Abiotic Stress Management for Resilient Agriculture

Paramjit Singh Minhas · Jagadish Rane
Ratna Kumar Pasala
Editors

Abiotic Stress Management for Resilient Agriculture

 Springer

Editors
Paramjit Singh Minhas
National Institute of Abiotic Stress
  Management
Indian Council for Agricultural Research
Baramati, Maharashtra, India

Jagadish Rane
National Institute of Abiotic Stress
  Management
Indian Council for Agricultural Research
Baramati, Maharashtra, India

Ratna Kumar Pasala
National Institute of Abiotic Stress
  Management
Indian Council for Agricultural Research
Baramati, Maharashtra, India

ISBN 978-981-10-5743-4   ISBN 978-981-10-5744-1   (eBook)
DOI 10.1007/978-981-10-5744-1

Library of Congress Control Number: 2017953015

Printed on acid-free paper

This Springer imprint is published by Springer Nature
The registered company is Springer Nature Singapore Pte Ltd.
The registered company address is: 152 Beach Road, #21-01/04 Gateway East, Singapore 189721, Singapore

Foreword

The global agriculture production has undergone drastic changes in recent years and is being seriously limited by various abiotic stresses. The Fourth Assessment Report of the Intergovernmental Panel on Climate Change (IPCC) revealed that by 2020 there could be a decline of agricultural yields of up to 50% in some countries in Africa as a result of climate change and variability. Moreover, a number of edaphic stresses, including chemical (nutrient deficiencies, excess of soluble salts, salinity, alkalinity), physical (high susceptibility to erosion, steep slopes, shallow soils, surface crusting and sealing, low water-holding capacity, impeded drainage, low structural stability, root restricting layer, high swell/shrink potential) and biological (low or high organic contents), have also emerged as major challenges for the production of crops, livestock, fisheries and other commodities.

This book addresses the management of soil-related abiotic constraints, stresses in drylands, heavy metal toxicities, salinity, water logging, high temperature and drought tolerance and presents mitigation strategies for immediate on-farm solutions with a special emphasis on approaches based on specific and potential plant bio-regulators for enhancing crop and water productivity in semi-arid regions. Special emphasis has been given to contextualizing the strategy for improving crop adaptations to climate change, biotechnological tools for improvement of tolerance and abiotic stress management in major food grains, commercial horticulture and vegetable crops and their production. This book also highlights livestock and their nutritional management during drought, mitigation options for GHG emissions from ruminants and mitigation of climatic change effects for sheep farming in arid environment. Overall, this volume covers a wide range of subjects that provide the readers a way forward in abiotic stress management to enable more productive agriculture.

I congratulate the editors for compiling this publication that will add a great deal to the global understanding of and implications for not only food security worldwide, but also the socioeconomic conditions of communities affected by climate change and management of abiotic stress for resilient agriculture.

ICAR   T. Mohapatra
Krishi Bhavan, New Delhi, 110 001, India

Preface

Several transformative changes, such as the growing population, changing lifestyles, expanding urbanization, accelerating land degradation, and climate change-induced abiotic stresses, are threatening the future food and nutrition security especially in low-income countries. The abiotic stress factors emerge mainly due to drought, extreme temperature (heat, cold chilling/frost), and floods in addition to edaphic settings, leading to chemical (ion/nutrient deficiencies/toxicities), physical (high erosion, hard pans/shallow soils, surface sealing/impeded drainage), and biological (low/high organic contents) constraints; these abiotic stress factors are also intrinsically linked to the production of crops, livestock, fisheries, and other commodities. Only 9% of the world's agricultural area is conducive for crop production, while 91% is afflicted by abiotic stresses which widely occur in combinations. While losses extending to more than 50% of agricultural production occur due to abiotic stresses, their intensity and adverse impact are likely to amplify manifold with climate change and overexploitation of natural resources. Fragile agroecosystems like the dryland areas are highly vulnerable to their disastrous impact.

Thus development of a strategic framework for inclusive, sustainable, and innovation-led agricultural growth is essential for these harsh agroecosystems afflicted by abiotic stresses. Multidisciplinary and holistic approaches to manage the stressed environments should aim at characterization of abiotically stressed environments; reoriented, novel, and scaled-up natural resource management (NRM) technologies for stress mitigation; improved adaptation to stressed environments; and task-oriented capacity building. Augmentation, integration, and promotion of the best available tools, approaches, and technologies should involve investments and incentives for breeding protocols, regional networks for exploring synergies, and dynamic policy support. Therefore, this book is an assemblage of 24 chapters by 68 experts in the area of abiotic stress tolerance/management, natural resource management, and strategic program of building resilience in crop, livestock, and policy implementation. Recent advances and prospects for understanding stress environments, adaptation and mitigation options in crops and animal husbandry, and policy support for abiotically stressed agroecosystems have been attempted. State-of-the art account of the information available has been synthesized in terms of challenges, scope and opportunities, coping strategies, and management of abiotic stresses using novel and new tools for resilient agriculture. Some of the chapters present management approaches for tackling specified stresses like edaphic constraints, stresses in drylands, heavy metal toxicities, salinity, waterlogging, high temperature and drought tolerance, and mitigation strategies for immediate on-farm solutions with a special emphasis on bio-regulators.

It is anticipated that this book will provide a practical update on our knowledge for improving management of abiotic stresses for resilient agriculture and allied sectors under changing global climate change conditions. This book establishes a set of principles based on current understanding on abiotic stresses and will be useful for different stakeholders, including agricultural students, scientists, environmentalists, policy makers, and social scientists.

We are extremely thankful to all the contributors for their efforts in providing comprehensive and cogent reviews.

Baramati, Maharashtra, India   Paramjit Singh Minhas
                                                                                      Jagadish Rane
                                                                                  Ratna Kumar Pasala

Contents

Contributors

Santanu Kumar Bal ICAR-National Institute of Abiotic Stress Management, Baramati, Pune, Maharashtra, India

R.M. Bhatt Division of Plant Physiology and Biochemistry, ICAR-Indian Institute of Horticultural Research, Bengaluru, Karnataka, India

Raghavendra Bhatta ICAR-National Institute of Animal Nutrition and Physiology, Bengaluru, Karnataka, India

Kiruba Shankari Arun-Chinnappa Centre for Crop Health, University of Southern Queensland, Toowoomba, QLD, Australia

Monika Dalal ICAR-National Research Centre on Plant Biotechnology, New Delhi, India

Kalyan De ICAR-Central Sheep and Wool Research Institute, Avikanagar, Rajasthan, India

K.A. Gopinath ICAR – Central Research Institute for Dryland Agriculture, Hyderabad, India

A.S. Kharub ICAR-Indian Institute on Wheat and Barley Research, Karnal, Haryana, India

Krishna Reddy Kakumanu International Water Management Institute, Hyderabad, India

S.S. Kukal Punjab Agricultural University, Ludhiana, India

Arvind Kumar Central Soil Salinity Research Institute, Karnal, Haryana, India

Arvind Kumar South Asia Breeding Hub, International Rice Research Institute, ICRISAT Campus, Hyderabad, Telengana, India

Davendra Kumar ICAR-Central Sheep and Wool Research Institute, Avikanagar, Rajasthan, India

Vishnu Kumar ICAR-Indian Institute on Wheat and Barley Research, Karnal, Haryana, India

N.P. Kurade ICAR-National Institute of Abiotic Stress Management, Baramati, Maharashtra, India

Chuni Lal ICAR-Indian Institute on Wheat and Barley Research, Karnal, Haryana, India

R.H. Laxman Division of Plant Physiology and Biochemistry, ICAR-Indian Institute of Horticultural Research, Bengaluru, Karnataka, India

P.K. Malik ICAR-National Institute of Animal Nutrition and Physiology, Bengaluru, Karnataka, India

H.M. Mamrutha ICAR-Indian Institute of Wheat & Barley Research, Karnal, Haryana, India

Paramjit Singh Minhas National Institute of Abiotic Stress Management, Indian Council for Agricultural Research, Baramati, Maharashtra, India

Varucha Misra ICAR-Indian Institute of Sugarcane Research, Lucknow, Uttar Pradesh, India

T. Mohanasundari Tamil Nadu Agricultural University, Coimbatore, Tamil Nadu, India

Raveendran Muthurajan Centre for Plant Molecular Biology & Biotechnology, Tamil Nadu Agricultural University, Coimbatore, Tamil Nadu, India

Vishnu V. Nachimuthu South Asia Breeding Hub, International Rice Research Institute, Hyderabad, Telangana, India

Ritu Nagdev ICAR-Indian Institute of Soil Science, Bhopal, Madhya Pradesh, India

Prakash S. Naik ICAR-Indian Institute of Vegetable Research, Varanasi, Uttar Pradesh, India

S.M.K. Naqvi ICAR-Central Sheep and Wool Research Institute, Avikanagar, Rajasthan, India

S. Naresh Kumar Centre for Environmental Science and Climate Resilient Agriculture, ICAR-Indian Agricultural Research Institute, New Delhi, India

A.V. Nirmale ICAR-National Institute of Abiotic Stress Management, Baramati, Maharashtra, India

K. Palanisami International Water Management Institute, Hyderabad, India

A.D. Pathak ICAR-Indian Institute of Sugarcane Research, Lucknow, Uttar Pradesh, India

S.S. Pawar ICAR-National Institute of Abiotic Stress Management, Baramati, Maharashtra, India

Kulasekaran Ramesh ICAR-Indian Institute of Soil Science, Bhopal, Madhya Pradesh, India

Lanka Ranawake Faculty of Agriculture, University of Ruhuna, Ruhuna, Sri Lanka

Jagadish Rane National Institute of Abiotic Stress Management, Indian Council for Agricultural Research, Baramati, Maharashtra, India

J.K. Ranjan ICAR-Indian Institute of Vegetable Research, Varanasi, Uttar Pradesh, India

Ratna Kumar Pasala National Institute of Abiotic Stress Management, Indian Council for Agricultural Research, Baramati, Maharashtra, India

G. Ravindra Chary ICAR-Central Research Institute Dryland Agriculture, Hyderabad, India

Robin Sabariappan Center for Plant Breeding and Genetics, Tamil Nadu Agricultural University, Coimbatore, Tamil Nadu, India

A. Sahoo ICAR-Central Sheep and Wool Research Institute, Avikanagar, Rajasthan, India

M.P. Sahu Swami Keshwanand Rajasthan Agricultural University, Bikaner, Rajasthan, India

B. Sajjanar ICAR-National Institute of Abiotic Stress Management, Baramati, Maharashtra, India

K.T. Sampath ICAR-National Institute of Animal Nutrition and Physiology, Bangalore, Karnataka, India

S. Sareen ICAR-Indian Institute of Wheat & Barley Research, Karnal, Haryana, India

Saman Seneweera Centre for Crop Health, University of Southern Queensland, Toowoomba, QLD, Australia

Kiruba A. Shankari Centre for Crop Health, University of Southern Queensland, Toowoomba, QLD, Australia

Arun K. Shanker ICAR – Central Research Institute for Dryland Agriculture, Hyderabad, India

P. Sharma ICAR-Indian Institute of Wheat & Barley Research, Karnal, Haryana, India

Parbodh C. Sharma Central Soil Salinity Research Institute, Karnal, Haryana, India

T.R. Sharma ICAR-National Research Centre on Plant Biotechnology, New Delhi, India

S. Sheoran ICAR-Indian Institute of Wheat & Barley Research, Karnal, Haryana, India

Arvind K. Shukla ICAR-Indian Institute of Soil Science, Bhopal, Madhya Pradesh, India

S.P. Shukla ICAR-Indian Institute of Sugarcane Research, Lucknow, Uttar Pradesh, India

S. Siddiqui ICAR-Indian Institute of Soil Science, Bhopal, Madhya Pradesh, India

C. Singh ICAR-Indian Institute of Wheat & Barley Research, Karnal, Haryana, India

G. Singh ICAR-Indian Institute of Wheat & Barley Research, Karnal, Haryana, India

Jogendra Singh ICAR-Indian Institute on Wheat and Barley Research, Karnal, Haryana, India

Major Singh ICAR-Indian Institute of Vegetable Research, Varanasi, Uttar Pradesh, India

Ch. Srinivasarao ICAR – Central Research Institute for Dryland Agriculture, Hyderabad, India

A.K. Shrivastava ICAR-Indian Institute of Sugarcane Research, Lucknow, Uttar Pradesh, India

Sangeeta Srivastava ICAR-Indian Institute of Sugarcane Research, Lucknow, Uttar Pradesh, India

Saumya Srivastava ICAR-Indian Institute of Soil Science, Bhopal, Madhya Pradesh, India

M. Swapna ICAR-Indian Institute of Sugarcane Research, Lucknow, Uttar Pradesh, India

R. Tiwari ICAR-Indian Institute of Wheat & Barley Research, Karnal, Haryana, India

V. Tiwari ICAR-Indian Institute of Wheat & Barley Research, Karnal, Haryana, India

T.V. Vineeth Central Soil Salinity Research Institute, Karnal, Haryana, India

S.M. Virmani Indian Resources Information and Management Technologies Ltd. (INRIMT), Hyderabad, India

Goraksha C. Wakchaure ICAR-National Institute of Abiotic Stress Management, Baramati, Maharashtra, India

About the Editors

Dr. Paramjit Singh Minhas has about four decades of diversified research experience on management of natural resources. His major research contributions have lead to better understanding of soil-water-plant interactions in saline and other edaphically harsh environments and the development of management strategies for deficit irrigation, use of low-quality waters, salinity afflicted and shallow basaltic soils for raising the production potential of crops, orchard/forestry plantations. By holding key research management positions in the Indian Council of Agricultural Research (viz., Project Coordinator, All India Coordinated Project on Management of Salt-Affected Soils and Use of Saline Water in Agriculture; Assistant Director General, Integrated Water Management; Director of Research, Punjab Agricultural University; Assistant Director General, Soil and Water Management; and Director, ICAR-National Institute on Abiotic Stress Management), he has demonstrated leadership skills in formulating and implementing research programs and providing vision and direction. His research endeavors have been recognized with several awards like Rafi Ahmed Kidwai Award, Swami Pranavananda Sarswati Award in Environmental Science and Ecology, Jain-INCID *Krishi Sinchai Vikas Puraskar* in 2005, Hari Om Ashram Trust Award, the CSSRI Excellence Award on Soil Salinity and Water Management, Su Kumar Basu Award, and the 12th International Congress Commemoration Medal. He is a fellow of National Academy of Agricultural Sciences, Indian Society of Soil Science, and Punjab Academy of Sciences, Associate Editor of *Agricultural Water Management*, and a member of the editorial board of Indian Journal of Agricultural Sciences.

Dr. Jagadish Rane Principal Scientist and Head of School of Drought Stress Management at ICAR-National Institute of Abiotic Stress Management (NIASM). He started his research carrier at ICAR-Indian Institute on Wheat and Barley Research in 1993. In 2007–2011, he led a plant phenotyping team for the evaluation of gene technology to improve drought tolerance in upland rice at International Center for Tropical Agriculture (CIAT). He has organized multi-location experiments across India to understand the plant traits and genes for resilience to abiotic stresses. He has established a plant phenomics platform and developed prototypes of low-cost phenotyping tools for field evaluation of germ plasm. He has standardized protocols for various screening procedures (viz., transgenic events in biosafety, identification of water-efficient genotypes, etc.).

Dr. Ratna Kumar Pasala Senior Scientist dealing with plant abiotic stress physiology at ICAR-Indian Institute of Oilseeds Research, Hyderabad, and a member of the editorial board of plant science journals and reviewer of many international and national journals. He has research experience in plant stress functional biology at ICAR-National Institute of Abiotic Stress Management (NIASM) and International Crop Research Institute for the Semi-Arid Tropics (ICRISAT). His major research expertise is in plant stress physiology and plant bio-regulators for the mitigation of abiotic stresses through redox-mediated mechanism. He is a recipient of R.D. Asana Gold Medal Award and Young Scientist Fellowship Award by FAO.

Part I

Advances and Prospects for Understanding Stress Environments

Abiotic Stresses in Agriculture: An Overview

Paramjit Singh Minhas, Jagadish Rane, and Ratna Kumar Pasala

Abstract

Agriculture production and productivity are vulnerable to abiotic stresses. These stresses emerge due to drought, temperature extremes (heat, cold chilling/ frost), radiation (UV, ionizing radiation), floods in addition to edaphic factors which include chemical (nutrient deficiencies, excess of soluble salts, salinity, alkalinity, low pH/acid sulfate conditions, high P and anion retention, calcareous or gypseous conditions, low redox, chemical contaminants—geogenic and xenobiotic), physical (high susceptibility to erosion, steep slopes, shallow soils, surface crusting and sealing, low water-holding capacity, impeded drainage, low structural stability, root-restricting layer, high swell/shrink potential), and biological (low or high organic contents) components. These stresses are the major challenges for production of crops, livestock, fisheries, and other commodities. Only 9% of the world's agricultural area is conducive for crop production, while 91% is under stresses which widely occur in combinations. While losses to an extent of more than 50% of agricultural production occur due to abiotic stresses, their intensity and adverse impact are likely to amplify manifold with climate change and over exploitation of natural resources. Fragile agroecosystems like the dryland areas are highly vulnerable to such disastrous impact. To mitigate the effects/impact of multiple stressors, proposed strategies include improved agronomic management, while the breeding of stress tolerant genotypes can enhance capacity for adaptation to stress environments. However, a holistic integrated multidisciplinary approach in systems perspectives is a need of the hour to get

P.S. Minhas (✉) • J. Rane • R.K. Pasala
National Institute of Abiotic Stress Management, Indian Council
for Agricultural Research, Baramati, Maharashtra, India
e-mail: minhas_54@yahoo.co.in; jagrane@hotmail.com; pratnakumar@gmail.com

© Springer Nature Singapore Pte Ltd. 2017
P.S. Minhas et al. (eds.), *Abiotic Stress Management for Resilient Agriculture*,
DOI 10.1007/978-981-10-5744-1_1

the best combination of technologies for a particular agroecosystem. Therefore, this compendium through different comprehensive chapters conveys relevant updates on trends in abiotic stresses and their impact in addition to scientific interventions for stress management through mitigation and adaptation options. The compendium also explains scope for modern science to mitigate abiotic stresses and improve adaptation through genetic improvement and some of the policy support endeavors. The way forward includes information on implementation of existing technologies and gaps to be filled through future research for abiotic stress management.

The food and nutritional security warrants the availability of adequate and quality food to meet the dietary and nutritional requirements for a healthy and productive life. The global food production has been increasing in line with and even sometimes ahead of demand in recent decades though many countries are still confronting problem of inadequacy of food supply. The world population is growing at an alarming rate and is anticipated to reach about 9.6 billion by 2050 from the present about 6 billion. Specifically the developing countries would be the major contributors to rise in population. For instance, India will be the most populous country by 2050, and its population is predicted to reach 1.6 billion from the present of about 1.3 billion. The predictions are that demand for food would increase by 70% and would even double in some low-income countries (FAO 2009). This of course is related with population rise, but the per capita food consumption coupled with its quality would also improve with growing economies. Therefore, enormous efforts are required to achieve the expected growth rate (44 million tons per annum) to ensure food security especially when agriculture is losing the productive lands due to urbanization and industrialization. Moreover, the past endeavors for improving agricultural productivity to meet the food demands were accompanied by land degradation and the impact of episodic climate variability; those have consequently increased the components, frequency, and magnitude of abiotic stresses. The abiotic stresses like drought, temperature extremes, floods, salinity, acidity, mineral toxicity, and nutrient deficiency have emerged as major challenges for production of crops, livestock, fisheries, and other commodities. These in fact are the principal causes of crop failure worldwide, dipping average yields for most major crops by more than 50% (Wang et al. 2007), mostly shared by high temperature (20%), low temperature (7%), drought (9%), and other forms of stresses (4%). Only 9% of the area is conducive for crop production, while 91% is under stresses in the world. Various anthropogenic activities are accentuating the existing stress factors. Thus, the abiotic stresses cause losses worth hundreds of million dollars each year due to reduction in crop productivity and crop failure. Specifically substantial agricultural land in tropics and subtropics, e.g., in India, is more challenged with penultimate combinations of abiotic stresses. Since these stresses threaten the very sustainability of agriculture, there is enhanced apprehension among the farmers, scientific communities, and policy makers regarding the intensity and adverse impacts that are likely to amplify manifold with climate change and due to over exploitation of natural resources.

For assessing the magnitude of the problems related to abiotic stresses, the initial section is dedicated for an overview and opportunities for alleviating the impacts of the atmospheric, drought, and edaphic factors on agricultural and horticulture crops, livestock including poultry, fisheries, etc. The environmental stresses like temperature (heat, cold chilling/ frost) and radiation (UV, ionizing radiation) are responsible for major reduction in agricultural production (Chap. 2). Moreover, the catastrophic events like droughts, floods, hailstorms, cyclones, etc. are occurring frequently and cause widespread land degradation apart from heavy monetary losses and serious setbacks to agricultural development. The nature and severity of impacts from extreme weather events like drought and hailstorms depend not only on their extreme features but also on duration of exposure to these stresses and extent of vulnerability of natural resources, livestock, and humans. Adverse impacts become disasters when they produce widespread damage and severely alter the routine functioning of communities or societies. Specifically the fragile dryland ecosystems, which support substantial population, are highly vulnerable to food insecurity since these are characterized by limited and erratic rainfall, intense pressure on resource use, and sensitivity to climatic shocks (Chap. 3). Emergence of widespread multiple nutrient deficiencies, depletion of organic carbon stocks, development of secondary salinity and waterlogging in canal-irrigated areas, low input use efficiencies (nutrient and water use), decreasing total factor productivity of fertilizers, etc. are all the consequences of application of component-based technologies for short-term yield gains (Chap. 4). Corrective measures and precautions both in short term and in long term should allow effective management of these situations which potentially limit the productivity of crops and dependent agricultural enterprises. Such efforts need community interventions particularly in cases of severely fragmented land holdings.

Development and promotion of strategies to minimize the impact of abiotic stresses are fundamental for sustaining agriculture. The proposed strategies include both the improved agronomic management and breeding novel genotypes with improved capacity for adaptation to stress environments. The major challenge is to enable accelerated adaptation and mitigation without threatening the sensitive agro-ecosystems that support livelihood of inhabitants striving to cope with abiotic stresses. The multi-thronged strategies are required to accomplish this task based upon the analysis of current situation and development and use of newer technologies including diversification of the production systems. Though the edaphic constraints are mostly coupled with intrinsic soil-forming processes, some of the anthropogenic causes like poor land management practices, e.g., overexploitation and lack of restorative measures aggravate these constraints. Except few, the processes leading to edaphic constraints are generally insidious and show up only gradually as the problem becomes more severe to cause yield declines. Farmers may ultimately be forced to either shift to less remunerative crops or soils can turn unfit for agriculture in extreme cases. Several resource conservation and cropping system-based strategies including conservation agriculture, integrated farming systems, watershed management, and other restorative measures are available to reduce the effects of edaphic stresses and for securing favorable soil conditions (Chaps. 5 and 6). Recent NRM technologies concentrate more on improving the whole farming systems at field or watershed level rather than only the productivity of specific commodities. Contingency crop plans have also been advocated for stabilizing agricultural production under the situation of weather aberrations especially the drought. Salinity and waterlogging have now emerged as global phenomena, which are adjunct of irrigated agriculture. The global annual cost of salt-induced land degradation in irrigated areas has been estimated to be US$ 27.3 billion because of lost crop production (Qadir et al. 2014). Soils are also getting increasingly polluted with toxic elements from geogenic or anthropogenic sources like sewage, industrial effluents, urban solid wastes, etc. The heavy metal toxicities not only impair productivity or agricultural crops, but these enter the food chain and become potential hazards to health of humans and animals. Remedial measures through engineering techniques and bioremediation have so far met with varying degrees of success (Chaps. 7 and 8). For promotion of growth and development under stress conditions, exogenous application of plant bio-regulators (PBRs) and other nutrient supplements has been tried under both controlled and actual field conditions (Chap. 10). These promote the ability of plants to cope with the stress conditions by mediating growth, development, nutrient allocation, and source sink transitions. Mainly the PBRs with thiol-groups, which are involved in redox signaling, can help alleviate stress for crops grown under drought, salinity, and heat stress (Chap. 11).

There are increasing evidences for climate change, which is happening and hence further global warming seems unavoidable (IPCC 2014). Agriculture sector is one of the most sensitive areas that are being afflicted by global warming and associated weather variability (Chaps. 2, 9 and 12). For instance, global production of maize, wheat, rice, and soybean is projected to decline up to 40–60% by 2090 (Rosenzweig et al. 2014). Climate change may impact the agricultural crops in four ways (Easterling and Apps 2005). First the agroecological zones may be altered with changes in temperature and precipitation. An increase in potential evapotranspiration and increased length of rainless periods are likely to intensify the drought stress especially in semiarid tropical and subtropical regions. The increased carbon dioxide is expected to have positive impact due to higher rate of photosynthesis and higher water use efficiency. The water availability (or runoff) is also a critical factor for determining the impact of climate change. Since the precipitation determines the length of growing season, the effect of climate change on total rainfall and interval between the rainfall events ultimately decide if the effects on agricultural crops are the positive or the negative. The losses in agriculture can result from climate variability, and there may be increase in frequency of extreme events like droughts, floods, etc.

The stress signal is first perceived at the membrane level by the receptors and then transduced in the cell to switch on the stress-responsive genes for mediating abiotic stress tolerance. Hence, deep insight into the mechanism of stress tolerance mediated by a plethora of genes involved in stress-signaling network is important for crop improvement (Chaps. 12 and 13). Each stress leads to multigenic responses of plant, and therefore it may result in alteration of a large number of genes as well as their products. A deeper understanding of the transcription factors regulating these genes, the products of the major stress-responsive genes, and cross talk between different signaling components will be an area of intense research activity in the future. The knowledge generated through these studies should be utilized in transforming the crop plants that would be able to tolerate stress condition without showing any growth and yield penalty. Attempts should be made to design suitable vectors for stacking relevant genes of one pathway or complementary pathways to develop durable tolerance. These genes should preferably be driven by a stress-inducible promoter to have maximal beneficial effects avoiding possible yield penalty during favorable season. Additionally, due importance should be laid on the physiological parameters such as the relative content of different ions as well as the water status in plant tissues for designing stress-resilient crop plants for the future.

Cereal food grains like rice and wheat have been the first priority for improvement by plant breeders since these continue to be the important staple food across the world. Therefore, these crops witnessed significant progress in improvement of germplasm, breeding lines, high-yielding cultivars, and yield stability. In addition, there have been ample efforts for diversification through cultivation of drought-tolerant crops like barley, sorghum, etc. With growing needs toward development of potentially resilient genotypes for emerging abiotic stresses (heat, drought, salinity, etc.), genetic alterations of these cereals are being attempted through genomics, bioinformatics, high-throughput phenomic tools, etc. (Chaps. 13, 14, 15, and 16). In the wake of climate change, recent developments in molecular genetics and biotechnology are also aiding acceleration of breeding process for adaptation in other crops like vegetables, sugarcane, etc. which are also an important component of human diets (Chaps. 17 and 19). Integration of proper crop management strategies with improved cultivars is essential to meet the goals of stress management in fruit crops in tropical and subtropical regions (Chap. 18).

Abiotic stresses also threaten the availability of feed resources for livestock from land-based production systems (Chap. 20). Specifically the drought conditions usually endanger the very sustainability of livestock in arid and semiarid regions. Moreover, the animal health is projected to be impacted by climate change via animal-related diseases with thermal stress and the extreme weather conditions. Therefore, vulnerability of ruminants and possibilities of improved nutrition and other management issues have been discussed (Chaps. 21 and 22).

The strategies that help to minimize the impacts of abiotic stresses include sound governmental policy and political will for post-disaster recovery and reconstruction for improving adaptive capacity. The under investment and market distortions especially in the regions having preponderance of abiotic stressors have been mainly responsible for poor R&D, weak institutions and infrastructure, and non-pragmatic pricing of inputs and natural resources. Hence, policies are identified which can shape development, dissemination, and marketing of technologies to sustain agricultural outputs using resource efficient methods in harsh agroecologies (Chaps. 23 and 24).

Keeping above in view, different chapters are compiled in a mode to bring out the latest developments on emerging techniques to tackle the complex problems related to abiotic stresses. Interventions of these technologies require appropriate knowledge of abiotic factors essential for developing preparedness measures suitable for socioeconomic and environmental conditions prevailing in the agroecosystem under consideration. So far the earlier compilations on abiotic stresses had focus on basic physiological and transgenic issues but not on problem-solving approaches and techniques; those are essential for inducing medium- to long-term resilience in production systems. Besides sustainable livelihood security to poor families, adaptation and mitigation strategies can provide an immense scope for ecosystem services. The authors trust that with the synthesis and integration of knowledge and experiences of experts from different disciplines, this book will open new vistas in the versatile field of abiotic stress management and will be useful for different stakeholders including agricultural students, scientists, environmentalists, policy makers, and social scientists.

References

Easterling W, Apps M (2005) Assessing the consequences of climate change for food and forest resources: a view from the IPCC. Climate Change 70:165–189

FAO (2009) How to feed the world in 2050. pp. 1–35. http://www.fao.org/fileadmin/templates/wsfs/docs/expert_paper/How_to_Feed_the_World_in_2050.pdf

IPCC (2014) Romero-Lankao P, Smith JB, Davidson DJ, Diffenbaugh NS, Kinney PL, Kirshen P, Kovacs P, Villers-Ruiz L (2014) North America. In: Barros VR, Field CB, Dokken DJ, Mastrandrea MD, Mach KJ, Bilir TE, Chatterjee M, Ebi KL, Estrada YO, Genova RC, Girma B, Kissel ES, Levy AN, MacCracken S, Mastrandrea PR, White LL (eds.) Climate Change 2014: Impacts, Adaptation, and Vulnerability. Part B: Regional Aspects. Contribution of Working Group II to the Fifth Assessment Report of the Intergovernmental Panel on Climate Change, Cambridge University Press, Cambridge, UK/New York, pp. 1439–1498

Qadir M, Quillérou E, Nangia V, Murtaza G, Singh M, Thomas RJ, Drechsel P, Noble AD (2014) Economics of salt-induced land degradation and restoration. Nat Res Forum 38:282–295

Rosenzweig C, Elliott J, Deryng D, Ruane AC, Müller C, Arneth A, Boote KJ, Folberth C, Glotter M, Khabarov N, Neumann K, Piontek F, Pugh TAM, Schmid E, Stehfest E, Yang H, Jones JW (2014) Assessing agricultural risks of climate change in the 21st century in a global gridded crop model intercomparison. Proc Natl Acad Sci 111(9):3268–3277

Wang W, Vinocur B, Altman A (2007) Plant responses to drought, salinity and extreme temperatures towards genetic engineering for stress tolerance. Planta 218:1–14

Atmospheric Stressors: Challenges and Coping Strategies

Santanu Kumar Bal and Paramjit Singh Minhas

Abstract

The basic principle of agriculture lies with how crop/livestock interacts with atmosphere and soil as a growing medium. Thus any deviation of external optimal atmospheric conditions affects the pathway through changes in atmospheric and edaphic/feed factors for crop/animal growth, development and/or productivity. Besides these, change and variability in atmospheric conditions have increased due to human activities to induce greenhouse gas emissions. In the continuation of current trend in carbon emissions, temperatures will rise by about 1 °C and 2 °C by the year 2030 and 2100, respectively. With warmer climate, frequency and severity of extreme weather events would increase as indicated by incidences of heat waves, extreme rains, hailstorm, etc. during recent years. Besides these, events like cloudburst, cyclone, sand/dust storm, frost and cold wave and deteriorated air quality are becoming regular events. However, the type and intensity of stress events will probably have varying impacts in different ecoregions. These events cause huge impact both in terms of mechanical and physiological on commodities across crop, livestock, poultry and fisheries. The quantum of impact on crops mainly depends on the type of stress and crop/animal/fish, its stage/age and mode of action of the stress. Management strategies for mitigation of these stresses require both application of current multidisciplinary knowledge, development of a range of technological innovations and timely interventions. It's high time to update our knowledge regarding existing

S.K. Bal (✉)
ICAR-National Institute of Abiotic Stress Management, Baramati, Pune 413115, Maharashtra, India
e-mail: bal_sk@yahoo.com

P. S. Minhas
National Institute of Abiotic Stress Management, Indian Council for Agricultural Research, Baramati, Maharashtra, India
e-mail: minhas_54@yahoo.co.in

© Springer Nature Singapore Pte Ltd. 2017
P.S. Minhas et al. (eds.), *Abiotic Stress Management for Resilient Agriculture*,
DOI 10.1007/978-981-10-5744-1_2

technologies and side by side explores new avenues for managing atmospheric stresses in agriculture. The first step for the scientific community will be to screen and identify species for tolerance to atmospheric stresses followed by complete insight of the biological processes behind the atmospheric stress response combined with emerging technologies in breeding, production, protection and postharvest which is likely to improve productivity and reduce losses. The type and level of stresses must be properly quantified through proper scientific planning for present as well as future references for finding mitigation and adaptation solutions. Keeping above in view, this chapter has been prepared which includes aspects covering atmospheric stresses, their challenges and coping strategies in various agricultural enterprises including crops, livestock, poultry and fisheries. This chapter will ignite the minds of all stakeholders including students and researchers to explore more in finding proper adaptation, and mitigation measures. This will pave the way for developing food and livelihood systems that will have greater economic and environmental resilience to risk.

2.1   Introduction

Agriculture is critical to development since the majority of the world's poorest and hungry people depend on it for their livelihoods. However, agriculture in turn depends on basic natural resources: biodiversity, soil and water and environmental factors. In spite large-scale development of soil-, water- and crop-based technologies to optimize and sustain crop productivity in the recent past, the latter continues to be affected significantly by number of climate variability factors. These factors like temperature, relative humidity, light, availability of water, mineral nutrients, CO_2, wind, ionizing radiation or pollutants determine plant growth and development (20). Effect of each atmospheric factor on the plant depends on their intensity and duration of act. For optimal growth, the plant requires a certain quantity of each of the environmental factors, and any deviation from such optimal conditions adversely affects its productivity through plant growth and development. These stress factors include extreme temperatures, too high or too low irradiation, extreme of water that induce drought or waterlogging, etc. (Fig. 2.1). Some of these are induced as a result of recurring features of climate variability, e.g. cold/heat waves, floods/heavy rain, cyclones/tidal waves, hail/thunder storms, etc., and these critical environmental threats are often referred as extreme weather events.

As climate change has become a reality, the implications of global warming for changes in extreme weather and climate events are of major concern for agrarian as well as civic society. However, since extreme events are typically rare events, therefore only limited observational data are available for their impacts (Lenton et al. 2008; Loarie et al. 2009; Sherwood and Huber 2010). Over the last couple of years, we experienced typical events, i.e. Kuwait reporting snow; the USA devastated by Hurricane Katrina and Paris sweltering in 40 °C heat; Mumbai sunk under 940 mm of heavy rainfall in a single day; Delhi froze with below 0 °C; Rajasthan had floods

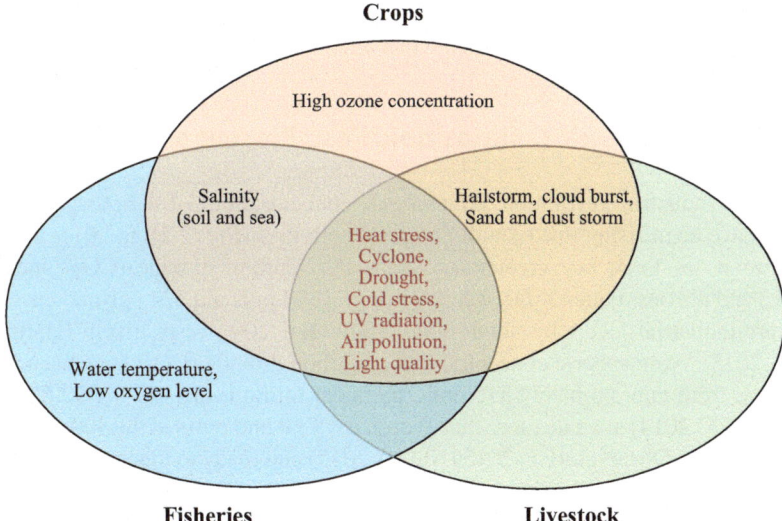

Fig. 2.1 Impact of atmospheric stresses across various agricultural commodities

twice in a year; across pan India there were unprecedented hailstorm events; and many more (Bal and Minhas 2016). The year 2016 will stand out in the historical record of the global climate in many ways. The average global temperature across land and ocean surface areas for 2016 was 0.90 °C above the twentieth century average ((NOAA 2015). This marks the fifth time in the twenty-first century a new record high annual temperature has been set and also marks the 40th consecutive year (since 1977) that the annual temperature has been above the twentieth century average (NOAA 2015). Heat waves were extremely intense in various part of the world including India and Pakistan leading to thousands of deaths. Similarly, extreme precipitation led to flooding that affected areas across Asia, South America, West Africa and Europe and dry conditions in southern Africa and Brazil which exacerbated multi-year droughts (WMO 2016). These events have signalled and forced us to accept the unusual change in the behaviour within our atmospheric. The Fifth Assessment Report of the IPCC has also reiterated that climate change is real and its impact is being felt across countries in the world, e.g. in many parts of India, and the number of rainy days and rain intensity have decreased and increased, respectively, which is counterproductive to the recharge of groundwater because of runoff of rainwater (IMD 2015). Warmer summer and droughts have also made agriculture nonsupportive. Global food production is gradually increasing, but the relative rate of increase especially for major cereal crops is declining (Easterling and Apps 2005; Fischer and Edmeades 2010). In these circumstances, what makes climate smart agriculture more important is ever-increasing demand for food, issues with climatic variability which makes farming more vulnerable to vagaries of nature. In this chapter, an attempt has been made to assess various atmospheric stressors: their current and expected future behaviour and damage potential in Indian agriculture along with possible management options.

2.2   Atmospheric Composition: Past, Present and Future

Since the industrial revolution, atmospheric concentrations of various greenhouse gases have been rising due to anthropogenic activities (Fig. 2.2). In 2011, the concentrations of three key greenhouse gases, viz. carbon dioxide (CO_2), methane (CH_4) and nitrous oxide (N_2O), were 391 ppm, 1803 ppb and 324 ppb and exceeded the pre-industrial levels by about 40%, 150% and 20%, respectively (Hartmann et al. 2013). Atmospheric concentrations of carbon dioxide (CO_2) have been rising steadily, from approximately 315 ppm (parts per million) in 1959 to 392.6 ppm in 2014 (IPCC 2014) at a rate more than 2 ppm per year and crossed the 400 ppm mark in Mauna Loa Observatory in 2015 (NOAA 2015) and ready to touch 410 ppm mark (NOAA 2017). If fossil fuel burning and other human interventions continue at a business as usual, CO_2 will rise to levels of 1500 ppm (NOAA 2015). Out of all GHGs, CO_2 is the most important GHG as more than 64% of total emission is CO_2, and other gases such as methane, nitrous oxide and fluorinated gases contribute 18, 6 and 12%, respectively.

Out of all anthropogenic sources of climate compounds, fossil fuel and biomass burning are the main ones. Atmosphere serves as a conduit for the transport of toxic material added sometimes inadvertently by the output of agricultural and industrial systems. When fossil fuels are burnt, carbon-based petrochemical products are broken up in combustion to form carbon dioxide (CO_2), carbon monoxide (CO), volatile organic compounds (VOCs), nitrogen oxides (NOx), sulphur oxides (SOx) and very fine particulates. However, chlorofluorocarbons (CFCs) a group of compounds

Fig. 2.2 Possible chain of events of the potential effect on climate and society on anthropogenic emissions (Source: Fuglestvedt et al 2003)

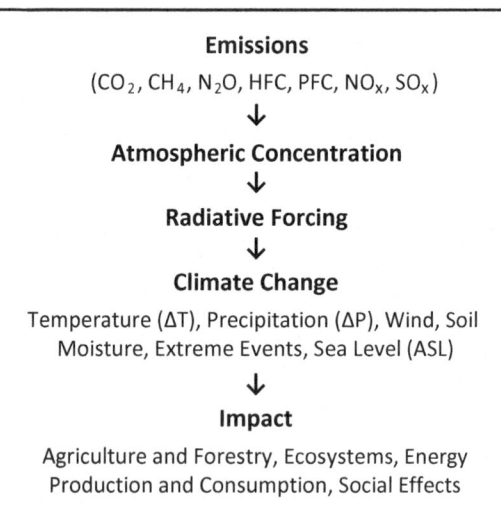

Emissions
(CO_2, CH_4, N_2O, HFC, PFC, NO_x, SO_x)
↓
Atmospheric Concentration
↓
Radiative Forcing
↓
Climate Change
Temperature (ΔT), Precipitation (ΔP), Wind, Soil Moisture, Extreme Events, Sea Level (ASL)
↓
Impact
Agriculture and Forestry, Ecosystems, Energy Production and Consumption, Social Effects

which contain the elements chlorine, fluorine and carbon are unlikely to have any direct impact on the environment in the immediate vicinity of their release whereas at global level have serious environmental consequences. Their long life in the atmosphere means that some end up in the higher atmosphere (stratosphere) where they can destroy the ozone (O_3) layer, thus reducing the protection it offers the Earth from the sun's harmful UV rays (SEPA 2016) besides contributing to global warming through the greenhouse effect. With the possible exception of CO_2 and black carbon, all compounds are likely to be affected by future climate change (Ramanathan and Carmichael 2008). In the atmosphere, when a sufficient concentration of sulphur and nitrogen oxides and hydrocarbons builds up and bombarded by sunlight, a complex series of chemical reactions takes place that creates more chemicals, including nitrogen dioxide (NO_2) and O_3 (Huthwelker et al. 2006). Methane (CH_4) is of particular importance since its climatic impact is strongly enhanced through atmospheric chemical interaction involving tropospheric O_3 and stratospheric H_2O. However, secondary aerosols like O_3 and sulphate are distinctly different from long-lived climate gases CO_2, CH_4, N_2O and halocarbons (Isaksen et al. 2014).

Intensification in agriculture has put tremendous pressure on the land and ecosystems. Approximately 508 million tons of on-field crop residues are generated per year in India, out of which 43 and 23% were rice and wheat, respectively (Koopmans and Koppejan, 1997). About 116 million tons out of these residues are burnt by farmers (Streets et al. 2003; Venkataraman et al. 2006). The primary carbon-containing gases emitted from biomass burning in order of abundance are CO_2, CO, and CH_4, which include two major greenhouse gases (Stockwell et al. 2014). In all, open burning of crop residue accounted for about 25% of black carbon, organic matter and carbon monoxide emissions, 9–13% of fine particulate matter (PM 2.5) and carbon dioxide emissions and about 1% of sulphur dioxide emissions.

2.2.1 Changes in Atmospheric Composition and its Effect on Climate

It is evident from scientific reports that global warming is most probably due to the man-made increases in greenhouse gas emissions, which ratifies the discernible human influence on the global climate. Emissions from fossil fuels, industry, land use and land-use change have increased greenhouse gas concentrations and led to almost 1 °C rise in global mean temperature on pre-industrial levels and also influenced the patterns and amounts of precipitation; reduced ice and snow cover, as well as permafrost; raised sea level; increased acidity of the oceans; increased the frequency, intensity and/or duration of extreme events; and shifted ecosystem characteristics (IPCC 2007a; IPCC 2013). The overall state of the global climate is determined by the balance between energy the Earth receives from the Sun and the energy which the Earth releases back to space, called the global energy balance, and how this energy balance is regulated depends upon the flows of energy within the global climate system. This process also causes the climate change to happen (Morshed 2013).

Earth's lowest part of the atmosphere has warmed up on an average by about 0.6 °C in the last 100 years or so (IPCC 2013). Global warming, in turn, has also triggered changes in the behaviour of many climatic parameters and occurrence of extreme events, more prominently in the last few decades. It is estimated that agriculture contributes around 10–15% of total anthropogenic emissions globally (IPCC 2007b). The greenhouse gases (GHG) trap heat radiated from the Earth and increase global mean temperature. The mean global temperature may increase by 0.3 °C per decade (Jones et al. 1999) with an uncertainty of 0.2–0.5% (Houghton et al. 1990) and would reach to approximately 1 °C and 2 °C above the present value by years 2030 and 2100, respectively, which would lead to global warming (IPCC 2013). Land-use and land-cover pattern over time in response to evolving economic, social and biophysical conditions has a serious impact on the atmospheric processes in addition to emission of heat-trapping greenhouse gases from energy, industrial, agricultural and other sources. Cities are warmer than surroundings because of greater extent of concrete area altering the surface albedo and affecting the exchange of water and energy between land and atmosphere (Lynn et al. 2009).

2.2.2 Climate Change and Extreme Events

While natural variability continues to play a key role in extreme weather, climate change has shifted the odds and changed the natural limits, making certain types of extreme weather especially cold wave, frost, hailstorm, thunderstorm, dust storm, heat wave, cyclone and flood more frequent and more intense. However, year-to-year deviations in the weather and occurrence of climatic anomalies/extremes have become a matter of concern. As the climate has warmed, some of the extreme weather events have become more frequent and severe in recent decades (IPCC 2014) and especially in warmer climate, the frequency of extremes such as heat waves increases (IPCC 2007b). Some of these events since 1990 are listed in Table 2.1, while the most prone areas in India are given in Table 2.2. During the period 1991–1999 (twentieth century), the Indian states – Jammu and Kashmir, Rajasthan and Uttar Pradesh – experienced 211, 195 and 127 number of cold waves, and Bihar, West Bengal and Maharashtra experienced 134, 113 and 99 heat waves, respectively (De et al. 2005). In addition, there is evidence (Holland and Webster 2007) that changes in distribution (e.g. tropical cyclone development occurring more equatorward or poleward of present day) have historically been associated with large changes in the proportion of major hurricanes. Earth is witnessing increases in extreme heat, severe storms, intense precipitation, drought and hailstorm (Allan 2011). The global temperature has increased by 0.74 °C during the past 100 years and could be increased by 2.6–4.8 °C by the end of this century (IPCC 2013). Heat waves are longer and hotter, and likely that has more than doubled the probability of occurrence of heat waves in some locations.

Heavy rains and flooding are more frequent. Patterns of precipitation and storm events, including both rain and snowfall, are also likely to change. However, some of these changes are less certain than the changes associated with temperature.

Table 2.1 Major atmospheric stress-related events in agriculture in India since 1990

Year	Type	Affected area	Loss of crops/livestock
Cyclone			
1994	Cyclone	Andhra Pradesh and Tamil Nadu	0.44 Mha
1996	Cyclone	Andhra Pradesh	0.10 Mha
1999	Super cyclone	Odisha	1.80 Mha
2006	Cyclone (Ogni)	Andhra Pradesh	Rice crop
2013	Cyclone (Phailin)	Odisha	0.50 Mha
30 Oct, 2014	Cyclone (Nilofar)	Gujarat	3 Mha cotton
			1.23 Mha groundnut
2014	Cyclone(Hudhud)	Andhra Pradesh	0.33 Mha
Cold wave/frost			
1994	Hailstorm	North West India, Andhra Pradesh, Maharashtra	0.46 Mha
2003	Frost	Punjab	Potato
2006, 2008	Frost	NW India especially in Punjab	
2011	Frost	Madhya Pradesh	0.007 Mha pigeon pea
2012	Cold wave	Madhya Pradesh and north Maharashtra	
2014	Hailstorm	Maharashtra	1.8 Mha
2015	Hailstorm and unseasonal rain	Northern and Central India	6.3 Mha
Heat wave			
1998	Heat wave	Odisha	>2000 human lives lost
2002	Heat wave	Southern India	>2002 human lives lost
2003	Heat wave	Andhra Pradesh	20 lakhs birds died
2015	Heat wave	Andhra Pradesh, West Bengal and Odisha	>2500 human lives lost
Extreme rain/cloudburst			
26 Jul. 2005	Extreme rain	Mumbai, India	57.0″ of rain in 10 hours
2008	Kosi floods	North Bihar	527 deaths, 19,323 livestock perished
5 Aug. 2010	Cloudbursts	Leh, Ladakh	9.8″ in 1 hour
29 Sep. 2010	Extreme rain	NDA, Pune, India	5.7″ of rain in 1 hour
4 Oct. 2010	Cloudbursts	Pashan, Pune	7.2″ in 1.5 hours
1 Jul. 2016	Cloudbursts	Pithoragarh	54.0″ in 24 hours

Source: http://www.mospi.gov.in

Table 2.2 Most probable areas of India affected by diverse atmospheric stressors

Type of natural disaster (season)	Affected regions in India
Cloudburst-induced landslides	North East Himalaya (Arunachal Pradesh, Assam, Manipur, Meghalaya, Mizoram, Nagaland and Tripura), including Darjeeling and Sikkim Himalaya
	North West Himalaya (Uttarakhand, Himachal Pradesh and Jammu and Kashmir)
	Western Ghats and Konkan hills (Tamil Nadu, Kerala, Karnataka, Goa and Maharashtra)
	Eastern Ghats of Araku area in Andhra Pradesh
Cyclones April to December, with peak activity between May and November	Eastern coast (Bay of Bengal): Calcutta, Andhra Pradesh, Orissa, Tamil Nad and West Bengal.
	Western coast (Arabian Sea): mainly strike Gujarat and less frequently, Kerala
Heat waves March to July with peak temperatures in April, May, late October	South Indian: Khammam and Ramagundam (Telangana), Kalburgi and Bangalore (Karnataka)
	Eastern states: Bankura and Kolkata (West Bengal) and Bhubaneswar, Titlagarh and Jharsuguda (Odisha)
	North India: Punjab, Allahabad and Lucknow (UP), Gaya (Bihar), Delhi
	West India: Vidarbha and Marathwada (Maharashtra), Churu (Rajasthan), Ahmadabad (Gujarat)
	Central India: Jashpur (Chattisgarh), Harda (Madhya Pradesh)
Cold waves from December to March	Northern states: Punjab, Haryana, New Delhi, Jammu and Kashmir, Himachal Pradesh
	Eastern states: Bihar and Tripura
	Western states: Rajasthan
	Northern and central MP, Maharashtra
Frost (Dec, Jan, Feb)	Punjab, Himachal Pradesh, Haryana, Uttarakhand
Hailstorm (Mar, Apr, May)	Himachal Pradesh, Maharashtra, Madhya Pradesh, Karnataka, Punjab, West Bengal
Sand and dust storm (May, June)	North-western states of Rajasthan, Delhi
Smog due to crop-residue burning	Punjab, Haryana, Uttarakhand, Western Uttar Pradesh (Oct–Nov)
	Maharashtra, Uttar Pradesh, Tamil Nadu (Dec–March)

Projections are that future precipitation and storm changes will vary by season and region. Some regions may have less precipitation, some may have more precipitation, and some may have little or no change. The amount of rain falling in heavy precipitation events is likely to increase in most regions, while storm tracks are projected to shift poleward (IPCC 2013). Besides these hailstorm has widened its horizon. In recent years, weather events especially deadly heat waves and devastating floods have necessitated in understanding the role of global warming in driving these extreme events. These events are part of a new pattern of more extreme weather across the globe shaped in part by human-induced climate change (IPCC 2013, 2014). All weather events are now influenced by climate change because all weather now develops in a different environment than before.

2.3   Impact of Atmospheric Stresses on Agriculture

The agricultural commodities include field crops, horticultural crops, livestock, poultry and fisheries. The basic principle of agriculture lies with how crop/livestock interacts with atmosphere and soil/food as a growing medium. The system acts as pathway which regulates the intake of water/feed, nutrients and gas exchanges. Thus any change in the quality and quantity of atmospheric variables will certainly affect the pathway through changes in atmospheric and edaphic/feed factors. Besides these, climate change is also adding salt to the wound by aggravating the extreme weather events. Agriculture production is sensitive to temperature, increasing carbon dioxide concentration as well as change in precipitation. Impacts of all these forces together imply that agriculture production will respond non-linearly to future climate change. The impacts are complex to understand. However, there is high level of agreement across studies that the impact in all probability is going to be negative for most crop categories. The volume of loss of agricultural produce due to extreme weather events will be more as compared to global warming effects (IPCC 2012). According to global assessment report, India's average annual economic loss due to natural disasters is estimated to be 10 billion dollar in which cyclone and flood accounts for 0.5 and 7.5%, respectively (UNISDR 2015). In a similar manner, other atmospheric stresses like hailstorm, heat wave, cold wave, frost, etc. cause huge losses to Indian agriculture. Since agriculture makes up roughly 14% of India's GDP, a 4.5–9% negative impact on production implies a cost of climate change to be roughly up to 1.5% of GDP per year (Venkateswarlu et al. 2013). Similarly a temperature increase of 3–4 °C can cause crop yields to fall by 15–35% and 25–35% in Afro-Asia and Middle East, respectively (Ortiz et al. 2008).

In quantitative and qualitative terms, effect of aberrant changes in atmospheric variables on agriculture can be conjectured, and most of them are estimated to be negative. The impact may be of direct (mechanical) or indirect (physiological) depending on the type of stress, type of crop, stage of crop and mode of action of the stress on the commodity. Thus the likely effect of climate change on crop production adds to the already complex problem as yield in some of the most productive regions of the world is approaching a plateau or even declining (Pathak et al. 2003).

2.4   Mechanical Damage

2.4.1   Hailstorm

Though hurricanes, tornadoes and lightning command more dramatic attention, but worldwide, hail ranks as one of the most dangerous and destructive of all severe weather phenomenons (Rogers 1996) causing a severe damage to standing crops in a very short span of time (Berlato et al. 2005). Hails can damage the field as well as fruit crops which depend on stage of the crop, size and texture of hailstones and speed and force of hail as it hits the ground. A high-velocity impact early in the season can also cause substantial damage, the impact of which may not be recovered with time (Rogers 1996). Damages in mature produce quickly become focal points for diseases like brown/grey rot smut, while the disease may initially be limited to damaged tissues but can quickly spread to intact plant parts in warm, humid conditions (Awasthi 2015). Hailstorm causes primary injuries due to direct impact of hails which causes heavy defoliation, shredding of leaf blades, breaking of branches and tender stems, lodging of plants, peeling of bark, stem lesions, cracking of fruits, heavy flower and fruit drop, etc. This is followed by secondary injuries which are nothing but the manifestations of primary injuries like dieback or wilting of damaged plant parts, loss of plant height, staining, bruises, discoloration of damaged parts like leaves and fruits affecting their quality and rotting of damaged fruits and/or tender stems and branches due to fungal and bacterial infections (Bal et al. 2014).

2.4.2 Extreme Rain/Cloudburst

A cloudburst is an extreme amount of precipitation, sometimes accompanied by hail and thunder that normally lasts no longer than a few minutes but is capable of creating flood conditions. Cloudbursts effect landslides which are resultant of shear failure along the boundary of moving soil or rock mass. The prolonged and intensive rainfall during cloudburst results in pore-water pressure variations, and seepage forces trigger for landslides (Surya 2012). The process leading to landslide is accelerated by anthropogenic disturbances such as deforestation and cultivation of crops lacking capability to add to root cohesion in steep slopes (Kuriakose et al. 2009). In hilly regions orchards are mainly established on the slopes of the mountains. Extreme rain can wash away the top productive soil layer and in extreme cases can wash away the whole orchard along with the landslide (Duran-Zuazo and Rodriguez-Pleguezuelo, 2008). High intensity of rains can damage the seedling planted on the ridges and also damages to ridges and furrow system.

2.4.3 Cyclone

The Indian Ocean is one of the six major cyclone-prone regions of the world. In India, cyclones from Indian Ocean usually occur between April and May and also between October and December. The frequency of tropical cyclones in the north Indian Ocean covering the Bay of Bengal and the Arabian Sea is the least in the world (7% of the global total); their impact on the east coast of India as well as the Bangladesh coast is relatively more devastating (Suma and Balaram 2014). Cyclones in coastal areas severely affect agriculture sector through direct damage by high-speed wind, rain and extensive flooding. High tides also bring in saline water and sand mass making the fields unsuitable for agriculture (Kumar et al. 2014). Fruit can also be stripped from the trees by the force of wind. Severe wind can cause lodging and damage to especially perennial orchard crops. Storm damage in crops is especially determined by storm severity and crop maturity. A severe storm close to harvest is more serious than a less severe storm earlier in the season (Iizumi and Ramankutty 2015).

During cyclone, feed and fodder become scarce and consequently many animals starve to death; veterinary dispensaries and livestock aid centres become dilapidated; and productivity of animals decreases. The indirect effects include infection and disease of farm animals, fish and crop plants. In addition, agricultural marketing is adversely affected due to lean season of animal, fish and crop production (Kumar et al. 2014).

In fisheries, sea grasses and mangroves which provide nurseries to many coastal fish species are vulnerable to increased rates of damage if storms and cyclones become more intense (Waycott et al. 2011). If storms become more intense, greater levels of damage to shrimp ponds is expected as waves will penetrate further inlands due to rise in sea level. Floods caused by cyclones and more extreme rainfall events are expected to be a threat to ponds constructed in low-lying areas or close to rivers. Flooding can cause damage to ponds and other infrastructure and the escape of fish through overtopping of pond dykes by rising waters.

2.4.4 Sand/Dust Storms

Sandstorm/dust storm adversely impacts the agricultural activities. There is a direct loss of plant tissue as a result of sandblasting by the sand and soil particles (Stefanski and Sivakumar 2009). With this loss of plant leaves, there is reduced photosynthetic activity and therefore reduced energy for the plant to utilize for growth, reproduction and development of grain, fibre or fruit. If the timing of the sand and dust storms is early enough in the agricultural season, the plant might be able to regrow the lost leaves, and the loss in the final crop yield could be relatively minor. However, even in this instance, any regrowth of leaves will still probably result in yield losses (Sivakumar and Stefanski 2007). Additionally these storms delay plant development, increase end-of-season drought risk, cause injury and reduced productivity of livestock, increase soil erosion and accelerate the process of land degradation and decertification, fill up irrigation canals with sediments and affect air quality. Additionally, the loss of energy for plant growth delays plant development, and in regions with short growing seasons, the plant damage reduces yield during grain development, and if it occurs at maturity but before harvest, there will be a direct harvest loss (Alavi and Sharifi 2015; Stefanski and Sivakumar 2009). If there is a large enough deposit of sand or soil material early in the season, the young plant would be buried and killed due to lack of sunlight for photosynthesis. The loss of topsoil increases soil erosion and accelerates the process of land degradation and decertification by removing the layer of soil that is inherently rich in nutrients and organic matter. Livestock not sheltered from the storm could be directly harmed, and any stress from the physical environment to livestock can reduce their productivity and growth (Stefanski and Sivakumar 2009; Starr 1980).

2.5 Physiological Damage

2.5.1 High Temperature Stress

2.5.1.1 Crops

Physiological processes in the plants are essentially affected by the alteration of surrounded environmental temperature. However, high temperatures beyond certain optimum level reduce plant growth by affecting the shoot net assimilation rates, and thus the total dry weight of the plant is collectively termed as heat stress (Wahid et al. 2007), which is one of the most important factors limiting crop production (Bita and Gerats 2013). In higher plants, heat stress significantly alters cell division and cell elongation rates which affect both leaf size and leaf weight (Prasad et al. 2008). Heat stress induces changes in rate of respiration and photosynthesis and leads to a shortened life cycle and reduced plant productivity (Barnabas et al. 2008). Exposure of plants to severe heat stress decreases the plant growth especially stem resulting in decreased plant height (Prasad et al. 2006).

As heat stress becomes more severe, a series of processes occurs in plants which affect rates of important metabolic processes, including photosynthetic CO_2 assimilation, dark respiration and photorespiration (Sicher 2015). As stress increases, closure of stomata slows down or stops CO_2 diffusion, consequent increase in photorespiration and ultimately inhibits growth processes of the plant. High heat-associated water uptake issue aggravates heat stress problem. There is a major slowdown in transpiration leading to reduced plant cooling and internal temperature increase. It also leads to inhibition of photosynthesis due to stomal closure (Lafta and Lorenzen 1995). At the cellular level, as stress becomes more severe, there is loss of membrane integrity, cell membrane leakage and protein breakdown, and finally, if stress is severe enough, there can be plant starvation and death of the plant (Bita and Gerats 2013). The heat stress also varies with the duration of exposure to high temperature, degree of heat and crop genotypes (Kim and Lee 2011). The most important effect of heat stress on plants is the reduction in the growth rate; however, coupled effect of heat stress along with drought stress had more detrimental effect on growth and productivity of crops as compared to the effect of the individual stress (Prasad et al. 2008). The most affected stage is the reproductive growth, and the affected process is pollen grain development (Bita and Gerats 2013). Sexual reproduction and flowering have been extremely sensitive to heat stress, which often results in reduced crop plant productivity (Hedhly et al. 2009; Thakur et al. 2010).

In general, higher temperatures are associated with longer and intense radiation and higher water use. C_3 plants generally have a greater ability for temperature acclimation of photosynthesis across a broad temperature range, CAM plants acclimate day and night photosynthetic process differentially to temperature, and C_4 plants are adapted to warm environments (Yamori et al. 2014). CAM plants can also strengthen the CO_2 fertilization effect and the CO_2 anti-transpiring effect of C_3 and C_4 plants to a considerable extent. However, higher night temperature may increase dark respiration of plants and diminishing net biomass production. The reduction in crop yield in response to high temperature is due to disturbance of relationship of source and sink for assimilation of photosynthates (Morita et al. 2005; Johkan et al. 2011).

The negative impact on yield of wheat and paddy in most part of India is due to increase in temperature, water stress and reduction in number of rainy days. For every 1 °C increase in temperature, yield of wheat, soybean, mustard, groundnut and potato are expected to decline by 3–7% (Agarwal 2009). In the year March 2004, temperatures were higher in the Indo-Gangetic plains by 3–6 °C, and as a result, the wheat crop matured earlier by 10–20 days. Presently, the Indian lowlands are the source of approximately 15% of global wheat production, but it is anticipated that climate changes will transform these into a heat-stressed, short-season production environment (Bita and Gerats 2013). In sorghum, heat stress reduces the accumulation of carbohydrate in pollen grains and ATP synthesis in the stigmatic tissue (Jain et al. 2007). In the IPCC scenario with the slowest warming trend, maize and soybean yields were predicted to decrease by 30–46% before the end of the century (Schlenker and Roberts 2009). In the year 2003, Europe experienced a heat wave in July, where temperatures rose by 6 °C above average and annual precipitation 50% below average and resulted in record crop yield reduction (Ciais et al. 2005). Many legumes and cereals show a high sensitivity to heat stress during flowering and severe reductions in fruit set, most probably as result of reduced water and nutrient assimilate transport during reproductive development (Young et al. 2004).

In horticultural crops, temperature is the most important factor. In onion and tomato, bulb initiation and formation, its bulb and fruit size and qualities are affected by sudden rise in temperature. In most fruit crops, generally higher temperature decreased the day interval required for flowering and cooler temperature though required more days for flowering, but the number of flowers produced increased proportionally at this temperature (Srinivasarao et al. 2016). Optimum temperature range in citrus is 22–27 °C, and temperatures greater than 30 °C increased fruit drop (Cole and McCloud, 1985). During fruit development when the temperatures exceed the optimum range of 13–27 °C with temperatures over 33 °C, there is a reduction in sugar content, acid content and fruit size in citrus (Hutton and Landsberg 2000). These effects can lead to change in choice of orchid crop and geographical shift in cultivation of particular crop.

2.5.1.2 Livestock

Under heat stress, a number of physiological and behavioural responses of livestock vary in intensity and duration in relation to the animal genetic makeup and environmental factors (Freeman, 1987). Heat stress is one of the most important stressors especially in hot regions of the world. In humid tropics along with extended periods of high ambient temperature and humidity, the primary non-evaporative means of cooling (viz. conduction, convection and radiation) becomes less effective with rising ambient temperature, and hence under such conditions, an animal becomes increasingly reliant upon evaporative cooling in the form of sweating and panting to alleviate heat stress (Kimothi and Ghosh 2005).

In north Indian condition, livestock begins to suffer from mild heat stress when thermal heat index (THI) reaches higher than 72, moderate heat stress occurs at 80 and severe stress is observed after it reaches 90 (Upadhyay et al. 2009). In Queensland, the first-service pregnancy rate decreased in dairy cattle, and number of services per pregnancy increased with THI above 72 which corresponds to temperature 25 °C and RH 50% (McGowan et al. 1996). Thermal stress lowers feed intake and reduces animal productivity in terms of milk yield, body weight and reproductive performance. Heat stress reduces libido, fertility and embryonic survival in animals. Enhanced heat dissipation during heat stress may also lead to electrolyte losses (Coppock et al. 1982). The poor reproductive performance in buffaloes especially during summer months is due to inefficiency in maintaining the thermoregulation under high environmental temperature and relative humidity (RH) as those have dark skin and sparse coat of body hair which absorb more heat along with poor heat dissipation mechanism due to less number of sweat glands (Marai and Haeeb 2010).

Weak symptoms of oestrous are exhibited in buffaloes during summer (Parmar and Mehta 1994) which results in reduction of luteinizing hormone secretion and oestradiol production in anoestrus buffaloes (Palta et al. 1997) leading to ovarian inactivity, and also the survival of embryo in the uterus is impaired due to the deficiency of progesterone in the hot season (Bahga and Gangwar 1988). This endocrine pattern may be partially responsible for the low sexual activities and low fertility in summer season in the buffaloes. The poor nutrition and high environmental temperature are the two major factors responsible for long anoestrous and poor reproductive performance in Murrah buffaloes (Kaur and Arora 1984). Similarly heat stress in lactating dairy cows causes significant loss of serum Na^+ and K^+ (West 1999) and also reduces birth weights of Holstein calves (Collier et al. 1982). High ambient temperature can adversely affect the structure and physiology of cells as well as functional and metabolic alterations in cells and tissues including cells of immune system (Iwagami 1996). Heat stress in lactating animals results in dramatic reduction in roughage intake, gut motility and rumination which alters dietary protein utilization and body protein metabolism (Ames et al. 1980). Temperature extremes can influence disease resistance in dairy calves (Stott et al. 1976; Olsen et al. 1980).

2.5.1.3 Poultry

Heat stress interferes with the broilers comfort and suppresses productive efficiency, growth rate, feed conversion and live weight gain (Etches et al. 1995; Yalcin et al. 1997) due to changes in behavioural, physiological and immunological responses. With rise in ambient temperature, the poultry bird has to maintain a balance between heat production and heat loss. This forces the bird to reduce its feed consumption by 5% to reduce heat from metabolism to a tune for every 1 °C rise in temperature between 32–38 °C (Sohail et al. 2012). In addition, heat stress leads to reduced dietary digestibility and decreased plasma protein and calcium levels (Bonnet et al. 1997; Zhou et al. 1998). Heat stress limits the productivity of laying hens, as reflected by egg production and egg quality, as the bird diverts feed metabolic energy to maintain its body temperature and also lower egg production and egg quality (Hsu et al. 1998; Tinoco 2001). The resulting hyperventilation decreases CO_2 blood levels, which may decrease eggshell thickness (Campos 2000). Plasma triiodothyronine and thyroxine, which are important growth promoters in animals, adversely affect heat-stressed broiler chickens (Sahin et al. 2001).

There are direct effects on organ and muscle metabolism during heat exposure which can persist after slaughter (Gregory, 2010); however, chronic heat exposure negatively affects fat deposition and meat quality in broilers (Imik et al. 2012). In addition, heat stress is associated with depression of meat chemical composition and quality in broilers (Dai et al. 2012). Chronic heat stress decreased the proportion of breast muscle, while increasing the proportion of thigh muscle in broilers (Lara and Rostagno, 2013) and protein content lower and fat deposition higher in birds subjected to hot climate (Zhang et al. 2012). Heat stress causes decrease in production performance, as well as reduced eggshell thickness and increased egg breakage (Lin et al. 2004). Additionally, heat stress has been shown to cause a significant reduction of egg weight, eggshell weight, eggshell per cent (Lara and Rostagno 2013) and all phases of semen production in breeder cocks (Banks et al. 2005). In hotter climate, immune-suppressing effect of heat stress is more on broilers and laying hens (Ghazi et al. 2012) and will alter global disease distribution (Guis et al. 2012) through changes in climate. This may also increase the insect vectors, prolong transmission cycles or increase the importation of animal reservoirs. Climate change would almost certainly alter bird migration and directly influence the virus survival outside the host (Gilbert et al. 2008).

2.5.1.4 Fisheries

Climate change will affect fisheries and aquaculture via acidification, changes in sea temperatures and circulation patterns, frequency and severity of extreme events and sea level rise and associated ecological changes (Nicholls et al. 2007). However, inland aquaculture will be affected by changing temperatures, water scarcity and salinization of coastal waters (Shelton 2014). Increased temperature may affect the distribution pattern of some fish species where some of them may be migrate to the higher latitude for cooler place (Barange and Perry 2009). Changes in temperature will have direct effects on swimming ability (Van-der-Kraak and Pankhurst 1997). Sea level rise due to glacier melting will destroy the mangrove forest as well as destroy the marine fish nursery ground. With rising temperature, the physiological activity of the fishes also increases with increase in oxygen demand, whereas the solubility of the oxygen in water is inversely related to temperature and salinity (Chowdhury et al. 2010). Thus, in pond culture system, critically low oxygen concentrations occur overnight when all aquatic organisms use the dissolved oxygen for respiration, and the decrease in dissolved oxygen-induced hypoxic condition results in reduction of growth and reproduction success of fishes (Weiss 1970). Temperature rise also increases the evaporation rate which will ultimately reduce the surface and volume of water in the fish ponds (Bhatnagar and Garg 2000).

With global warming, tropical and subtropical areas will experience more reduction in ecosystem productivity than temperate and polar ecosystems (Shelton 2014).

Marine fisheries sector is already overexploited due to overfishing (Hilborn et al. 2003) and inland fisheries already affected due to pollution, habitat alteration and introduction of alien species/culture fish (Allan et al. 2005). The effects of increasing temperature on freshwater and marine ecosystems where temperature change has been rapid are becoming evident, with rapid poleward shifts in distributions of fish and plankton (Brander 2007). High temperature can cause stratification leading to algae blooms and reduced levels of dissolved oxygen. Tilapia can tolerate dissolved oxygen concentration as low as 0.1–0.5 mg L^{-1} but only for a limited period (Bell et al. 2011). Though tilapia and carp are considered hardy fish, repeated or prolonged exposure to extreme temperature and low dissolved oxygen, especially at higher stocking densities, increase the stress and the susceptibility of the fish to other physiological complications and diseases. Reproduction of fish is highly sensitive to fluctuation in temperature. For example, the fish spp. tilapia can tolerate temperature up to 42 °C, whereas exposure to high temperature results in more deformities in early larval stage and sex ration skewed towards male. In river Ganga, an increase in annual mean minimum water temperature by 1.5 °C has been recorded in the upper cold-water stretch of the river and by 0.2–1.6 °C in the aquaculture farms in the lower stretches in the Gangetic plains (Vass et al. 2009). This change in temperature has resulted in a perceptible biogeographically distribution of the Gangetic fish fauna (Menon 1954).

2.5.2 Extreme Rain and Cloudburst

2.5.2.1 Crops

Extreme rain or cloudburst causes severe waterlogging in poorly drained areas and can impact crop growth in both the short and long term through oxygen deprivation or anoxia through a number of biological and chemical processes. Germinating seeds and emerging seedlings are very sensitive to waterlogging as their level of metabolism is comparatively higher (Tuwilika 2016). The first symptom of flooding damage is stomata closure, which affects not only gas exchange but also decreases the passive absorption of water, which is also negatively influenced by anaerobic conditions in the rhizosphere (Kozlowski and Pallardy 1997). A decrease in transpiration leads to leaf wilting and early senescence and finally resulting in foliar abscission (Ashraf 2012).

There are large differences in plant tolerance to flooding and insufficient aeration of root media among herbaceous species (Das 2012). Waterlogged conditions reduce root growth and can predispose the plant to root rots. Therefore, plants having experienced waterlogging display nitrogen and phosphorus deficiencies due to restricted root development (Postma and Lynch 2011). In light-textured soils, waterlogging impacts growth in cereals by affecting the availability of nitrogen in the soil through excessive leaching of nitrate nitrogen beyond the rooting zone. In heavier soils, nitrate nitrogen can be lost through denitrification (Aulakh and Singh 1997). The amount of loss depends on the amount of nitrate in the, soil temperature and length of time that the soil is saturated (Tuwilika 2016).

2.5.3 Frost and Cold Wave

2.5.3.1 Crops

Cold or low temperature stress comprises of chilling (<20 °C) and freezing temperatures (<0 °C) those that hamper the plant growth and development. Generally, exposure to cold temperature affects developmental events in the shoot apex which directly determine the differentiation of the panicle and hence potential yield and spikelet fertility resulting in fewer grains. In addition, photosynthesis is impaired which reduces growth and results in indirect yield loss because there is less carbohydrate available for grain production (Takeoka et al. 1992).

In India frost mainly occurs in the translunar plans and elevated hilly areas. It is a phenomenon when gross minimum temperature drops to below 0 °C in a short time and plants suffer injury. In plants conversion of cellular water into ice is a major reason for cell rupture in cold stress (Mckersie and Bowley 1997; Olien and Smith 1997). In Punjab, potato crop suffered heavily during 2006 and 2008 when the timely planted crop (mid-Sept. to mid-Oct.) performed better than late-planted crop (late Oct. to mid-Nov.). In late-planted crop, yield reduced by 30–60% (Arora et al. 2010). The temperature at which frost damage occurs in a crop depends on the species and type of cultivar. In potato, frost damage is likely to occur when the temperature drops to −2 °C or lower, and it can cause partial or complete loss of leaf area leading to a reduction in photosynthesis and hence yield (Carrasco et al. 1997). According to Angadi et al. (2000), temperatures below 10 °C result in slower and reduced growth and premature stem elongation in rapeseed and mustard. At early stage of plant growth, various phenotypic symptoms occur in response to chilling stress; however, at flowering stage low temperature may cause delayed flowering, bud abscission and sterile or distorted flowers, while at grain filling the source-sink relation is altered, kernel filling rate is reduced, and ultimately small-sized, unfilled or aborted seeds are produced (Jiang et al. 2002; Thakur et al. 2010). Low temperatures affect not only normal heading but also panicle exertion and prevent the normal elongation of internodes of rice (Farrell et al. 2006). In case of fruit crop, damage to trees was relatively more in low-lying areas where cold air settles and remains for longer time on the ground. Frost during early flowering and ear emergence can result in partial or complete sterility of florets and spikelets and therefore reduced grain number and yield (Al-Issawi et al. 2012).

2.5.3.2 Livestock

Despite the absence of a challenge to homeothermy in cattle, there are marked seasonal fluctuations in the cattle's level and efficiency of production which probably arise from hormonal and adaptive changes occurring as a consequence of mild cold stress (Young 1981). In cold stress, cow's requirement for nutrient and energy intake increases due increased metabolism rate. Freezing of fodder crops results in changes their metabolism and composition that can be toxic to livestock (FAO 2016). Here, two problems need to be considered – prussic acid poisoning and bloat. Prussic acid is not normally present in plants, but under certain conditions, forage can accumulate large quantities of cyanogenic glycosides which can convert to prussic acid

(Robson 2007) a potent, rapidly acting poison, which enters the bloodstream of affected animals and is transported through the body. It inhibits oxygen utilization by the cells, and the animal dies from asphyxia. Prussic acid poisonings occurs in areas associated with light frost. Cold condition stimulates appetite of animals, which may be slightly beneficial for production but the same may reduce utilization efficiency of dietary energy (Young 1981). Cold environment increases the whole-body glucose turnover and glucose oxidation, thus resulting in less production of ketones and resulting increased metabolic rate (Ravussin et al. 2014).

2.5.3.3 Poultry

Cool temperatures are the primary triggers for the accumulation of fluid causing abdominal swelling (ascites) during commercial broiler production (Wideman, 2001) which accounts for losses of about US$ 1 billion annually worldwide (Maxwell and Robertson 1997). The incidences are higher in the colder environmental temperatures (Wideman 1988; Shlosberg et al. 1992; Yahav et al. 1997), because cold ambient temperatures increase the cardiac output, oxygen requirement and blood flow and result in increased pulmonary arterial pressure overload on the right ventricle (Julian et al. 1989). In white Leghorn hens, a reduction in environmental temperature from 20 to 2 °C almost doubles the oxygen requirement (Gleeson, 1986). During the development of ascites, birds exhibit classic haematological changes. Haematocrit, haemoglobin and red blood cell counts (RBC) all increase dramatically (Cueva et al. 1974; Maxwell et al. 1986, 1987; Yersin et al. 1992). Birds exposed to cold stress have severely injured liver and affects thyroid hormones which play a key role in energy expenditure and body temperature homeostasis (Nguyen et al. 2016). Moderate cold exposure during early post-hatching period causes long-term negative effect on growth performance of chicken (Baarendse et al. 2006).

2.5.3.4 Fisheries

Depth of water plays an important role in fish growth and development. When water depth in fish pond is less than 0.5 m, rapid lowering of temperature in a shallow depth reduces dissolved oxygen level, which is detrimental to the fish's survival (Changi and Ouyang 1988). Higher air temperatures and incoming solar radiation increase the surface water temperatures of lakes and oceans (Hader et al. 2015). Foggy and cloudiness and lack of solar radiation or photosynthesis by aquatic vegetation also raise carbon dioxide levels and deplete oxygen levels in water bodies. Under acute cold condition, when water temperature is around 10–12 °C in the month of December 2002 and January 2003, bacterial septicaemia and fungal infections were reported in some ponds (Declercq et al. 2003). Stray incidences of fish/prawn mortality (3–5%) take place of fishponds. This may probably due to emaciated growth because of low feed intake during extreme winter month.

2.5.4 Air Quality and Pollution

2.5.4.1 Crops

Among the major greenhouse gases are methane and tropospheric ozone, which are both of concern for air quality (West et al. 2007). Primary air pollutants, such as sulphur dioxide, nitrogen oxide and particulates, are emitted directly into the atmosphere. These are generally present in high concentrations in urban areas or close to large point sources, such as around thermal power stations have large effects on local farming communities. Secondary pollutants like tropospheric (ground level) ozone are formed by subsequent chemical reactions in the atmosphere and have increased historically at the surface in industrialized regions and in the global background troposphere (Vingarzan 2004). Ozone is considered to be the most powerful pollutant for its impacts on crops (Fuhrer et al. 1997). Ozone symptoms characteristically occur on the upper surface of affected leaves and appear as bleaching of the leaf tissues. Ozone exposure has a strong linear relationship to yield loss in wheat grown in field experiments (Fuhrer et al. 1997).

At lower atmosphere, chances of encountering radiation-absorbing aerosols or chemical substances such as ozone (O_3) and sulphur dioxide are more than it does at lower elevations. With regard to plants, UV-B (280–320 nm) impairs photosynthesis in many species; while rice is considered to be sensitive to elevated levels of UV-B (Teramura et al. 1991), cotton is known to be sensitive to O_3 (Temple 1990). India comes under region with high photochemical smog and the condition most likely to aggravate in the future (Krupa and Kickert 1993). UV-B radiation damages DNA, proteins, lipids and membranes and increases plant susceptibility to disease (Hidema and Kumagai 2006). Excess UV-B can directly affect plant physiology and cause massive amounts of mutations. However, it has not yet been ascertained whether an increase in greenhouse gases would decrease stratospheric ozone levels. When the concentration of sulphates and nitrates increases in the atmosphere, very fine acidic particles are formed, and when nitrogen oxides and reactive organic gases combine, especially on sunny, still days, a photochemical (ozone) smog is formed. NO_2 is the pollutant closely associated with ozone (Cho et al. 2011) because of the role that NO_2 plays as a precursor of O_3 in polluted air.

Sulphur dioxide enters the leaves mainly through the stomata, and acute injury is caused when leaves absorb high concentrations of sulphur dioxide in a relatively short time. Newly expanded leaves usually are the most sensitive to acute sulphur dioxide injury. And different plant species and varieties of the same species may vary considerably in their sensitivity (McCormac and Varney 1971). The effects of SO_2 on crops are influenced by other biological and environmental factors such as plant type, age, sunlight levels, temperature and humidity. Thus, even though sulphur dioxide levels may be extremely high, the levels may not affect vegetation (Cohen et al. 1981). The burning of sulphur-containing fuels release SO_2, and when the atmosphere is sufficiently moist, the SO_2 transforms into tiny droplets of sulphuric acid and causes acid rain (Ahrens 2015). Acid rain damages the protective waxy coating of leaves and allows acids to diffuse into them, which interrupts the evaporation of water and gas exchange so that the plant can no longer breathe (Izuta

2017). Elevated CO_2 stimulates photosynthesis leading to increased carbon (C) uptake and assimilation, thereby increasing plant growth (Kant et al. 2012). However, as a result of differences in CO_2 use during photosynthesis, plants with a C_3 photosynthetic pathway often exhibit great growth response relative to those with a C_4 pathway (Poorter 1993; Rogers et al. 1997).

Particulate matter such as cement dust, magnesium-lime dust and carbon soot deposited on vegetation can inhibit the normal respiration and photosynthesis mechanisms within the leaf (Thara et al. 2015). Cement dust causes chlorosis and death of leaf tissue by the combination of a thick crust and alkaline toxicity in wet weather (Griffiths 2003), and dust coating affects the normal action of pesticides and other agricultural chemicals applied as sprays to foliage (Ravichandra 2013). In addition, accumulation of alkaline dusts in the soil can increase soil pH to levels adverse to crop growth (Thara et al. 2015). Air pollution injury to plants can be evident in several ways. There may be a reduction in growth of various portions of a plant, or plants may be killed outright. Acute symptoms of injury from various pollutants in different horticultural and agronomic groups are visible on the affected plant. Symptom expressions produced include chlorosis, necrosis, abscission of plant parts and effects on pigment systems (Taylor 1973).

2.5.4.2 Livestock

Air contains fine suspended particles, bacteria, pollens, which cause allergies and respiratory diseases in animals (Takizawa 2011). Pig and poultry are kept in indoor facilities for a variable part of their life. However, for dairy cattle, goat and sheep; the housing facilities are quite open and air quality is to a certain extent comparable with the outdoor air quality. Also in many piggeries, high levels of ammonia, airborne dust, endotoxin and microorganisms can be found (Wathes et al. 1998). Ammonia is considered as one of the most important inhaled toxicants in agriculture, and long-term low-level exposure causes mucosal damage, impaired ciliary activity and secondary infections in laboratory animals (Davis and Foster 2002).

Dust particles within a livestock farming environment consist of up to 90% organic matter (Aarnink et al. 1999; Heber et al. 1988), which provides opportunities for bacteria and odorous components to adhere themselves to these particles. The contaminated air is dissipated into the external environment via ventilation; however, the concentration of airborne contaminants can be still higher within livestock building (Arogo et al. 2006). Dust particles that can potentially harm livestock is grouped as inhalable dust particles-PM100 (less than100 microns in diameter), thoracic dust particles-PM10 (less than 10 microns in diameter) and repairable dust particles-PM 5 (less than 5 μ in diameter). PM5 can enter the smallest cavities of the lung, the alveolithus making them the most hazardous causing shortness of breath, chronic bronchitis, asthma and other lung diseases (Choiniere 1993; Banhazi 2009). CO_2 is considered to be a potential inhalation toxicant and of CO_2 in the blood caused acidosis in animal (Hernandez et al. 2014). Similarly fluoride-contaminated forage that is eaten by cattle or sheep may cause fluorosis.

2.5.4.3 Poultry

In poultry houses, birds and their wastes generate different forms of air pollution, namely, ammonia, carbon dioxide, methane, hydrogen sulphide and nitrous oxide gases, as well as the dust (Kocaman et al. 2005). Inadequate ventilation in the poultry shed accumulates gases such as carbon dioxide, ammonia and methane and reach toxic levels (David et al. 2015). Most ammonia originates from the decomposition of the nitrogen-containing excretion from the kidneys and the gut of the bird (Groot-Koerkamp and Bleijenberg 1998). Poor environments normally don't cause disease directly, but they do reduce the chickens' defence mechanism, making them more susceptible to existing viruses and pathogens (Quarles and Kling 1974). Aerial ammonia in poultry facilities is usually found to be the most abundant air contaminant. Ammonia concentration varies depending upon several factors including temperature, humidity, animal density and ventilation rate of the facility. Chickens exposed to ammonia shows reductions in feed consumption, feed efficiency, live weight gain, carcass condemnation and egg production (Reece and Lott 1980). The presence of dust in animal housing adversely affects health, growth and development of animals and increases disease transfer within flocks (Feddes et al. 1992).

2.5.4.4 Fisheries

Oceans absorb approximately 25% of anthropogenic CO_2 (Logan 2010), and these dissolved CO_2 reacts with seawater to form weak carbonic acid, causing pH to decline and reducing the availability of dissolved carbonate ions which is required by many marine calcifying organisms to build their shells or skeletons (Orr et al. 2005). Till date, average alkalinity of ocean has declined from 8.2 to 8.1 (IPCC 2007c), equivalent to a 30% increase in acidity. Havenhand et al. (2008) reported that expected near-future levels of ocean acidification reduce sperm motility and fertilization success of the sea urchin and suggest that other broadcast spawning marine species may be at similar risk. Similarly, impacts on oxygen transport and respiration systems of oceanic squid make them particularly at risk of reduced pH (Portner and Langenbuch 2005). Fish embryos and larvae are more sensitive to pH change than juvenile and adults (Brown and Sadler 1989). Extremely high or low pH values in water cause damage to fish tissues, especially the gills, and haemorrhages may occur in the gills and on the lower part of the body and cause secondary infection (Declercq et al. 2013). The toxicity of ammonia is affected by the amount of free CO_2 in the water as the diffusion of respiratory CO_2 at the gill surface reduces the pH of the water (Svobodova et al. 1993). Radiation plays an important role in aquatic ecosytem. UV-B radiation (280–320 nm) impairs growth and photosynthesis in many species, especially in phytoplankton.

2.6 Management Strategies to Cope Atmospheric Stress Events

Global climate change will have significant impacts on future agriculture, and therefore climate change mitigation for agriculture is a global challenge. In a country like India, one of the most vulnerable countries owing to its large agricultural sector, vast population, rich biodiversity, long coastline and high poverty levels will be severely affected by climate change if new strategies for amelioration are not devised (Nelson 2009; Fischer and Edmeades 2010). For this a thorough understanding is required for various physical, physiological, metabolic and biochemical processes that occurs in normal as well as stresses environments so as to form the basis for developing climate smart mitigation strategies. Nevertheless, plant and livestock responses to high temperatures clearly depend on genotypic parameters, as certain genotypes are more tolerant (Prasad et al. 2006; Challinor et al. 2007). Though plants adapt to various stresses by developing more appropriate morphological, physiological and biochemical characteristics, analysing plant phenology in response to heat stress often gives a better understanding of the plant response and facilitates further molecular characterization of the tolerance traits (Wahid et al. 2007). As far as atmospheric stresses are concerned, a complete insight of the biological processes behind the atmospheric stress response combined with classical and emerging technologies in production, breeding and protection engineering is likely to make a significant contribution to improved productivity and reduction in losses. The following section contains some of the adaptation and management options available to mitigate those atmospheric stresses.

2.6.1 High Temperature Stress

2.6.1.1 Crops

High temperature stress could be avoided by agronomic/crop management practices. The selection of type of tillage and planting methods play an important role in emergence and growth of a crop. The methods of planting may improve the plants tolerance against heat stress through the soil moisture. The presence of crop residues/mulch on the soil surface keeps soil temperature lower than ambient during the day and higher at night. Mulches also conserves soil moisture (Geiger et al. 1992), thereby making it available to crop for a longer period which augments transpiration, keeps the canopy cool and protects the crop from terminal heat and decline in yield. The selection of cultivar as per the agroecological conditions and probable temperature variability scenario during the crop growth period is important to get better yield under high temperature stress conditions. Although periods of elevated temperature may occur at any period during the growing season of the crop, earlier sowing, use of earlier maturing cultivars and adopting heat stress tolerant cultivars are the best options (Krishnan et al. 2011). Selecting optimum planting time helps in avoiding high temperature stress during anthesis and grain filling so that crop escapes the hot and desiccating wind during grain-filling period which is one of the critical stages.

Water management is critical from the view point that damage potentiality due to high temperature stress is commonly associated with water stress, and plants can tolerate heat stress until crop keeps on transpiring. Continuous supply of water to heat-stressed crop helps to sustain grain-filling rate, duration and size of grain (Dupont et al. 2006). Water-stressed plants attempt to conserve water by closing their stomata; as a consequence evaporative cooling diminishes, leaf temperatures increase drastically, and at that point, the metabolic activity in plants stops. Proper irrigation scheduling as per soil type, crop type, stage of the crop as per the available water and weather may help in mitigating the effects of heat stress on crop.

Chemicals having potential to protect the plants against high temperature (Hasanuzzaman et al. 2012) is another option. Proline accumulates to high concentration in cell cytoplasm under stress conditions without interfering with cellular structure or metabolism. Exogenous proline guarantees the protection of vital enzymes of carbon and antioxidant metabolism and improved water and chlorophyll content as the basis of heat tolerance in chickpea plants (Kaushal et al. 2011). Activities of different antioxidant enzymes are temperature sensitive, and activation varies with different temperature ranges, tolerance or susceptibility of different crop varieties, their growth stages and growing season (Chakraborty and Pradhan 2011). In wheat, spray of indole acetic acid, gibberellic acid and abscisic acid significantly improves grain yield under high temperature stress (Cai et al. 2014).

Phytohormone/bioregulators like salicylic acid also play significant role in the regulation of plant growth and development. Salicylic acid improves the plant growth and yield of maize by enhancing photosynthetic efficiency (Khan et al. 2003). In mustard, the same with lower concentration as foliar application increased the H_2O_2 level and reduced the catalase which amplifies the potential of plants to withstand the heat stress (Dat et al. 1998). Abscisic acid (ABA) as a plant growth hormone regulates stomata opening, root hydraulic conductivity and application of α-tocopherol and SA decreased consumption of photosynthates and increased membrane stability which aided in transport of photosynthates, which thereby induce tolerance to different abiotic stresses such as drought, salinity and temperature stresses increased yield (Farooq et al. 2008).

Application of nitrogen, phosphorus and potassium improves plant growth under moderate heat stress (Dupont et al. 2006). Under high temperature stress, foliar application of thiourea promotes root growth by enhancing assimilate partitioning to root at seedling and pre-anthesis growth stages (Anjum et al. 2011). Potassium involves in several physiological processes, i.e. photosynthesis, translocation of photosynthates into sink organs, maintenance of turgidity and activation of enzymes are increased under stress conditions (Mengel and Kirkby, 2001), and the deficiency of potassium resulted in decrease in photosynthetic CO_2 fixation and impairment in partitioning and utilization of photosynthates. Application of KNO_3 and zinc can also improve heat tolerance in wheat (Graham and McDonald 2001). Calcium is found to control guard cell turgor and stomatal aperture (Webb et al. 1996). The foliar application of potassic fertilizer, urea and zinc may help in improving crop yield by alleviating the ill effects of high temperature.

Due to incomplete understanding of tolerance mechanisms and accurate phenotyping for stress responses, the conventional and molecular plant-breeding efforts have been limited (Bita and Gerats 2013). In addition, generating high-yielding and stress-tolerant crops requires not only a thorough understanding of the metabolic and developmental processes involved in stress responses but also in energy regulation (Hirayama and Shinozaki, 2010). High temperature stress is detrimental to cereal crop productivity, and the existence of genetic variability in heat stress toler-ance is an indispensable factor for the development of more tolerant cultivars. If limited variability for tolerance to heat is available within a crop species, wild germ-plasm can also be used as tolerance source (Pradhan et al. 2012).

Agroforestry as an integral component of conservation agriculture can play a major role as trees can buffer climate extremes that affect crop performance. In particular, the shading effects of trees can buffer soil as well as canopy temperature. In addition it also sublimes atmospheric saturation deficit and reduces exposure to supra-optimal temperatures, of which physiological and developmental processes and yield become increasingly vulnerable (Challinor et al. 2005). Trees in farms bring favourable changes in field microclimate by influencing radiation flux (radia-tive and advective processes), air temperature, wind speed and saturation deficit of understory crops all of which will have a significant impact on modifying the rate and duration of photosynthesis, transpiration, etc. (Monteith et al. 1991).

2.6.1.2 Livestock

In animal husbandry, physical modifications of environment, genetic development of heat tolerant breeds and nutritional stress management are the three major key components to sustain production in hot environment (Beede and Collier, 1986). Scientifically designed sheds provide comfortable environment to animals. However, in hot and humid areas though shade reduces heat accumulation, it hardly reduces air temperature or relative humidity, and additional cooling is necessary for farm animals (St-Pierre et al. 2003). Genetic variation exists among animals for cooling capability and stress tolerance. Heat shock proteins (HSP) play important role in stress responses of animals. It was observed that genotype differences exist in HSP genes of indigenous breed and cross-bred dairy cattle indicating the relative heat stress tolerance phenotype of native indigenous cattle (Sajjanar et al. 2015). These results indicate that more heat-tolerant animals can be selected genetically or cross-breeding programme (Kimothi and Ghosh 2005). As adaptation to heat stress requires the physiological integration of many organs and systems, viz. endocrine, cardiorespiratory and immune system (Altan et al. 2003), the administration of anti-oxidants has proved useful for improvement of several immune functions (Victor et al. 1999).

Antioxidants, both enzymatic and non-enzymatic, provide necessary defence against oxidative stress as a result of thermal stress (Rahal et al. 2014). Both vitamin C and vitamin E have antioxidant properties. Vitamin C and E having antioxidant property along with electrolyte supplementation was found to ameliorate the heat stress in buffaloes (Sunil Kumar et al. 2010). Zinc and other trace elements like cop-per and chromium act as typical antioxidants as they work indirectly as normal copper levels are necessary to maintain the structural integrity of DNA during oxi-dative stress (Rahman 2007). Additional supplementation of electrolytes (Na, K and Cl) is one among the nutritional strategies which has beneficial effects in heat-stressed dairy cows in terms of milk yield, acid base balance and lower temperature (Sanchez et al. 1994). Lactating cows and buffaloes have higher body temperature of 1.5–2 °C than their normal temperature. Therefore to maintain thermal balance, they need more efficient cooling devices to reduce thermal load (Upadhyay et al. 2009). Provision of sprinklers and fans to mitigate heat stress facilitates the buffalo heifers to reduce the heat load and increased the time of lying down (Tulloch 1988).

2.6.1.3 Poultry

In hot and humid environment, poultry shades should be designed as an open style house with proper shading for adequate air movement, grass cover on the ground surface to reduce sunlight reflection and shiny surface roof for more reflection of solar radiation. During hot periods, lower-protein diets supplemented with limiting amino acids should be replaced with high-protein diet (Pawar et al. 2016). Glucose in drinking water helps in alleviating the influence of heat stress on whole-blood viscosity and plasma osmolarity (Zhou et al. 1998). Aviaries should be equipped with overhead sprinkler systems, which cool the air and reduce the chances of heat injuries. Addition of ammonium chloride and potassium chloride to drinking water is desired to maintain carbon dioxide and blood pH under control. Vitamin E supplementation is beneficial to the egg production of hens at high temperature and associated with an increase in feed intake (Kirunda et al. 2001).

2.6.1.4 Fisheries

Genetic variation also exists among fish species for thermal tolerance against increasing diurnal water temperature and induced hypoxia condition in which heat shock proteins (HSP) play important role in stress responses and determinant for critical thermal maxima and critical thermal minima of each fish species which is also dependent upon fish habitat and adaptation. Brahmane et al. (2017) identified that tilapia, *Oreochromis mossambicus* juveniles, responds to constant rearing temperatures with significantly higher growth achieved at 30 °C as compared to 25 °C. Besides genetic screening, few adaptive measures may be taken as restoration of mangrove forest can protect shorelines from erosion and provide breeding ground for fish while sequestering carbon (Daw et al. 2009). In addition, deepening of the stock area, raising bund height, adjusting crop calender and planting shade tress in and around the stock area would counter the higher temperature. (FAO 2012).

2.6.2   Extreme Rain and Cloudburst

2.6.2.1 Crops

Conservation agriculture provides alternatives that can address challenges posed by erosion due to intense rain events. Soil surface covered by plant residues not only increases water infiltration and cuts down soil erosion and runoff but also mitigates some of the challenges presented by climate change (FAO 2009). On the other hand, CA practices increase carbon accumulation (sequestration) through recycling of crop residues and increase the nutrient supply and turnover capacity of soils and resulted in significant changes in the physio-biological properties of the soils. Zero tilled fields covered with organic mulch result in enhanced erosion check, increased water-holding capacity of soils and ability to withstand longer dry spells during crop growing period (Bhattacharyya et al. 2015).

2.6.3 Frost and Cold Wave

2.6.3.1 Crops

Long-term counter measures for frost protection mainly include selection of favourable topography, breeding of frost-tolerant variety, non-use and overuse fertilizer and adjusting the sowing time (Arora et al. 2010). The fruit crops under frosty weather can be protected by creation of hot air or smoke, covering small fruit plant areas with straw, dry grass, etc. (Rathore et al. 2012). Sprinkler irrigation releases latent heat of fusion by releasing heat into the surrounding air and maintains soil moisture in the soil profile. Growth regulator and other growth-promoting chemicals also enhance resistance to cold stress (Colebrook et al. 2014). Windbreaks or shelter belts if raised around the plantations prevent the convective frost or cold wave damage (Rathore et al. 2012). In the places where frost is a recurrent phenomenon, new orchard growers should select low temperature-tolerant fruit species and varieties depending upon their leaf structure, succulency and concentration of solutes as these reutilize water formed after melting of crystals for initiating metabolic activities in the cells (Rathore et al. 2012).

2.6.3.2 Livestock

The problem of fodder poisoning can be prevented or at least minimized with proper management of the fodder field and feeding pattern (Vough 1978). Removing livestock from pastures for several days after a frost is the best preventative management strategy to reduce prussic acid poisoning in case of Sudan grass and sorghum-Sudan grass pastures (Lemengar and Johnson 1997). In periods of cold weather, provision of windbreak is to be made enabling the cows to take shelter in the leeward side of the cold wave and preferably reducing the walking area so that animals stand in a group to stay warm. Appropriate nutritional supplementation is the key to managing cold stress.

2.6.3.3 Poultry

Poultry house made up of low-heat-conducting materials like bamboo and wood helps to maintain optimum night temperature in the shelter. In addition, with the help of light bulbs and physical barriers, movement of chicks should be restricted nearer the heat source. The orientation of poultry house should be east-west alignment for proper ventilation and to gain maximum solar energy during winter. The surface of poultry house should be covered with a bedding material called litter; it maintains uniform temperature and also absorbs moisture and promotes drying (Banday and Untoo 2012). In winter season, energy-rich source like oil/fat should be added to the diet so that the requirement of other nutrients will be reduced.

2.6.3.4 Fisheries

Adoption of hardy fish species like tilapia and carp may be taken up. Though the optimum range for growth of common carp is similar to tilapia, i.e. 23–30 °C, carp is much more cold tolerant than tilapia (Bell et al. 2011). To escape the cold stress due to lowering of temperature, deep water ponds (100–200 cm) may be adopted compared to shallow water ponds (<50 cm). Though the optimal temperature range for growth of common carp is similar to tilapia, i.e. 23–36 °C, carp is much more cold tolerant than tilapia. Intermittent aeration of water would probably help to alleviate respiratory stress to fishes.

2.6.4 Hailstorm

2.6.4.1 Crops

Shelter belts and windbreaks around orchard are always recommended to avoid heavy damage to the main crop, which also lowers the water requirements and other related stresses. In areas with higher probability of hailstorm occurrence, shade nets can be a good option especially for high-value crops along with nylon nets used for protection against bird damage. Though little information is available on measures for faster recovery in hail-damaged plants, application of additional nitrogen encourages new growth (Patel and Rajput 2004; Badr and Abou El-Yazied 2007). In case of orchard crops, broken branches and twigs should be removed followed by spraying of recommended chemicals to avoid secondary fungal and bacterial infections. The fallen fruits should be removed to reduce the spread of disease and pest during their decay. Water-based paint should be applied on large wounds on trunks and branches to avoid desiccation and disease infection. Fruit thinning by removal of hail-damaged fruits improves yield and quality of remaining fruit. Bud-breaking chemicals and growth/bioregulators may be applied to induce the vegetative growth in orchard crop along with fertilizers (Bal et al. 2014; Boyhan and Kelley 2001). Proper drainage facilities are to be provided to avoid waterlogging and to avoid secondary infection of diseases. Near-maturity bulb crops like onion and garlic may be harvested to avoid rotting.

2.6.5 Cyclone

2.6.5.1 Crops

Pre-cyclone preventive measures must be taken care of as per the nature of urgency. Windbreak reduces the wind speed; hence promotion of several tiers of windbreak plants near the sea coast should be encouraged (Kumar et al. 2014). In case of damage in early-growth stage of the crop, replanting/transplanting of short-duration field crops is one of the strategies to reduce the loss. If the crop is in reproductive phase, by doing partial removal of the canopy although there will be yield reduction, partial yield can be harvested. If the cyclone forewarning is well before time, partial or complete removal of canopy is possible. Stakes made of bamboo often overcomes moderate wind speed. Banana plantations use of healthy suckers for replanting after cyclone damage is recommended, and survived banana plants having banana fingers in fruit-filling stage may be provided with additional nutrition through bunch tip (Kumar et al. 2014). However, quick harvest, threshing and drying the grains before the cyclonic system through early warning, is the best option. Availability of covered threshing floor cum drying yard and polythene sheets for covering the produce so that grains do not get moist are possible options to prevent grain damage.

2.6.5.2 Livestock

Animals should be shifted to a safer place as soon as warnings for impeding cyclones are received. Animal sheds should be constructed with lightweight timber, coconut and palm tree leaves that will not harm the animals even if they collapse during cyclonic winds. Preserving the fodder resource should be on higher priority for meeting the demand during the crisis period. Molasses can be used as an alternate fodder during cyclone periods and periods of non-availability of fodder to protect the animals from disaster impacts and also from scarcity of food/fodder. To provide clean and unpolluted water to the animals, it may be treated with chlorine or bleaching powder before giving it to the animals. Animals must be vaccinated against communicable diseases during emergency periods, and the carcasses of animals must be buried in a pit over which lime will be sprinkled. In addition, nutritional supplements must be provided to improve immunity in animals (NDMA 2008).

2.6.5.3 Poultry

Birds need special care after storm events. Restocking of birds and arrangement of feeds and medicines should be arranged for poultry revival (Kumar et al. 2014). There should be sufficient supply of quality water preferably chlorinated for prohibiting the growth of bacteria. In cooler areas, birds may be kept warm using heaters to reduce stress (Sen and Chander 2003). Moisture-infected feed is to be replaced with dry feed. Birds which die due to diseases must be treated separately and cremated or buried in a deep pit and lime sprinkled over it along with other anti-infectants (NDMA 2008).

2.6.5.4 Fisheries

Fish farmers should be provided with fish fingerlings and feed, boats and nets as per need (Kumar et al. 2014). There is need to improve on the awareness of the fishers knowledge to climate change by involving them in the disaster preparedness and planning process. Restocking of fish fingerlings should be done immediately. Physical protection of inlet and outlet of aquaculture farms and ponds must be ensured to reduce the migration of the fishes (Kumar et al. 2014).

2.6.6 Sand/Dust Storm

Though the environmental and health hazards of storms cannot be reduced permanently, its impact can be reduced by taking appropriate measures.

2.6.6.1 Crops

Surface crop residues help stabilize the soil, reduce erosive force of wind and reduce number of saltating particles. Appropriate control of dust raising factors such as increasing the vegetation cover must be taken to stabilize soil and sand dunes by acting as windbreaks (Speer 2013; Shivakumar 2005). Especially standing stems decrease the wind energy available for momentum transfer at soil surface (Hagan and Armbrust 1994) as geometry of crop alters the soil microclimate, impacts the degree of soil protection and also conserves soil moisture (Nielsen and Hinkle 1994). Specified tillage operation increases cohesion of soil particles by mechanical action but needs to be taken as per soil type and topography specifications. In arid and semiarid areas, summer tillage is discouraged and a limited tillage recommended after the first monsoon showers (Gupta et al. 1997).

2.6.6.2 Livestock

The main purpose for implementing control methods for airborne pollutants in the livestock building is to ensure that production efficiency gets maximized (Banhazi et al. 2009). To control the dust, we need to understand the livestock environment as in the confines of a building, the air quality depends directly on building management, feeding and manure handling, ventilation system and the overall cleanliness (Choiniere 1993). This can be achieved through improved configuration and management of livestock buildings, provision of adequate ventilation, decreased stocking density and management of the animals contained in these buildings (Banhazi et al. 2009).

2.7   Way Forward

Since unusual atmospheric events are becoming usual events, to make future agriculture remunerative, risk-free and sustainable, first the dynamic characteristics of atmospheric stressors have to be understood. The scientific community must respond to the need of credible, objective and innovative scientific alternatives to tackle the stress impacts. There are, however, ways by which the adverse impacts can be mitigated and agriculture can be adapted to changing scenarios. The first step is to form integrated interdisciplinary research partnerships as atmospheric stressors already pose and will continue to pose challenges for agriculture and managed ecosystems.

As every year is becoming warmer than the previous year and unprecedented heat or cold wave conditions have become a common phenomenon, adaptation and mitigation options have to be explored either using field management practices or by applying genetic improvement tools for developing tolerant varieties. The use of chemicals like plant bioregulators and growth hormones having minimal or no residual effects and rescheduling of sowing/planting periods need to be explored. Regular events like extreme rain and hailstorm have necessitated focusing on developing protective structures. Especially research on designing low-cost shade net or poly-structures for high-value horticultural crops is the need of the hour as horticultural production in India has already surpassed the food grain production. Stresses arising out of increased atmospheric aerosol and decrease in available light need extra attention as change in land-use pattern and crop residue burning has changed the way we have been dealing these aspects in the past. Lastly abrupt changes in the magnitude and periodicity of atmospheric variables will have impact on distribution of disease vectors and pest dynamics which needs extra attention.

With unprecedented increase in demand for animal proteins, our prime focus must be towards developing low-cost environment suitable animal shelter structures for improved animal production and wellness. Future research in livestock must be done keeping a balance between competition for natural resources and projected atmospheric anomalies as unlike crop production where there will be a vertical growth, however livestock production is expected to have a horizontal growth. A large agenda of work still remains concerning the robust prediction of animal growth, body composition, feed requirement and waste output in future climate. To cater the target for lowering the emission level, interdisciplinary research approach must be undertaken to lower the methane emission from animal sector. Lastly the use of biotechnology can't be ignored if we want to impart heat-tolerant traits in high meat and milk yielding breeds.

Though the country has achieved food grain security for entire population, we are yet to solve the problem of poor nutrition, especially protein- and mineral-associated health issues. As poultry meat is the cheapest source of proteins, this sector has potential to provide food and livelihood securities to major chunk of Indian population. With emergence of heat stress as one of the major problems in Indian poultry industry, our primary area of focus should be to explore innovative approaches, including genetic marker-assisted selection of poultry breeds for increased heat tolerance and disease resistance for better productivity. Application of modern molecular techniques in poultry breeding has great potential to improve poultry productivity in a sustainable manner. Simultaneously, the possibilities of heat stress mitigation must be explored in terms of designing of suitable poultry housing for hot regions. Nutrition being one of the major factors in mitigating heat stress, study of the nutrient supplementation and feeding practices should be given priority.

Future research on fisheries must be oriented to understand the implications of greenhouse gas emission-induced sea acidification on fisheries ecosystem productivity and habitat quality and quantity especially on habitat of phytoplankton. Secondly, biotechnological intervention in terms of identification of heat and hypoxia tolerance traits in fish, identifying the molecular pathways and marker-assisted selection of fast-growing and tolerant species.

Though a lot of scientific advances have been made related to understanding of physical and physiological aspect of various atmospheric stressors, much work remains to be done regarding quantifying its impacts and long-term implications on agriculture. Thus only option left before us is to fight it in our own way. Firstly before doing so, the types and level of stresses must be properly quantified for future references. Secondly researches on finding mitigation and adaptation measures need to be scientifically planned so as to make it economically viable. By doing so, these scientific measures can be successfully adopted by growers to make future agriculture economically sustainable and less risky against the atmospheric anomalies.

References

Aarnink AJA, Roelofs PFMM, Ellen H, Gunnink H (1999) Dust sources in animal houses. In: Proceedings of international symposium on dust control in animal production facilities Aarhus, Denmark, pp. 34–40

Agarwal PK (2009) Global climate change and Indian agricultural Case studies from ICAR network project, ICAR pp 148

Ahrens CD (2015) Essentials of meteorology: an invitation to the atmosphere. LENGAGE Learning Publishers

Alavi M, Sharifi M (2015) Experimental effects of sand-dust storm on tolerance index, percentage phototoxicity and chlorophyll a fluorescence of *Vigna radiata* L. Proc Int Acad Ecol Environ Sci 5(1):16–24

Al-Issawi M, Rihan HZ, El-Sarkassy N, Fuller MP (2012) Frost hardiness expression and characterisation in wheat at ear emergence. J Agron Crop Sci 199:66–74

Allan RP (2011) Climate change: human influence on rainfall. Nature 470(7334):344–345

Allan JD, Abell R, Hogan ZEB, Revenga C, Taylor BW, Welcomme RL, Winemiller K (2005) Overfishing of inland waters. Biocontrol Sci 55:1041–1051

Altan O, Pabuccuoglu A, Alton A, Konyalioglu S, Bayraktar H (2003) Effect of heat stress on oxidative stress, lipid peroxidation and some stress parameters in broilers. Br Poult Sci 44(4):545–550

Ames DR, Brink DR, Willms CL (1980) Adjusting protein in feedlot diet during thermal stress. J Anim Sci 50(1):1–6

Angadi SV, Cutforth HW, Miller PR, McConkey EMH, Brabdt SA, Volkmar KM (2000) Response of three Brassica species to high temperature stress during reproductive growth. Can J Plant Sci 56:693–701

Anjum F, Wahid A, Farooq M, Javed F (2011) Potential of foliar applied thiourea in improving salt and high temperature tolerance of bread wheat (*Triticum aestivum*). Int J Agric Biol 13:251–256

Arogo J, Westerman PW, Heber AJ, Robarge WP, Classen JJ (2006) Ammonia emissions from animal feeding operations. In: Rice JM, Caldwell DF, Humenik FJ (eds) Animal agriculture and the environment: national center for manure and animal waste management white papers. ASABE, St Joseph, pp 41–88

Arora RK, Singh RK, Gulati S (2010) Managing loss from ground frost a major constraint to potato production in north western plans. Potato J 37(1–2):73–74

Ashraf MA (2012) Waterlogging stress in plants: A review. Afr J Agr Res 7:1976–1981

Aulakh MS, Singh B (1997) Nitrogen losses and fertilizer N use efficiency in irrigated porous soils. Nut Cyc Agroecosyst 47:197–212

Awasthi LP (2015) Recent advances in the diagnosis and management of plant diseases. Springers India, p 285

Baarendse PJ, Kemp B, Van Den Brand H (2006) Early-age housing temperature affects subsequent broiler chicken performance. Br Poult Sci 47(2):125–130

Badr MA, Abou El-Yazied AA (2007) Effect of fertigation frequency from sub-surface drip irrigation on tomato yield grown on sandy soil. Aust J Basic Appl Sci 1(3):279–285

Bahga CS, Gangwar PC (1988) Seasonal variations in plasma hormones and reproductive efficiency in early postpartum buffalo. Theriogenology 30:1209–1223

Bal SK, Minhas PS (2016) Managing abiotic stresses in agricultural field: ICAR-NIASM initiative. In: Proceedings of the IES international conference on "natural resource management: ecological perspectives" held at SKUAS&T (Jammu), J&K, India during 18–20 Feb 2016, p 44

Bal SK, Saha S, Fand BB, Singh NP, Rane J, Minhas PS (2014) Hailstorms: causes, damage and post hail management in agriculture. Tech Bull No. 5 National Institute of Abiotic Stress Management, Baramati 413115, Pune, Maharashtra (India) pp. 44

Banday T, Untoo M (2012) Adverse season and poultry farming: management of poultry in extreme weather. Available on: www.en.engormix.com

Banhazi TM (2009) User-friendly air quality monitoring system. App Engr Agric 25:281–290

Banhazi TM, Currie E, Reed S, Lee I-B, Aarnink AJA (2009) Controlling the concentrations of airborne pollutants in piggery buildings. In: Aland A, Madec F (eds) Sustainable animal production: the challenges and potential developments for professional farming. Wageningen Academic Publishers, Wageningen, pp 285–311

Banks S, King SA, Irvine DS, Saunders PTK (2005) Impact of a mild scrotal heat stress on DNA integrity in murine spermatozoa. Reproduction 129(4):505–514

Barange M, Perry RI (2009) Physical and ecological impacts of climate change relevant to marine and inland capture fisheries and aquaculture. In: Climate change implications for fisheries and aquaculture overview of current scientific Knowledge, FAO Fisheries and Aquaculture Technical Paper No. 530. FAO, Rome, pp 7–106

Barnabas B, Jager K, Feher A (2008) The effect of drought and heat stress on reproductive processes in cereals. Plant Cell Environ 31(1):11–38

Beede DK, Collier RJ (1986) Potential management strategies for intensively managed cattle during thermal stress. J Anim Sci 62(2):543–554

Bell JD, Johnson JE, Ganachaud AS, Gehrke PC, Hobday AJ, Hoegh-Guldberg O, Le Borgne R, Lehodey P, Lough JM, Pickering T, Pratchett MS, Waycott M (2011) Vulnerability of tropical fisheries and aquaculture to climate change. pp 665

Berlato MA, Farenzena H, Fontana DC (2005) Association between El Nino southern oscillation and corn yield in Rio Grande do Sul State. Pesq Agrop Brasileira 40:423–432

Bhatnagar A, Garg SK (2000) Causative factors of fish mortality in still water fish ponds under sub-tropical conditions. Aquaculture 1(2):91–96

Bhattacharyya R, Ghosh BN, Mishra PK, Mandal B, Srinivasa Rao C, Sarkar D, Das K, Kokkuvayil Sankaranarayanan K, Lalitha M, Hati KM, Franzluebbers AJ (2015) Soil degradation in India: challenges and potential solutions. Sustain For 7:3528–3570. doi:10.3390/su7043528

Bita CE, Gerats T (2013) Plant tolerance to high temperature in a changing environment: Scientific fundamentals and production of heat stress-tolerant crops. Front Plant Sci 4:273. (online)

Bonnet S, Geraert PA, Lessire M, Carre B, Guillaumin S (1997) Effect of high ambient temperature on feed digestibility in broilers. Poult Sci 76(6):857–863

Boyhan GE, Kelley WT, (2001) Onion production guide, Bulletin 1198. College of Agricultural and Environmental Sciences, University of Georgia, Georgia, pp. 56

Brahmane MP, Sajjanar B, Kumar N, Pawar SS, Bal SK, Krishnani KK (2017) Impact of rearing temperatures on Oreochromis mossambicus (Tilapia) growth, muscle morphology and gene expression. J Environ Biol

Brander KM (2007) Global fish production and climate change. Proc Natl Acad Sci U S A 104:19709–19714

Brown DJA, Sadler K (1989) Fish survival in acid waters. In: Morris R, Taylor EW, DJA B, Brown JA (eds) Acid toxicity and aquatic animals. Cambridge University Press, Cambridge, UK, pp 31–44C

Cai T, Xu H, Peng D, Yin Y, Yang W, Ni Y, Chen X, Xu C, Yang D, Cui Z, Wang Z (2014) Exogenous hormonal application improves grain yield of wheat by optimizing tiller productivity. Field Crop Res 155:172–183

Campos EJ (2000) Avicultura: razoes, fatos e divergencias. FEP-MVZ Escola de Veterinaria da UFMG, Belo Horizonte, 311 p

Carrasco E, Devaux A, Garcla W, Esprella R (1997) Frost tolerant potato varieties for the Andean Highlands. In: Program Report 1995–1996. Int Potato Center, Lima pp, pp 237–232

Chakraborty U, Pradhan D (2011) High temperature-induced oxidative stress in Lens culinaris, role of antioxidants and amelioration of stress by chemical pre-treatments. J Plant Interact 6:43–52

Challinor AJ, Wheeler TR, Slingo TM, Hemming D (2005) Quantification of physical and biological uncertainty in the simulation of yield of a tropical crop using present-day and doubled CO_2 climates. Philos Trans R Soc B 360:2085–2094

Challinor A, Wheeler T, Craufurd P, Ferro C, Stephenson D (2007) Adaptation of crops to climate change through genotypic responses to mean and extreme temperatures. Agric Ecosyst Environ 119(1–2):190–204

Changi WYB, Ouyang H (1988) Dynamics of dissolved oxygen and vertical circulation in fish ponds. Aquaculture 74:263–276

Cho K, Tiwari S, Agrawal SB, Torres NL, Agrawal M, Sarkar A, Shibato J, Agrawal GK, Kubo A, Rakwal R (2011) Tropospheric ozone and plants: absorption, responses and consequences. Rev Environ Contam Toxicol 212:61–111

Choiniere YMJ (1993) Farm workers health problems related to air quality inside livestock barns. Ministry of Agriculture and Food Factsheet, 4, 3

Chowdhury MTH, Sukhan ZP, Hannan MA (2010) Climate change and its impact on fisheries resource in Bangladesh (www.benjapan.org/iceab10/22.pdf)

Ciais P, Reichstein M, Viovy N, Granier A, Ogee J, Allard V, Aubinet M, Buchmann N, Bernhofer C, Carrara A et al (2005) Europe-wide reduction in primary productivity caused by the heat and drought in 2003. Nature 437:529–533

Cohen SS, Gale J, Poljakoff-Mayber A, Shmida A, Suraqui S (1981) Transpiration and the radiation climate of the leaf on Mt. Hermon: a Mediterranean mountain. Aust J Ecol 69:391–403

Cole P, McCloud P (1985) Salinity and climatic effects on the yields of citrus. Aust J Exp Agric 25:711–717

Colebrook EH, Thomas SG, Phillips AL, Hedden P (2014) The role of gibberellin signalling in plant responses to abiotic stress. J Exp Biol 217:67–75

Collier RJ, Beede DK, Thatcher WW, Israel LA, Wilcox LS (1982) Influences of environment and its modification on dairy animal health and production. J Dairy Sci 65:2213–2227

Coppock CE, Grant PA, Portzer SJ (1982) Lactating dairy cow responses to dietary sodium, chloride, bicarbonate during hot weather. J Dairy Sci 65(4):566–576

Cueva S, Sillau H, Valenzuela A, Ploog H (1974) High altitude induced pulmonary hypertension and right heart failure in broiler chickens. Res Vet Sci 16:370–374

Dai SF, Gao F, Xu XL, Zhang WH, Song SX, Zhou GH (2012) Effects of dietary glutamine and gamma aminobutyric acid on meat colour, pH, composition, and water-holding characteristic in broilers under cyclic heat stress. Br Poult Sci 53(4):471–481

Das HP (2012) Agrometeorology in extreme events and natural disasters. BS Publications, Hyderabad

Dat JF, Lopez-Delgado H, Foyer CH, Scott IM (1998) Parallel changes in H_2O_2 and catalase during thermotolerance induced by salicylic acid or heat acclimation in mustard seedlings. Plant Physiol 116(4):1351–1357

David B, Mejdell C, Michel V, Lund V, Moe RO (2015) Air quality in alternative housing systems may have an impact on laying hen welfare. Part II—Ammonia. Animals (Basel) 5(3):886–896

Davis MS, Foster WM (2002) Inhalation toxicology in the equine respiratory tract (Last Updated: 28-Feb-2002). In: Lekeux P (ed) Equine respiratory diseases. International Veterinary Information Service, Ithaca. www.ivis.org, 2002; B0319.0202

Daw T, Adger WN, Brown K, Badjeck MC (2009) Climate change and capture fisheries: potential impacts, adaptation and mitigation. In: Cochrane K, De Young C, Soto D, Bahri T (eds) Climate change implications for fisheries and aquaculture: overview of current scientific knowledge, FAO Fisheries and Aquaculture Technical Paper. No. 530. FAO, Rome, pp 107–150

De US, Dube RK, Prakasa Rao GS (2005) Extreme weather events over India in last 100 years. J Indian Geophys Union 9:173–187

Declercq AM, Haesebrouck F, Broeck WV, Bossier P, Decostere A (2013) Columnaris disease in fish: a review with emphasis on bacterium-host interactions. Vet Res 44(1):27

Dupont FM, Hurkman WJ, Vensel WH, Tanaka C, Kothari KM, Chung OK, Altenbach SB (2006) Protein accumulation and composition in wheat grains: effects of mineral nutrients and high temperature. Eur J Agron 25(2):96–107

Duran-Zuazo VH, Rodriguez-Pleguezuelo CR (2008) Soil-erosion and runoff prevention by plant covers. A review. Agron Sustain Dev 28:65–86

Easterling W, Apps M (2005) Assessing the consequences of climate change for food and forest resources: a view from the IPCC. Climate Change 70(1):165–189

Etches RJ, John TM, Verrinder Gibbins AM (1995) Behavioural, physiological, neuroendocrine and molecular responses to heat stress. In: Daghir NJ (ed) Poultry production in hot climates. CAB International, Wallingford, pp 31–65

FAO (2009) The state of food and agriculture 2009. Rome, FAO

FAO (2012) The state of world fisheries and aquaculture 2012. Rome. pp. 209

FAO (2016) Climate change and food security: risks and responses. Rome (www.fao.org/3/a--i5188e.pdf)

Farooq M, Basra SMA, Rehman H, Saleem BA (2008) Seed priming enhances the performance of late sown wheat (*Triticum aestivum* L.) by improving chilling tolerance. J Agron Crop Sci 194:55–60

Farrell TC, Fukai S, Williams RL (2006) Minimizing cold damage during reproductive development among temperate rice genotypes. I. Avoiding low temperature with the use of appropriate sowing time and photoperiod-sensitive varieties. Aust J Agric Res 57:75–88

Feddes JJR, Koberstein BS, Robinson FE, Ridell C (1992) Misting and ventilation rate effects on air quality, and heavy from turkey performance and health. Can Agric Eng 34(2):177–181

Fischer R, Edmeades GO (2010) Breeding and cereal yield progress. Crop Sci 50:S-85–S-98

Freeman BM (1987) The stress syndrome. Worlds Poult Sci J 43(1):15–19

Fuglestvedt JS, Berntsen TK, Godal O, Sausen R, Shine KP, Skodvin T (2003) Metrics of climate change: assessing radiative forcing and emission indices. Clim Chang 58(3):267–331

Fuhrer J, Skarby L, Ashmore MR (1997) Critical levels for effects of ozone on vegetation in Europe. Environ Pollut 97(1–2):91–106

Geiger SC, Manu A, Bationo A (1992) Changes in a sandy Sahelian soil following crop residue and fertilizer additions. Soil Sci Soc Amer J 56(1):172–177

Ghazi SH, Habibian M, Moeini MM, Abdol Mohammadi AR (2012) Effects of different levels of organic and inorganic chromium on growth performance and immunocompetence of broilers under heat stress. Biol Trace Elem Res 146(3):309–317

Gilbert M, Slingenbergh J, Xiao X (2008) Climate change and avian influenza. Rev Sci Tech 27(2):459–466

Gleeson M (1986) Respiratory adjustments of the unanaesthetized chicken, gallus domesticus, to elevated metabolism elicited by 2,4 dinitrophenol or cold exposure. Comp Biochem Physiol 83(2):283–289

Graham AW, McDonald GK (2001) Effect of zinc on photosynthesis and yield of wheat under heat stress. In: Proceedings of the 10th Australian Agronomy Conference Hobart, January 29–February 1, 2001, Australian Society of Agronomy, Hobart, Tasmania, Australia

Gregory NG (2010) How climatic changes could affect meat quality. Food Res Int 43:1866–1873

Griffiths H (2003) Effects of Air Pollution on Agricultural Crops. Factsheet Available on www.omafra.gov.on

Groot-Koerkamp PW, Bleijenberg R (1998) Effect of type of aviary, manure and litter handling on the emission kinetics of ammonia from layer houses. Br Poult Sci 39:379–392

Guis H, Caminade C, Calvete C, Morse AP, Tran A, Baylis M (2012) Modelling the effects of past and future climate on the risk of bluetounge emergence in Europe. J R Soc Interface 9(67):339–350

Gupta JP, Kar A and Faroda AS (1997) Desertification in India: problems and possible solutions. Yojana (Independence Day Issue on Development and Environment): 55–59

Hader D, Williamson CE, Wangberg SA, Rautio M, Rose KC, Gao K, Helbling EW, Sinha RP, Worrest R (2015) Effects of UV radiation on aquatic ecosystems and interactions with other environmental factors. Photochem Photobiol Sci 14:108–126

Hagan LJ, Armbrust DV (1994) Plant canopy effects on wind erosion saltation. Trans ASAE 37(2):461–465

Hartmann DL, Klein Tank AMG, Rusticucci M, Alexander LV, Brönnimann S, Charabi Y, Dentener FJ, Dlugokencky EJ, Easterling DR, Kaplan A, Soden BJ, Thorne PW, Wild M, Zhai PM (2013) Observations: atmosphere and surface. In: Stocker TF, Qin D, Plattner GK, Tignor M, Allen SK, Boschung J, Nauels A, Xia Y, Bex V, Midgley PM (eds) Climate change 2013: the physical science basis, Contribution of working group I to the fifth assessment report of the intergovernmental panel on climate change. Cambridge University Press, Cambridge, UK/New York

Hasanuzzaman M, Nahar K, Alam MM, Fujita M (2012) Exogenous nitric oxide alleviates high temperature induced oxidative stress in wheat (*Triticum aestivum* L.) seedlings by modulating the antioxidant defense and glyoxalase system. Aust J Crop Sci 6(8):1314–1323

Havenhand JN, Buttler F, Thorndyke MC, Williamson JE (2008) Near-future levels of ocean acidification reduce fertilisation success in a sea urchin. Curr Biol 18:R651–R652

Heber AJ, Stroik M, Faubion JM, Willard LH (1988) Size distribution and identification of aerial dust particles in swine finishing buildings. Trans ASAE 31(3):882–887

Hedhly A, Hormaza JI, Herrero M (2009) Global warming and sexual plant reproduction. Trends Plant Sci 14:30–36

Hernandez J, Benedito JL, Abuelo A, Castillo C (2014) Ruminal acidosis in feedlot: from aetiology to prevention, The Scientific World J, ID 702572 http://dx.doi.org/10.1155/2014/702572

Hidema J, Kumagai T (2006) Sensitivity of rice to ultraviolet-B radiation. Ann Bot 97(6):933–942

Hilborn R, Branch TA, Ernst B, Magnusson A, Minte-Vera CV, Scheuerell MD, Valero JL (2003) State of the world's fisheries. Annu Rev Environ Resour 28:359–399

Hirayama T, Shinozaki K (2010) Research on plant abiotic stress responses in the post genome era: past, present and future. Plant J 61(6):1041–1052

Holland GJ, Webster PJ (2007) Heightened tropical cyclone activity in the North Atlantic: natural variability or climate trend? Phil Trans the Royal Soc A 365(1860):2695–2716

Houghton JT, Collander BA, Ephraums JJ (eds) (1990) Climate change – the IPCC scientific assessment. Cambridge University Press, Cambridge, p 135

Hsu JC, Lin CY, Chiou PW (1998) Effects of ambient temperature and methionine supplementation of a low protein diet on the performance of laying hens. Anim Feed Sci Technol 74(4):289–299

Huthwelker T, Ammann M, Peter T (2006) The uptake of acidic gases on ice. Chem Rev 106(4):1375–1444

Hutton RJ, Landsberg JJ (2000) Temperature sums experienced before harvest partially determine the post-maturation juicing quality of oranges grown in the Murrumbidgee Irrigation Areas (MIA) of New South Wales. J Sci Food Agric 80:275–283

Iizumi T, Ramankutty N (2015) How do weather and climate influence cropping area and intensity? Global Food Sec 4:46–50

Imik H, Ozlu H, Gumus R, Atasever MA, Urgar S, Atasever M (2012) Effects of ascorbic acid and alpha-lipoic acid on performance and meat quality of broilers subjected to heat stress. Br Poult Sci 53(6):800–808

India Meteorological Department (2015) Ministry of Earth Science, Govt of India

IPCC (2007a) In: Solomon S, Qin D, Manning M, Chen Z, Marquis M, Averyt KB, Tignor M, Miller HL (eds) Climate change 2007: the physical science basis, Contribution of working group I to the fourth assessment report of the intergovernmental panel on climate change. Cambridge University Press, Cambridge, UK/New York. 996 pp

IPCC (2007b) Fourth assessment report of the intergovernmental panel on climate change: the impacts, adaptation and vulnerability (Working Group III). Cambridge University Press, New York

IPCC (2007c) Summary for policymakers. In: Solomon S, Qin D, Manning M et al (eds) Climate change 2007: the physical science basis, Contribution of the working group I to the fourth assessment report of the intergovernmental panel on climate change. Cambridge University Press, Cambridge, UK, p 22

IPCC (2012) In: Field CB, Barros V, Stocker TF, Qin D, Dokken DJ, Ebi KL, Mastrandrea MD, Mach KJ, Plattner GK, Allen SK, Tignor M, Midgley PM (eds) Managing the risks of extreme events and disasters to advance climate change adaptation, A special report of working groups I and II of the intergovernmental panel on climate change. Cambridge University Press, Cambridge, UK/New York. pp 582

IPCC (2013) Climate change 2013: the physical science basis. Contribution of working group I to the fifth assessment report of the intergovernmental panel on climate change, Stocker TF, Qin D, Plattner GK, Tignor M, Allen SK, Boschung J, Nauels A, Xia Y, Bex V, Midgley PM (eds.) Cambridge University Press, Cambridge, UK/New York

IPCC (2014) Climate change 2014: impacts, adaptation, and vulnerability, Regional aspects. Contribution of working group II to the fifth assessment report of the IPCC. Cambridge University Press, Cambridge, UK/New York

Isaksen ISA, Berntsen TK, Dalsoren SB, Eleftheratos K, Orsolini Y, Rognerud B, Stordal F, Amund Sovde O, Zerefos C, Holmes CD (2014) Atmospheric ozone and methane in a changing climate. Atmos 5:518–535

Iwagami Y (1996) Changes in the ultrasonic of human cells related to certain biological responses under hyperthermic culture conditions. Hum Cell 9(4):353–366

Izuta T (2017) Air pollution impacts on plants in East Asia. Springers, Japan. doi 10.1007/978-4-431-56438-6

Jain M, Prasad PV, Boote KJ, Hartwell AL Jr, Chourey PS (2007) Effects of season-long high temperature growth conditions on sugar to starch metabolism in developing microspores of grain sorghum (*Sorghum bicolor* L. Moench). Planta 227(1):67–79

Jiang QW, Kiyoharu O, Ryozo I (2002) Two novel mitogen-activated protein signalling components, *OsMEK1* and *OsMAP1*, are involved in a moderate low-temperature signaling pathway in Rice. Plant Physiol 129(4):1880–1891

Johkan M, Oda M, Maruo T, Shinohara Y (2011) Crop production and global warming. In: Casalegno S (ed) Global warming impacts: case studies on the economy, human health, and on urban and natural environments. InTech, Rijeka, pp 139–152

Jones PD, New M, Parker DE, Martin S, Rigor IG (1999) Surface air temperature and its changes over the past 150 years. Rev Geophys 37:173–199

Julian RJ, McMillan I, Quinton M (1989) The effect of cold and dietary energy on right ventricular hypertrophy, right ventricular failure and ascites in meat-type chickens. Avian Pathol 18(4):675–684

Kant S, Seneweera S, Rodin J, Materne M, Burch D, Rothstein SJ, Spangenberg G (2012) Improving yield potential in crops under elevated CO2: integrating the photosynthetic and nitrogen utilization efficiencies. Front Plant Sci 3(162). doi:10.3389/fpls.2012.00162

Kaur H, Arora SP (1984) Annual pattern of plasma progesterone in normal cycling buffaloes (*Bubalus bubalis*) fed two different levels of nutrition. Anim Reprod Sci 7:323–332

Kaushal N, Gupta K, Bhandhari K, Kumar S, Thakur P, Nayyar H (2011) Proline induces heat tolerance in chickpea (*Cicer arietinum* L.) plants by protecting vital enzymes of carbon and antioxidative metabolism. Physiol Mol Biol Plants 17(3):203–213

Khan NM, Qasim F, Ahmed R, Khan AK, Khan B (2003) Effects of sowing date on yield of maize under Agro climatic condition of Kaghan Valley. Asian J Plant Sci 1(2):140–147

Kim YG, Lee BW (2011) Relationship between grain filling duration and leaf senescence of temperate rice under high temperature. Field Crop Res 122(3):207–213

Kimothi SP, Ghosh CP (2005) Strategies for ameliorating heat stress in dairy animals. Dairy Year Book pp 371–377

Kirunda DFK, Scheideler SE, Mckee SR (2001) The efficiency of Vitamin E (DL-α-tocopheryl acetate) supplementation in hens diets ti alleviate egg quality deterioration associated with high temperature exposure. Poult Sci 80:1378–1383

Kocaman B, Yaganoglu AV, Yanar M (2005) Combination of fan ventilation system and spraying of oil-water mixture on the levels of dust and gases in caged layer facilities in Eastern Turkey. J Appl Anim Res 27(2):109–111

Koopmans A, Koppejan J (1997) Agricultural and forest residues; Generation, utilization and availability. Paper presented at the Regional Consultation on Modern Applications of Biomass Energy 6: 10

Kozlowski TT, Pallardy SG (1997) Physiology of woody plants. Academic Press, San Diego

Krishnan P, Ramakrishnan B, Reddy R, Reddy V (2011) High temperature effects on rice growth, yield, and grain quality. Adv Agron 111:87–206

Krupa SV, Kickert RN (1993) The effects of elevated ultraviolet (UV)-B radiation on agricultural production. Report submitted to the formal commission on 'Protecting the Earth's Atmosphere' of the German Parliament, Bonn, Germany. pp. 432

Kumar A, Brahmanand PS, Nayak AK (2014) Management of cyclone disaster in agriculture sector in coastal areas. Directorate of Water Management NRM Division (ICAR) Chandrasekharpur, Bhubaneswar, p 108

Kuriakose SL, Sankar G, Muraleedharan C (2009) History of landslide susceptibility and a chorology of landslide-prone areas in the Western Ghats of Kerala, India. Environ Geol 57(7):1553–1568

Lafta AM, Lorenzen JH (1995) Effect of high temperature on plant growth and carbohydrate metabolism in potato. Plant Physiol 109:637–643

Lara LJ, Rostagno MH (2013) Impact of heat stress on poultry production. Animals 3:356–369

Lemengar R, Johnson K (1997) Frost-damaged forages can be deadly http://www.ansc.purdue.edu/beef/articles/FrostDamagedForages.pdf

Lenton T, Held H, Kriegler E, Hall J, Lucht W, Rahmstorf S, Schellnhuber H (2008) Tipping elements in the Earth's climate system. Proc Natl Acad Sci 105(6):1786–1793

Lin H, Mertens K, Kemps B, Govaerts T, De Ketelaere B, Baerdemaeker D, Decuypere J, Buyse J (2004) New approach of testing the effect of heat stress on eggshell quality: mechanical and material properties of eggshell and membrane. Br Poult Sci 45(4):476–482

Loarie S, Duffy P, Hamilton H, Asner G, Field C, Ackerly D (2009) The velocity of climate change. Nature 462(7276):1052–1055

Logan CA (2010) A review of ocean acidification and America's response. Biocontrol Sci 60(10):819–828

Lynn BH, Carlson TN, Rosenzweig C, Goldberg R, Druyan L, Cox J, Civerolo K (2009) A modification to the NOAH LSM to simulate heat mitigation strategies in the New York City Metropolitan Area. J Appl Meteorol Climatol 48(2):199–216

Marai IFM, Haeeb AAM (2010) Buffalo's biological functions as affected by heat stress – a review. Livest Sci 127:89–109

Maxwell MH, Robertson GW (1997) World broiler ascites survey 1996. Poult Int 36:16–19

Maxwell MH, Robertson GH, Spence S (1986) Studies on an ascitic syndrome in young broilers. 1. Hematology and pathology. Avian Pathol 15(3):511–524

Maxwell MH, Tullett SG, Burton FG (1987) Haemotology and morphological changes in young broiler chicks with experimentally induced hypoxia. Res Vet Sci 43(3):331–338

McCormac BM, Varney R (1971) Introduction to the scientific study of atmospheric pollution. D. Reidel Publishing Company, Holland

Mckersie BD, Bowley SR (1997) Active oxygen and freezing tolerance in transgenic plants. In: Li PH, Chen THH (eds) Plant cold hardiness, molecular biology, biochemistry and physiology. Plenum Press, New York, pp 203–214

Mengel K, Kirkby EA (2001) Principles of plant nutrition, 5th edn. Kluwer Academic Publishers, Dordrecht

Menon AGK (1954) Fish geography of the Himalayas. Zoological Survey India, Calcutta 20(4):467–493

Monteith JL, Ong CK, Corlett JE (1991) Microclimatic interactions in agroforestry systems. For Ecol Manag 45(1–4):31–44

Morita S, Yonermaru J, Takahashi J (2005) Grain growth and endosperm cell size under high night temperature in rice (Oryza sativa L.) Ann Bot 95(4):695–701

Morshed AMA (2013) Root reasons behind the unusual behaviours of the Earth climate thus the causes of natural disasters. J Environ Sci Toxicol Food Technol 7(2):05–07

NDMA (2008) National disaster management guidelines, Government of India, Management of Cyclones. NDMC, Centaur Hotel, New Delhi, pp. 91–92

Nelson GC (2009) Climate change: impact on agriculture and costs of adaptation. Int Food Policy Res Institute, Washington

Nguyen PH, Greene E, Donoghue A, Huff G, Clark FD, Dridi S (2016) A new insight into cold stress in poultry production. Adv food Technol Nutr Sci open J 2(1):1–2

Nicholls RJ, Wong PP, Burkett VR, Codignotto JO, Hay JE, McLean RF, Ragoonaden S, Woodroffe CD (2007) Coastal systems and low-lying areas. In: Parry ML, OF Canziani, Palutikof JP, Linden PJ, Hanson CE (eds) Climate change 2007: impacts, adaptation and vulnerability, Contribution of working group II to the Fourth Assessment Report of the Intergovernmental Panel on Climate Change. Cambridge University Press, Cambridge, UK, pp 315–356

Nielsen DC, Hinkle SE (1994) Wind velocity, snow and soil water measurements in sunflower residues. Great Plains Agric Council Bull 150:93–100

NOAA (2015) National centres for environmental information, State of the Climate: Global Analysis for Annual Report 2015.; https://www.ncdc.noaa.gov/sotc/global/201511

NOAA (2017) Monthly mean CO2 level at Mauna Loa observatory, Earth System Research Laboratory, National Oceanic and Atmospheric Administration, USA. www.esrl.noaa.gov/gmd/ccgg/trends/

Olien CR, Smith MN (1997) Ice adhesions in relation to freeze stress. Plant Physiol 60(4):499–503

Olsen DP, Paparian CJ, Ritter RC (1980) The effects of cold stress on neonatal calves. ll. Absorption of colostral immunoglobulins. Can J Comp Med 44(1):19–23

Orr JC, Fabry VJ, Aumont O, Bopp L, Doney SC, Feely RA, Gnanadesikan A et al (2005) Anthropogenic ocean acidification over the twenty-first century and its impact on calcifying organisms. Nature 437(681–686)

Ortiz R, Braun HJ, Crossa J, Crouch JH, Davenport G, Dixon J (2008) Wheat genetic resources enhancement by the International Maize and Wheat Improvement Center (CIMMYT). Genet Resour Crop Evol 55(7):1095–1140

Palta P, Mondal S, Prakash BS, Madan ML (1997) Peripheral inhibin levels in relation to climatic variations and stage of estrous cycle in Buffalo (*Bubalus bubalis*). Theriogenology 47:989–995

Parmar AP, Mehta VM (1994) Seasonal endocrine changes in steroid hormones of developing ovarian follicles in Surti buffaloes. Indian J Anim Sci 64:111–113

Patel N, Rajput TBS (2004) Fertigation a technique for efficient use of granular fertilizer through drip irrigation. J Agric Eng 85(2):50–54

Pathak H, Ladha JK, Aggarwal PK, Peng S, Das S, Singh Y, Singh B, Kamra SK, Mishra B, Sastri ASRAS, Aggarwal HP, Das DK, Gupta RK (2003) Trends of climatic potential and on-farm yields of rice and wheat in the Indo Gangetic Plains. Field Crop Res 80(3):223–234

Pawar SS, Sajjanar B, Lonkar VD, Kurade NP, Kadam AS, Nirmal AV, Brahmane MP, Bal SK (2016) Assessing and mitigating the impact of heat stress on poultry. Adv Anim Vet Sci 4(6):332–341

Poorter H (1993) Interspecific variation in the growth response of plants to an elevated ambient CO_2 concentration. Vegetatio 104(1):77–97

Portner HO, Langenbuch M (2005) Synergistic effects of temperature extremes, hypoxia, and increases in CO_2 on marine animals: from Earth history to global change. J Geophys Res 110:C09S10

Postma JA, Lynch JP (2011) Root cortical aerenchyma enhances the growth of maize on soils with suboptimal availability of nitrogen, phosphorus, and potassium. Plant Physiol 156(3):1190–1201

Pradhan GP, Prasad PVV, Fritz AK, Kirkham MB, Gill BS (2012) Response of *Aegilops* species to drought stress during reproductive stages of development. Funct Plant Biol 39(1):51–59

Prasad PVV, Boote KJ, Allen LH Jr (2006) Adverse high temperature effects on pollen viability, seed-set, seed yield and harvest index of grain-sorghum (*Sorghum bicolor* L. Moench) are more severe at elevated carbon dioxide due to higher tissue temperatures. Agric Forest Metero 139(3–4):237–251

Prasad PVV, Staggenborg SA, Ristic Z (2008) Impacts of drought and/or heat stress on physiological, developmental, growth, and yield processes of crop plants. In: Ahuja LH, Saseendran SA (eds) Response of crops to limited water: understanding and modeling water stress effects on plant growth processes, Adv Agric Sys Model Series, vol 1. ASA-CSSA, Madison, Wisconsin, pp 301–355

Quarles CL, Kling HF (1974) Evaluation of ammonia and infectious bronchitis vaccination stress on broiler performance and carcass quality. Poult Sci 53(4):1592–1596

Rahal A, Kumar A, Singh V, Yadav B, Tiwari R, Chakraborty S, Dhama K (2014) Oxidative stress, prooxidants, and antioxidants: the interplay. Biomed Res Int. doi:10.1155/2014/761264

Rahman K (2007) Studies on free radicals, antioxidants, and co-factors. Clin Interv Aging 2(2):219–236

Ramanathan V, Carmichael G (2008) Global and regional climate changes due to black carbon. Nat Geosci 1:221–227

Rathore AC, Raizada A, Jaya Prakash J, Sharda VN (2012) Impact of chilling injury on common fruit plants in the Doon Valley. Curr Sci 102(8):1107–1111

Ravichandra NG (2013) Fundamentals of plant pathology. PHI Learning Private Limited, New Delhi

Ravussin Y, Xiao C, Gavrilova O, Reitman ML (2014) Effect of intermittent cold exposure on brown fat activation, obesity, and energy homeostasis in mice. PLoS One 9(1):e85876. https://doi.org/10.1371/journal.pone.0085876

Reece FN, Lott BD (1980) The effect of ammonia and carbon dioxide during brooding on the performance of broiler chickens. Poult Sci 59:1654–1661

Robson S (2007) Prussic acid poisoning in livestock. PRIMEFACT 417, Regional Animal Health Leader, Animal and Plant Biosecurity, Wagga

Rogers S (1996) Hail damage: physical meteorology and crop losses. Proc Fla State Hort Soc 109:97–103

Rogers HH, Runion GB, Krupa SV, Prior SA (1997) Plant responses to atmospheric CO_2 enrichment: implications in root-soil microbe interactions. In: Allen LH Jr, Kirkham MB, Olszyk DM, Whitman CE (eds) Advances in carbon dioxide effects research, ASA Special Publication No. 61. ASA, CSSA, and SSSA, Madison, pp 1–34

Sahin N, Sahin K, Kucuk O (2001) Effects of vitamin E and vitamin A supplementation on performance, thyroid status and serum concentrations of some metabolites and minerals in broilers reared under heat stress (32°C). Vet Med 46(11–12):286–292

Sajjanar B, Deb R, Singh U, Kumar S, Brahmane M, Nirmale A, Bal SK, Minhas PS (2015) Identification of SNP in HSP90AB1 and its association with relative thermotolerance and milk production traits in Indian dairy cattle. Anim Biotechnol 26(1):45–50

Sanchez WK, Beede DK, Cornell JA (1994) Interactions of Na^+, K^+ and Cl^- on lactation, acid-base status and mineral concentrations. J Dairy Sci 77:1661–1675

Schlenker W, Roberts MJ (2009) Nonlinear temperature effects indicate severe damages to U.S. crop yields under climate change. Proc Nat Acad Sci USA 106:15594–15598

Sen A, Chander M (2003) Disaster management in India: the case of livestock and poultry. Rev Sci Tech Off Int Epiz 22(3):915–930

SEPA (2016) Scottish pollutant release inventory (2016) Scottish environment protection agency. SEPA Stirling Office Strathallan House Castle Business Park, Stirling, FK9 4TZ, pp. 39

Shelton C (2014) Climate change adaptation in fisheries and aquaculture compilation of initial examples, FAO Fisheries and Aquaculture Circular No. 1088. Rome, FAO, p 34

Sherwood SC, Huber M (2010) An adaptability limit to climate change due to heat stress. Proc Natl Acad Sci 107(21):9552–9555

Shivakumar MVK (2005) Impact of sand storms/dust storms on agriculture. Natural Disasters and Extreme Events in Agriculture. Publisher-Springer eBook, pp 159–177

Shlosberg A, Pano G, Handji J, Berman E (1992) Prophylactic and therapeutic treatment of ascites in broiler chickens. Br Poult Sci 33(1):141–148

Sicher RC (2015) Temperature shift experiments suggest that metabolic impairment and enhanced rates of photorespiration decrease organic acid levels in soybean leaflets exposed to supra-optimal growth temperatures. Meta 5:443–454

Shivakumar MVK, Stefanski R (2007) Climate and land degradation: an overview. In: Shivakumar MVK, Ndiangui N (eds) Climate and land degradation. Springer, Berlin, p 623

Sohail MU, Hume ME, Byrd JA, Nisbet DJ, Ijaz A, Sohail A, Shabbir MZ, Rehman H (2012) Effect of supplementation of prebiotic mannanoligosaccharides and probiotic mixture on growth performance of broilers subjected to chronic heat stress. Poult Sci 91(9):2235–2240

Speer MS (2013) Dust storm frequency and impact over Eastern Australia determined by state of Pacific climate system. Weather Climate Extremes 2:16–21

Srinivasarao NK, Shivashankara RH, Laxman RH (2016) Abiotic physiology of horticultural crops. Springers (India) Pvt. Ltd. pp. 12

Starr JR (1980) Weather, climate, and animal performance. WMO Technical Note No. 190. WMO Publication No. 684. WMO, Geneva, Switzerland

Stefanski R, Sivakumar MVK (2009) Impacts of sand and dust storms on agriculture and potential agricultural applications of a SDSWS. IOP Conf Series Earth Environ Sci 7(2009):012016

Stockwell CE, Yokelson RJ, Kreidenweis SM, Robinson AL, DeMott PJ, Sullivan RC, Reardon J, Ryan KC, Griffith DWT, Stevens L (2014) Trace gas emissions from combustion of peat, crop residue, domestic biofuels, grasses, and other fuels: configuration and Fourier transform infrared (FTIR) component of the fourth Fire Lab at Missoula Experiment (FLAME-4). Atmos Chem Phys 14:9727–9754

Stott GH, Wiersma F, Mevefec BE, Radwamki FR (1976) Influence of environment on passive immunity in calves. J Dairy Sci 59(7):1306–1311

St-Pierre NR, Cobanov B, Schnitkey G (2003) Economic losses from heat stress by U.S. livestock industries. J Dairy Sci 86:52–77

Streets D, Yarber K, Woo J, Carmichael G (2003) Biomass burning in Asia: Annual and seasonal estimates and atmospheric emissions. Glob Biogeochem Cycles 17(4):1099–1108

Suma M, Balaram PS (2014) Perspective study of coastal disaster management at Andhra Pradesh 16(12): 55–60

Sunil Kumar BV, Singh G, Meur SK (2010) Effects of addition of electrolyte and ascorbic acid in feed during heat stress in buffaloes. Asian Aust J Anim Sci 23(7):880–888

Surya P (2012) Training module on comprehensive landslides risk management. National Institute of Disaster Management, New Delhi, p 282

Svobodova Z, Lloyd R, Máchova J, Vykusova B (1993) Water quality and fish health, EIFAC Technical Paper. No. 54. FAO, Rome, 59 p

Takeoka Y, Mamun AA, Wada T, Kanj BP (1992) Reproductive adaptation of rice to environmental stress. Japan Sci Soc Press Tokyo, Japan, pp 8–10

Takizawa H (2011) Impact of air pollution on allergic diseases. Korean J Intern Med 26(3):262–273

Taylor OC (1973) Acute responses of plants to aerial pollutants. Adv Chemother 122:9–20

Temple PJ (1990) Growth form and yield responses of four cotton cultivars to ozone. Agron J 82:1045–1050

Teramura AH, Ziska LH, Sztein AE (1991) Changes in growth and photo-synthetic capacity of rice with increased UV-B radiation. Physiol Plant 83:373–380

Thakur P, Kumara S, Malika JA, Bergerb JD, Nayyar H (2010) Cold stress effects on reproductive development in grain crops: an overview. Environ Exp Bot 67(3):429–443

Thara SB, Hemanthkumar NK, Shobha J (2015) Micro-morphological and biochemical response of Muntingia calabura L. and Ixora coccinea L. to air pollution. J Res Plant Biol 5(4):11–17

Tinoco IFF (2001) Avicultura industrial: novos conceitos de materiais, concepcoes e tecnicas construtivas disponiveis para galpoes avicolas brasileiros. Rev Bras Cienc Avic 3(1):1–25

Tulloch DG (1988) The importance of the wallow to the water buffalo (*Bubalus bubalis* L.) Buffalo J 4:1–8

Tuwilika SV (2016) Impact of flooding on rural livelihoods of the Cuvelai Basin in Northern Namibia. J Geo Regio Plan 9(6):104–121

UNISDR (2015) Making development sustainable: the future of disaster risk management. Global Assessment Report on Disaster Risk Reduction. Geneva, Switzerland: United Nations Office for Disaster Risk Reduction (UNISDR), pp. 266

Upadhyay RC, Ashutosh, Raina VS, Singh SV (2009) Impact of climate change on reproductive functions of cattle and buffaloes. In: Aggarwal PK (ed) Global climate change and Indian agriculture: case studies from the ICAR network project. ICAR Publication, New Delhi, pp 107–110

Van-Der-Kraak G, Pankhurst NW (1997) Temperature effects on the reproductive performance of fish. In: Wood CM, McDonald DG (eds) Global warming: implications for freshwater and marine fish. Cambridge University Press, Cambridge, pp 159–176

Vass KK, Das MK, Srivastava PK, Dey S (2009) Assessing the impact of climate change on inland fisheries in River Ganga and its plains in India. Aquat Ecosyst Health Manage 12(2):138–151

Venkataraman C, Habib G, Kadamba D, Shrivastava M, Leon J, Crouzille B, Boucher O, Streets D (2006) Emissions from open biomass burning in India: integrating the inventory approach with high-resolution Moderate Resolution Imaging Spectroradiometer (MODIS) active-fire and land cover data. Global Biogeochem Cycles 20(2):1–12

Venkateswarlu B, Maheswari M, Srinivasa Rao M, Rao VUM, Srinivasa Rao Ch, Reddy KS, Ramana DBV, Rama Rao CA, Vijay Kumar P, Dixit S, Sikka AK (2013) National Initiative on Climate Resilient Agriculture (NICRA), Research Highlights (2012–13). Central Research Institute for Dryland Agriculture, Hyderabad

Victor VM, Guayerbas N, Garrote D, Del Rio M, De La Fuente M (1999) Modulation of murine macrophage function by N-Acetyl cytosine in a model of endotoxic shock. Biofactors 5:234

Vingarzan R (2004) A review of surface ozone background levels and trends. Atmos Environ 38:3431–3442

Vough L (1978) Preventing prussic acid poisoning of livestock. Extension Circular 950. Oregon State University Extension Service. http://forages.oregonstate.edu/fi/topics/pasturesandgrazing/grazingsystemdesign/preventingprussicacidpoisening

Wahid A, Gelani S, Ashraf M, Foolad M (2007) Heat tolerance in plants: an overview. Environ Exp Bot 61(3):199–223

Wathes CM, Phillips VR, Holden MR, Sneath RW, Short JL, White RPP, Hartung J, Seedorf J, Schroder M, Linkert KH, Pedersen S, Takai H, Johnsen JO, Groot Koerkamp PWG, Uenk GH, Metz JHM, Hinz T, Caspary V, Linke S (1998) Emissions of aerial pollutants in livestock buildings in Northern Europe: overview of a multinational project. J Agril Eng Res 70(1):3–9

Waycott M, McKenzie LJ, Mellors JE, Ellison JC, Sheaves MT, Collier C, Schwarz AM et al (2011) Vulnerability of mangroves, sea grasses and intertidal flats in the tropical Pacific to climate change. In: Bell JD, Johanna EJ, Hobday AJ (eds) Vulnerability of tropical pacific fisheries and aquaculture to climate change. Secretariat of the Pacific Community, Noumea, pp 297–368

Webb AAR, McAinsh MR, Taylor JE, Hetherington AM (1996) Calcium ions as intercellular second messengers in higher plants. Adv Bot Res 22:45–96

Weiss RF (1970) The solubility of nitrogen, oxygen and argon in water and sea water. Deep-Sea Res 17:721–735

West JW (1999) Nutritional strategies for managing the heat stressed dairy cows. J Anim Sci 77(2):21–35

West JJ, Fiore AM, Naik V, Horowitz LW, Schwarzkopf MD, Mauzerall DL (2007) Ozone air quality and radiative forcing consequences of changes in ozone precursor emissions. Geophy Res Lett 34(6). doi:10.1029/2006GL029173

Wideman RF (1988) Ascites in poultry. Monsanto Nutr Update 6:1–7

Wideman RF (2001) Pathophysiology of heart/lung disorders: pulmonary hypertension syndrome in broiler chickens. World's Poult Sci 57(3):289–307

WMO (2016) Hotter, drier, wetter. Face the future. WMO Bulletin 65(1):2016

Yahav S, Straschnow A, Plavnik I, Hurwitz S (1997) Blood system response of chickens to changes in environmental temperature. Poult Sci 76(4):627–633

Yalcin S, Settar P, Ozkan S, Cahaner A (1997) Comparative evaluation of three commercial broiler stocks in hot versus temperate climates. Poult Sci 76(7):921–929

Yamori W, Hikosaka K, Way DA (2014) Temperature response of photosynthesis in C_3, C_4, and CAM plants: temperature acclimation and temperature adaptation. Photosynth Res 119(1–2):101–117

Yersin AG, Huff WE, Kubena LF, Elissalde MH, Harvey RB, Witzel DA, Giror LE (1992) Changes in haematological, blood gas, and serum biochemical variables in broilers during exposure to simulated high altitude. Avian Dis 36(2):189–196

Young BA (1981) Cold stress as it affects animal production. J Anim Sci 52(1):154–163

Young LW, Wilen RW, Bonham-Smith PC (2004) High temperature stress of *Brassica napus* during flowering reduces micro and mega gametophyte fertility, induces fruit abortion, and disrupts seed production. J Exp Bot 55(396):485–495

Zhang ZY, Jia GQ, Zuo JJ, Zhang Y, Lei J, Ren L, Feng DY (2012) Effects of constant and cyclic heat stress on muscle metabolism and meat quality of broiler breast fillet and thigh meat. Poult Sci 91(11):2931–2937

Zhou WT, Fijita M, Yamamoto S, Iwasaki K, Ikawa R, Oyama H, Horikawa H (1998) Effects of glucose in drinking water on the changes in whole blood viscosity and plasma osmolality of broiler chickens during high temperature exposure. Poult Sci 77(5):644–647

Agriculture Drought Management Options: Scope and Opportunities

3

Jagadish Rane and Paramjit Singh Minhas

Abstract

Drought is one of the recurring features of Indian agriculture especially in the rainfed areas. Severity of its impacts depends upon its nature (chronic and contingent), its duration and frequency, and the extent of area afflicted. Drought not only impacts production at farm level vis-à-vis national food security but also causes miseries to human life and livestock. The present drought management strategies, however, are skewed toward crisis management rather than risk management. The latter needs enhanced insight into even the minute features of the agroecologies for viable solutions to mitigate drought stress. These solutions are determined by capacity to reduce soil moisture deficit, minimizing the impact of drought and accelerating the recovery. However, the key technologies for drought proofing are watershed management, *in situ* water conservation, and integrated farming systems that include resilient crops, contingent crop plans, etc. Drought stress management further needs shaping through modern tools for characterization of agroecosystem, stress mitigation options, and genetic modification of crops for drought tolerance. In this context, the present review attempts to look at various options being offered by advances in drought management.

3.1 Introduction

Drought is an integral part of farmers' life in 68% of cultivated area in India that is vulnerable to water deficits resulting from failure of rains and lack of access to stored water in natural or artificial reservoirs above or below the ground. Definitions and different perspectives of this natural disasters have been elaborated in different

J. Rane (✉) • P.S. Minhas
National Institute of Abiotic Stress Management, Indian Council for Agricultural Research, Baramati, Maharashtra, India
e-mail: jagrane@hotmail.com; minhas_54@yahoo.co.in

© Springer Nature Singapore Pte Ltd. 2017
P.S. Minhas et al. (eds.), *Abiotic Stress Management for Resilient Agriculture*,
DOI 10.1007/978-981-10-5744-1_3

51

reviews and reports (Kramer 1983; Wilhite and Glantz 1985; NAAS 2011). While some of these definitions are conceptual others are operational or disciplinary (Table 3.1). The management of drought has now become an essential feature of national disaster management system that has evolved as an integrated institutional mechanisms for holistic, proactive, multi-disaster-oriented and technology-driven strategy through various approaches for prevention, mitigation, preparedness, and response (GoI 2009; Rathore et al. 2014). Till recently, the focus of drought management strategy was primarily on crisis management during the drought, and there is now an increasing awareness on management of risks of drought for the human and livestock population as well as agricultural crops. Though the reasons can be traced ultimately to atmospheric and hydrological droughts, the agricultural drought is unique in the sense that it impacts the society through its adverse effect mainly on crop plants including forages, which then affect livestock. By definition, agricultural drought is the situation when crop plants face deficit of soil moisture for their growth and development which consequently lead to losses in productivity. Here drought stress refers to inadequate water availability in quantity and distribution during the life cycle of the crop, which is the most important production risk for many crops worldwide (Beebe et al. 2013). In this context, the intervention of modern science for management of agricultural drought assumes immense significance.

Countries like India with major agroecologies being rainfed have been witnessing frequent drought episodes, but their frequency has increased in the recent past. Though this cannot be exclusively attributed to climate change, the predictions are now warranting enhanced efforts to explore drought adaptation and mitigation options to ensure food demand of about 1.6 billion population expected by 2050. Even in the absence of climate change, diminishing water supplies, urbanization, shifting diets, and the additional demand on cereals like maize for fodder and fuel pose challenges for agricultural sector (Hubert et al. 2010). In countries like India where the two-fifth of irrigated agriculture contributes about half the production and the rest, which is prone to drought, contributes three-fifth of the production, the rest three-fifth comprising of rainfed ecosystems contribute only about two-fifth of production.

Despite reduced contribution of agriculture to national GDP since independence, about half the population depends on agriculture, and about half of that are vulnerable to impacts of drought. Drought-prone agricultural lands cannot be neglected as the food production needs to be doubled by 2050 particularly when climate change events are likely to amplify adverse effects of natural disasters. Lessons learnt in the past have helped evolving institutional mechanisms for reducing the impact of drought though holistic solution are yet to be evolved for this recurring problem. In addition to the focus on staple food crops, increased understanding of response of horticultural crops and livestock production system to drought is essential.

Table 3.1 Concepts and definitions of drought

Type	Description
Conceptual	A long period with no rain, especially during a planting season (American Heritage Dictionary 1976)
	An extended period of dry weather, especially one injurious to crops (Random House Dictionary 1969)
Operational	An operational definition, for example, would be one that compares daily precipitation values to evapotranspiration (ET) rates to determine the rate of soil moisture depletion and expresses these relationships in terms of drought effects on plant behavior at various stages of crop development (Wilhite and Glantz 1985)
Disciplinary	Defined according to disciplinary perspectives (Subrahmanyam 1967) 1. Meteorological/atmospheric drought
	Meteorological drought is classified based on rainfall deficiency w.r.t. long-term average – 25% or less is normal, 26–50% is moderate, and more than 50% is severe
	In India meteorological drought is classified based on rainfall deficiency w.r.t. long-term average – 25% or less is normal, 26–50% is moderate, and more than 50% is severe
	Palmer drought severity index (PDSI)-based definition: The palmer drought severity index (PDSI), developed in 1965 by W. C. Palmer, is the widely referred meteorologic drought definition in the United States and is well known internationally
	The PDSI relates drought severity to the accumulated weighted differences between actual precipitation and the precipitation requirement of evapotranspiration (ET). Although commonly referred to as a drought index, the PDSI is actually used to evaluate prolonged periods of abnormally wet or abnormally dry weather
	2. Hydrological drought
	Hydrological drought is best defined as deficiencies in surface and subsurface water supplies leading to a lack of water for normal and specific needs. Such conditions arise even in times of average (or above average) precipitation when increased usage of water diminishes the reserves
	Definitions of hydrologic drought are concerned with the effects of dry spells on surface or subsurface hydrology, rather than with the meteorological explanation of the event. For example, Linsley et al. (1975) considered hydrologic drought a "period during which stream flows are inadequate to supply established uses under a given water management system"
	3. Agricultural drought
	Agricultural drought definitions link various characteristics of meteorological drought to agricultural impacts, focusing, for example, on precipitation shortages, departures from normal or numerous factors such as evapotranspiration
	A plant's demand for water is dependent on prevailing meteorological conditions, biological characteristics of the specific plant, its stage of growth, and the physical and biological properties of the soil. An operational definition of agricultural drought should account for variable susceptibility of crops at different stages of crop development. For example, deficient subsoil moisture in an early growth stage will have little impact on final crop yield if topsoil moisture is sufficient to meet early growth requirements. However, if the deficiency of subsoil moisture continues, a substantial yield loss would result

(continued)

Table 3.1 (continued)

Type	Description
	4. Socioeconomic drought
	Definitions which express features of the socioeconomic effects of drought in addition to features of meteorological, agricultural, and hydrological drought. For example, socioeconomic drought is said to be occurring when the supply of particular commodity drastically falls short of demand due to less precipitation and harms the progress of society for that particular year or season

3.2   Intensity of Drought and Losses

On an average, severe drought occurs once in 5 years in most of the tropical countries, though often these may occur during successive years causing huge losses to agriculture and misery to human life and livestock. Almost every year, varying intensities of drought affect one or the other region of the country. Almost two-thirds of the geographic area of India receive low rainfall (<1000 mm), which is also characterized by erratic and uneven distribution. Technical Committee on Drought Prone Area Programme and Desert Development Programme identified about 120 million hectares of the country's area, covering 185 districts (1173 development blocks) in 13 states as drought prone. Based on the historical records, about 130 droughts/famines have been reported in one or other part of the country between 1291 and 2009. During the twentieth century alone, droughts of varied intensities occurred during 28 years in India (NAAS 2011). The loss in production of food grains due to drought averaged over in 1970–1996 has been estimated to be 1.8 billion year^{-1}, which was equivalent to 8% of the value of food grain production in the region. This needs to be revisited taking into consideration the worst droughts that occurred till recently. In 1972 and 2009 the nationwide rainfall deficits were 24% and 23%, respectively. Rainfall was 19% below normal during the droughts of 1979, 1987, and 2002. Food production was declined by an average of 10% year-on-year in a drought year. In 1987, the delayed onset of the monsoon in certain parts and the prolonged dry spells in most parts of the country severely affected agricultural operations in 58.6 M ha of cropped area in 263 districts in 15 states and 6 Union territories. About half of the area was not sown at all. In Gujarat and Rajasthan the drought was third to fourth in succession and even Punjab and Haryana received less than 50% of normal rainfall. In addition to acute water shortage for 54,000 villages of Rajasthan, there was severe deficit of fodder for livestock. Drought in 1987 affected about 16.8 million cattle across the country (Anonymous 1991).

In terms of magnitude, the drought of the year 2002 ranked fifth among the severest droughts India faced since 1875. The intensity of aridness in July at 51% rainfall deficiency surpassed all previous droughts. The impact of drought spreads over 56% of the landmass threatening livelihood of about 300 million people in 18

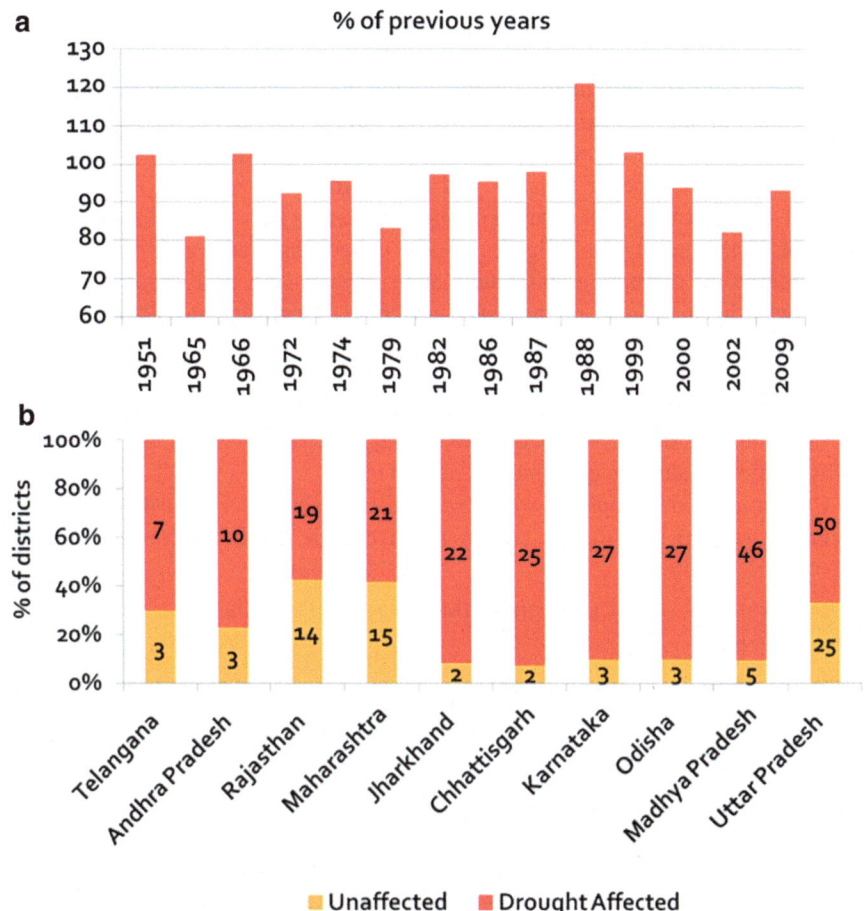

Fig. 3.1 Losses in food production during drought years (**a**) and drought-affected districts in different states of India during 2015–2016 (**b**) (Based on DOAC (2016), Gov India data)

states (NAAS 2011). This drought reduced the sown area to 112 million hectares from 124 million hectares and the food grain production to 174 million tons from 212 million tons, thus leading to a 3.2% decline in agricultural GDP (Murthy et al. 2010). Most of the drought-prone areas lie in the arid (19.6%), semiarid (37%), and subhumid (21%) areas of the country that occupy 77.6% of its total land area of 329 million hectares (Anonymous 2012). In 2015–2016, more than 40% of the area of 10 states was severely affected by drought (Fig. 3.1) indicating that drought will continue to be the major constraint in ensuring the food security that encompasses availability, access, utilization, and stability of a healthy food supply (FAO 1996). This necessitates an overview of progress made so far and scope ahead for further gains from scientific advances.

3.3   Achievements and Challenges

The growth of crops and the food production in the country are strongly influenced by total rainfall during *Kharif* season (Venkateswarlu et al. 2012). Though not completely solved by persistent efforts during the last six decades for developing improved practices, better logistics, and timely interventions, there is evidence to support the view that such drought-proofing measures have lowered the impact of droughts of the recent years (Fig. 3.2). These impacts have emerged from technologies focused for prevention, mitigation, and risk management. Providing access to water was the main goal of integrated watershed management and other in situ water conservation measures, which were successful in many places but yet to cover hitherto unattended regions. The success of these interventions was more conspicuous in regions that receive erratic or optimum rains. However, the rainfall abrasions during southwest monsoons continue to be major factors contributing toward instability of food production especially in rainfed areas.

Crop improvement options were explored extensively for almost all the crops including rainfed rice that is known to have high water requirement. Several

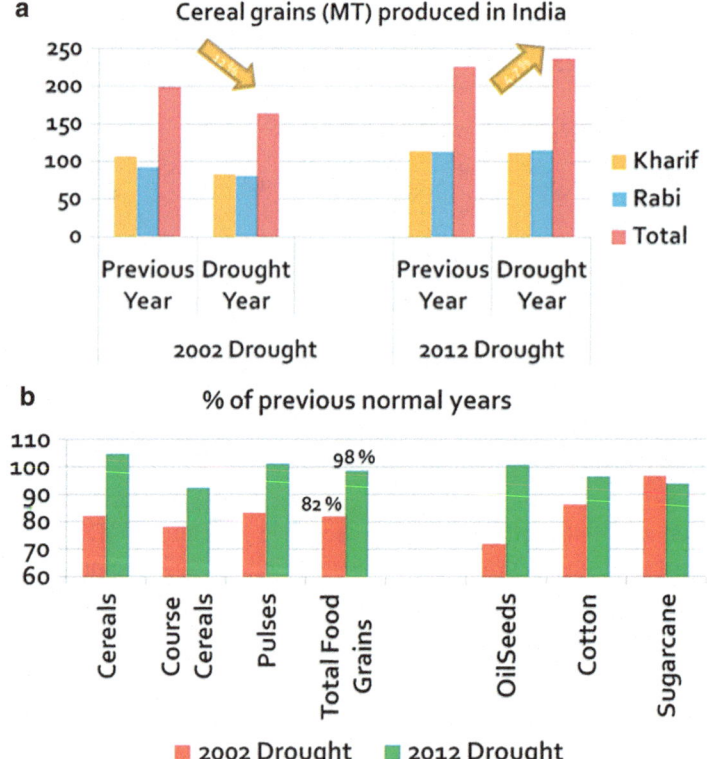

Fig. 3.2 Comparison between 2002 and 2012 drought years with respect to cereal production (**a**) and percent reduction in different food crops (**b**) (Based on DOAC (2016))

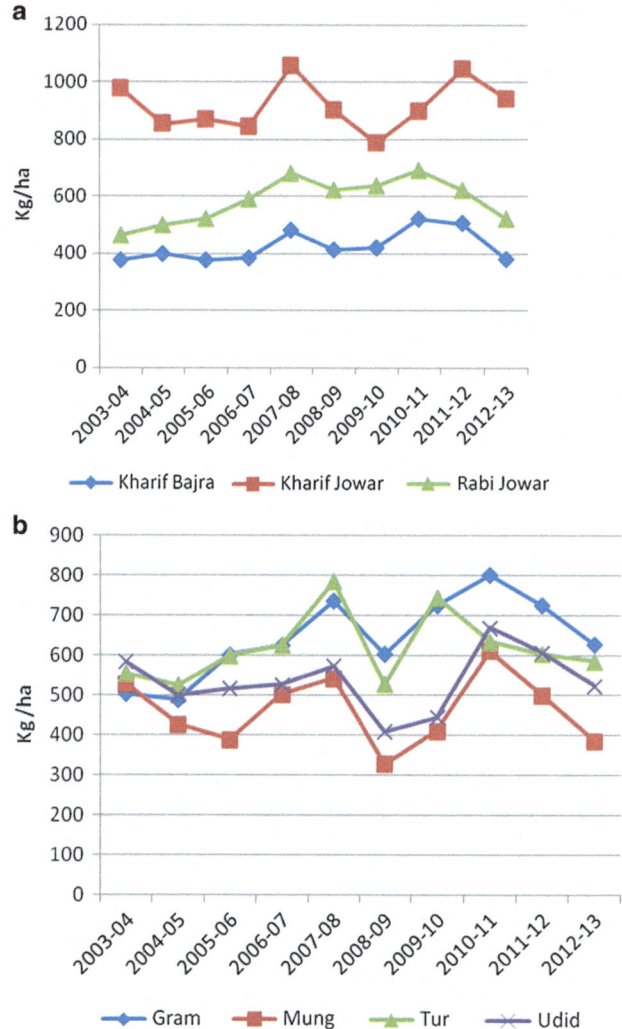

Fig. 3.3 Trend in productivity of cereal (**a**) pulse (**b**) crops in Maharashtra (Based Maharashtra Agriculture Statistics)

high-yielding varieties have been released for cultivation in rainfed and drought-prone areas by national (AICRPs for different crops) and international institutes. In addition, improved production practices have been developed to get maximum yield from these varieties. However, these technologies remain ineffective when the drought is featured by extreme delay and total failure of rains (Fig. 3.3).

Integrated farming system featured by small ruminants in livestock was perceived as viable solution for assured income under drought situation; however, the fodder crisis during drought usually limits the potential of such ventures in large scale. Unlike vast well-managed grasslands in the developed and scarcely populated rural landscapes of the west, Indian livestock production continues to depend on village common lands and largely on crop residues. Dryland fruit crops though do not require large quantities of water; the acute shortage during the dry spell can devastate their orchards as has been observed during the recent droughts in Vidarbha and Marathwada regions of Maharashtra.

Soil and water conservation measures in drought-prone areas are also given due attention as the absence of rains and consequent lack of crop cover lead to substantial soil loss contributing to land degradation. However, the distressed farmers pay little attention to this fact unless provided with support from concerned public and private agencies. Dire necessity for contingency plans to tackle the drought was emphasized in various forums, and as a consequence the contingency plans have been prepared for 614 districts. Persistent efforts are being made to scale up the best practices for rainwater and soil management through linking on-station and on-farm research (Venkateswarlu et al. 2008; Rao et al. 2014). Focus is also on alternate and diversified land use systems that can contribute to agricultural drought management. Remote sensing has emerged as robust tools for assessment, monitoring, and forewarning the drought episodes, and thereby efforts are on strengthening the forecasts for early warning, vulnerability, and assessment of drought.

3.4   Scope for Technological Interventions

Unlike other natural disaster, the occurrence and impact of drought is gradual and hence allows sufficient scope for managing it through technological interventions. Agriculture persists in drought-prone areas because the drought years are followed by normal years. For example, deviations from normal rainfall from 1998 to 2015 in Solapur District of Maharashtra (Fig. 3.4) reveal cyclic nature of drought with phase of normal years of 3 to 4 years followed by drought years of almost similar duration. While beginning of the cycle the district received 70% higher than the normal, the end of the cycle witnessed 40–60% less rains. There were intermediate years when the rain deficits were not severe. Such trends are very common particularly in rainfed areas. In this context, drought cycle management is crucial for enhancing the benefit for farmers with approaches that encompasses prevention of losses, mitigation, and risk management (Table 3.2). Such approaches should consider the short- and long-term profit without technological footprint on the environment. This should be based on weather predictions and land use plans suited to soil type and other features of agroecologies.

Through the drought cycle management, farmers can get higher productivity from improved crops and resilient livestock, production technologies, and protection technologies particularly during favorable years, while risk cover can take care of him during harsh years. At the same time during normal as well as severe drought periods, attention is needed for protecting and developing the agroecosystems through appropriate land degradation prevention efforts.

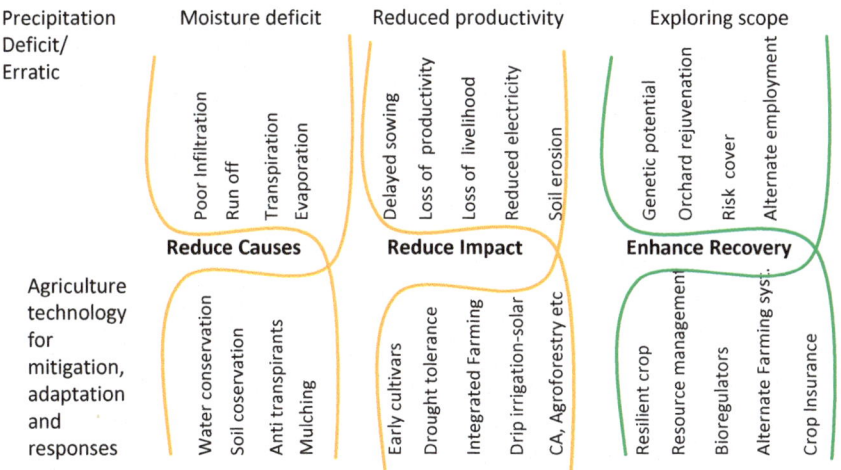

Fig. 3.4 Conceptual framework for managing drought stress through agricultural technology interventions

Table 3.2 Technology intervention for drought cycle management

Beneficiary	Source	Technologies and tools	Drought cycle
Farmer	More productivity	Improved crops and resilient livestock	Favorable years
		Improved production technologies	
		Improved protection technologies	
	Less reduction in losses	Robust weather predictions	Unfavorable years
		Resource management	
		Resilient crop with environmental plasticity	
	Continued income	Risk cover	Transition period
		Protection of biennial and perennial crops to ensure recovery	
		Sustainable farming systems	
Natural ecosystem	Prevention of land degradation	Soil conservation technologies	Unfavorable years
	Enrichment of natural reservoirs	Watershed technology	Favorable years
		Ground water recharge	

While drought management in broad sense including even nonagricultural sector has been given due emphasis for crisis management, it is suggested that the three key components (Fig. 3.5) should be integrated in conceptual framework for agricultural drought management. The conceptual framework considers precipitation deficit and erratic rainfall with long spells of dry weather as primary cause of soil

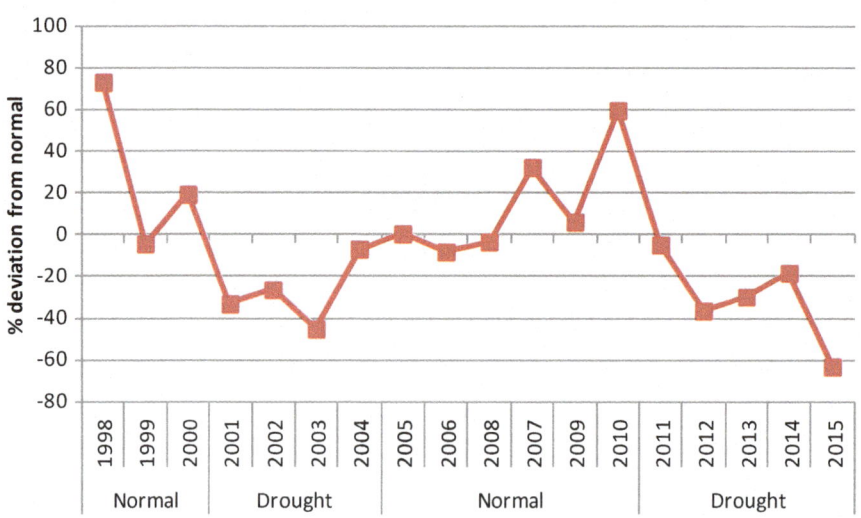

Fig. 3.5 Drought cycle at Solapur District of Maharashtra depicting deviation from normal rainfall (Based on Maharashtra Agricultural Statistics)

moisture deficit. However, it also recognizes features of soil as factor that allows determination of options for reducing the causes of soil moisture deficit. While reduced agricultural productivity is the immediate impact, it hopes on technologies to reduce these impacts with resilient commodities or stress-mitigating natural management options. Further, the framework considers recovery from stress as one of the major component for management of drought particularly with respect to perennial horticultural crops, livestock, and also farmer. There are several proposals, policies, and institutional mechanisms which cover prevention, mitigation, stress relief, and recovery in broader perspectives including socioeconomic aspects. However, the present review focuses largely on scope and options offered by science for reducing drought risks for agriculture.

3.4.1   Reducing the Causes of Soil Moisture Deficit

There are four major factors that lead to soil moisture deficit, viz., loss due to poor infiltration of water, run-off, high evaporation, and transpiration. These are largely determined by climate, soil properties and topography of land, and laws of water dynamics (Unger et al. 2010; Kirkham 2011). These causal factors also determine the type, intensity, and duration of agricultural drought and can be successfully addressed through modern approaches for integrated watershed management (Molden 2007; Joshi et al. 2008; Bhan 2013; Nagaraja and Ekambaram 2015) and conservation agriculture (Hudson 1987) involving mulching (Shekour et al. 1987; Choudhary 2016) and intercultural practices. Antitranspirants have been demonstrated to improve crop yield in sorghum (Fuehring 1975), maize (Fuehring and Finkner 1983), and potato (Pavlista 1995) but are cost prohibitive.

3.4.2 Minimizing the Impact of Drought

Major impact of drought on agriculture is manifested mainly through damage to crops and gradually the livestock dependent on fodder production and water. The prolonged drought and incessant rains can cause heavy losses by degradation of land. The agriculture once crippled by drought can affect livelihood and hence income of rural mass which consequently face severe economic constraints for farmers who often depend on loans. The nature of damage to crops depends on drought intensity, which is determined by its timing, duration, crop growth stages, and also the soil type and landscape at a particular location. Technological interventions to minimize the losses can emerge from natural resource management with focus on conservation and input use efficiency, genetic improvement with focus on tolerance to stress, and policy support research with focus on adaptation of technology and market strategies for enhancing profit for farmers. When farmers get engrossed with immediate relief from the drought, the damage to land due to soil degradation is often neglected. The community efforts supported by government and voluntary agencies can help prevent this by employing appropriate strategies involving integrated watershed concepts. Already released varieties of crops for drought-prone areas if matched with resources and agroecosystem features can minimize the losses due to drought. Contingency plans have been developed for minimizing the losses but needs to be extended to drought-prone regions. In addition to resource management technologies, improvement in resilience to soil moisture stress can be achieved by adopting drought-tolerant crops. This needs genetic improvement efforts. New omic technologies, specifically genomics and phenomics, have opened new avenues to support conventional breeding approaches for developing crop varieties tolerant to soil moisture deficit at different locations. The yield component-based selection procedure is gradually leaning toward trait-based approaches for imparting drought tolerance in crops.

3.4.3 Enhancing the Recovery from Drought

The recovery from agricultural drought should be attended by taking into consideration the necessity to recover the annual crops from mid- and early season drought, perennial horticultural plant after long dry spell, recovery of livestock, and recovery of farmers. This can be accomplished by technological interventions such as application of bio-regulators, real-time contingency plan for midseason drought, fodder banks, and risk cover through crop insurance. The foremost factor that enables recovery from drought is resumption of precipitation while technologies can enhance the recovery of crops damaged by deficit soil moisture. Technologies can benefit the crops if drought occurs at mid-, early, or late season if not throughout the crop growth stage particularly in case of annual staple food crops. The role of bioregulators have been demonstrated to alleviate the abiotic stresses including soil moisture stress in some of the crops as illustrated in other chapters. This option has to be explored for all the drought tolerant cultivars and different drought situations.

Resource management technologies for real-time contingency have been successfully demonstrated but needs to be scaled up to suit location-specific requirements. Crop improvement technologies have resulted in crop cultivars specific to rainfed, arid, and semiarid regions which experience drought; however, they need to be tailored for updated agronomic practices and nature of drought that varies widely across the location. There is a scope for development of technologies to address delayed onset to maximum of three weeks from normal date for the given region, early onset and sudden breaks, early withdrawal of monsoon, and delayed withdrawal or extended monsoon which all contribute to drought faced by the crops. There is a scope to test these technologies in real-time situation.

3.5 Reorientation of Approaches

Some of the evolved technologies are not being adopted due to constraints at the field level. These gaps can be bridged through a deep insight into the nature of drought in location-specific context, mechanisms, and traits that enable crop plants to survive and perform in soil moisture-deficit environment. Varieties of crops with high and stable yield, production technologies that can reduce gap between potential and realizable yield, and protection technology that employ bio-regulators can become integral components of drought management plans. The technologies that were evolved during the past decades may not provide much dividend as any improvement in production and productivity should not have environmental footprints.

3.5.1 Insight into Agroecosystems Facing Drought

Variability in drought ecologies exists across the country, and strategies to manage these cannot be the same though adequate access to water is ultimate solution for development. However, it is possible to outline common options as well as specific options for mitigation and adoption by soil type and the climate-dependent choice of agriculture enterprises. For example, in Maharashtra, where about 80% of cropped area is rainfed, the intensity of drought and its impact vary widely on deep/medium/shallow black soils due to deficient rains at early stages, midseason, or/and at terminal stages of crops (Fig. 3.6). So the management options should match the location-specific requirement for improvement in agricultural productivity. This also demands close link between departments of agriculture, agriculture extension agencies, research institutes, and academic institutes.

3.5.2 Effective Contingency Plans

Contingency planning is a management process that analyzes specific potential events or emerging situations that might threaten society or the environment and establishes arrangements in advance to enable timely, effective, and appropriate

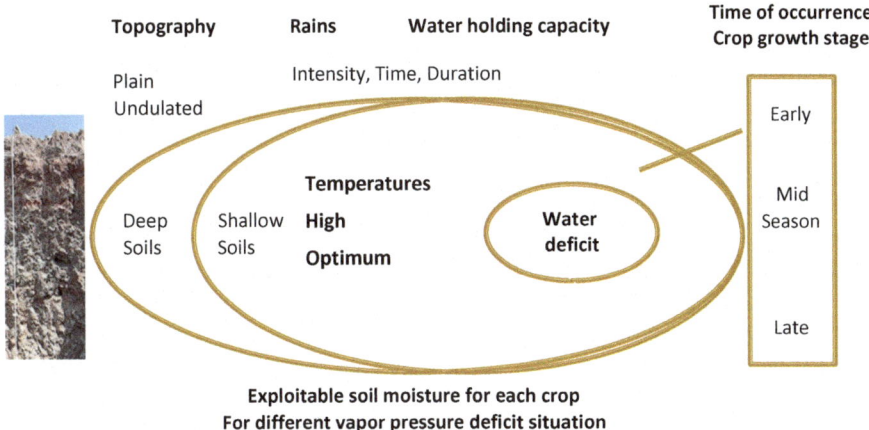

Fig. 3.6 Factors contributing variability and nature of drought

responses to such events and situations. It refers to a feasible strategy for implementation to mitigate the losses due to stress when problem occurs. The fundamental issue of prediction of drought in advance is addressed by meteorological divisions while the task of minimizing the losses to agriculture in general and crops in specific from the next drought depends on existing technologies. Hence, the contingency plan necessarily encompasses technological interventions to minimize the loss. The effectiveness of the contingency plan is determined by the scale at which it is implemented and the emphasis on the diversity in nature of drought to fit the best available technology. Crop contingency plans have been prepared for 614 districts that have now become the integral part of drought management action plan for the nation. In addition, some real-time contingency plans in the event of failed or deficient rains have been successfully demonstrated (Srinivasarao et al. 2013).

3.5.3 Drought Cycle Management

Drought is a cyclic process and therefore actions to be taken have to be phase dependent (Pantuliano and Wekesa 2008; Lesukat 2012). When drought is inevitable, seeding or planting can be avoided instead of losing crop. Vast areas remain uncultivated and become vulnerable to wind and excess rains responsible for soil erosion. However, the limited rains can sustain some cover crop that may be useful in fixing nitrogen and enriching soil with organic matter during the dry spell. This can help in improving the productivity of subsequent crops as benefit of cover crops are increasingly evident (Mahama et al. 2016). Farming system approach with new technological options for crop and livestock suited to specific agroecologies can minimize the losses due to drought. Emphasis on livestock during the predicted drought years and crop production during the favorable years needs to be optimized at each smallest unit of agriculture sector.

Farmers have learnt about the nature of drought by experiencing it and are usually hesitant to adopt costly technologies unless convinced about the benefits of emerging ventures. The technologies which are optimized for small and marginal holdings will have more acceptances. For enhancing the acceptance of these technologies, it is necessary to combine them with weather-based risk cover through crop insurance in the event of drought. Availability of quality seeds of crops and other essential inputs become major constraints for re-sowing if early secession of monsoon spoils the crop. Similarly, failure of crop during the drought year places substantial pressure on seed supply chain for the subsequent year. Hence, drought cycle management approach should include production and supply of seeds and other inputs like seed treatment formulation/microbial cultures for ensuring benefits from favorable season in drought-prone area. Further, activities essential for prevention of degradation of land and water resources can be beyond the capacity of farmers when crop cultivation is totally hampered by drought. Many of the soil health improvement approaches can be carried out during the drought periods for the benefit of crops during ensuing season with the support of disaster management agencies.

3.5.4 Integrated Models for Soil–Crop–Atmosphere Interaction

For drought cycle management, diversity-based contingency plan can be effectively implemented with an improved insight into the soil–crop–atmosphere continuum. Performance of crops under a given environment is influenced by several production factors which are embedded in soil and atmosphere, and these factors contribute to the complexity in the assessment of stress-prone areas. Crop modeling and simulation facilitate the decomposition of complex nonlinear interactions among production factors. When combined with biology, environment, and policy sciences, it strengthens agriculture decision-making (Adlul et al. 2014). Several models are being used to predict climate change, impact under different scenarios, and the crop performance. International Agricultural Model Intercomparison and Improvement Project (AgMIP) is looking forward to filling the gaps in current generation of simulation models that are inadequately account for soil–crop–atmosphere interaction responses to the wide variability of temperature and precipitation (Hatfield et al. 2011) Hence, it seeks to replace them with the most advanced and robust crop simulation models to project future crop production and to enhance development of adaptation strategies to cope with climate change (Rosenzweig et al. 2013). Further, crop modeling-based management of limited water sources (Islam et al. 2014) may be immensely useful for drought-prone regions. This is highly crucial as India is facing water scarcity as evident from per capita availability of water declining sharply from 5177 m^3 in 1951 to 1544 m^3 in 2011 (CWC 2013). It is projected to reduce further to 1465 and 1235 m^3 by the year 2025 and 2050, respectively, under high population growth scenarios (Kumar et al. 2005).

3.5.5   Emphasis on Genetics × Environment × Management Interaction

Only the genotype and environment interaction used to be the focal point for genetic improvement technologies till recently with the assumption that yield potential improvement can take care of the rest of aspects. At present there is drastic change in the way crops are raised. Conventional agronomic practices are now making ways for conservation and precision agriculture. Since our ability to expand the available land resources are not a viable option, actual yield increases and overall productivity should come from resource management for increasing land productivity. In this context, development of methods of screening genotypes for a variety of responses to combinations of environmental and management scenarios is highly essential. This needs transdisciplinary teams to represent each component of the G × E × M interaction (Hatfield and Walthall 2015). Closing the yield gap, defined as the difference between the farmer's yield and potential yield of a crop variety, is one of the best options as the gap can be as high as 50% under rainfed conditions (Lobell et al. 2009; Hatfield and Walthall 2015). Further, linking water use efficiency (WUE) with radiation use efficiency (RUE) will provide an opportunity for improvement in adaptation to drought environment (Hatfield and Walthall 2015). The proposals to consider concept of effective use of water rather than WUE may also change the way we select the crop for drought stress environments (Blum 2005).

Choice of a crop for a particular soil moisture stress environment is not necessarily governed by yield potential of a variety. For example, Lok 1 a wheat variety released 29 years ago is widely grown in central and peninsular India despite new cultivars released in the recent years. Traditional crop varieties outperform new drought-resistant varieties because of differing soil management practices (Hatfield and Walthall 2015). Recent analysis of field-level data revealed that the sensitivity of maize yields to drought stress associated with high vapor pressure deficits has increased from 1995 to 2006 though yields have increased in absolute value under all levels of stress for both soybean and maize. The agronomic changes tend to translate improved drought tolerance of plants to higher average yields but not to decreasing drought sensitivity of yields at the field scale (Lobell et al. 2014). Further, new options such as plastic mulch on ridge and bare furrows have advantage only in sandy soils and not in heavy soils (Xianju et al. 2014). Hence, it is essential to explore G × E × M approaches for further improvement in productivity of crops. Use of microbial consortium or microbial product in agriculture (Asghar et al. 2015: Xiao-Min and Huiming 2015) can add new dimension to G × E × M approach.

3.5.6   Mining Genetic Resources

There is an ample opportunity to search for stress-tolerant genes in wide range of cultivated and wild species of plants in genebanks. About 1750 plant genebanks have been established worldwide, with 7.4 million accessions maintained as ex situ collections in either seed banks, field collections, and in vitro and cryopreservation conditions (FAO 2010). Countries like India have as many as 346,000 germplasm of various crops, and these germplasm can be treasure of desirable traits for drought tolerance. In addition there are sets of mutant lines with induced genetic variation which exist for different crop species, and new techniques are emerged to create new genetic variants for gene function studies by employing molecular tools. A Robust CRISPR/Cas9 System for convenient, high-efficiency multiplex genome editing in plants is emerging and can be helpful in studying the genes associated with stress tolerance (Ma et al. 2016). Such high-end genomics and marker-aided selection can get support of emerging phenomics techniques (Furbank and Tester 2011; Yang et al. 2013; Klukas et al. 2014; Rahaman et al., 2015) to screen these large set of genotypes with great precision to discover traits and genes relevant to drought tolerance.

While the tremendous progress during the green revolution has been attributed to germplasm exchanged for crop improvement, recent trade-related issues and biodiversity ownerships have largely restricted the free flow of germplasm, which has become the cause of concerns (Aseffa and Welch 2015). At the same time, new inventions have made a way for several patents that have delayed the translation of science for benevolence of humanity. For example, about 70 patents delayed benefit of highly innovative and nutrient-rich golden rice for the targeted population having no access to sufficient quantity and quality of food. In this context, the Public Sector Intellectual Property Resource for Agriculture (PIPRA; www.pipra.org) is trying to establish "best practices" encouraging the greatest commercial application of publicly funded research without compromising the rights of public institutions to fulfill their responsibilities toward the public at large. With hundreds of inventions related to drought tolerance in plants patented, significant utility is expected from PIPRA's efforts to establish a database providing an overview of IPR currently held by public institutions, with up-to-date information on the licensing status of these IPRs. Further the crop improvement for drought-prone regions can get a boost if freedom to operate (FTO) for use of new invention is facilitated through "technology packages" of complementary patents. (Atkinson et al. 2003; Gepts 2004). Efforts should be made to avoid complex restrictions and obligations that may hinder the exchange of genetic materials, public research, and further innovation (Seyoum and Welch 2015).

While the focus is on genetic variability for drought improvement in staple food crops, the opportunity to cultivate other crop species is yet to gain momentum. The human population today derives most of its calories from a very narrow set of crops, with only about 30 species providing 95% of the global food energy (Prescott and Prescott 1990; FAO 2010) and just three major crops, viz., wheat, maize, and rice, provide over 60% of our food supply. On the contrary, over 7000 species are known as edible and are either partly or fully domesticated, suggesting that a large share of potential food sources is underutilized (Rehm and Espig 1991; Wilson 1992). These underutilized crops carry unrealized potential to contribute to human welfare, in particular for income generation for the world's poor, food security and nutrition, and reduction of "hidden hunger" caused by micronutrient deficiencies resulting from uniform diets. It is increasingly felt that challenges of global climate change and food security can be addressed by rescuing and using more diversity in agricultural and food production systems, both in terms of crop as in terms of varieties within any given crop (PAR 2010; FAO 2011). Diverse agroecosystem, crop diversification, and dietary diversification should be assessed in specific context of drought for integrating qualitative and quantitative knowledge on underutilized crops for decision support system. Any efforts to choose the best combination of unexplored crops and the well-characterized drought-prone agroecosystem can greatly contribute to the drought management in agriculture.

3.5.7 Dryland Horticulture Technologies

Extensive studies have been conducted to understand the mechanisms underlying plant responses to soil moisture deficit. However, implementation of technologies except for micro irrigation is not gaining momentum. Plants strategy to survive and perform during and after stress needs to be exploited for those horticultural crops where information is available. There is a need to generate such information for other crops for productive use of scarce and poor quality water though these plants are apparently adapted to dry region. Future orchard management practices for dryland horticulture should consider the emerging basic knowledge that mechanisms of tolerance to drought in young leaf are different from that in mature leaves (Claeys and Dirk 2013). It is well know that pollination is critical to fruit production, but the interactions of pollination with plant resources on a plant's reproductive and vegetative features are largely overlooked (Klein et al. 2015). There is an indication that thermal imaging can explain the responses of plant to soil moisture levels (Lima et al. 2016). There is clear demonstration of benefits of simple techniques such as mulching with straw or plastic films in conserving soil moisture and enhancing water use efficiency in crops like peach in semiarid region (Wanga et al. 2015). Deficit irrigation techniques optimized and method adopted to optimize the same in orchard crops like grapes (Lanaria et al. 2015), papaya (Nunes de Limaa et al. 2015), and pomegranate (Catola et al. 2016; Parvizi et al. 2016) need to be extended to other fruit crops to save water without compromising the quality of fruits. Such optimizations are likely to be largely determined by the type of soil and precipitation events at a given agroecology. However, site preparation techniques, careful species selection, planting, and care during the initial and other critical stages of development are the versatile tools for sustaining orchards during drought period (Minhas et al. 2015). Recently the information on abiotic stress physiology of different horticulture crops has been compiled (Srinivasa Rao et al. 2016) which can further serve as guide for translation of scientific knowledge into management practices for better harvest and income for farmers of dry and drought-prone areas.

3.5.8   Efforts Beyond the Usual Business

Some of the unusual approaches such as artificial rains can help to cope during severe water scarcity periods. Managing clouds to get rains have been tried by human being since time immemorial, and cloud seeding has been demonstrated for enhancing precipitation. Recently, Chinese scientists have perfected this art of managing clouds. The science of cloud seeding involves introduction of small particles for promoting condensation of moisture from cloud by serving as a nuclei to start with and precipitate down as snow or rain. Since the world is at the verge of facing severe water crisis, such technologies are likely to evolve fast. If the art and science are made prefect for managing cloud to get precipitation at desired time, at desired place, and in amount optimum for agricultural activity, our strategy to manage drought-like situation should be reoriented.

Introduction of genetically engineered crops warrant altogether different crop production approaches to comply with the norms of biosecurity. In addition, these crops need high input use efficiency-driven resource management practices as the cost of seeds may be higher than usual traditionally bred crop varieties. On the other hand, introduction of new crops may need altogether different agronomic practices. Future research for drought-prone agriculture should consider these aspects. If inbuilt constraints such as economically unviable size of land holdings are addressed through institutional mechanisms, every piece of land can get benefit of appropriately integrated modern science and viable conventional approaches emerging from traditional wisdom.

3.5.9   Village Transformation Into Drought-Ready Community

To make community-level drought planning more widespread, there should be close linkages between key stakeholders, planners, researchers, and villagers. This needs additional resources to work with organizations and communities. This can be achieved by wider publicity of success stories emerged from model villages. To be qualified as drought ready, the villages should get a start well in time, gather information on drought, establish monitoring, involve in public awareness and education, and prepare for response to drought and recovery. Advances in information technology can be effectively employed to facilitate many of the components of drought-ready community. For example, it is often seen that the farmers face lack of sufficient seeds of desired quality following the drought episodes though the likelihood of better weather offer them to recover from the drought shocks. Such situations can be predicted and the provision can be made to arrange for seeds and other inputs from other places which remain unaffected by drought. Alternatively, there should be access to technology and infrastructure to preserve the seeds during drought phase of drought cycle. Similarly, there must be ways to save drying orchards by bringing water from other places wherever possible. The cost of such ventures will be lesser for the community relative to individual.

3.6   Conclusion

Both the natural resource management and crop improvement research have played significant role in improving the productivity of crops in drought-prone areas as evident from reduced impact of recent drought on overall agricultural production of the country. Implicitly, we cannot neglect regional adverse impact of drought that tends to continue miseries of farmer in drought-prone area. It is also recognized that the achievement so far in developing technologies is not sufficient to meet the future demand for food, improved livelihood, and for stable and descent income of the marginal farmers. Hence, there is a need for accelerated efforts to develop technologies that can reduce the cause of soil moisture deficit, minimize the impact of drought, and accelerate the recovery to normal after the drought event. There is a scope for enhanced capacity to understand drought-prone environment facilitated by remote sensing and GIS and instrumentation and automation to monitor moisture in soil and plant tissues. Our understanding of impact of conservation agriculture as well as advances in genomics, phenomics, and metabolomics for genetic improvement can provide new opportunities for developing novel mitigation and adaptation technologies for drought-prone agriculture. In addition to crop-based agriculture, integrated farming approaches involving livestock and agroforestry with improved mechanization of agricultural operation can transform agricultural landscapes in drought-prone areas and also the remunerations for the farmers. Such ventures should give due consideration for land use plan options. With improvement in monsoon forecast, these technologies can provide robust solutions for management of drought at smallest level of administration. While the scientific community will continue to advance the understanding about the science of tolerance to drought in crop plants, it will better serve crop improvement scientists by providing a more in-depth and clear understanding of the adaptation and mitigation solutions. The focus of search for solutions should be a major component of translational research by involving multidisciplinary teams. With the state-of-the-art facilities for high throughput genomics, phenomics, proteomics spread across the institution, interinstitutional collaboration can bring the expertise on one platform where communications will be robust enough to avoid duplication and to promote synergies in developing scientific solutions for management of agricultural drought.

References

Adlul I, Paresh BS, Soora NK, Nataraja S, Sikka AK, Aggarwal PK (2014) Modeling water management and food security in India under climate change: practical applications of agricultural system models to optimize the use of limited water. In: Ahuja LR, Ma L, Lascano RJ (eds) Advances in agricultural systems modeling 5. Practical applications of agricultural system models to optimize the use of limited water, pp 267–316

Anonymous (1991) In: Misra DC (ed) The drought of 1987: response and management, vol I. DOAC, GOI

Anonymous (2012) Crisis management plan: drought. Department of Agriculture and Cooperation, GOI

American Heritage Dictionary (1976) Houghton Mifflin, Boston

Aseffa S, Welch EW (2015) Ex-post use restriction and benefit- sharing provisions for access to non-plant genetic materials for public research. Appl Econ Perspect Pol 37(4):667–691

Asghar HN, Zahir ZA, Akram MA, Ahmad HT, Hussain MB (2015) Isolation and screening of beneficial bacteria to ameliorate drought stress in wheat. Soil Environ 34(1):100–110

Atkinson RC, Beachy RN, Conway G, Cordova FA, Fox MA, Holbrook KA, Klessig DF, McCormick RL, McPherson PM, Rawlings HR III et al (2003) Intellectual property rights: public sector collaboration for agricultural IP management. Science 301:174–175

Beebe SE, Rao IM, Blair MW, Acosta-gallegos JA (2013) Phenotyping common beans for adaptation to drought. Front Physiol 4:1–20. doi:10.3389/fphys.2013.00035

Bhan S (2013) Land degradation and integrated watershed management in India. Intern Soil Water Cons Res 1:49–57

Blum A (2005) Drought resistance, water-use efficiency, and yield potential–are they compatible, dissonant, or mutually exclusive? Crop Pasture Sci 56:1159–1168

Catola S, Marino G, Emiliani HT, Musayev M, Akparov Z et al (2016) Physiological and metabolomic analysis of *Punica granatum (L.)* under drought stress. Planta 243:441–449

Choudhary VK (2016) Response of land configuration and mulches on maize–frenchbean–toria cropping system. Agron J 108(5):2147–2157

Claeys H, Dirk I (2013) The agony of choice: how plants balance growth and survival under water-limiting conditions. Plant Physiol 162:1768–1779

CWC (2013) Water and related statistics. Water Resources Information System Directorate, Information System Organization, Water Planning & Project Wing, Central Water Commission, New Delhi. http://www.cwc.nic.in/main/downloads/Water%20and%20Related%20Statistics-2013.pdf. Accessed 28 July 2013

DOAC (2016) Agricultural statistics at a glance-2015. Department of Agriculture and Cooperation (http://eands.dacnet.nic.in/PDF/Agricultural_Statistics_At_Glance-2015.pdf

FAO (1996) Rome declaration on World food security and World food summit plan of action. World Food Summit 13–17 November 1996. Rome

FAO (2010) Report on the state of the World's plant genetic resources for food and agriculture. FAO, Rome

FAO (2011) Save and grow: a policymakers' guide to the sustainable intensification of smallholder crop production. FAO, Rome. http://www.fao.org/docrep/014/i2215e/i2215e.pdf

Fuehring HD (1975) Yield of dryland grain sorghum as affected by antitranspirant, nitrogen and contributing micro-watershed. Agron J 67:255–257

Fuehring HD, Finkner MD (1983) Effect of folicote antitranspirant application on field grain yield of moisture–stressed corn. Agron J 75(4):579–582

Furbank RT, Tester M (2011) Phenomics–technologies to relieve the phenotyping bottleneck. Trends Plant Sci 16:635–664

Gepts P (2004) Who owns biodiversity and how should the owners be compensated. Plant Physiol 134:1295–1307

GoI (2009) National Policy on disaster management. Ministry of Home Affairs Government of India http://nidm.gov.in/PDF/policies/ndm_policy2009.pdf. Last Accessed: 20 Apr 2014

Hatfield JL, Boote KJ, Kimball BA, Ziska LH, Izaurralde RC, Ort D, Thomson AM, Wolfe D (2011) Climate impacts on agriculture: implications for crop production. Agron J 103:351–370

Hatfield JL, Walthall CL (2015) Meeting global food needs: realizing the potential via genetics × environment × management interactions. Agron J 107:1215–1226

Hubert B, Rosegrant M, Boekel MA, Van JS, Ortiz R (2010) The future of food: scenarios for 2050. Crop Sci 50:S-33–S-50. doi:10.2135/cropsci2009.09.0530

Hudson NW (1987) Soil and water conservation in semi-arid areas. Soil resources, management and conservation service. FAO Land and Water Development Division. Food and Agriculture Organization of the United Nations, Rome

Islam A, Shirsath PB, Soora KN, Nataraja S, Lascano RJ (eds) (2014) Advances in agricultural systems modeling 5. Modeling water management and food security in India under climate change: practical applications of agricultural system models to optimize the use of limited water practical applications of agricultural system models to optimize the use of limited water, 267–316

Joshi PK, Jha AK, Wani SP, Sreedevi TK, Shaheen FA (2008) Impact of watershed program and conditions for success: a meta-analysis approach. Global theme on Agroecosystems, Report 46. International crops research Institute for the Semi-Arid Tropics and National Centre for agricultural economics and policy research

Kirkham MB (2011) Water dynamics in soils. In: Hatfield JL, Sauer TJ (eds) Management: building a stable base for agriculture, pp 53–65

Klein AM, Hendrix SD, Clough Y, Scofield A, Kremen C (2015) Interacting effects of pollination, water and nutrients on fruit tree performance. Plant Biol 17:201–208

Klukas C, Chen D, Pape JM (2014) Integrated analysis platform: an open-source information system for high-throughput plant phenotyping. Plant Physiol 165:506–518

Kramer P (1983) Water relations of plants. Academic Press, New York

Kumar R, Singh RD, Sharma KD (2005) Water resources of India. Curr Sci 89:794–811

Lanaria V, Alberto P, Paolo S, Stanley HG, Oriana S (2015) Optimizing deficit irrigation strategies to manage vine performance and fruit composition of field-grown 'Sangiovese' (*Vitis vinifera* L.) grapevines. Sci Hort 179(2014):239–247

Lesukat M (2012) Drought contingency plans and planning in the greater horn of Africa a desktop review of the effectiveness of drought contingency plans and planning in Kenya, Uganda and Ethiopia. UNON, Published by United Nations International Strategy for Disaster Reduction (UNISDR) Regional Office for Africa

Lima RSN, García-Tejero I, Lopes TS, Costa JM, Vaz M, Durán-Zuazo VH, Campostrini E (2016) Linking thermal imaging to physiological indicators in *Carica papaya L.* under different watering regimes. Agric Water Manag 164:148–157

Linsley RK, Kohler MA Jr, Paulhus JLH (1975) Hydrology for engineers, 2nd edn. McGraw Hill/Kogukusha, Tokyo

Lobell DB, Cassman KG, Field CB (2009) Crop yield gaps: their importance, magnitudes, and causes. Ann Rev Environ Resour 34:179–204

Lobell DB, Roberts MJ, Schlenker W, Braun N, Little BB, Rejesus RM, Hammer GL (2014) Greater sensitivity to drought accompanies maize yield increase in the U.S. Midwest. Science 344:516–519

Ma X, Zhu Q, Chen Y, Liu Y (2016) CRISPR/Cas9 platforms for genome editing in plants: developments and applications. Mol Plant 9:961–974

Mahama GY, Vara Prasad PV, Kraig L et al (2016) Rice cover crops, fertilizer nitrogen rates and economic return of grain sorghum. Agron J 108(1):1–16

Minhas PS, Bal SK, Sureshkumar P, Singh Y, Wakchaure, GC, Ghadge SV, Nangare DD, Taware PB (2015) Turning basaltic terrain into model research farm: chronicle description. NIASM technical bulletin 8, ICAR -National Institute of Abiotic Stress Management, Baramati, Pune, Maharashtra(India), p.64.

Molden D (2007) In: Molden D (ed) Water for food, water for life: a comprehensive assessment of water Management in Agriculture. International Water Management Institute/Earthscan, London/Colombo

NAAS (2011) Drought preparedness and mitigation. Policy Paper No. 50, National Academy of Agricultural Sciences, New Delhi 22

Nagaraja B, Ekambaram G (2015) A critical appraisal of integrated watershed management programme in India. IOSR-JHSS 20:17–23

Nunes de Limaa RS, Fábio A, Mazzei MA, Amanda OM, Bruna Corrêa SD, Tiago MF, de AG M, Elias FS, David MG, Eliemar C (2015) Partial rootzone drying (PRD) and regulated deficit irrigation (RDI) effects on stomatal conductance, growth, photosynthetic capacity, and water-use efficiency of papaya. Sci Horti 183:13–22

Pantuliano S, Wekesa M (2008) Improving drought response in pastoral areas of Ethiopia Somali and afar regions and Borena zone of Oromiya region. ODI, London

PAR (2010) Biodiversity for food and agriculture: contributing to food security and sustainability in a changing world; Outcomes of an Expert Workshop held by FAO and the Platform on Agrobiodiversity Research: Rome, Italy, 14–16 April; 2010

Parvizi H, Ali RS, Seyed HA (2016) *Punica granatum* (L.) Rabab under partial root zone drying and deficit irrigation regimes. Agric Water Manag 163:146–158

Pavlista AD (1995) Paraffin enhances yield and quality of the potato cv. Atlantic J Prod Agric 8:40–42

Prescott AR, Prescott AC (1990) How many plants feed the world? Conserv Biol 4:365–374

Random House Dictionary (1969) Random House, New York

Rahaman MM, Chen D, Gillani Z, Klukas C, Chen M (2015) Advanced phenotyping and phenotype data analysis for the study of plant growth and development. Front Plant Sci 6:619. doi:10.3389/fpls.2015.0061

Rathore BMS, Sud R, Saxena V, Rathore LS, Rathore TS, Subrahmanyam VG, Roy MM (2014) Country Report prepared for the regional workshop for Asia Pacific as part of the UN water initiative. "Capacity Development to Support National Drought Management Policies" organized from 6–9th May 2014 in Hanoi, Vietnam Page 1 of 7 Drought Conditions and Management Strategies in India

Rehm S, Espig G (1991) The cultivated plants of the tropics. Verlag Josef Margraf and CTA, Weikersheim

Rosenzweig C, Jones JW, Hatfield JL, Ruane AC, Boote KJ, Thorburn P, Antle JM, Nelson GC, Porter C, Janssen S, Asseng S, Basso B, Ewert F, Wallach D, Baigorria G, Winter JM (2013) The agricultural model Intercomparison and improvement project (AgMIP): protocols and pilot studies. Agric For Meteorol 170:166–172

Seyoum A, Welch EW (2015) Ex-post use restriction and benefit-sharing provisions for access to non-plant genetic materials for public research. Appl Econ Persp Pol 37(4):667–691

Shekour GM, Brathwaite RAI, McDavid CR (1987) Dry season sweet corn response to mulching and antitranspirants. Agron J 79:629–631

Srinivasarao CH, Ravindrachary G, Mishra PK, Subba Reddy G, Shankar GRM, Venkateshwaralu B, Sikka AK (2014) Rainfed farming: a profile of doable technologies AICRPDA-CRIDA. Tech Bull:194

Srinivasarao NK, Shivashankara KS, Laxman RH (2016) Abiotic stress physiology of horticultural crops. Springer, New Delhi, p 368

Srinivasarao Ch, Ravindra Chary G, Mishra PK, Nagarjuna Kumar R, Maruthi Sankar GR, Venkateswarlu B, Sikka AK (2013) Real time contingency planning: initial experiences from AICRPDA. All India Coordinated Research Project for Dryland Agriculture (AICRPDA), Central Research Institute for Dryland Agriculture (CRIDA), ICAR, Hyderabad 500 059, India, pp 63

Subrahmanyam P (1967) Incidence and spread of continental drought, WMO/IHD Report No. 2, Geneva, Switzerland: WMO

Unger PW, Kirkham MB, Nielsen DC (2010) Water conservation for agriculture. In: Zobeck TM, Schillinger WF (eds) Soil and water conservation advances in the United States, SSSA Spec Publ, vol 60. SSSA, Madison, pp 1–45

Venkateswarlu B, Kokate KD, Gopinath KA, Srinivasarao C, Anuradha B, Dixit S (eds) (2012) Coping with climate variability: technology demonstration on farmers' fields in vulnerable districts. Central Research Institute for Dryland Agriculture, Hyderabad, p 160

Venkateswarlu B, Ramakrishna YS, Subba Reddy G, Mayande VM, Korwar GR, Prabhakar M (2008) Rainfed farming: a profile of doable technologies. CRIDA. Technical Bulletin, pp 48

Wanga H, Chenbing WB, Xiumei Z, Falin W (2015) Mulching increases water use efficiency of peach production on the rainfed semiarid loess plateau of China. Agric Water Manag 154:20–28

Wilhite DA, Glantz MH (1985) Understanding the drought phenomenon: the role of definitions. Water Intern 10:111–120

Wilson EO (1992) The diversity of life. Penguin, London

Xianju L, Zizhong L, Qingguo B, Dongjian C, Wenxiao D, Zenghui S (2014) Sun effects of rainfall harvesting and mulching on corn yield and water use in the corn belt of Northeast China. Agron J 106(6):2175–2184

Xiao-Min L, Huiming Z (2015) The effects of bacterial volatile emissions on plant abiotic stress tolerance. Front Pl Sci 6:774

Yang W, Duan L, Chen G, Xiong L, Liu Q (2013) Plant phenomics and high-throughput phenotyping: accelerating rice functional genomics using multidisciplinary technologies. Curr Opin Plant Biol 16:180–187

Edaphic Stresses: Concerns and Opportunities for Management

4

Paramjit Singh Minhas

Abstract

Sustainable intensification of soil resources is inevitable to maintain food and nutritional security with ever-increasing demographic pressures. However, the major gains in agricultural productivity in the recent past have led to severe land degradation and the resultant edaphic constraints. These include chemical stresses like emerging nutrient deficiencies (93, 91, 51, 43% soils rated low in N, P, K, Zn, respectively) with mining (8–10 million Mg of NPK annually) along with acidity (pH < 5.5 in 17.93 M ha), salinity (6.73 M ha), and pollutants, while severe soil erosivity (water 82.47 M ha and wind 12.40 M ha), shallow soils (26.4 M ha), soil hardening (21.4 M ha), and low water-holding capacity (13.75 M ha) constitute the major physical constraints. The Indian soils are inherently low in low organic carbon, and climate change is further impacting the farming systems. Even the conservative estimates are that the edaphic stresses cause about two-third loss of agricultural production. Several land and water management practices have been put forward to minimize the impact of these stresses including conservation agriculture, rainwater harvesting, integrated nutrient management, integrated farming systems, etc. However, to alleviate the effects of multiple stressors, a holistic approach to build up systems perspectives is a need of the hour. The new tools emerging especially in the areas of resource-conserving farming systems, conservation agriculture, precision irrigation technologies, biotechnology and omic sciences, etc. are opening up new opportunities for tackling these stresses. The database compiled here should allow for better focus of research to develop the best combination of agroecosystem-based technologies. Moreover, these should improve awareness among decision makers to evolve policy guidelines and take up measures toward "sustainable development goals."

P.S. Minhas (✉)
National Institute of Abiotic Stress Management, Indian Council for Agricultural Research, Baramati, Maharashtra, India
e-mail: minhas_54@yahoo.co.in

© Springer Nature Singapore Pte Ltd. 2017
P.S. Minhas et al. (eds.), *Abiotic Stress Management for Resilient Agriculture*,
DOI 10.1007/978-981-10-5744-1_4

4.1   Introduction

The food and nutritional security warrants the availability of adequate and quality food to meet the dietary and nutritional requirements for a healthy and productive life. The Indian agriculture has registered a phenomenal growth during the past four decades. However, it is still facing the challenge of ensuring food security for the ever-increasing population amidst constraints such as deterioration of soil quality, reduction in per capita land availability, forecasted water scarcities, climate change, etc. Nevertheless, most observers still agree that food security must be achieved through more productive use of land resources and that this must occur without further degradation. With increased pressures on land from urban agglomerations, horizontal expansion in agriculturally suitable land seems no longer a feasible option and rather the cultivated area has since long been static around 141 million hectares (DAC 2015). Thus the only way possible is considerable improvements in soil productivity (Yadav 2007; Singh 2010). But the past endeavors in meeting the demands of ever-increasing population have accompanied with land degradation and the impact of episodic climate change, those that have consequently increased number of edaphic stresses and their varied levels for crop production (NRAA 2011). This has happened because the focus was more on exploitation of soil and water resources and less on improving, restoring, reclaiming, and enhancing their productivity and sustainability (NIASM 2015). As a result, the agricultural productivity has been witnessing stagnation even in irrigated areas. Emergence of widespread multiple nutrient deficiencies, depletion of organic carbon stocks, pollutants/contamination by toxic substances, soil sealing and subsoil hardening, development of secondary salinity and waterlogging in canal-irrigated areas, low-input use efficiencies (nutrient and water use), and decreasing total factor productivity of fertilizers, etc. are all the consequences of application of component-based technologies for short-term yield gains. Additionally, the fragile dryland ecosystems are characterized by limited and erratic rainfall, intense pressure on resource use, and sensitivity to climatic risks and populations that are highly vulnerable to food insecurity (WG-NRMRF 2011). Therefore, sustenance of agricultural growth would require processes that would help us to meet current and long-term societal needs for food, fiber, and other resources, while maximizing benefits through conservation of natural resources and maintenance of ecosystem functions.

4.2   Processes and Extent of Soil Edaphic Stresses

The edaphic constraints are either natural or anthropogenic in origin and afflict almost all global land resources to variable degrees. Some stresses like soil salinity/acidity and erosion, for example, may have both origins. The major stresses as described in Friedrich et al. (2008) are given in Table 4.1. Natural stresses may be either intrinsic (i.e., a consequence of characteristics specific to the soil and its internal processes) or extrinsic (i.e., a consequence of ambient conditions in the external environment of the soil). A further point regarding these edaphic stresses is that

Table 4.1 Edaphic constraints to food production

Category	Examples of stress factors
(a) Natural stresses	
(i) Stresses caused by internal soil processes	
Chemical conditions	Nutrient deficiencies; excess of soluble salts; salinity and alkalinity; low base saturation; low pH; aluminum and manganese toxicity; acid sulfate condition; high P and anion retention; calcareous or gypseous conditions, low redox (hydromorphic conditions)
Physical conditions	High susceptibility to erosion; steep slopes; shallow soils; surface crusting and sealing; low water-holding capacity; impeded drainage; low structural stability; root-restricting layer; high swell/ shrink potential
Biological conditions	Low or high organic matter content
Ecosystem conditions	Low soil resilience; natural soil degradation
(ii) Stresses caused by external conditions and processes	
Climate-controlled conditions	Extreme climatic regimes; extreme high or low temperature; insufficient length of growing season; waterlogging; excessive nutrient leaching; global warming
Biological conditions	Pests and diseases; high termite population
Catastrophic events	El Niño and extreme storm events/cyclones; floods; droughts; landslides; earth quake activity; volcanic eruption
Ecosystem conditions	Impaired ecosystem functions and services; loss of soil quality and soil health
(b) Anthropic stresses	
Chemical conditions	Acidification by acid rain and acidifying fertilizers; drainage of wetlands; exposure to mine wastes; contamination with toxins
Physical conditions	Accelerated soil erosion; soil compaction; subsidence of drained organic soils
Biological conditions	Diminished biodiversity; high incidence of pests and diseases; allelopathy; loss of predators

Source: Friedrich et al. (2008)

rather than singly, these commonly exist in combinations. A typical example is of salinity and aridity (water stress) which are known to occur together. Also the shallow basaltic soils with low moisture retention capacity are prone to water stress in addition to having constraints like nutrient deficiencies, surface crusting, etc. Often the severity of stress controls the degree of adverse impacts in terms of decline in ability of the soils to support plants and animals. This obviously occurs due to reduction in the capacity to retain and supply adequate moisture and nutrients required for optimal growth of crops. As an example, the typical characteristics of selected soil types of India leading to different edaphic constraints are listed in Table 4.2, and area under each soil order is as given in Velayutham et al. (1999).

So far the concepts of edaphic stresses have been generally applied, and the knowledge on their assessment and management is limited and diverse. Still there are many perceptions on how these vary over time and space. But the edaphic stresses often lead to situations where cost-effective production is not feasible under a given set of site conditions and cultivation practices. Therefore, based upon the

Table 4.2 Edaphic constraints of some major soil groups in India

Soil group	Area[a] (M ha)	Characteristics leading to edaphic constraints
Vertisols	25.9 (8.5%)	Very hard when dry and very plastic, sticky and untrafficable when wet; deep cracking; narrow workable soil moisture range; low infiltration and poor internal drainage; salinity and alkalinity; macro/micronutrient deficiencies
Inceptisols	122.79 (39.7)	Vulnerable to erosion and degradation; structural instability, low soil–water retention; rapid surface sealing after rain and crusting with subsequent drying and hardening; limited soil moisture range for optimal soil tilth; sodicity and *Kankar* pan to restrict tree root proliferation
Entisols/A ridisols	98.37 (32.3%)	Low water retention capacity and unsaturated movement; poor structure and high erodibility; low nutrient reserves and high leaching losses; shallow depth; secondary salinization; high wind erosion
Alfisols	41.87 (13.5%)	Surface crusting and sealing; low water retention; low fertility; acidity; high erosivity due to undulating/rolling topography

Source: [a]NBSS&LUP Staff (2002)

soil's edaphic limitations that lead to the marginality of agricultural production, the soils are often classified as degraded lands, marginal soils, underutilized lands, unproductive soils, wastelands, etc. Usually the edaphic constraints are used as synonymous to these terms and antonymous to the soil quality. Though authenticity and accuracy of various estimates are questionable, it is reported that about 2 billion ha (23% of the world's usable land) is affected by degradation resulting in edaphic stresses to a degree sufficient to reduce their productivity (FAO 2011). About 5.8 M ha of land gets degraded annually and overgrazing, deforestation, agricultural activities, overexploitation of vegetation, and industrial activities contribute 35, 39, 27, 7, and 1%, respectively. Similarly 56, 28, 12 and 4% of land area is prone to degardation with water erosion, wind erosion, chemical and physical processes, respectively.

Land degradation estimates by different organizations have also been at variance in India. This is attributed to different approaches, methodologies, and criteria followed for assessment. Nevertheless, a large land area is under degradation due to various factors. These datasets were harmonized in GIS environment in 2010 (ICAR and NAAS 2010). The estimates are that 36.5% (120.72 M ha) of the total geographical area are degraded. The soil erosion due to water and wind are the major cause of soil degradation (94.97 M ha) and this is followed by chemical degradation (24.68 M ha). The latter includes decline in soil fertility as indicated by nutrient deficiencies and loss of organic matter, acidity, salinity/alkalinity, pollutants, etc. The waterlogged and flood-prone area is about 12 M ha.

A matrix which was earlier used by Friedrich et al. (2008) to define 25 stress classes in world soils is given in Table 4.3. The most limiting has been considered to be continuous moisture stress, while high shrink/swell potential is the least limiting factors. Nine inherent land quality classes (the land quality prior to human

Table 4.3 Land area of major edaphic stress classes and land quality classes

Quality class	Class code	Kind of land resource stress	Criteria for assigning stress	World (M ha)[a]	India (M ha)[b]
IX	25	Continuous moisture stress	Aridic SMR, rocky land, dunes	3648.0	11.46
VIII	24	Continuous low temperatures	Gelisols	2177.6	5.43
	23	Steep lands	Slopes >32%	48.4	0.91
VII	22	Shallow soils	Lithic subgroups, root-restricting layers <0.3 m	735.8	26.40
	21	Salinity/alkalinity	"Salic, halic, natric" categories	306.8	6.73
VII	20	High organic matter content	Histosols	122.1	NA
VI	19	Low water-holding capacity/high permeability	Sandy, gravelly, and skeletal families	336.3	13.75
	18	Low moisture and nutrient status	Spodosols, ferritic, sesquic and oxidic families, aridic subgroups	346.2	14.06
	17	Acid sulfate conditions	"Sulf" great groups and subgroups	11.2	0.03
	16	High anion retention	Anionic subgroups, acric great groups, oxidic families	249.8	NA
	15	Low nutrient-holding capacity	Loamy families of ultisols, oxisols	778.8	36.72
V	14	Excessive nutrient leaching	Soils with udic, perudic SMR, but lacking mollic, umbric, argillic	447.1	13.75
	13	Calcareous, gypseous conditions	With calcic, petrocalcic, gypsic, petrogypsic horizons; carbonatic and gypsic families	247.1	NA
	12	High exchangeable aluminum	pH < 5.5 in 0.3 m and Al saturation > 60%	406.2	17.93
	11	Seasonal moisture stress	Ustic or xeric suborders but lacking mollic or umbric epipedon, argillic or kandic horizon	1034.2	14.30
IV	10	Impeded drainage/marshy	Aquic suborders, "gloss" great groups	282.9	1.16
	9	High anion exchange capacity	Andisols	91.3	NA
	8	Low structural stability/soil hardening	Loamy soils and entisols except fluvents	136.9	21.57

(continued)

Table 4.3 (continued)

Quality class	Class code	Kind of land resource stress	Criteria for assigning stress	World (M ha)[a]	India (M ha)[b]
III	7	Seasonal low temperatures	Cryic or frigid STR	300.9	15.20
	6	Minor root-restricting layer/subsurface hard pan	Soils with plinthite, fragipan, duripan, densipan, petroferric, placic <100	151.7	11.31
	5	Seasonal excess water/temporary waterlogging	Recent terraces, aquic subgroup	136.2	6.24
II	4	High temperatures	Isohyper and/or isomegathermic STR excluding mollisol and alfisol	250.6	14.30
	3	Low organic matter content	With ochric epipedon	310.1	111.5
	2	High shrink/swell potential	Vertic subgroups	92.5	28.00
I	1	Few constraints	Other soil	408.8	NA

Sources: [a]Friedrich et al. (2008); [b]Compiled by Minhas and Obi-Reddy (2017)

interference) were established using the stress factors as a key. Class I has the most and Class IX the least favorable attributes. The criteria used in soil taxonomy to define these stress classes and land quality classes by Eswaran et al. (2003) have been included in this matrix. The areas falling in Classes VII (60–80% risk for sustainable agriculture), VIII, and IX (>80% risk) is over half the global land (54%), i.e., it shows severe or prohibitive constraints for agriculture. Another nearly a third of the land area (30%) falls under Classes V and VI (40–60% risk), i.e., having serious limitations. Only meager land (13%) is under Classes II, III, and IV (20–40% risk), i.e., with moderate limitations. Land with no or few constraints (< 20%; Class I) make up a minuscule 3%. About 52% of the land area falls under classes (25, 24, and 11) where stresses are mainly controlled by climate. The above classification was basically intended to focus on the inherent ability of the soil to produce grain crops on sustainable basis but the impact of some other management factor including irrigation and climate were not considered. A similar attempt was made by Minhas and Obi-Reddy (2017) to equate and estimate areas under different stress classes in India (Table 4.3). Though, the estimates need to be harmonized with respect to overlaps, e.g., most of the sandy terrains and dunes of Northwest India fall under aridic zone with continuous moisture stress and low water- and nutrient-holding capacities (Hipher arid Soil Moisture Regime, i.e., AER 2.1 and 2.2). Similarly salinity and aridity go together. Anyhow the overall situation seems to be a little better than the world averages. It is since about half the soils are grouped under classes V and VI (40–60% risk) and the rest is equally distributed under severe and moderate limitations. Even the conservative estimates are that all these stresses are limiting the agricultural production to one-third of the potential. Low organic carbon and thereby nutrient deficiencies are the most limiting factors followed by shallowness of soils in peninsular India. Among the processes of degradation, water erosivity is afflicting the most land area (about 25% of the geographic area) followed by aridity (13%).

The soil-forming factors like parent material, climate, relief, organisms, and time basically define the intrinsic stress conditions, e.g., intense leaching of nutrients in humid tropics may lower soil fertility in the long run. Further the landscape factors like steep terrain, boulders, undulations, and sand dunes may add to these constraints. Unfavorable climatic conditions like droughts, cyclones, hailstorms, heat/cold waves, etc. also play a role by exuberating edaphic stresses, e.g., landslides and episodes of soil erosion may be triggered by the intense rainfall during cyclonic events especially in coastal areas. These conditions are essentially beyond human control, but increasing population pressures, poor land management practices, and the emerging inequities in land and resource access have been the major driving forces for these stresses since the past about half a century (since the green revolution era). Further the mismatch between land use and its attributes, lack of restorative management, and deforestation to meet ever-increasing fuel and energy demands have aggravated the constraints. However, the emerging concerns about edaphic constraints are those being caused by inappropriate agricultural practices. Some of these are discussed below along with possible remediation measures.

4.3 Chemical Stress Conditions

The chemical constraints in Indian agriculture occur mainly due to mining of nutrients, loss of organic matter, and through buildup of salinity and pollutants.

4.3.1 Nutrient Deficiencies and Mining

Nutrient mining has increased with cropping intensity during the post green revolution period and multiple nutrient deficiencies are emerging due to low inherent fertility of most of Indian soils (Sanyal et al. 2014). Of the soil samples analyzed so far, about 93, 91, and 51% for N, P, and K fall under low to medium category (IISS 2015). Unfortunately, with continued low-input agriculture (current annual fertilizer consumption being only 154.6 kg ha^{-1}), the adequate replenishment of nutrients is not happening, and there is a net negative balance of about 8–10 million Mg of NPK. The most mined nutrient from soils is K with its removal and additions being about 7 and 1 million Mg (Sharma and Singh 2012). The continued mining of soil nutrients is impairing soil health and crop productivity. Once a nutrient becomes limiting, the overall fertilizer responses vis-à-vis crop productivity is lower since it does not allow for the full expression of other nutrients. Benbi et al. (2006) reported a drop in partial factor productivity of NPK for food grain production from 81 in 1966–1967 to 16 kg grain per kg of NPK in 2003–2004. The irrigated areas are afflicted more, e.g., the response was 13.4 kg grain per kg NPK in 1970 which has declined to 3.7 kg grain per kg NPK in 2005. In other words, to produce around 2 Mg ha^{-1} in 1970, the requirements for NPK fertilizers were only 54 kg NPK ha^{-1} was while about 218 kg NPK ha^{-1} are being now used for the same yield. Evidences have been generated that unfertilized exhausted soils can produce wheat, rice, and maize yields only in the ranges of 0.8–1.1, 1.5–1.6, and 0.3–0.7 Mg ha^{-1}, respectively (Swarup and Wanjari 2000). The yields can be increased and also sustained at higher levels of 4.0–5.0, 2.4–4.6, and 2.5–3.0 Mg ha^{-1} for rice, wheat, and maize, respectively, if deficient nutrients are brought to sufficiency levels through chemical fertilizers. Integrated use of various types of organic sources of nutrients like FYM, crop residues, and green manures along with fertilizers (INMS) not only maintain the yield but also reduce input cost, enhance profit, and improve soil health (Yadav et al. 2000). Yields of crops could be sustained with substitution of 50% nutrients by FYM, crop residues, or green manure.

Moreover, the deficiencies of micronutrient like zinc, boron, molybdenum, iron, manganese, and copper have been reported to exist in 49, 33, 13, 12, 5, and 3% of soils, respectively (Sharma and Singh 2012). Widespread deficiencies of zinc occur in alluvial soils of Indo-Gangetic plain, black soils of Deccan Plateau, and red and other associated soils. About 86 and 73% of soils in Maharashtra and Karnataka have been reported to be deficient in zinc, respectively. Similarly, the secondary nutrients like sulphur (41% samples) are becoming increasingly deficient. The boron deficiency is found in red and lateritic, acidic, coarse-textured alluvial and highly calcareous soils. Iron deficiency has been noticed in the paddy on coarse-textured alluvial soils while that of manganese are appearing on wheat when cultivated after paddy on coarse-textured alkaline soils These deficiencies are also being translated in terms of human and animal health.

4.3.2 Depleting Soil's Organic Carbon (SOC)

Soil organic carbon (SOC) governs soil productivity by influencing physicochemical-biological environment of the soils and is already low in more than one-third of Indian soils and has declined by 30–60% as a consequence of cultivation (Swarup and Wanjari 2000). This is being further negatively impacted with imbalanced use of fertilizers, removal and burning of crop residues, and reduced use of FYM and other organics. The reduced productivity of the rice–wheat system in IGP plain has been linked with declining soil organic matter contents. The earlier report on declining productivity in Haryana and Punjab (Sinha et al. 1998) had hinted at decrease in SOC from 0.5% in the 1960s to 0.2% in 1998 in major rice–wheat regions of Northwestern India. However, the recent evaluation of large soil test data for 25 years (1981–1982 to 2005–2006) has brought out that the SOC rather increased by 38%, i.e., 2.9 g kg^{-1} in 1981–1982 to 4 g kg^{-1} in 2005–2006 that has been ascribed to higher root biomass, stubble, and rhizo-depositions with increase in rice–wheat productivity from 5.9 to 8.1 Mg ha^{-1} by Benbi and Brar (2009). The fear of IGP losing SOC under intensive agriculture were also annulled by Milne et al. (2006) who predicted SOC to increase by about 4.5% from 1990 to 2000. Prediction also shows that SOC can be maintained at the base level of 0.45% in alfisols with conjunctive use of chemical fertilizers and FYM, while it would reduce to 0.30 and 0.36% under no fertilizer and chemical fertilizer treatments, respectively (IISS 2015). Adaptation of management practices like conservation agriculture/residue recycling, diversification/intensification of crops, integrated nutrient management, and agroforestry have been shown to enhance possibilities of carbon sequestration. The increase in crop yield ranges between 15–33 kg ha^{-1} with increase of 1 Mg C ha^{-1} SOC (Benbi and Chand, 2007), while in case of rainfed crops of rice, pearl millet, groundnut, lentil, finger millet, and soybean, the increase was 160, 170, 13, 18, 101, and 145 kg ha^{-1}, respectively (Srinivasarao et al. 2014).

4.3.3 Soil Contaminations

4.3.3.1 Geogenic Toxic Elements

The geogenic sources and the anthropogenic sources like sewage water, industrial effluents, urban solid wastes, fertilizers, etc. are polluting soils. The pollutants are becoming a potential hazard to the health of humans and animals since these enter into their food chain. The soils in about 34,000 km^2 of area in Malda, Murshidabad, Nadia, North and South 24 Parganas, and Bardman districts of West Bengal adjoining Bhagirathi river are contaminated with arsenic (Sanyal et al. 2012). With consumption of contaminated food, nearly 30 million people inhabiting these areas are exposed to arsenic poisoning. Similarly, high levels of fluoride occur in about 200 districts in India. The soils of seleniferous region of Northeastern Punjab suffer due to problem of selenium toxicity. Since the selenium content of crops grown on these soils is many times higher than the permissible limit of 5 mg Se kg^{-1}, chronic selenosis (selenium poisoning) in animals and human beings is caused with the intake of contaminated food. Nitrate pollution occurs in intensively irrigated and high productivity regions due to excessive use of chemical fertilizers in India, especially in states like Punjab, Haryana, and western Uttar Pradesh.

4.3.3.2 Urban Effluents Toxicities

The sewage generated is estimated to be 38,354 million liters per day (MLD) in major cities of India, while the wastewater generated by industries is about 13,468 MLD (CPCB 2015). Since the conventional treatment technologies for wastewaters are often cost prohibitive, only about one-fifth of this water is treated that too until secondary treatment level. However, with increased quantities of wastewater being generated with urbanization and industrial development, their disposal is becoming a major problem (Minhas and Samra 2004). Thus, land disposal in peri-urban areas seems the most logical sink due to restrictions in their disposal into surface streams. In addition to irrigation, the other benefits of sewage use include the additions of nutrients especially N and P and also the organic matter (Minhas et al. 2015). It is estimated that with a potential to annually irrigate about 2.5 M ha, sewage waters can add about 1 million Mg of nutrients and generate employment of 220 million man-days. However, their disposal is associated with accumulation of salts, toxic constituents, and heavy metals in addition to pathogenic contaminations. Therefore for minimizing sewage water-related bio-transfer of heavy metals and pathogenic contaminations, guidelines and regulatory measures have been advocated. Among treatment technologies, the low cost include the biological mechanisms like phy-toremediation, etc. Disposal in tree plantations around urban areas to create green belts is another alternative, but their loading rates need standardization (Yadav et al. 2016). Though the astute management practices can minimize the risks of wastewa-ters, the national water development strategies so far have not come out with the policies for promotion of environmentally safe use of these waters. National poli-cies have to be based on reasonably accurate estimates of the quality and quantity and the expected damage associated with their uncontrolled use/disposal. The other techno-economic issues relate to treatment standards, pricing/subsidies, and mar-kets for non-conventional water use in addition to creating awareness of risk man-agement through educational and political campaigns.

4.3.3.3 Sewage Sludge Toxicities

About 57 million Mg of urban solid wastes are generated per annum from over 450 class I and class II cities in India (Sharma and Singh 2012) and this would increase to 107 million Mg by 2025. Though a rich source of organic matter and nutrients, it has toxic substances of wastewater concentrated in the solid phase. Thus in the absence of any proper management system, these wastes could become a potential source of soil pollution. Such contaminations could be avoided by segregating the industrial waste containing heavy metals and toxic chemicals from the biodegradable waste. The biodegradable component of wastes is, generally, rich source of nutrients and organic carbon. Its conversion into compost is desired to serve the twin objec-tives of cleaning the environment and augmenting the supplies of organic manure. There should, therefore, be growing interest in development of cost-effective and eco-friendly composting technologies both in public and private sectors.

4.3.3.4 Brackish Water Use

Development of irrigation facilities has played a major role in enhancing agricul-tural productivity, and these areas now contribute about three-fifths of the country's total food production. Especially the groundwater withdrawals have increased man-ifold during the second half of the last century, and groundwater use soared from $10–20 \text{ km}^3$ in 1950 to $240–260 \text{ km}^3$ during 2000. However, groundwater in about 19.3 M ha area has been rated brackish in states like Rajasthan, Haryana, Punjab, and others. The indiscriminate use of these waters in the absence of proper soil–water–crop management strategies poses grave risks to soil health and environment (Minhas and Gupta 1992; Minhas 2012). It is since the emerging salinity, sodicity, and toxicity problems in soils and their impact on crop productivity and quality. For their effective use, not only the typical characteristics of the waters but factors like soil texture, rainfall, and crop tolerance have to be considered. However, if their use is not viable for agricultural crops, a shift to alternative uses like salt-tolerant forest/fruit trees and other high-value crops should be equally remunerative (Dagar and Minhas 2016).

4.3.4   Soil Salinity

About 6.73 M ha of land is afflicted with salinity (2.96 M ha) and alkalinity (3.77 M ha) mainly in Gujarat, Uttar Pradesh, Rajasthan, West Bengal, and Andhra Pradesh (ICAR and NAAS 2011). The expansion of the canal works has led to the occurrence of waterlogging and secondary salinity in their irrigation commands. The area under salinity has been increasing at the rate of 3000 to 4000 ha per annum. This is mainly because the emphasis has been on their water distribution and the associated drainage and efficient water use usually have been sidelined. The major achievement with respect to their reclamation has been on alkali soils where about 45,000 ha are being put under cultivation using the gypsum-based technology. A total of about 1.8 M ha of alkali land has been reclaimed so far, contributing 12–15 million Mg of paddy-wheat annually and changed the socioeconomic profile of more than 9 million people residing in rural India (CSSRI 2015). The reclamation of saline soils does not require any amendment. The leaching of salts is induced by providing subsurface drainage system, and the effluent is disposed (Minhas and Sharma 2002). Subsurface drainage technology has been demonstrated successfully on farmers' fields in Haryana, Gujarat, and Rajasthan, reclaiming nearly 60,000 ha of waterlogged saline lands (Kamra 2015). However, the problems associated with its operation, maintenance, and appropriate disposal of drainage water in addition to high investment for installation (Rs. 75 and 100,000 for alluvial and black soils, respectively) are the major bottlenecks for large-scale expansion of drainage networks. The alternatives are also being proposed in terms of agroforestry/bio-drainage, but combinations of agronomic, engineering, and agroforestry measures, as suited to the site conditions, are now considered to be more realistic (Minhas and Dagar 2016).

4.3.5   Soil Acidity

About 17.93 M ha of acidic soils (pH < 5.5) suffer from deficiencies as well as toxicities of certain nutrients and have very low productivity (< 1 Mg ha^{-1}). Depending upon the level of acidity, type of soil, crop grown, and climatic conditions, acid soils can reduce productivity by 10–50% (Sharma and Sarkar 2005). The reduction is attributed to low base saturation (20–25%); deficiency of calcium, magnesium, molybdenum, boron, and zinc; low cation exchange capacity of kaolinitic clay and poor nutrient retention; poor organic matter build up and nitrogen availability; high P fixation and its low availability; and excess/toxicity of iron, aluminum, and manganese. For reclamation via neutralizing acidity, these soils must receive lime and adequate supplies of fertilizers. The liming at 0.2–0.4 Mg ha^{-1} (one-tenth of lime requirement) in furrows along with recommended fertilizers was quite effective in realizing higher and economic yields. Cheaper liming materials include basic slag, lime sludge and low-grade lime stones, etc.

4.4   Physical Stress Conditions

4.4.1   High Erosivity

4.4.1.1 Water Erosion

The water erosion takes away productive topsoil to cause decline in crop productivity, erosion of biodiversity, flash floods, and siltation into rivers and other water bodies. About 5.3 billion Mg of soil gets eroded annually that carries along with it about 8 million Mg of plant nutrients (Sharda 2011). The erosion rates monitored among different land resource regions are 23.7–112.5, 80, 27–40, and 2.1 Mg ha^{-1} for the black soil, Shiwalik, northeastern region with shifting cultivation and the north Himalayan forest regions, respectively. While 61% soil gets moved and deposited at unwanted locations with the result of increased off-site costs, nearly 29% is transported to the sea. The remaining 10% was deposited in multipurpose reservoirs reducing their holding capacity by 1 to 2% per annum (Dhruvanarayana and Ram-Babu 1983). The annual loss in production of major crops due to soil erosion has been estimated to vary from 7.2 million Mg (UNEP 1993) to 13.5 million Mg (Bansil 1990). The loss in production for 11 major crops varied from 1.7 to 4.1% of total production (Brandon et al. 1995). Experimental studies in lower Himalayan region indicated that removal of 1 cm of topsoil caused 76 kg ha^{-1} reduction in maize grain yield and 236 kg ha^{-1} in straw yield (Khybri et al. 1988). The reduction was observed to be 103 kg ha^{-1} in Shiwalik region of Punjab (Sur et al. 1998). Soil loss from sorghum, pearl millet, and castor bean was computed to be 138, 84, and 51 kg ha-cm^{-1}, respectively (Vittal et al. 1990). Recent computation on the loss of productivity of food grains due to soil erosion was 1.34 billion Mg from alluvial, black, and red of which the contribution of cereals, oilseeds, and pulses was 68.3, 20.9, and 12.8%, respectively. Among oilseed crops, groundnut, and soybean occupying 26.5 and 36.5% of total area contribute 12.3% and 10.4% to total monetary loss (Sharda et al. 2010). Similarly, in pulse crops, gram (6.4%) and pigeon pea (2.9%) are the main contributors to total monetary loss as compared to other crops. The appropriate soil and water conservation measures have been recommended for reducing the losses due to soil erosion. For this integrated watershed management has been advocated as an excellent strategy to achieve optimal production. This involves water harvesting through construction of check dams along gullies, land leveling, bench terracing, contour bunding, and planting of grasses (Palanisami et al. 2002). Thereby, substantial investments on integrated soil and water conservation programs over the past about five decades have yielded local successes and the development of participatory watersheds models. Area under wasteland got decreased but that under irrigation to horticultural/fodder/fuel plantations increased. The grazing of animals changed to stall feeding and seasonal migration stopped. A meta-analysis of 311 case studies on watershed management programs showed a mean benefit–cost ratio, and the internal rate of return of programs was 2.14 and 22% while these performed best in rainfall zone of 700–1000 mm (Joshi et al. 2008).

4.4.1.2 Wind Erosion

Wind erosion causes decrease in land productivity at both the sites from where the finer particles are blown away and at sites where they are deposited. Wind erosion is prevalent in arid and semiarid regions of the country covering an area of about 28,600 km^2 in the states of Rajasthan, Haryana, Gujarat, and Punjab (ICAR and NAAS 2011). About 68% of the affected area is covered by sand dunes and sandy plains. It has been estimated that out of 2,08,751 km^2 mapped area of Western Rajasthan, 30% is slightly affected by land degradation, while 41% is moderately, 16% severely, and 5% very severely affected (Narain and Kar 2006). Decrease in rainfall gradient and increase in wind strength from east to west are responsible for the spatial variability in sand reactivation pattern. According to recent estimates, about 75% area of Western Rajasthan is affected by wind erosion hazard of different intensities (Narain et al. 2000) besides 13% area under water erosion and 4% under waterlogging and salinity/alkalinity. The spatial extent of the problem is increasing in the recent decades, especially due to increased cultivation and grazing pressures on the erstwhile stable sandy terrain leading to depletion of vegetation cover. However, under desert development program (DDP) and watershed development projects, affected areas are being rehabilitated along with stabilization of sand dunes through appropriate soil conservation measures. The harmonized area statistics on land degradation in the country shows that 12.4 M ha area on arable lands is affected by wind erosion of more than 10 Mg ha^{-1} yr.$^{-1}$ (Maji 2007).

4.4.2 Waterlogging

The area prone to waterlogged and floods has been estimated to be 12 M ha. In high rainfall areas of Central India, the *vertisols* are kept fallow during rainy season, and only one crop is possible during post-rainy season. However, their productivity can be enhanced by raised–sunken bed system that allows to cultivate soybean on raised beds while paddy in sunken beds (Tomar et al. 1995). The cropping is also possible in sunken bed during post-rainy season. Likewise, by growing *Rabi* legumes on raised beds under raised–sunken bed system, the productivity of rice fallows in Eastern India could be increased. The water stagnation occurs for more than 6 months in the alluvial plains of eastern India and the conditions allow for only one paddy crop that has a very low yield potential (<1.0 Mg ha^{-1}). The low productivity of the land resource is the prominent cause of poverty in the region. There is a need for switch over to integrated farming systems that could convert threat of water abundance into opportunities for enhancing income and employment. One such farming system model involving aquaculture enhanced water productivity by 136% and cost–benefit of the system was 1.52-fold over the farmer's practice of Rs. 3.3 m^{-3} of water (DAC 2014).

4.4.3   Other Soil Physical Constraints

About 89.5 M ha suffers from one or another form of physical constraint in the country (Painuli and Yadav 1998). These include shallow depth, soil hardening, slow and high permeability, subsurface compacted layer, surface crusting, and temporary waterlogging. Maximum area is affected by shallow depth (26.4 M ha) followed by soil hardening (21.6 M ha) and the least by temporary waterlogging (6.25 M ha). Soil compaction is a management problem resulting from movement of heavy machinery and repeated tillage operations accompanied with reduction in organic matter and destruction of soil aggregates. Compaction causes deterioration in soil structure and impedes root growth and biological activity besides generating high amount of runoff during intense storms. Soil scaling due to surface hardening and crust formation together affects 31.82 M ha area. The technologies for treating the soil affected by subsurface mechanical impedance and compaction include chiseling, chiseling plus amendment application, construction of ridges, and raised and sunken bed technology. The impact assessment of technologies developed for soils having subsurface mechanical impedance under field conditions show spectacular increase in production of major crops varying from 12 to 63%. The raised and sunken bed (RSB) technology for vertisols having high clay content has been found to be highly remunerative on sustainable basis (Painuli et al. 2002). The technology is equally effective in subsurface compacted soils as effective rooting depth is increased by 0.30 m in raised beds. The technologies for checking subsurface mechanical impedance and compaction also help in conserving soil and water besides increasing productivity. For example, the conservation agriculture (CA) has now emerged as a new paradigm to attain sustainability and overcome soil physical constraints induced by rice–wheat systems in the IG Plains (Hobbs et al. 2008). The CA technologies becoming popular in irrigated systems include zero-tillage/bed planting with residue recycling, direct drilling of wheat in to paddy residues, direct seeding of paddy, and laser-assisted precision land leveling. The other practices that alleviate these physical constraints include integrated nutrient management, cropping systems to include legumes, optimal tillage, mulching and amendment use, etc.

4.5   Extrinsic Stresses

4.5.1   Climate Variability Impacts

There is increasing evidence that climate change-related elements are contributing to accelerated resource degradation and the resultant edaphic stresses. The average increase in temperature in India during 1901 and 2005 has been 0.51 °C compared to 0.74 °C at global level. The increase was in the order of 0.03 °C per decade during 1901–1970 while it was around 0.22 °C per decade for the period from 1971 to 2004 indicating greater warming in the recent decades. Increase in the twenty-first century is projected to vary between 3 to 6 °C with southern regions registering 2–4 °C increase, while the increase (> 4 °C) would be more pronounced in the northern states and eastern peninsular region. The resultant heat stress would have serious impact on agriculture, water resources, forests, national ecosystems, fisheries, and energy sectors, e.g., a simulation study on the impact of high temperature on irrigated wheat in North India indicated that grain yield can decrease by 17% if the temperature increased by 2 °C (Aggarwal et al. 2001). Moreover, the increased temperature would result in reduced quantity and quality of soil's organic matter which is already low in Indian soils. The OSC loss per degree of warming may increase by 8–9% in areas with temperatures of 10–15 °C, while only 2% for areas with 35 °C. The N mineralization would increase but its availability may decline with gaseous losses via volatilization and denitrification.

The normal season rainfall from the period 1961–1990 has been projected to increase in India by 15 to 40% till the end of the twenty-first century. Monthly rainfall data for all the 36 subdivisions of the country indicate that it is exhibiting an increasing trend in June and August, while the July rainfall showed a decreasing trend (Guhathakurta and Rajeevan 2006). Analysis of long-term rainfall data for over 1100 stations across India shows pockets of deficit rainfall over Eastern Madhya Pradesh, Chhattisgarh, and Northeast region in Central and Eastern India (Subba-Rao et al. 2007), especially around Jharkhand and Chhattisgarh. In contrast trends indicate increase in rainfall (10–12%) along the west coast, northern Andhra Pradesh and parts of NW India (NAPCC 2008). In the southern peninsular region, a shift in peak monthly rainfall by 20–25 days from September to October is recorded. Further, the intensification of hydrologic cycle due to global warming may result in higher intensity rains, frequent floods and droughts, shift of rainy season toward winter, and receding glaciers causing higher flow during few decades followed by substantial reductions thereafter. Analysis of rainfall data with intensities of 10, 100, and above 100 mm revealed that in the recent period, the frequency of rain events of more than 100 mm intensity have increased, while the frequency of moderate events over Central India has significantly decreased during 1951 to 2000 (Goswami et al. 2006). Thus high-intensity storms would cause high erosion losses leading to severe land degradation problems.

The climate change would disturb the water balance in different parts and ground water quality due to intrusion of seawater in coastal areas. Thermal expansion of seawater due to global warming coupled with melting of glaciers and snowfields would result in the rise of sea level by 0.1–0.5 m by the middle of the twenty-first century (IPCC 2013). It is expected that by the end of the century, 68 to 77% of the forest areas are likely to experience shift in forest types with corresponding reduction in forest produce and livelihood prospects. Coastal wetlands would have serious impact due to change in the composition of plant species and expected sea level rise. The marine and aquatic life would be impacted due to rise of seawater temperature and sea level resulting in their migration to favorable regions, thus affecting livelihood of coastal people. The energy requirements in summers in plains would increase more than being compensated by saving in energy due to increased temperature in winter in northern mountainous regions. The demand for energy would also increase for irrigation needs due to high evaporative demands in cropped areas.

4.5.2 Catastrophic Events

Occurrence of floods, droughts, and other weather extremes are a common feature in many parts of the country. These natural disasters cause widespread land degradation apart from heavy monetary losses and a serious setback to economic development of the country. It has been estimated that eight major river valleys spread over 40 M ha area of the country covering 260 M population are affected by floods. Besides environmental degradation, poverty and marginalization are other major factors which force the poor to live in threatened and exposed conditions. About 60% of total flood-prone area in the country lies in Indo-Gangetic basin, which supports 40% of India's population with 60 M ha of cultivable land. The Brahmaputra basin is also critical as it experiences several floods within a year thus seriously affecting all developmental activities. The incidence of floods is not restricted to humid and subhumid regions but has also caused extensive damage in the desert districts of Rajasthan and Gujarat in the recent years. Average flood damage to houses, crops, and public utilities during 1953–2002 has been estimated as Rs. 13.76 billion affecting an area of 7.38 M ha and a population of 32.97 million (CESI 2014). Human and cattle loss has been put at 1560 and 91,555 affecting 3.48 M ha of cropped area. The maximum damage to area, human and livestock population, crops, and public utilities occurred during the years 1977, 1978, 1979, 1988, and 1998.

The impact of drought on the techno-economic and socioeconomic aspects of agricultural development and growth of the nation is severe, and droughts often result in huge production and monetary losses (Samra et al. 2006). About 68% of total sown area and 23% of land area covering a total of 183 districts and 12% of population are accounted as drought prone. In a state like Rajasthan (arid), about 56% of the total area and 33% of the total population are chronic drought-prone-affected areas followed by Andhra Pradesh, Gujarat, and Karnataka with corresponding figures as 30 and 22%, 29 and 18%, and 25 and 22%, respectively. Except Kerala, Punjab, and northeastern region, every state has one or more drought-prone areas. Apart from floods and droughts, cyclones frequently occur in the entire 5700 km long coastline of Southern and Peninsular India besides the Islands of Lakshadweep and Andaman and Nicobar Islands affecting 10 M population. Nearly 56% of the total area of the country is susceptible to seismic disturbances affecting 400 M people.

India has experienced 40 major droughts during the 200 years between 1801–2002 with 10 years under severe drought category (> 39.5% area affected) and 5 years under phenomenal drought (> 47.7% area affected) (Subbareddy et al. 2008). Since Independence, India has experienced 15 droughts out of which 3 were of severe, 7 of moderate and 5 of slight intensity affecting 13.3–49.2% of total geographical area of the country (DAC 2009). Drought-prone areas are more vulnerable to land degradation. In a good or normal rainfall year, they substantially contribute to agriculture production particularly for groundnut, millets, and sorghum crops where they account for one-third to one-fourth of the total national production.

Similarly, one-sixth to one-tenth of other important crops like ragi, maize, and cotton and 12% of rice production is realized from these areas besides sizeable contribution to the production of pulses and oilseeds.

4.6 Anthropic Stresses

4.6.1 Biological Processes

Deforestation, overgrazing, mismanagement of agricultural land, overexploitation, and bio-industrial activities are the main anthropogenic causes of loss of organic matter and soil's biodiversity. Farm level practices which sustain carbon on soils such as integrated nutrient management, reuse of crop residues via conservation tillage, diversified cropping, etc. are not being followed adequately and thus imbalanced use of fertilizers, removal of crop residues or in situ burning, intensive cropping, etc. are causing depletion of organic matter and in turn result in loss of a soil's biological population.

A decline of carbon stock especially of soils is resulting in various soil constraints in soil. It is since disruption of carbon cycle between pedosphere and atmosphere as a result of deforestation, desertification, and soil erosion. India is the lowest contributor of the GHG compared to North America and many other industrial and developed countries (0.29 Mg per capita consumption compared to 5.37 and 4.63 by the USA and Australia at 1996 level). However, with growing industrialization and economic development, India may become the second fastest growing GHG contributor in the world (increase in per capita consumption to 1.02 Mg by 2004) next to China (NAPCC 2008). While the CO_2 emissions at 1997 level had been 237 million Mg, it is projected to increase to 775 million Mg by the end of the century if coal consumption continues at the present rate (Ravi Sharma 2007).

4.6.2 Vegetation Degradation

Widespread vegetation degradation is occurring in pasture lands and open forests affecting the biodiversity. The permanent pastures and grazing lands exist in major portion, e.g., 32, 10, 6, 5.1, 5, 4.5, and 4% of land in states like Himachal Pradesh, Sikkim, Madhya Pradesh (including Chhattisgarh), Karnataka, Rajasthan, Gujarat, and Maharashtra, respectively. In India, with about 500 million livestock population, more than 50% of fodder demand is met from grasslands. The grazing intensity in India is about 42 animals per ha of land, while the threshold is 5 animals per ha. About 100 million cow units graze in forest lands while sustainable level is 31 million per annum and thereby 78% of India's forests are being affected. Per capita forest area in India is only 0.07 ha which is far below the world average of 0.8 ha. Dense forests are losing their crown density and productivity continuously, the current productivity being one-third (0.7 Mg ha^{-1}) of the actual potential. The combined availability of green fodder from pasture lands and grazing lands, agricultural lands, and forest (899.3 million Mg) is far short of the actual demand of 1820 million Mg (DAC 2009). It causes indiscriminate grazing on forest lands leading to large-scale degradation thereby seriously affecting natural regeneration of forests. The present forest cover of 20.6% (Indiastat 2011–2012) is far below the 33% cover recommended by National Forest Policy of 1988, the proportion being 60% in the hill regions and 20% in the plains. The net annual loss of forest areas is put at 74000 ha which is mainly attributed to overgrazing and over-extraction of firewood from 78% of the forest areas and fire hazards in 71% of forest lands.

4.7   Epilogue

As stated above, the Indian soils are being degraded leading to multiple edaphic constraints that are threatening the sustainability of agricultural production. Except few, the processes leading to these constraints are generally insidious and show up only gradually as the problem becomes more severe to cause yield declines. Farmers may ultimately be forced to either shift to less remunerative crops or, in extreme cases, soils can turn unfit for agriculture. The situation is further going to worsen with global warming when edaphic stressors are expected to show greater impacts. Even by conservative estimates, mitigation of about half the edaphic stresses can raise the food production level to about two-fold. Thus development and promoting strategies to minimize the edaphic constraints and improving the quality and health of soils are fundamental to sustained agriculture and food security of the country. The coping strategies for minimizing the impacts of edaphic constraints include (i) mitigation through improved methods of soil and land management, (ii) adaptation though selection of crops those are tolerant to the specific constraints or develop tolerant cultivars as a result of bioengineering, and (iii) shifting to alternative uses for the land. In fact the options to be adopted are defined by the typical edaphic factors and available opportunities. Moreover, the research and policy strategies should aim at both to negate the impacts of edaphic stresses and also to prevent their further spread. Some of these are listed below:

- Several research and developmental organizations are in fact working on edaphic stressors, but their efforts are too inadequate considering the magnitude of the problem. Looking at the past scenario, it comes out that these organizations have been working in isolation and within their disciplinary boundaries. But to alleviate the effects of multiple stressors, a holistic multidisciplinary approach to build up systems perspectives is a need of the hour to get best combination of technologies for a particular agroecosystem that are often featured by multiple stressors and that needs to be defined with greater precision. For this, national networks should be built up on priority.
- Geo-referenced information system needs to be created for edaphic stresses using remote sensing, geographic information system (GIS), and scientifically designed indicators. The prognosis of hot spots for various edaphic stresses

would help in prioritization of action plans for developing integrated frameworks to alleviate the stressors.

- New tools have emerged from decades of research in the areas of conservation agriculture, precision irrigation technologies, biotechnology, remote sensing, geospatial technologies, information technology, nanotechnology and polymer sciences, etc., which have opened up new opportunities for tackling edaphic stresses. Therefore, it is of national importance to not only initiate high-quality research programs, which are of global standard in this important area, but also to capture, synthesize, adopt, and apply the technological advances taking place within and outside the country.
- Crop diversification and other measures in land use for improving soil quality such as conservation agriculture, integrated nutrient management, and efficient water management through modern irrigation and drainage methods are needed. The approaches like agroforestry and integrated crop–livestock systems can further improve soil health, biodiversity, and ecosystem services. All these need up-scaling with the ultimate goal to develop "Sustainable Agricultural Systems" those that are less vulnerable to shocks and stresses.

References

Aggarwal PK, Roetter R, Kalra N, Hoanh CT, Van Keulen H, Laar HH (2001) Land use analysis and planning for sustainable food security, with an illustration for the state of Haryana. Indian Agricultural Research Institute, New Delhi, International Rice Research Institute, Philippines and Wageningen University, Netherlands

Bansil PC (1990) Agricultural statistical compendium. Techno-Economic Research Institute, New Delhi

Benbi DK, Brar JS (2009) A 25-year record of carbon sequestration and soil properties in intensive agriculture. Agron Sustain Dev 29:257–265

Benbi DK, Chand M (2007) Quantifying the effects of soil organic matter on indeginous N suppy and wheat productivity in semi-arid subtropical India. Nutr Cycl Agroecosyst 79:103–112

Benbi DK, Nayyar VK, Brar JS (2006) Green revolution in Punjab: the impact on soil health. Indian J Fert 2:57–62

Brandon C, Homman K, Kishor NM (1995) The cost of inaction: valuing the economy- wide cost of environmental degradation in India. The World Bank, New Delhi

CESI (2014) Compendium of environment statistics India, Central Statistical Organization, Ministry of Statistics and Programme Implementation, Government of India (Website: https//www.mospi.gov.in)

CPCB (2015) Status of water supply, wastewater generation and treatment in class I cities and class II towns of India. Series: CUPS/70/2009-10. Central Pollution Control Board, India

CSSRI (2015) Vision-2050, Central Soil Salinity Research Institute, Karnal, Haryana, India

DAC (2009) Manual for drought management. Department of Agriculture & Cooperation, Ministry of Agriculture, Government of India, New Delhi

DAC (2014) State of Indian agriculture 2012–13. Department of Agriculture & Cooperation, Ministry of Agriculture, Government of India, New Delhi

DAC (2015) Agricultural statistics at a glance-2014. New Delhi, Department of Agriculture & Cooperation, Directorate of Economics and Statistics, Ministry of Agriculture, Government of India/Oxford University Press

Dagar JC, Minhas PS (2016) Agroforestry for Management of Waterlogged Saline Soils and Poor-Quality Waters, Adv Agrofor Series 13. Springer, New Delhi

Dhruvanarayana VV, Ram-Babu (1983) Estimation of soil erosion in India. J Irrig Drain Eng 109(4):419–434

Eswaran H, Beinroth FH, Reich PF (2003) A global assessment of land quality. In: Wiebe K (ed) Land quality, agricultural productivity, and food security: biophysical processes and economic choices at local, regional, and global levels. Publ. Edward Elgar, Northampton, pp 111–132

FAO (2011) State of World's land and water resources: managing systems at risk. Food and Agricultural Organization/Earthscan, Rome/London

Friedrich HB, Hari-Eswaran, Reich PF (2008) Edaphic constraints on food production. In: Chesworth W (ed) Encyclopedia of soil science. Springer, Dordrecht, pp 202–206

Goswami BN, Venugopal V, Sengupta D, Madhusoodanan MS, Prince KX (2006) Increasing trend of extreme rain events over India in a warming environment. Science 314:1442–1445

Guhathakurta P, Rajeevan M (2006) Trends in the rainfall pattern over India, National climate centre research report. India Meteorological Department, Pune

Hobbs PR, Ken-Sayre, Gupta RK (2008) The role of conservation agriculture in sustainable agriculture. Phil Trans R Soc A 363:543–555

ICAR and NAAS (2010) Degraded and wastelands of India; status and spatial distribution. Indian Council of Agricultural Research & National Academy of Agricultural Science, New Delhi

IISS (2015) Vision 2050, Indian Institute of Soil Science, Bhopal: India.

Indiastat (2011–12) http://www.indiastat.com/table/agriculture, Ministry of Agriculture, Government of India. Accessed 15 May 2015

IPCC (2013) Climate change 2013: the physical science basis. In: Stocker TF, Qin D, Plattner GK, Tignor M, Allen SK, Boschung J, Nauels A, Xia Y, Bex V, Midgley PM (eds) Contribution of working Group I to the fifth assessment report of the intergovernmental panel on climate change. Cambridge University Press, Cambridge, UK/New York

Joshi PK, Jha AK, Wani Suhas P, Sreedevi TK, Shaheen FA (2008) Impact of watershed program and conditions for success: a meta-analysis approach. Global theme on agro-ecosystems. Report no. 46. Patancheru 502 324, Andhra Pradesh, India; International Crops Research Institute for the Semi-Arid Tropics, Hyderabad

Kamra SK (2015) An overview of sub-surface drainage for management of saline and waterlogged soils in India. Water Energy Int 6(09):46–53

Khybri ML, Prasad SN, Sewa R (1988) Effect of top soil removal on the growth and yield of rain-fed maize. Indian J Soil Conserv 8(2):164–169

Maji AK (2007) Assessment of degraded and wastelands of India. J Indian Soc Soil Sci 55(4):427–435

Milne E, Bhattacharya T, Pal DK, Easter M, Williams S (2006) A system for estimating carbon organic stocks and changes at regional scale: a case study from Indo-Gangetic Plains of India, Proc. International Conference on Soil, Water and Environ. Quality, Indian Society of Soil Science, New Delhi, pp 303–313

Minhas PS (2012) Sustainable management of brackish-water agriculture. Soil water and agronomic productivity. (eds) Lal R, Stewart BA. Adv Soil Sci 19: 289–327

Minhas PS, Dagar JC (2016) Synthesis and way forward: agroforestry for waterlogged saline soils and poor-quality waters. In: Dagar JC, Minhas PS (eds) Agroforestry for management of waterlogged saline soils and poor-quality waters, Adv Agrofor Series 13. Springer, New Delhi

Minhas PS, Gupta RK (1992) Quality of irrigation water – assessment and management. Information and Publication Section, Indian Council of Agricultural Research, New Delhi

Minhas PS, Obi-Reddy GP (2017) Edaphic stresses and agricultural sustainability: an Indian perspective. Agric Res 6:8–21

Minhas PS, Samra JS (2004) Wastewater use in Peri-urban agriculture: impacts and opportunities. Bull. No. 2/2004, Central Soil Salinity Research Institute, Karnal, India

Minhas PS, Sharma OP (2002) Management of soil salinity and alkalinity problems in India. J Crop Prod 7:181–230

Minhas PS, Yadav RK, Khajanchi-Lal, Chaturvedi RK (2015) Impact of long-term irrigation with domestic sewage and nutrient rates I. Performance, sustainability and produce quality of peri-urban cropping systems. Agric Water Manag 156:100–109

NAPCC (2008) National action plan on climate change. Prime Minister's Council on Climate Change, Government of India: New Delhi

Narain P, Kar A (2006) Desertification. In: Chadha KL, Swaminathan MS (eds) Environment and agriculture. Malhotra Publication House, New Delhi

Narain P, Kar A, Ram B, Joshi DC, Singh RS (2000) Wind erosion in Western Rajasthan, Central Arid Zone Research Institute, Jodhpur, India

NBSS&LUP Staff (2002) Soils of India, NBSS Publ. 94, National Bureau of Soil Survey and Land Use Planning, Nagpur, India

NIASM (2015) Vision-2050. National Institute of Abiotic Stress Management, Baramati, Pune, India

NRAA (2011) Challenges of food security and its management, National Rainfed Area Authority, NASC Complex, New Delhi

Painuli DK, Yadav RP (1998) Tillage requirement of Indian soils. In: Singh GB, Sharma BR (eds) Fifty years of natural resource management research in India. ICAR, New Delhi, pp 245–262

Painuli DK, Tomar SS, Temba GP, Sharma SK (2002) Raised-sunken Technology for Vertisols of high rainfall areas. Technical bulletin, AICRP on soil physical constraints and their amelioration for sustainable crop production. Indian Institute of Soil Science, Bhopal

Palanisami K, Suresh-Kumar D, Chanderashekhran B (2002) Watershed management; issues and policies for 21st century, Associated Publishing Company, New Delhi, India

Ravi-Sharma (2007) India: status of National Communications to the UNFCCC. Web address http://www.whre.org/policy/climate_change/alapdf/ala-08-india.pdf,

Samra JS, Singh G, Dagar JC (2006) Drought management strategies in India. Indian Council of Agricultural Research, New Delhi, p 227.

Samra JS, Narain P (2006) Soil and water conservation. In: Handbook of agriculture. Indian Council of Agricultural Research, New Delhi, pp 230–253

Sanyal SK, Jeevan-Rao K, Sadana US (2012) Toxic elements and other pollutants. In Soil Science in the Service of Nation, Proceedings of the Platinum Jubilee Symposium, ISSS, New Delhi, pp 266–291

Sanyal SK, Majumdar K, Singh VK (2014) Nutrient management in Indian agriculture with special reference to nutrient mining – a relook. J Indian Soc Soil Sci 62:307–325

Sharda VN (2011) Strategies for arresting land degradation in India. In: Sarkar D, Azad AK, Singh S, Akter N (eds) Strategies for arresting land degradation in South Asian countries. SAARC Agriculture Centre, Dhaka, pp 77–134

Sharda VN, Dogra P, Prakash C (2010) Assessment of production losses due to water erosion in rainfed areas of India. J Soil Water Conserv 65:79–91

Sharma PD, Sarkar AK (2005) Managing acid soils for enhancing productivity. Tech. Bull. NRM Division, Indian Council of Agricultural Research, New Delhi

Sharma PD, Singh MV (2012) State of health of Indian soils. In: Soil Science in the Service of Nation, Proceedings of the Platinum Jubilee Symposium, Indian Society Soil Science, New Delhi, pp 191–213

Singh RB (2010) Towards a food secure India and South Asia: making hunger history. Bangkok, Asia-Pacific Association of Agricultural Research Institutes

Sinha SK, Singh GB, Rai M (1998) Decline in crop productivity in Haryana and Punjab: myth or reality. Indian Council of Agricultural Research, New Delhi, India

Srinivasarao C, Lal R, Kundu S, Prasad MBB, Venketswarlu B, Singh AK (2014) Soil carbon sequestration in rainfed production systems in semi-arid tropics of Indi. Sci Total Env 487:587–603

Subba-Rao AVM, Choudary SB, Manipandan N, Rao VUM, Rao GGSN, Ramakrishna YS (2007) Rainfall trends, periodicities and vulnerable areas to climate change over India. National Conference on "Impact of Climate Change with Particular Reference to Agriculture", August 22-24, 2007, TNAU, Coimbatore

Subbareddy G, Reddy YVR, Vittal KPR, Thyagaraj CR, Rama-Krishna YS, Somani LL (2008) Dryland agriculture. Agrotech Publishing Academy, Udaipur

Sur HS, Singh R, Malhi SS (1998) Influence of simulated erosion on soil properties and maize yield in north-western India. Commun Soil Sci Plant Anal 29:2647–2658

Swarup A, Wanjari RH (2000) Three decades of all India research project on long-term fertilizer experiments to study changes on soil quality, crop productivity and sustainability. Indian Institute of Soil Science, Bhopal

Tomar SS, Tambe GB, Sharma SK, Tomar VS (1995) Studies on some land management practices for increasing agricultural production in Vertisols of central India. Agric Water Manag 30:91–106

UNEP (1993) Land degradation in South Asia: its severity, causes and effects upon the people. World soils research report no. 78, FAO, Rome

Velayutham M, Mandal DK, Mandal C, Sehgal JL (1999) Agro-ecological subregions of India for development and planning. NBSS Publication no. 35. National Bureau of Soil Survey and Land Use Planning, Nagpur, India

Vittal KPR, Vijayalakshmi K, Rao UMB (1990) The effect of cumulative erosion and rainfall on sorghum, pearl-millet and castor-bean yields under dry farming conditions in Andhra Pradesh, India. Exp Agric 26:429–439

WG-NRMRF (2011) Report of Working Group on Natural Resource Management and Rainfed Farming for XII Plan, Planning Commission, Government of India, New Delhi

Yadav JSP (2007) Soil productivity enhancement: prospect and problems. J Indian Soc Soil Sci 55:455–463

Yadav RL, Diwedi BS, Kanta-Prasad, Tomar DA, Shurpali NJ, Pandey PS (2000) Yield trends and changes in soil organic carbon and available NPK in a long-term rice- wheat system under irrigated use of manures and fertilizers. Field Crop Res 68:219–246

Yadav RK, Minhas PS, Khajanchi-Lal, Dagar JC (2016) Potential of wastewater disposal through tree plantations. In: Dagar JC, Minhas PS (eds) Agroforestry for management of waterlogged saline soils and poor-quality waters, Adv Agrofor Series 13. Springer, New Delhi

Part II

Adaptation and Mitigation Options

Soil-Related Abiotic Constraints for Sustainable Agriculture

5

S.S. Kukal, G. Ravindra Chary, and S.M. Virmani

Abstract

Currently, natural resources, land (soil, air, water) and biologicals (biodiversity) are in a state of serious decline. The factor productivity of monetary and non-monetary inputs is impaired both in the irrigated agricultural regions and in the dryland ecologies. Groundwater tables are lowering at 1 m per annum in most areas. There is little recharge of the free aquifer. The soil health is also in a state of decline, where only major nutrients (NPK) were applied to obtain high yields in the green revolution areas; today apparent from NPK, the deficiencies of Zn, Fe, S, Cu and Mo, amongst others, are showing up. The groundwaters are becoming toxic due to increased salinity, alkalinity and the contamination due to arsenic, fluorides, chlorides and in some cases uranium. The content of CO_2 in our atmosphere has almost doubled (since the industrial revolution about a century ago), which is leading to global warming; $1-1.5°C$ rise in ambient environmental temperature is already evident. The circulation of monsoon is setting new normals. It is in a state of flux. The occurrence of unseasonal rains, droughts and floods, soil erosion and loss of soil organic matter has increased with rise in temperature, thus seriously impairing soil health and its quality for producing sustainable yields.

S.S. Kukal (✉)
Punjab Agricultural University, Ludhiana 141001, India
e-mail: sskukal@rediffmail.com

G.R. Chary
ICAR-Central Research Institute Dryland Agriculture, Hyderabad 500 059, India
e-mail: rcgajjala@crida.in

S.M. Virmani
Indian Resources Information and Management Technologies Ltd. (INRIMT), Hyderbad 500003, India
e-mail: virmanism1938@gmail.com

© Springer Nature Singapore Pte Ltd. 2017
P.S. Minhas et al. (eds.), *Abiotic Stress Management for Resilient Agriculture*,
DOI 10.1007/978-981-10-5744-1_5

5.1   Introduction

Soil is not only a store house of essential nutrients to plants apart from providing plant anchorage, but is also crucial for ecological balance of the earth's environment. However, the soils, one of the important natural resources, are being threatened by their degradation in terms of physical, chemical and biological health mainly due to their intensive cultivation coupled by disturbed natural carbon cycle. The ratio of land to population especially in countries like India, having 2.4% of the world's arable land supporting 17.5% of the global population, is a matter of concern for meeting future food demands. The mismanagement and misuse of soils have resulted in their degradation in the form of accelerated erosion, salinization (both primary and secondary), depletion of soil organic matter, elemental imbalance, deficiencies of essential nutrients, soil compaction, surface sealing, etc.

The disturbance of natural carbon cycle, whereby the plant residues should have found their way back into the soils, has led to severe depletion of soil organic matter (SOM), thereby affecting not only the chemical but physical and biological health as well. The greater reliance of soil fertility on chemical fertilizers has led to alternate disposals of crop residues mainly open burning in the fields itself. The degradation of soils due to different reasons leads to abiotic and biotic constraints to plant growth and yield, the effect being differential under irrigated and dryland conditions. Also the nutrient imbalance in soils, leading to deficiency and toxicity of some elements, could seriously affect the human health through soil-plant system. This chapter aims to discuss the effects and management of soil-related abiotic constraints in irrigated and dryland regions.

5.2   Constraints in Irrigated Agriculture

5.2.1   Irrigated Agriculture Scenario in India

There has been a steady increase in area under irrigation in India over the last few decades. The net irrigated area increased by 24% during the period from 1980–1981 to 1990–1991, and, thereafter, the growth rate fell to around 15% each in the next two decades (Anonymous 2016). The biggest challenge to India's future irrigation is that of ever-increasing population, which currently stands at 1321 million, being 18.4% of the world population. The population of the country is projected to be 1700 million by 2050, and agriculture is expected to dominate as the primary source of livelihood in rural areas (Anonymous 2016). The proportion of agriculture-dependent population in India to the total population has decreased over the last few decades, even though the total agriculture-dependent population has been increasing but at a decreasing rate of 1.1% (1980s) and 1.0% (1990s). The increase in area under irrigation has been mainly due to groundwater exploitation. The higher groundwater irrigation expansion is due to the fact that the surface irrigation has not expanded over the years due to the slowdown in public investments in large-scale irrigation infrastructure coupled with noncompletion of ongoing irrigation projects. The most severe problem facing Indian canal irrigation is the rapid deterioration of the structures and maintenance being woefully neglected, leading to poor capacity utilization, rising incidence of water logging and salinity and lower water use efficiency (WUE). Thus, the surface irrigation is threatening to become unsustainable both physically and environmentally (Gulati et al. 1999). In the absence of new large-scale surface irrigation schemes, and the availability of low cost electric and diesel pumps coupled with little or no electricity charges, the groundwater has been a major driver in the irrigated area expansion. The FAO reports indicate higher yields (30–50%) in groundwater irrigated areas than in areas with surface irrigation. The irrigated area in the country is about 40%, with Punjab leading (98%), followed by Haryana (83%) and UP (68%). All other states have less than half area under irrigation. Because of the highest levels of irrigation in Punjab, Haryana and western Uttar Pradesh, these areas are contributing 98% of wheat and 66% of rice to the central pool of India.

The demand for food grains in the country is estimated to be 291 Mt. by 2025, which may increase to 377 Mt. by 2050. On the other hand, the total production is estimated to be 292 Mt. by 2025 and 385 Mt. by 2050 (Venkateswarlu and Prasad 2016). It is also predicted that by 2050, the agricultural production in states like Punjab might not be sustainable even at the present level if major steps are not taken to arrest the fast-depleting groundwater in the region. Overpumping is not only leading to fall in water table levels and failure of tube wells, but pumping costs are increasing due to higher energy consumption. The fall in fresh waters near the coastal areas has led to ingression of sea water, with serious environmental implications. In sustaining agricultural production, the northwestern states have already depleted their good- quality groundwater resources (GWR), which were conserved across more than 100 years with the introduction of a mighty canal network. The agriculture in these states is presently facing not only the problem of a declining groundwater table coupled with increased energy cost for pumping but also the deterioration of groundwater quality. Under such situation, the changes in water quality are adversely affecting agriculture and vice versa. Inefficient and/or over-use of chemical fertilizers and pesticides in agriculture and untreated disposal of industrial and urban wastes are increasingly contaminating the surface as well as groundwater by such elements as lead, zinc, copper, chromium and cadmium especially in areas with high industrial activity, e.g. in districts of Ludhiana, Faridabad, Kanpur, Varanasi, etc. An increase in arsenic content reported in several of the districts of West Bengal is attributed to the lowering of groundwater table due to excessive groundwater withdrawal for irrigation apart from other causes. This is leading to serious and widespread toxicity symptoms adversely affecting the health of people in the region. The reverse flow of water from southwest Punjab (with shallow and poor quality waters) to the central Punjab (with deep and sweet water) is being observed to hamper the water quality in central Punjab.

5.2.2   Soil-Related Constraints and Their Impacts

5.2.2.1 Soil Physical Constraints

Plants, apart from nutrients in soil, require air and water in definite proportions for their proper growth and development. In addition, the roots of many crop plants (except rice) respire independently in soils. This requires an optimum balance of air and water in soil pores. The availability of right proportion of air and water depends on the physical conditions of the soil. The soils with imbalance of air and water may lead to reduced uptake of nutrients by the plants despite of their sufficient presence in the soils. Such soils are designated as physically constrained soils. Many a times, despite of optimum input applications, best crop management practices and favourable weather conditions, the crops may not yield as per the expectations. This could again be due to unfavourable soil physical conditions. The soil physical characteristics, viz. supporting power and bearing capacity, tillability, moisture storage capacity and its availability to plants, drainage, ease of root penetration and favourable temperature conditions are all interrelated. The physical constraints in irrigated areas may appear due to excessive soil erosion (both by water and wind), surface and subsurface soil compaction, inappropriate proportions of primary soil particles, etc.

Detachment of Aggregates and Soil Erosion

The detachment of surface soil aggregates by wind energy, raindrops or moving irrigation water especially in low organic carbon or coarse-textured soils could lead to decreased infiltration of water due to clogging of soil pores by finer soil particles. In addition, the lesser permeable layer may lead to formation of soil crust upon drying, thereby hampering seed germination in case of freshly sown crop. Moreover, the runoff water during excessive rains may move from one field to another carrying along with it the nutrients to the lower most fields, thereby reducing the fertility of such fields situated on the higher side and leading to reductions in crop yields. The loss of soil fertility and topsoil makes the land unsuitable for biomass production. Studies (Hadda and Sur 1989) have indicated nutrient loss of 106 kg ha^{-1} in lower Shiwaliks region of northwest India, thereby indicating that erosion is a major threat to the crop yields in the region.

Surface Compaction and Crusting

The problem of surface soil crusting-sealing and hard setting of coarse- and medium-textured soils, particularly in the subhumid to semi-arid tropics in India, appears to be the most serious. The aggregates are easily dispersed under raindrop impact in soils with low organic matter, thereby creating a thin layer of dispersed soil particles (clay) on the surface, which is compact with higher strength on drying leading to soil crusting. The formation of crust layer reduces the exchange of gases between soil and atmosphere and also affects the germination of sown seeds. Surface sealing impedes seedling emergence, because of the strength needed to break through the crust, and it forms an oxygen-deficient layer immediately below the crust. The formation of crust on the surface by rain showers within 2 h of pearl millet sowing has been reported to reduce the seedling emergence to 40%. The seeding emergence increased from 42 to 67% with increase in soil moisture from 11.1 to 17.9% at the time of seeding (Agrawal and Sharma 1980). The alfisols in semi-arid tropics of India are highly prone to soil crusting leading to lower infiltration rates and higher runoff losses on gentle slopes.

Water Logging

Water logging has been categorized as surface water logging, where water stagnates on the land surface and subsurface water logging where groundwater table rises within the root zone, both having an adverse impact on the physico-chemical properties of the soil profile. While the surface stagnation is a perpetual problem in the humid regions, the problem is also encountered in the arid and the semi-arid regions, resulting from heavy rainstorms during the monsoon season or the mismanagement of irrigation water or both. Lands under rice-wheat system, prone to development of a plough sole/hard pan at the bottom of the plough layer, alkali lands under reclamation, irrigated lands with poor quality waters, especially with high SAR or alkali waters with high RSC, areas with inadequate drainage and areas with poor on-farm development or practicing flood irrigation are also prone to short-term surface stagnation of water. The subsurface water logging is quite widespread in irrigation commands and coastal regions of India. Most irrigation projects, being designed without adequate drainage provision, cause water table to rise, for example at a rate of 0.6 m y^{-1} in Bhakra canal command at Bhuna in Haryana (1975–2000) and 1 m y^{-1} in the initial years of the Indira Gandhi *Nahar Pariyojna*, Phase I. As soon as the water table reaches within 2 m of the soil surface, the root zone available to plants becomes restricted, and salts rise to the surface by capillary action, resulting in soil salinization. The districts of Muktsar, Fazilka, Bathinda and Faridkot in southwestern Punjab are presently facing severe problem of water logging mainly due to over-irrigation from the canals. The preliminary area statistics of degraded and wastelands of India put the figures at 1.66 M ha under surface and 4.75 M ha under subsurface water logging (ICAR and NAAS 2010).

The problem of short-term water stagnation in crop lands leads to yield decline of 2–8% for a water stagnation of 1 day to 20–48% when water stagnates for 6 days (Gupta et al. 2004). In several water logging sensitive crops, complete failure has been reported with a water stagnation of 2–4 days (Gupta 2014). A survey in various irrigation commands of India revealed that paddy, wheat, cotton and sugarcane suffered yield losses of 38 to 77% due to water logging (Joshi and Agnihotri, 1984). In the Indira Gandhi *Nahar Pariyojana*, wheat yields in areas with water table at less than 70 cm were only about 25% of the optimum yield with water table at or around 120 cm (Anonymous 2002). Besides decline in yield, water logging adversely impacts cropping intensity, restricts the choice of crops, impairs product quality and has several socio-economic and environmental implications. The water logging and salinity cause a loss of Rs. 12–27 billion annually.

Subsurface Compaction

The excessive use of machinery for cultivation and harvesting leads to compaction of soil with consequent reduction in porosity, available soil moisture content and higher resistance to root penetration. The higher bulk density of subsoils under rice-wheat cropping system commonly observed in Indo-Gangetic Plains (IGP) leads to aeration stress to crop roots. Intensive cultivation in conjunction with lower organic input is also responsible for loss of soil structure and consequent compaction. The area under hard pan soils in the country is 10.63 M ha.

Excessive dry and wet tillage (puddling), use of rotavators and freewheeling of tractors and harvesters (combined) in the IGP have in general, resulted in gradual subsoil compaction, increased bulk density and led to the formation of hard pan at 15–22 cm depth due to compaction and migration of silt and clay particles from the upper layers to this layer (Kukal and Aggarwal 2003a, b; Patil et al. 2005). Subsurface compaction can have a number of negative effects on soil quality and crop production including reduction in the size of pores, percolation rate of soils, water- and air-holding capacity of soils and crop yields. Various studies (Kukal and Aggarwal 2003b) have shown subsurface compaction to have detrimental effects on the upland crops to the extent of 5–15%. Reduced soil aeration due to temporary water logging conditions in the root zone can lead to aeration stress, and the plant roots are unable to breathe properly due to lack of oxygen. Kukal and Aggarwal (2003b) reported a 7–8% decrease in wheat yield under aeration stress caused by temporary water logging conditions during the first irrigation in central Punjab. The yellowing of wheat due to nitrogen deficiency after heavy irrigation coupled with rain showers is a common feature in rice soils with subsurface compaction.

Slowly Permeable Soils

The high clay content in soils (9.43 M ha) results in dominance of microporosity. The micropores, being water-retaining pores, lead to lower permeability of such soils. Such soils are difficult to cultivate with small window period for cultivation. It takes a longer time for these soils to reach field capacity moisture content after a pre-sowing irrigation. These soils if cultivated under higher moisture conditions can lead to puddling, whereas cultivation under dry conditions leads to breakdown of aggregates to produce dust. Moreover, a heavy irrigation in these soils can lead to aeration stress conditions for the younger plants as most of the dominating micropores are filled with water, and there is little air for the plant roots to breathe. Also the diffusion of gases between atmosphere and soil is slow and may lead to higher concentration of carbon dioxide in such soils. During higher rainfall years, the crops suffer from temporary water logging and later from salinity due to rise in ground-water table. The flat relief condition creates problems of drainage, and water remains stagnant at the surface for a long time making agricultural operations difficult. In low rainfall years or on drying, the soils become hard and dry and develop deep and wide cracks resulting into loss of stored moisture during maturity stage of the crop.

Highly Permeable Soils

The soils (10.77 Mha), containing higher amount of sand particles, are dominating in macrospores, which are water-conducting pores. This results in immediate movement of water below the root zone. This necessitates more frequent irrigation in such soils. Most of the coarse- and medium-textured soils in Punjab are being cultivated for rice. The percolation rate of such soils being higher leads to consumption of higher amounts of irrigation water (Kukal and Aggarwal 2003a) to grow rice crop in comparison to fine-textured soils. The leaching losses of nutrients in these soils are a common feature.

Shallow Soils

Limited soil depth and volume in shallow soils restrict root growth and hence restrict supply of water and nutrients to the crop in required amounts. This leads to reduced crop yields.

5.2.2.2 Soil Chemical Constraints

Salinization and Sodification

Soil salinization is a slow build-up process, which may lead to reduction of crop yields initially to the final abandonment of the land as it is not economically viable to cultivate such lands. These soils are commonly found in arid and semi-arid regions where evaporation exceeds the rainfall, apart from being dominant in the coastal regions. The salinity leads to reduced plant growth by restricted availability of water to the plants as a result of increased osmotic stress provoking withering of plants, preferential absorption of one ion that might retard the absorption of other essential plant nutrients and excess uptake of some of the salt constituents to cause toxicity of specific ions in the plants.

Higher accumulation of sodium ions (Na^+) on the soil exchange complex results in soil sodification. The sodification leads to alteration of soil physical properties, viz. increased soil dispersion, destruction of the soil structure and formation of crust on the soil surface that together hinders air/water movement in the soil. The soils with both the excess salts and higher content of exchangeable sodium percentage are designated as saline-alkali soils. Besides the adverse impacts on chemical and physical processes, soil salinization/sodification has a direct negative effect on soil biology resulting in reduced crop productivity. An area of 6.73 M ha area is saline/sodic, 2.96 M ha being saline and 3.77 Mha sodic (Sharma et al. 2004). Extreme events, climatic aberrations and anthropogenic interventions are likely to further aggravate the aerial extent of these soils, with the predictions indicating the extent of salt-affected soils may treble to 20.0 M ha by 2050 (CSSRI 2014). The annual rise of groundwater in the command areas of Gandak and Indira Gandhi canal has been recorded to be 43–83 cm. About 60% of these command areas will develop drainage and salinity problems in 100 years as per an estimate by Central Arid Zone Research Institute (CAZRI), Jodhpur.

High amounts of salts in water bodies are being released by many industries through effluents. Due to higher mobility of these salts (particularly Na^+ and Cl^-) within the soil, the groundwater in and around these industrial areas gets contaminated. These salt-loaded effluents when used for irrigation (in the adjoining lands) due to higher nitrate contents, contaminate agricultural land with heavy metals, thereby, degrading soil structure and decreasing crop productivity.

Chemical degradation of soil ultimately leads to reduced crop yields in the range of 30–63%. The annual losses due to water logging and soil salinity in 11 irrigation projects in India were estimated at US\$ 200 million (Joshi and Agnihotri 1984). Datta and de-Jong (2002) estimated the potential annual loss of about Rs. 1669 million from waterlogged saline areas in Haryana, while it has been estimated at INR 250–3000 million per year in crop production alone in the Tungabhadra command (Anonymous 2002).

Pollutant Chemicals

The soils often get polluted with chemicals which enter into the soil body through emissions from industries, power plants, vehicles, radioactive and toxic chemical leaks during disasters, etc. In India, about 100 million tonnes of gaseous pollutants are being added to the atmosphere every year through burning of fossil fuels and emissions from industries leading to air pollution. The thermal power plants release 100–110 t Hg $year^{-1}$, which finally gets precipitated in the soil body. Similarly, the pollutants in the surface and groundwater add harmful chemicals into the soils when such polluted waters are used for irrigation purposes. The potential carcinogenic and noncarcinogenic persistent organic pollutants (POP) like polycyclic aromatic hydrocarbons (PAH), polychlorinated biphenyls (PCB) and other organic pollutants from contaminated soil may affect humans through their ingestion and inhalation or dermal (skin) exposure to contaminated soil/dust or during tillage of polluted dry soil.

The pollutants from extensively used agrochemicals, like fertilizers (e.g. Cd through phosphatic fertilizer) and pesticides (organic pollutants), also pollute the soils. High concentration of heavy metals (Cd, Cr, Cu, Pb, Ni and Zn) could be observed in composts prepared from mixed municipal solid wastes in many cities of India. The release of pollutant-loaded industrial effluents on the agricultural land has destroyed soil fertility around several industrial units in the country.

Higher concentration of pollutants like As in the groundwater and selenium in soils in many parts of India are a common feature. Higher levels of NO_3 in groundwater have been recorded at some places with higher N fertilizer inputs indicating probable contamination from injudicious use of N fertilizers.

Nutrient Imbalance

The long-term imbalanced use of fertilizer nutrients has been responsible for soil degradation. Although overall nutrient use ($N:P_2O_5:K_2O$) of 4:2:1 is considered ideal for Indian soils, the present use ratio of 6.8:2.8:1 is far off the mark. The imbalanced nutrient use has increased the gap between crop removal and fertilizer application. Long-term experiments in India have indicated decreasing P and K in soils with dominance of N application alone. The deficiencies of micro- and secondary nutrients in crops are becoming a common feature in different parts of the country being just one deficient nutrient in 1950 to nine in the year 2005–2006, which might further increase if the practice of imbalanced fertilization continues. On the other hand, relatively higher concentration of phosphorus has been reported in some intensively cultivated irrigated areas like Punjab in India.

5.2.3 Managing Soil-Related Constraints for Sustained Agricultural Production

5.2.3.1 Soil Physical Constraints

Highly Permeable Soils
The coarse-textured soils, being highly permeable can be managed through compaction with repeated passes of a tractor-drawn heavy roller at optimum soil moisture content to attain the desired level of compaction. This leads to conversion of macropores to micropores, thereby, helping in retention of water in such soils. Alternately, the addition of clay @ 2% in red sandy loam of Andhra Pradesh has been reported to increase crop yield by more than 10%. This technology is viable only in regions where fine-textured soil is available either from the ponds or nearby fields.

Subsurface Mechanical Impedance
The subsurface compaction leads to mechanical impedance, which restricts plant root growth and movement of air, water and nutrients, thereby affecting crop yields. The subsurface compaction can be ameliorated by chiselling, chiselling plus amendment or ridge cultivation (Painuli and Yadav 1998). The subsurface high bulk density layer can be broken down by deep tillage/chiselling, thereby facilitating vertical and horizontal growth of roots. The chiselling has been recommended up to a depth of 30–50 cm depending upon soil and crop requirements. The effect of chiselling has been observed till seventh successive crop in red soils, but the effect lasted for smaller time in light-textured soils. Thus, chiselling is recommended every *kharif* season in light-textured soils and once in 2–3 years in red soils. Addition of amendments like gypsum at 5 t ha^{-1} or FYM at 20 t ha^{-1} can reduce the rate of compaction. The ill effect of compaction can also be overcome by sowing the crops on ridges or raised beds with an aim to increase the rooting volume above the compacted layer.

Slowly Permeable Soils
These soils with frequent aeration stress immediately after rainfall or irrigation can be managed by various tillage and land form treatments, viz. ridges and furrows, broad bed and furrow and raised and sunken beds especially in black soils of low and other high rainfall areas so as to avoid aeration stress during rainy season (Painuli and Yadav 1998).

Shallow Soils

The shallow soils can be effectively used for crop growing by constructing 10 cm high ridges above the soil surface as these were found to improve the root growth. Addition of clay or paddy husk can further improve the soil physical condition and crop growth. In the slopy red soils of Andhra Pradesh, the farmers are faced with the twin problems of shallow soil depth and soil erosion by water. The formation of ridges and furrows along the contours with vetiver barrier at a vertical interval of 1.0 m can reduce runoff and soil loss by 88 and 92%, respectively. This can also help in retaining higher moisture in the soil during crop growth, which is beneficial for crop yields.

Crusting Soils

The crusting can be managed by seed line mulch technology (Nagarajarao and Gupta 1996). The application of FYM @ 3 t ha^{-1} or chopped wheat straw on the soil surface after sowing can help in dissipating the kinetic energy of raindrops, thereby preventing the disintegration of aggregates and dispersion of soil particles leading to crusting. In addition, higher soil water (3%) can be retained in the upper 5 cm layer of the crusted soil, thereby helping in better seedling emergence.

5.2.3.2 Soil Chemical Constraints

Salt-Affected Soils

The sodic soils can be reclaimed by replacing a large part of the exchangeable sodium with calcium by using a suitable amendment. The work at the Central Soil Salinity Research Institute, Karnal shows that gypsum requirement sufficient to meet 50% of replacement is required to start with rice as the first crop. The combined application of FYM @ 20 t ha^{-1} and gypsum (25% gypsum requirement) proved to be more effective than application of gypsum equivalent to 50% gypsum requirement. On the other hand, saline soils can be reclaimed by leaching of excess soluble salts from the root zone, lowering of water table below the critical depth, selection of suitable crops and their varieties and agronomic practices.

5.3   Abiotic Constraints in Rainfed Agroecosystems

Rainfed agriculture in India accounts to 53% of total net cultivated area (74 Mha) and contributes to about 40% of total food grain production and livelihoods of the majority of small and marginal farmers. However, the productivity from rainfed agriculture continues to be <1 t ha^{-1} because of inherent biophysical and socio-economic constraints. Drought has been a recurring feature of rainfed agriculture affecting agricultural production in India. The recurrence interval of drought years in rainfed areas indicates 3–4 drought years in every 10-year period, of these 2–3 are of moderate and 1 of severe intensity. Due to such drought in 2012, 5.68 Mha could not be sown during *kharif* season, which led to a loss of about 12.76 million tonnes of *kharif* food grain production, and the food grain production declined by 4.66% to 252.68 million tonnes (MT). In view of the challenges posed by twin problems of unabated land degradation and climate change/variability and nation's expectations of second green revolution from rainfed areas, it is essential to address natural management issues, particularly soil and moisture management to enhance productivity to at least 1 t ha^{-1} in such areas.

The major soil constraints that are limiting productivity of rainfed crops are shallow depth, low plant available water capacity (PAWC), subsoil hard pans, very low subsoil saturated hydraulic conductivity, imperfect soil/land drainage, subsoil gravelliness, calcareousness, low soil organic carbon, multiple nutrient deficiencies, etc. (Table 5.1).

5.3.1 Managing Soil-Related Constraints for Rainfed Agriculture

5.3.1.1 Soil Physical Constraints for Enhanced Moisture Conservation

The sustainable management of soil physical environment has a significant role in water and nutrient uptake and losses, pollutant transport and also emission of greenhouse gases from soil. In rained agriculture, managing optimum soil physical environment is essential not only for sustainable management of soil and water resources but also for realizing yield potential of crops. In broad sense, the soil physical environment in rainfed regions can be improved through (a) building *in situ* moisture reserves to tide over the recurring drought spells and (b) preventing loss of stored soil moisture.

The strategies to manage soil moisture availability should include *in situ* water harvesting to tide over the recurring drought spells, preventing subsequent soil moisture losses through various practices. These practices are agroecology-specific, i.e. physiography, rainfall pattern, soil type and crop. Summer tillage coinciding with pre-monsoon showers increases rainwater infiltration, thereby recharging the soil profile and controlling weeds. This results in increased yields of rainfed crops in alfisols (Table 5.2). In the regions with unimodal type of rainfall, in semi-arid regions with shallow alfisols, crop sowing across the slope and ridging later can prove to be useful due to better rainwater absorption and control of pernicious weeds.

Tillage in combination with compartmental bunding is the most effective soil management practice for *in situ* moisture conservation in Vertic Inceptisols. Superficial scraping to eliminate weeds can conserve moisture, which otherwise could have been consumed by the weeds.

Reducing runoff losses through tillage operations across the slope could help sustain crop growth for another 2 weeks in case of intermittent failure of rains. This can increase the crop yields to the extent of 30–40%. Ridges and furrows along the contours are required to be laid on a minor gradient of 0.3–0.4% to prevent temporary water logging. However, this practice is not feasible for closely spaced

Table 5.1 Major soil constraints in rainfed agro-ecologies

Climate/Agroecological-subregion/Growing period	Major soil type(s) (Soil subgroups)	Major soil constraints	
		Physical	Chemical
(a) Cold arid			
1.1, 1.2 AERs <90 days	Skeletal soils (Typic Cryorthents, Typic Cryorthids)	Shallow depth, sandy texture, inadequate leaching, very low PAWC	Calcareousness, alkaline, low to medium organic matter
(b) Hot arid			
2.1,2.2,2.3, 2.4 AESRs <90 days	Desert and saline soils (Typic Camborthids, Typic Torripsamments, Typic Calciorthids, Typic Natrargids, Typic Salorthids, Ustochreptic Camborthids, Typic Paleorthids	Sandy texture, very low PAWC	Moderately calcareousness, alkaline, low organic matter
3.0 AESR <90 days	Mixed red and black soils (Typic Rhodustalfs, Typic Paleusterts)	Subsoil hard pans, low to medium PAWC	Slightly acidic, high sub soil density, alkaline, subsoil sodicity
(c) Semi-arid			
4.2, 4.4, 6.1, 14.1 AESRs 90–150 days	Alluvial soils (Typic Ustochrepts, Fluventic Ustochrepts, Typic Eutrochrepts, lithic Ustorthents)	Coarse soil texture, low to moderate PAWC	Multiple nutrient deficiencies
5.1,5.2,5.3,6.1,6.2,6.3,6.4 AESRs 90–150 days	Black soils (VerticUstochrepts, Typic Chromusterts, Entic Pellusterts, Vertic Haplaquepts, Vertic Halaquepts, Typic Pellusterts)	Swell-shrink potential, very low subsoil saturated hydraulic conductivity	Slightly to strongly alkaline, calcareousness, N, P, Zn. S and B deficiency
7.1,7.2,7.3 AESRs 90–150 days	Mixed black and red soils (Udic Rhodustalfs, Typic Pellusterts, Typic Chromusterts, Typic Haplustalfs)	Red soils: Gravelliness, low PAWC, high sub soil bulk density. Black soils: Very low saturated hydraulic capacity, imperfectly soil drainage	Low N, P, Zn, strongly to very strongly alkaline, calcareousness

(continued)

Table 5.1 (continued)

Climate/Agroecological-subregion/Growing period	Major soil type(s) (Soil subgroups)	Major soil constraints	
		Physical	Chemical
8.1,8.2,8.3 AESRs	Red loamy soils (Typic Haplustalfs, Vertic Ustropepts, Oxic Paleustalfs, Oxic Rhodustalfs, Typic Rhodustalfs)	Gravelliness, low to Medium PAWC, fine sub soil texture	Moderately alkaline, low CEC, multiple nutrient deficiencies (N,P,Zn)
(d) Subhumid			
10.1,10.2,10.3,10.4 AESRs 150–210 days	Black soils (Vertic Ustochrepts, Udic Chromusterts, Entic Chromusterts, Typic Chromusterts)	High shrink-swell potential, slowly permeability, high subsoil compaction	Calcareousness, alkaline, N,P,Zn deficiency
10.4 AESR 150–210 days	Red soils (Typic Plinthustalfs)	Subsoil gravelliness, low PAWC	Acidic, P fixation
11.0 AESR 150–210 days	Black soils (Entic Chromusterts)	Imperfectly drained, very slow saturated hydraulic conductivity in subsoil, *gilgai* micro-relief, high shrink-swell potential	Alkaline, calcareousness
	Red and yellow soils (Udic Rhodustalfs)	Gravelly subsoil, moderate permeability	Mildly acidic, low in N, P, B and Zn
12.1,12.2,12.3 AESRs	Red and lateritic soils (Typic Haplustalfs, Typic Plinthustalfs, Typic Haplustults)	Shallow depth, gravelly like quality subsoil, low PAWC, high bulk density of subsoil	Moderate to mild acidic, low CEC, moderate to high P fixation, low in N, P, Zn, B, ca, mg, S, Mo, toxicity of Fe, al, Mn
	Tarai soils (Aquic Hapludolls)	Weak soil structure, coarse to granular A-horizon, imperfect drainage	Calcareousness, alkaline, deficiency of N, P, Zn

(continued)

Table 5.1 (continued)

Climate/Agroecological-subregion/Growing period	Major soil type(s) (Soil subgroups)	Major soil constraints	
		Physical	Chemical
(e) Humid to per-humid			
14.2,14.3,14.4,14.5 AESRs >210 days	Brown forest and Podzolic soils (Typic Haplustalfs, Mollic Haplaquepts, Dystric Eutrochrepts, lithic Udorthents, Typic Hapludolls)	Weak soil development, Imperfectly drained, low PAWC	Mildly acidic to mildly alkaline
15.1,15.2,15.3,15.4,16.1 AESRs >210 days	Alluvial soils (Aeric Fluvaquents, Fluventic Eutrochrepts, Typic Dystrochrepts, Aeric Haplaquepts, Typic Haplumbrepts, Umbric Dystrochrepts)	Medium to excess leaching of bases	Mildly alkaline to strongly acidic, low to medium base saturation, moderate to low CEC, P fixation
16.2,16.3,17.1,17.2 AESRs >210 days	Red and lateritic hill soils (Typic Arguidolls, Cumulic Hapludolls, Typic Haplumbrepts, Typic Udorthents, Ultic Hapludalfs, Typic Paleudalfs)	Limiting soil depth, loamy skeletal soils, excessive leaching	Moderately to strongly acidic, low available P

Source: Velayutham et al. (1999)

Table 5.2 Effect of off-season tillage on the yield of sorghum (three rainy seasons) and castor bean (two rainy seasons) on alfisols at Hyderabad and barley (three post-rainy seasons) on Inceptisols at Varanasi

Practice	Sorghum (q ha^{-1})	Barley (q ha^{-1})	Castor bean (q ha^{-1})
Without off-season tillage (farmers practice)	18.7	13.7	1.32
With off-season tillage (improved system)	26.0	15.7	1.31

Source: AICRPDA (2003)

row-sown crops (row spacing <30 cm). In such cases, deep furrows (popular as 'conservation/dead furrows') can be created after eight to ten crop rows (~ 3 m interval) for capturing the rain/runoff water. The ridges and furrows with water surplus were found useful in regions with unimodal and low rainfall (~600–700 mm). The serration of the soil surface is an efficient way, which provides permitting more time for water entry into the soil profile.

In the regions with unimodal and medium to high rainfall (mean annual rainfall <800 mm), the raised and sunken beds are most appropriate to provide simultaneous drainage and storage of runoff water, whereas in the regions with bimodal and low rainfall (mean annual rainfall <750 mm), scooping, compartment bunding and tied ridging were found to be effective for *in situ* water conservation. Dividing the fields into sectors of 3 m × 3 m by compartment bunds in Vertisols can increase crop yields up to 50%. Tillage not only helps conserving moisture in the surface but also helps in conserving it in the seed zone (Acharya et al. 1988). Mechanical shattering of hard pans by chiselling or mould board ploughing helps in improving infiltration of water into the soils. In regions with medium rainfall and Vertisols, the broad bed and furrow system (popularly known as ICRISAT technology), combining benefits of good aeration and safe disposal of excess rainwater is useful.

Addition of manures and inorganic fertilizers leads to improvement of the aggregation status of soil (Rehana-Rasool et al. 2007, 2008). The bulk density of soil amended with higher inorganic fertilizer and farmyard manure normally shows a negative linear relationship (Bandyopadhyay et al. 2010). The application of manures improves water retention by increasing micropores and inter-aggregate pores. The specific surface area increases by the addition of organic manures, thereby resulting in increased water-holding capacity at higher tensions. The long-term experiments indicated improvement in infiltration and water retention capacity of soils with the addition of farmyard manure (FYM), groundnut shells and other crop residues including the green leaf manuring. The highest soil organic carbon (SOC) stock was observed with 50% recommended dose of fertilizer (RDF) + 4 t groundnut shells ha^{-1} (47.2 t SOC ha^{-1}) followed by 100% RDF (36.2 t SOC ha^{-1}) and lowest in control (32.2 t SOC ha^{-1}), which in turn helped in good soil moisture retention and better groundnut pod yield.

Mulching improves soil hydrothermal regime, soil aggregation, infiltration rate, soil water storage, retardation of erosion and controls evaporation losses from soil surface. An innovative technology, i.e. seed line mulch, involves application of FYM @ 3 t ha^{-1} or chopped wheat straw on the seeded rows immediately after sowing prevents the disintegration of aggregates and dispersion of soil (Nagarajarao and Gupta 1996). In crusting soils, application of straw mulch on seed lines increases the emergence of cotton finger millet, soybean, cowpea and horse gram by reducing raindrop impact. Increased plant biomass produced by fertilizer application, results in increased return of organic material to soil in the form of decaying roots litter, leaf fall and crop residues. The frequent cultivation between the crop rows to create dust mulch is useful, particularly breaking soil crusting in red soils. In post-rainy season-cropped heavy black soils (mean annual rainfall of 550–600 mm), vertical mulch increases water intake and reduces runoff.

5.3.1.2 Tactical Recycling of Harvested Runoff

The best practice that can be done in terminal drought is to provide life-saving irrigation to the rainfed crop. During severe drought at Arjia, protective irrigation increased yields by 377–1000 kg ha^{-1} in maize (out of seven experimental seasons, there was a total failure of crop in 1 year and drought occurred in 3 seasons). At

Agra, the protective irrigation increased the yield by 633–696 kg ha^{-1} in pearl millet (out of seven experimental seasons, the drought occurred only in 3 seasons). Under moderate drought at Varanasi, the protective irrigation to rice increased yields by 320–1190 kg ha^{-1} (out of seven experimental seasons, drought occurred in 3 seasons). Single supplemental irrigation gave an advantage usually to the tune of 200 kg grain ha^{-1} cm^{-1} water applied.

For sustainable use of the harvested rainwater, strict budgeting needs to be followed. For example, the crops like sugarcane, rice and wheat with higher evapotranspiration (ET) requirements should be avoided, and the crops like pulses and oilseeds with lower ET requirements be preferred. Various rainwater harvesting systems, viz. Khadin in Western Rajasthan, Nadi system in Southern Rajasthan and Bandh in Baghelkhand region of Madhya Pradesh, are some of the indigenous techniques, which need to be improved on scientific lines. Farm ponds need to be an essential component in the rainfed regions to harvest excess rain/runoff water for its subsequent use in crop production. A pond of 250 m^3 ha^{-1} catchment is recommended. Even smaller ponds can be dug to harvest excess rain/runoff water in arid ecosystem. Irrigation during initial 2 years is essential for trees and fruit trees.

5.3.1.3 Resilient Crops and Cropping Systems for Drought Mitigation

Since rainfed agriculture is risk prone due to weather aberrations such as delayed onset of monsoon and in-season drought/long dry spells, selection of crops and varieties and adoption of intercropping systems based on the biophysical environment (rainfall pattern and soil type) of a specific location are essential prerequisite for drought mitigation and enhanced productivity of rainfed crops. Based on long-term analysis of climate data in terms of probability of onset and withdrawal of the monsoon and occurrence of dry spells, effective cropping seasons have been worked out for various rainfed regions in the country. In arid regions where annual/crop seasonal rainfall is < 500 mm and length of growing period is < 90 days, short duration drought-tolerant pulses such as mothbean and cereals of 10–12 weeks duration such as pearl millet and minor millets are suggested. Suitable varieties of predominant rainfed crops for weather aberrations were identified, e.g. for delayed onset of monsoon, the sorghum varieties recommended were CSH-1 at Akola, M-35-1 at Bellary, S-1049 at Dantiwada and CSH-6 at Udaipur (AICRPDA, 2003).

The cropping intensity could be increased considerably depending on the soil type and moisture availability period. However, the duration of the crop cultivars influence the selection of a cropping system. Hence, the research in this area clearly brought out that in the high rainfall (> 1000 mm) regions of Orissa, Eastern Uttar Pradesh and Madhya Pradesh, a second crop could be grown in the residual moisture after a 90 days duration variety of upland rice than 120 days duration, similarly in the Vertisols of Malwa (Madhya Pradesh) and Vidarbha (Maharashtra), a change of 140 or 150 days sorghums to about 90 or 100 days cultivars provided an opportunity to grow chickpea or safflower in sequence. Double cropping was possible only in areas receiving more than 750 mm rainfall with a soil moisture storage capacity of more than 200 mm.

The low productivity of ICSs prior to 1970 was due to the replacement series, while the work by AICRPDA suggested additive series was most successful, i.e. out of 59 experiments on replacement and additive series with base crops as sorghum, maize, pearl millet, pigeon pea, safflower and wheat; in 54 experiments the LERs of the additive series were greater (with average of 23% more) than replacement series with multiple benefits of higher output and returns, spread labour peaks, maintenance of soil fertility (with inclusion of legume) and stability in production. The performance of the ICSs were strongly correlated with the amount of seasonal rainfall; when it was above normal, optimum productivity was achieved; under normal rainfall conditions, fairly high values of land equivalent ratios were achieved; and under low rainfall conditions, one of the two crops reasonably yielded providing an insurance against the risk due to weather aberrations (Ravindra-Chary et al. 2012). Further, the ICSs over the years and across locations in India proved that *kharif* season is more favourable to intercropping than *rabi*, probably due to replenishment of soil moisture (Chetty 1983). Pigeon pea either as base crop or intercrop performed better, particularly in sorghum, cotton and pearl millet-based intercropping systems (AICPRDA 2003).

5.3.1.4 Soil Fertility Management

Soil fertility in drought-prone regions is generally low especially being deficient in nitrogen. Most of these soils are also deficient in available phosphorus, sulphur and micronutrients (mainly Zn and Fe). There is, thus vast potential for increasing crop yields through fertilization, especially N fertilizers, across the contrasting rainfed environments, leading to favourable cost-benefit ratio. A water stressed crop can also recover faster if it is fertilized immediately after relief from stress. Integrated nutrient management studies have established the fact that green manuring is a dependable source of several plant nutrients, typically able to meet half of the N requirement of a crop. Inclusion of legumes in a rotation has been reported to benefit the succeeding crop equivalent to 10–30 kg N ha^{-1}. An integration of FYM (10 t ha^{-1}) + recommended NPK at Bangalore not only stabilized productivity and improved sustainability but also improved economics of production. The increase in urease, phosphatase and dehydrogenase activity increased with integrated nutrient management. In most of the situations, the yield sustainability was higher when the recommended dose of fertilizer was applied. Available nitrogen, organic carbon and phosphorus content in soil increased with organic fertilizer application. Application of crop residues in combination with chemical fertilizer can result in higher sustainable yield and maintain higher levels of nitrogen, phosphorus and organic carbon.

5.3.1.5 Soil Carbon Management

Conjunctive use of chemical fertilizers and organic manure resulted in higher sustainable yield index (SYI) over unfertilized control and sole application of either chemical fertilizers or organic manures. The average soil organic carbon (SOC) sequestration rate measured with different management treatments ranged between $0.32–1.26$ Mg C ha^{-1} $year^{-1}$. The mean annual C input were recorded to be maximum in soybean system followed by that in rice and groundnut systems. The carbon footprints (Tg CE ha^{-1} $year^{-1}$) were higher in cereal cropping systems followed by oilseed and pulse systems. The carbon footprints per unit amount of yield were higher for rice (2.88)-lentil (6.15) sequence in Inceptisols.

Crop residue management and carbon sequestration: The organic C concentration in the surface soil (0–15 cm) largely depends on the total input of crop residues remaining on the surface or incorporated into the soil. Generally, farmers burn crop residues like stalks of pigeon pea and cotton without recycling them. The shredding of stalks of pigeon pea, cotton and green biomass into small pieces facilitates easy mixing of the residue in the soil or compost/vermicompost pits. Higher yields with crop residue retention or incorporation are due to increased infiltration and improved soil properties, higher soil organic carbon and earthworm activity and improved soil structure after a period of 4–7 years. Permanent crop cover with recycling of crop residues is a prerequisite and integral part of conservation agriculture. The new variants of zero-till seed-cum-fertilizer drill/planters such as PAU Happy Seeder (Dr J S Gill, Personal Communication) and rotary-disc drill have been developed for direct drilling of seeds even in the presence of surface residues (loose and anchored up to 8 t ha^{-1}). Annual inputs of biomass-C as crop residues significantly increases SOC sequestration and stock, following an asymptotic relationship between the SOC stock and the magnitude of the inputs of biomass-C (Srinivasarao et al. 2012).

The practice of no-tillage can lead to higher SOC due to reduced soil disturbance, decreased fallow period and incorporation of cover crops in the rotation cycle. Eliminating summer fallowing in arid and semi-arid regions and adopting no till with residue mulching improves soil structure, lowers bulk density and increases infiltration capacity. It is well established that ecosystems with higher biodiversity absorb and sequester more C than those with lower biodiversity. Conservation agriculture promotes soil C sequestration by tipping the balance in favour of C inputs relative to C outputs. In conservation agriculture system, the soil organic carbon (SOC) was higher (0.31–0.45%) than conventional system (0.29–0.42%).

Agroforestry systems and carbon sequestration: Inclusion of trees in the agricultural landscapes often improves productivity of the system, while providing opportunities to create carbon (C) sinks. The above-ground carbon sequestration, the assimilation of carbon in to the plant matter, is highly variable in some major agroforestry systems around the world ranging from 0.29–15.2 Mg ha^{-1} $year^{-1}$. In general, the AFS in the arid and semi-arid and degraded regions have a lower carbon sequestration potential than those on fertile humid sites; and the temperate agroforestry systems have relatively lower sequestration potential compared with the tropical systems. The tree-based systems also add substantial quantities of litter to the soil every year, which is added to the soil and contributes towards the soil carbon sequestration.

5.3.1.6 Integrated Watershed Management (IWM)

Watershed management is a holistic approach towards optimizing the use of land, water and vegetation to alleviate drought, moderate floods, prevent soil erosion, and improve water availability and increase fuel, fodder, fibre and agricultural production on a sustained basis. Integrated watershed management (IWM) is the key to conservation and efficient utilization of natural resources of soil and water particularly in rainfed agriculture where water is the limiting factor for agricultural production. An analysis of the watershed programmes across the diverse rainfed agroecosystems in India revealed that these programmes benefitted farmers through increased area under irrigation (33.5%), increased cropping intensity (63%), reduced soil loss (0.8 t ha^{-1}) and runoff (13%) and improved groundwater availability. Economic assessment showed that these watershed programmes were beneficial and viable with a cost-benefit ratio of 1:2.14 (Joshi et al. 2005).

5.3.1.7 Rainfed Land use Planning

The land use planning on scientific lines in drought-prone regions is one of the rational approaches for drought mitigation. The cadastral level soil site-specific cropping systems centred on land use modules have been developed/identified for different regions (Ravindra-Chary et al. 2008a). A drought-resilient, less risk-prone farming system based on the land requirements and farmers' capacities can be developed to mitigate the drought and to address the unabated land degradation and imminent climate change. The SCUs are basically for soil and water conservation prioritized activities to mitigate drought. The SQUs are to address soil resilience and improve soil organic carbon, problem soils amelioration and wastelands treatment and linked to various schemes and programmes in operation like National Horticultural Mission (NHM), *Rashtriya Krishi Vikas Yojana* (RKVY), etc. The LMUs would be operationalized at farm level for taking decisions on arable, nonarable and common lands for cropping, agroforestry, agrohorticulture, etc. and, further, for levying the most fragile land parcels for ecorestoration. Rainfed land use planning modules should be based on these units for risk minimization, enhanced land productivity and income, finally, for drought proofing.

5.3.1.8 Delineation of Rainfed Agro-economic Zones

We need to focus on resource conservation and management, increased productivity and profitability, making rainfed agriculture as dependable and improving the livelihoods of the farmers/people in these areas. This requires to delineate core 'Rainfed Agro-Economic Zones' (RAEZs) in a district or part of a region in a state

(Ravindra-Chary et al. 2008b). The important criteria for delineation of these RAEZs could be a predominantly rainfed region with predominant rainfed production system, source and percentage net irrigated area and livelihoods majorly dependent on rainfed agriculture. Here, the rainfed farmers' livelihoods improvement and sustaining the land resources would be focal, wherein all the issues related to production through processing, profitability, improved livelihoods in harmony with conservation and maintenance of land resources.

5.4 Way Forward

The food requirements of the country are continuously increasing due to rising populations, increased gross national product, changing food habits (protein rich diets) and substantial migration of populations from rural to the urban areas. But the productivity increase is not able to match the rise in population, even in irrigated 'green revolution regions'. The productivity gains have slowed down considerably or in some cases are on decline due to the emergence of widespread multiple nutrient deficiencies, depletion of soil organic carbon stocks, development of secondary salinity and water logging in canal irrigated ecologies. The other abiotic stresses like changes in the environmental temperatures, e.g. heat, cold, chilling (frost, hail, etc.) and radiation (UV, ionizing radiation) are also responsible for major reduction in agricultural production.

The 70% of the cultivated lands are cropped under fragile dryland agriculture. These ecosystems are exposed to erratic rainfall, high volume intense rains, which occur only on 20–30 days during the 4 months of monsoon season, thus leading to sheet erosion of surface soils and low productivity. The rainfed/drylands are also characterized by intense grazing pressure on limited pastures. Both human and cattle population in these areas suffer from serious food (fodder) and water insecurity. The catastrophic events due to deterioration of natural resources combined with climate change are on the rise. Some agendas for future research are given below.

In irrigated agriculture, the soil-related abiotic stresses, both physical and chemical, have to be managed for achieving sustainable productivity. Light-textured soils have to be managed through compaction, while subsurface mechanical impedance has to be managed through chisel technology, chisel plus amendment technology and ridge technology. Water logging, particularly in highly productive black soil regions with intensive cropping systems has to be managed through site-specific landform treatments. For managing crusting problem, there needs to be focused research on integrated approach with tillage, soil and crop management strategies. Management and reclamation of sodic soils, reclamation of saline soils in the changing cropping patterns and rainfall variability are a challenge and need to be addressed.

In rainfed agroecosystems, managing rainwater, either *in situ* or *ex situ* for higher water productivity and sustainable conservation rainfed agriculture, is the prime challenge, particularly in the present context of climate change/variability in diverse rained agro-ecologies. Location-specific research on *in situ* moisture conservation practices coping with in-season drought is need of the hour. Similarly, rainwater harvesting and efficient utilization strategies to be developed are based on potential runoff, catchment-command area relationships and quantification of stored rainwater in structures like farm pond with quantified information on critical irrigations specific to rainfed crops. Carbon management strategies specific to application of locally available material is needed. Resilient cropping systems research, particularly intercropping systems, to be focused on developing practices that are amenable to *in situ* moisture conservation practices and mechanization with resilient crops and genotypes.

Scientific and participatory land use planning is a buzzword for achieving the different goals of the various stakeholders. In stressed ecosystems like rainfed where in the major crop- based production systems are established as best land use planning over a period of time, no single land use or single criteria have sustained the land productivities, incomes, ecosystem and finally the livelihoods, the reasons being highly complex situations of risk, diverse socio-economic settings and subsistence agriculture. Thus, land use planning in drought-prone areas should aim at increased land productivity in totality through various means from annuals to perennials (integration with animal component) to cope with drought and also to address inherent unabated land degradation. The final aim is to build a biodiverse mixed farming system model for individual farmer to sustain the farming system and achieving the goals of food, nutritional, economic and ecological securities with complimentary benefits of drought mitigation or drought proofing and sustainable land management as a buffer to impact of land use change.

References

Acharya CL, Bisnoi SK, Yaduvanshi HS (1988) Effect of long-term application of fertilizers and organic and inorganic amendments under continuous cropping on soil physical and chemical properties in an Alfisol. Indian J Agric Sci 58:509–516

Agrawal RP, Sharma DP (1980) Management practices for improving seeding emergence of pearl millet (*Pennisetumglaucum* L.) under surface crusting. J Agron Crop Sci 149:398–405

AICRPDA (2003) Annual reports 1971–2001. Eldoscope electronic document.All India Co-ordinated Research Project for Dry land Agriculture (AICRPDA), Central Research institute for Dryland Agriculture (CRIDA), Hyderabad, India. pp. 6357

Anonymous (2002) Recommendations on water logging and salinity control based on pilot area drainage research. CSSRI, Karnal and ALTERRA-ILRI, Wageningen, pp. 100

Anonymous (2016) World population prospects: the 2015 revision. United Nations, Department of Economic and Social Affairs, Population Division

Bandyopadhyay KK, Misra AK, Ghosh PK, Hati KM (2010) Effect of integrated use of farmyard manure and chemical fertilizers on soil physical properties and productivity of soybean. Soil Tillage Res 110:115–125

Chetty CKR (Ed.) (1983) Research on intercropping systems in dry lands – a review of Decade's work (1971–72 to 1980–81). All India Co-ordinated Research Project for Dry land Agriculture, Hyderabad, India, pp. 124

CSSRI (2014) CSSRI vision 2050. IACR-Central Soil Salinity Research Institute, Karnal, India. pp. 29

Datta KK, de-Jong C (2002) Adverse effect of water logging and soil salinity on crop and land productivity in north-west Haryana, India. Agric Water Manag 57:223–238

Gulati A, Meinzen-Dick R, Raju KV (1999) From top down to bottoms up: institutional reforms in Indian Canal irrigation. Institute of Economic Growth, Delhi, India

Gupta SK (ed) (2014) Agricultural land drainage in India. Agrotech Publishing Academy, Udaipur. 316 p

Gupta SK, Sharma DP, Swarup A (2004) Relative tolerance of crops to surface stagnation. J Agric Engg 41:44–48

Hadda MS, Sur HS (1989) Effect of land modifying measures on soil erosion and nutrient losses. J Res (PAU) 26:37–47

ICAR and NAAS (2010) Degraded and wastelands of India – status and spatial distribution. Indian Council of Agricultural Research, New Delhi. 158 p

Joshi PK, Agnihotri AK (1984) An assessment of adverse effects of canal irrigation in India. Indian J Agric Econ 38:528–536

Joshi PK, Jha AK, Wani SP, Joshi L, Shiyani RL (2005) Meta analysis to assess impact of watershed program and people's participation, comprehensive assessment research report no 8, IWMI, Colombo, Sri Lanka, pp. 18

Kukal SS, Aggarwal GC (2003a) Puddling depth and intensity effects in rice-wheat system on a sandy loam soil. I. Development of subsurface compaction. Soil Tillage Res 72:1–8

Kukal SS, Aggarwal GC (2003b) Puddling depth and intensity effects in rice-wheat system on a sandy loam soil. II. Water use and crop performance. Soil Tillage Res 74:37–45

Nagarajarao Y, Gupta RP (1996) Soil physical constraints and its ameliorative measures. In: Biswas TD, Narayanasamy G (eds) Soil management in relation to land degradation and environment, Bulletin 17. Indian Society Soil Science, New Delhi, pp 66–73

Painuli DK, Yadav RP (1998) Tillage requirements of Indian soils. In: Singh GB, Sharma BR (eds) 50 years of natural resource management research. Natural Resource Management Division, ICAR, New Delhi

Patil BN, Patil SG, Hebbara M, Manjunatha MV, Gupta RK, Minhas PS (2005) Bio-ameliorative role of tree species in salt-affected Vertisols of India. J Trop For Sci 17:346–354

Ravindra-Chary G, Maruthi-Sankar GR, Subba-Reddy G, Ramakrishna YS, Singh AK, Gogoi AK, Rao KV (2008a) District-wise promising technologies for rainfed cereals based production systems in India. All India Coordinated Research Project for Dryland Agriculture, Central Research Institute for Dryland Agriculture (CRIDA), Indian Council of Agricultural Research (ICAR), Hyderabad, AP, India, 204 pp

Ravindra-Chary G, Venkateswarlu B, Maruthi-Sankar GR, Dixit S, Rao KV, Pratibha G, Osman M, Kareemulla K (2008b) Rainfed Agro-Economic Zones (RAEZs): a step towards sustainable land resource management and improved livelihoods. Lead Paper, National Seminar on Land Resource Management and Livelihood Security, the Indian Society of Soil Survey and Land Use Planning, 10–12 September 2008, Nagpur, India, p. 70

Ravindra-Chary G, Venkateswarlu B, Sharma SK, Mishra JS, Rana DS, Ganesh-Kute (2012) Agronomic research in dryland farming in India: an overview. Indian J Agron 57:157–167

Rehana-Rasool, Kukal SS, Hira GS (2007) Soil physical fertility and crop performance as affected by long term application of FYM and inorganic fertilizers in rice-wheat system. Soil Tillage Res 96:64–72

Rehana-Rasool, Kukal SS, Hira GS (2008) Soil organic carbon and physical properties as affected by long term application of FYM and inorganic fertilizers in maize-wheat system. Soil Tillage Res 101:31–36

Sharma RC, Rao BRM, Saxena RK (2004) Salt affected soils in India-current assessment. In: Advances in Sodic Land Reclamation, International Conference on Sustainable Management of Sodic Lands, Feb 9–14, Lucknow, India. pp. 1–26

Srinivasarao C, Deshpande AN, Venkateswarlu B, Lal R, Singh AK, Kundu S, Vittal KPR, Mishra PK, Prasad JVNS, Mandal UK, Sharma KL (2012) Grain yield and carbon sequestration potential of post monsoon sorghum cultivation in Vertisols in the semi arid tropics of central India. Geoderma 175–176:90–97

Velayutham M, Mandal DK, Mandal-Champa, Sehgal JL (1999) Agro-ecological subregions of India for planning and development. NBSS & LUP, Nagpur, India, NBS Publ. 35, pp. 372

Venkateswarlu B, Prasad JVNS (2016) Carrying capacity of Indian agriculture: issues related to rainfed agriculture. Curr Sci 102:882–888

Developments in Management of Abiotic Stresses in Dryland Agriculture

Ch. Srinivasarao, Arun K. Shanker, and K.A. Gopinath

Abstract

Abiotic stress is one of the important consequences of climate change that will have a telling effect on crop growth and productivity in the near future. The impact of abiotic stress on crop production has emerged as a major research priority during the past decade. Several forecasts for the coming decades project increase in atmospheric CO_2 and temperature and changes in precipitation, resulting in more frequent droughts and floods, cold and heat waves and other extreme events. The key aspect of sustainable development in agriculture involves resource conservation-based strategies, cropping system-based strategies and exploitation of genetic resources. Soil degradation should be prevented by practices and techniques, such as no-till sowing of crops, drip irrigation, crop rotation and leaving land fallow. Suitable farming systems that have potential to increase food production and promote soil conservation for each agroecological zones should be identified. Holistic land management and soil health restoration are also one of the key aspects of suitable development towards managing food security. In addition, maintenance of soil life and soil quality through practices such as organic fertilizer supplementation and judicious use of chemical fertilizers and pesticides should be a priority. Judicious management of water on a watershed basis should be undertaken to make maximum use of available water. Biotechnological improvement of crops and evolution of crop varieties suitable for climate change also form the key for sustainable development. Tapping of genetic resources to identify varieties and new crops that can adapt and cope with climate risk also forms a key approach in sustainable development to tackle the problem of food security. Adaptation to climate change requires long-term investments in strategic research and new policy initiatives that put climate

C. Srinivasarao (✉) • A.K. Shanker • K.A. Gopinath
ICAR – Central Research Institute for Dryland Agriculture, Hyderabad, India
e-mail: cherukumalli2011@gmail.com; arunshank@gmail.com; gopinath@crida.in

© Springer Nature Singapore Pte Ltd. 2017
P.S. Minhas et al. (eds.), *Abiotic Stress Management for Resilient Agriculture*,
DOI 10.1007/978-981-10-5744-1_6

change adaptation in planning. A comprehensive understanding of abiotic stress, especially the mechanism and tolerance aspects for adaptation strategies, across the full range of warming scenarios and regions is essential for preparing for climate change. Therefore, a judicious mixture of basic and applied research outlooks has been presented on developments in management of abiotic stresses in dryland agriculture.

6.1   Introduction

Rainfed agriculture constitutes 80% of global agriculture and plays a critical role in achieving global food security. However, growing world population, water scarcity and climate change threaten rainfed farming through increased vulnerability to different abiotic stresses like droughts and other extreme weather events. Drought, salinity, temperature, radiation and heavy metal stresses are among the major stresses, which adversely affect plant's growth and productivity. Various forms of abiotic stresses limit agricultural production on most of the world's 1.4 billion cultivated hectares. Irrigation will not be a practical solution as water becomes scarcer, and irrigation already in place will lead to soil salinization (Richards 1996). High and low temperatures, acid soils and soils with high levels of metal ions reduce crop productivity over large tracts of land and will remain a major challenge for the foreseeable future. Solutions to the problem will be as complex as the problem itself. Abiotic stresses do not generally come in isolation, and many stresses occur simultaneously thus severely affecting crop yields. In response to these stress signals, nature has developed diverse pathways for combating and tolerating them. These pathways act in cooperation to alleviate stress.

All the abiotic stresses have profound impact on agricultural systems. Among these, water stress is the predominant stress which causes huge loss in agricultural production, more so because water stress is usually accompanied in varying degrees by other stresses, viz. salinity, high temperature, high radiation and nutrient stress. The management of natural resources in dryland areas has a role in determining food security for the growing population and in reducing poverty in the coming decades (Rockstrom et al. 2007). Enhancing the efficiency and sustainability of natural resource management (NRM) is important not only because the livelihoods of millions of rural poor (>500 million) are directly dependent on these areas but also because these areas will continue to play a crucial role in improving the livelihoods of farmers. The potential for technically based NRM interventions for management of different abiotic stresses varies across countries. For example, in India, a number of proven technologies have been developed for dryland farming over eight decades due to a huge investment in research since 1930s. Location-specific technologies including in situ moisture conservation, rainwater harvesting in farm ponds and its efficient utilization, integrated nutrient management, resilient crops and cropping systems, improved planting methods and contingency crop plans have been developed for management of abiotic stresses including drought and improve the productivity and profitability of dryland systems (Srinivasarao et al. 2016a).

Conventional plant breeding approaches have yielded limited results so far mainly because of the lack of understanding on the traits that is required for abiotic stress tolerance (Ceccarelli 1996). The aim of plant breeding is also to ensure that abiotic stresses are managed well by increasing the tolerance mechanism of crops. It has the potential to produce high-yielding varieties but requires identification of major traits and their incorporation into high-yielding varieties using conventional or biotechnological tools. Conventional plant breeding in tropical regions so far only increased the yield of crops grown under several abiotic stresses at about half the rate achieved for crops grown in temperate regions. Crop response to stress is dependent on numerous traits many of which are constitutive and expressed irrespective of the presence of stress stimuli, but such constitutive traits may also be modified by stress (Firbank 2005).

Molecular genetics works primarily through insertion of genetic material, although gene insertion must also be followed up by selection (Dunwell 2000). The potentials for increasing tolerance of crops to abiotic stresses by molecular approaches are vast, although it is well known that the actual production of transgenic crop varieties with improved abiotic stress tolerance is slow. Identification of differentially expressed genes in various cells or under different conditions is one of the main areas which will decisively pave way for the release of new crop varieties which are not only adapted to dryland conditions but are also cross adapted to the accompanying stresses in drylands.

6.2 The Stresses Afflicting Dryland Agriculture

Stress is defined as 'any environmental variable, which can induce a potentially injurious strain in plants'. Since crop plants cannot control environmental conditions, they have evolved two major strategies for surviving adverse environmental conditions – they either avoid the stress or tolerate it. Plants lack the avoidance mechanism like mobility; they have evolved intricate mechanical methods to avoid stress, the best example of which is altering life cycle period in such a way that a stress-sensitive growth period is before or after the occurrence of the stress. On the other hand, tolerance mechanisms mainly involve biochemical and metabolic means which are in turn regulated by genes to counteract, nullify or tolerate the given environment.

6.2.1 Drought

Drought has been a frequent feature of agriculture in India. In the past, India experienced 24 large-scale droughts, with increasing frequencies during the periods 1891–1920, 1965–1990 and 1999–2012 (NRAA 2013). Rainfed areas in India experience 3–4 years of drought in every 10-year period. Of these, two to three are in moderate and one or two may be of severe intensity (Srinivasarao et al. 2013a). Droughts in India have periodically led to famines, including the Bengal famine of 1770, in which up to one third of the population in affected areas died and the 1876–1877 and 1899 famines in which over 5 million and more than 4.5 million people died, respectively.

6.2.2 Floods

India's vulnerability to floods can be visualized from the flood damages at current prices during 1953–2010 of Rs. 8.12 trillion. About 49.8 M ha land (15.2% of geographical area) is flood prone, and about 10–12 M ha is actually flooded each year. Floods occur in almost all river basins in India. The main causes of floods are heavy rainfall, inadequate capacity of rivers to carry high flood discharge and inadequate drainage to carry away the rainwater quickly to streams/rivers. Flash floods occur due to high rate of water flow and poor permeability of the soil. Most of the floods occur during rainy season (June–September) and are usually associated with cyclones and active monsoon conditions (Sikka et al. 2016).

Crop plants require a free exchange of atmospheric gases for photosynthesis and respiration. The most common barrier to gas diffusion is water that saturates the root zone in flooded soils or that accumulates above soil due to floods (Bennett and Freeling 1987). Prolonged flooding shifts the soil microbial flora in favour of anaerobic microorganisms that use alternative electron acceptors to oxygen. As a consequence, the soil tends to accumulate more reduced and phytotoxic forms of mineral ions such as nitrite and ferrous ions, and few plants are adapted to grow in these soils. Short-term anaerobic stress to an adult plant that is caused by poor drainage or periodic flooding reduces oxygen levels around the root and influences root development directly, whereas changes in shoot development may follow as a result of metabolic alterations in the roots (Boru et al. 2001).

6.2.3 High Temperature

Heat stress often is defined as where temperatures are hot enough for sufficient time that they cause irreversible damage to the plant's function or development (Hall 2000). Extreme positive departures from the normal maximum temperature result in heat wave in different parts of India. The maximum heat waves occur over east Uttar Pradesh followed by Punjab, east Madhya Pradesh and Saurashtra and Kutch in Gujarat (Raghavan 1967). During the decade 1991–2000, a significant increase in the frequency, persistency and spatial coverage of heat wave/severe heat wave was observed in comparison to that during the earlier decades 1971–1980 and 1981–1990 (Pai et al. 2004).

A few districts in Jammu and Kashmir, Himachal Pradesh, Punjab, Haryana, Uttar Pradesh, Uttarakhand, Madhya Pradesh and northeastern states of India are likely to experience more frequent hot days compared to the baseline. In a recent study, sensitivity of wheat yields to minimum temperature during post-anthesis period was quantified, and it was found that wheat yields in India for the period 1980–2011 declined by 7% (204 kg ha^{-1}) for a 1 °C rise in minimum temperature. Exposure to minimum temperature exceeding 12 °C for 6 days and to maximum temperature exceeding 34 °C for 7 days during post-anthesis period are thermal constraints in achieving high productivity levels in wheat (Bapuji Rao et al. 2015).

Extreme temperatures can cause premature death of plants. Among the cool-season annuals, pea (*Pisum sativum*) is very sensitive to high day temperatures with death of the plant occurring when air temperatures exceed about 35 °C for sufficient duration, whereas barley (*Hordeum vulgare*) is very heat tolerant, especially during grain filling. For warm-season annuals, cowpea can produce substantial biomass when growing in one of the hottest crop production environments on earth (maximum daytime air temperatures in a weather station shelter of about 50 °C), although its vegetative development may exhibit abnormalities such as leaf fasciations (Thiaw and Hall 2004).

6.2.4 Cold

The temperature below which chilling injury can occur varies, ranging from 0 to 4 °C for temperate fruits, 8 °C for subtropical fruits and about 12 °C for tropical fruits such as banana. Chilling injury is the physical and/or physiological changes that are induced by exposure to chilling temperatures. The physiological changes may be considered primary or secondary. The primary injury is the initial rapid response that causes a dysfunction in the plant but is readily reversible if the temperature is raised to non-chilling conditions (Kratsch and Wise 2000). Secondary injuries are dysfunctions that occur as a consequence of the primary injury and that may not be reversible. The characteristic visual symptoms are the consequence of secondary chilling injuries. Physiological age, seedling development and preharvest climate can also influence chilling sensitivity. The severity of injury to chill-sensitive tissues tends to increase with decreasing temperatures and with length of low temperature exposure.

In India, the maximum of cold waves generally occur in Rajasthan followed by Jammu and Kashmir and Uttar Pradesh. The frequency of events over different time periods indicates that in recent years the state of Rajasthan is experiencing more cold waves and Jammu and Kashmir are experiencing a few (Sikka et al. 2016). Depending upon the time of occurrence, they are either beneficial or harmful to the field and orchard crops. Cold wave conditions that prevailed during the winter of 2010–2011 and 2011–2012 coincided with the flowering and seed formation stage of wheat in Punjab resulting in good yields (Samra et al. 2012). Frost and cold waves greatly impact legume crops. During flowering stages, these crops are likely to be adversely affected at temperature of −2 to −3 °C. Those in pod formation stage are a bit more tolerant but are likely to be damaged at a temperature of −3 to −4 °C.

6.2.5   Salinity

In India, the total degraded area is estimated at 120.7 Mha, of which 104.2 M ha (86.3%) is arable land and 16.5 Mha (13.7%) is open forest land. Of the total degraded land area, 73.3 Mha (60.7%) is caused by water erosion, 12.4 Mha (10.3%) by wind erosion, 5.4 Mha (4.5%) by salinization and 5.1 M ha (4.2%) by soil acidification. Some areas are affected by multiple degradation processes (Maji 2007). In many states in India, between 40 and 80% of the land area is classified as degraded by one or other processes. High salt concentrations decrease the osmotic potential of soil solution creating water stress in plants. Secondly, they cause severe ion toxicity, since Na^+ is not readily sequestered into vacuoles as in halophytes. Finally, the interactions of salts with mineral nutrition may result in nutrient imbalances and deficiencies (Flowers and Yeo 1995). The complexity of plant salt tolerance in part comes from the fact that salinity imposes not only ionic stress but also osmotic stress (Munns 1993). The ionic stress is primarily caused by sodium toxicity to plants. Some plant species are also sensitive to chloride toxicity. In certain saline soils, the ion toxicity is further aggravated by alkaline pH.

6.2.6   Heavy Metals

Metal contamination issues are becoming increasingly common in cultivated areas. Plant responses to metals are dose dependent. For essential metals, these responses cover the phases from deficiency through to sufficiency/tolerance to toxicity. For non-essential metals, only the tolerance and toxicity phases occur (Reichmann 2002). The presence of different heavy metals like Cd, Cu, Mn, Bi, Zn, etc. in certain level in soils is natural, but their enhanced level is an indicator of the degree of pollution load in that specific area. The heavy metals at higher levels have strong snowballing properties and toxicity due to which they have a hazardous effect not only on crop plants but also on human health.

6.3   Metabolic Alterations and Stress Signalling Pathways

One of the most crucial functions of plant cells is their ability to respond to fluctuations in their environment. Understanding the connection between a plant's initial responses and the downstream events that constitute successful adjustment to its altered environment is one of the next grand challenges of plant biology (Wendehenne et al. 2004). Oxidative stress which results from almost all the abiotic stresses involves the formation of reactive oxygen species (ROS) in plant cells. In order to understand the implications of ROI in plants under stress, one should have a clear idea of the nature of damage it can cause to the plants. Lipids are the most important molecules that are attacked by ROI (Jonak et al. 2006). Proteins and DNA are the other targets of ROI. The mechanisms by which oxygen radicals damage membrane lipids are well accepted, and consequently oxidative damage is often exclusively

Fig. 6.1 Signalling
cascade during abiotic
stress

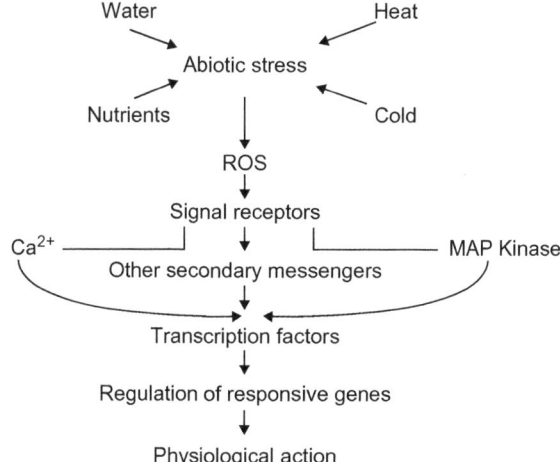

associated with these peroxidation reactions in membrane lipids. Lipid peroxidation reactions in plant membranes would selectively degrade unsaturated fatty acids and accumulate aldehydes, hydrocarbons and cross-linked products. When examining the effects of environmental stresses on plant membranes, many studies measured the products of lipid peroxidation, such as malondialdehyde and/or ethane, and concluded that oxygen free radicals are involved in these stress responses (Fig. 6.1).

Inter- and intracellular signalling is one of the important methods why plants initiate a defence response to the above stresses. Low molecular weight molecules that primarily regulate the protective responses of plants against multiple abiotic stresses via synergistic and antagonistic actions are referred to as signalling cross-talk. In plants, the mitogen-activated protein kinase (MAPK) cascade plays a crucial role in various biotic and abiotic stress responses and in hormone responses that include ROS signalling (Moon et al. 2003). Protein phosphorylation and dephosphorylation are perhaps the most common intracellular signalling modes (Fig. 6.1). They regulate cellular processes such as enzyme activation, assembly of macromolecules and protein localization and degradation.

In plants, many protein kinases and phosphatases are thought to be involved in environmental stress responses on the basis of several studies. A continuously growing number of genes coding for protein kinases in plants have been reported. The MAPK cascades are the major component downstream of receptors or sensors that transduce extracellular stimuli into intercellular responses (Viswanathan and Zhu 2004). All plant MAPKs have a Thr-Glu-Tyr activation motif, except members of subfamily V, where Glu is replaced by Asp. Recently, a MAPK kinase 2 (MKK2) from *Arabidopsis*, specifically activated by cold and salt stress and by the stress-induced MAPK kinase MEKK1, was found to increase freezing and salt tolerance in transgenic plants, suggesting the importance of MAPK cascades in plant's responses to multiple stresses (Kasuga et al. 1999). Studies of transcriptional activation of some stress-responsive genes have also led to the identification of cis-acting

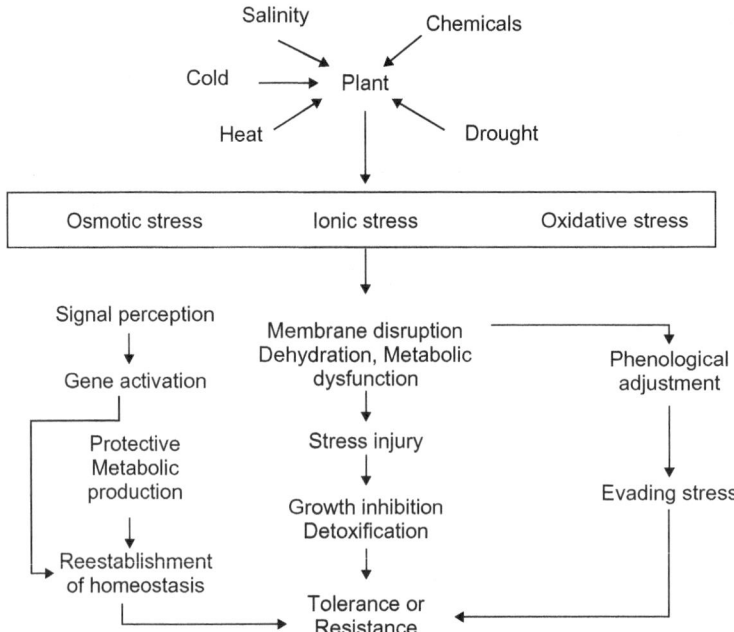

Fig. 6.2 Physiological and biochemical events resulting in tolerance or resistance to abiotic stress

elements ABA-responsive element (ABRE) and dehydration-responsive element (DRE)/C-repeat (CRT) that function in ABA-dependent and ABA-independent gene expression in response to stress, respectively (Arnholdt-Schmidt 2005).

Gene expression profiling has also identified many potential signalling molecules that change in expression in response to stresses. In addition to this, it has been found that plant hormone signalling pathways are among a set of core pathways that are used repeatedly in many different developmental contexts. Both physiological studies of stress adaptation and molecular analysis of diverse stress gene regulation patterns have suggested a network of multiple signalling pathways that mediate multiple stress responses in plants (Babu et al. 2003). The similarities among intermediate signalling molecules used by diverse stresses imply the existence of intracellular networks rather than linear pathways. The sequence of events after signalling due to stress throws light on the mechanism of tolerance in plants affected by abiotic stress (Fig. 6.2).

If a limited number of signalling intermediates can interact in a combinatorial fashion, such networks could allow specific cellular responses to numerous, potentially conflicting, signals. Some of the constituents of MAP kinase cascades are activated by cold, drought, salinity, H_2O_2, heat, shaking, wounding, pathogens, elicitors, ABA, salicylic acid and ethylene, suggesting that they might function as complex networking molecules. Only recently, the molecular components of these pathways have begun to be identified (Bohnert et al. 2006). However, understanding signal pathway crosstalk will become increasingly important for our understanding of complex signalling networks.

6.4   Natural Resource Management-Based Strategies

India is facing the biggest challenge of meeting the food demands by increasing the production (per unit land) simultaneously without degrading the soil and water resources and maintaining a favourable ecological balance (Srinivasarao et al. 2015b). Production potential of crops particularly under rainfed conditions depends on the resource endowments of the region and the management practices adopted. Hence, scientific management of natural resources plays crucial role for adaptation to different abiotic stresses and enhancing productivity and profitability of rainfed farming systems. Improved water storage through in situ moisture conservation and stored runoff is the basic for bringing resilience to drought or moisture stress conditions often encountered by the rainfed crops. Other strategies under natural resource management for bringing resilience are through site-specific nutrient management, foliar sprays, watershed management and efficient recycling of farm wastes/residues.

6.4.1   Soil Carbon Sequestration

Soils hold the key to productivity and resilience to climate vagaries in dryland agriculture. Higher organic matter in soil improves soil aggregation which in turn improves soil aeration and soil water storage, reduces soil erosion, improves infiltration and generally improves surface and groundwater quality. However, the soil organic carbon (SOC) which is the seat of major soil processes and functions, is only <5 g kg^{-1} in rainfed soils, while the desired level is 11 g kg^{-1}. Maintaining or improving soil organic matter is a prerequisite to ensure soil quality, productivity and sustainability (Srinivasarao and Gopinath 2016).

Organic and inorganic and total C stocks in soil vary between and within soil types. Vertisols and associated soils contain higher C stocks, followed by inceptisols, alfisols and aridisols. In general, concentration of SOC is more than that of soil inorganic C (SIC) in alfisols and aridisols, while SIC is more than SOC in vertisols and inceptisols. The SOC stocks (Mg ha^{-1}) range from 26.7 to 59.7 with a mean of 43.7 in inceptisols, 23.3 to 49.8 with a mean of 30.8 in alfisols, 28.6 to 95.9 with a mean of 46.4 in vertisols and 20.1 to 27.4 with a mean of 23.7 in aridisols (Srinivasarao et al. 2013b).

The carbon-positive nutrient management strategies in rainfed production systems were developed in long-term experiments at different All India Coordinated Research Project for Dryland Agriculture (AICRPDA) centres. The average SOC sequestration rate (kg C ha^{-1} $yr.^{-1}$) measured with different management treatments were (1) 570 for 50% recommended dose of fertilizer (RDF) + groundnut shells (GNS) at 4 Mg ha^{-1}, (2) 570–720 with application of farmyard manure (FYM) at 10 t ha^{-1} + 100% NPK, (3) 650 for 25 kg N ha^{-1} (sorghum residue) + 25 kg N (*Leucaena* clippings), (4) 240 for 50% recommended dose of nitrogen (RDN) through fertilizer +50% RDN through FYM, (5) 790 for 6 Mg ha^{-1} FYM + 20 kg N + 13 kg P and (6) 320 for 100% organic (FYM). The critical level of C input requirements for maintaining SOC at the antecedent level ranged from 1 to 3.5 Mg C ha^{-1} yr.$^{-1}$ and differed among soil type and production system. The critical level of C input was higher in soybean system and lower in winter sorghum system and increased with increase in mean annual temperature from humid to semiarid to arid ecosystems. For each ton of soil organic carbon improvement, productivity enhancement ranging from 50 to 300 kg ha^{-1} was recorded among different agro-ecoregions (Srinivasarao et al. 2013b).

In soybean-safflower cropping sequence at Indore (Madhya Pradesh; hot-dry, semiarid vertisols), combination of FYM and chemical fertilizer increased the profile SOC stock (69.9 Mg ha^{-1}) and overall SOC build-up (37.1%) and also sequestered high amount of SOC (11.9 Mg C ha^{-1} or 0.79 Mg C ha^{-1} yr.$^{-1}$) compared with control and chemical fertilizer alone. Higher seed yield (2.10 and 1.49 Mg ha^{-1} of soybean and safflower, respectively) was obtained with application of FYM at 6 Mg ha^{-1} + N20 + P13. For every t C ha^{-1} increase in the root zone, there was 0.145 and 0.059 t ha^{-1} increase in seed yield of soybean and safflower, respectively. In case of groundnut-finger millet rotation, the SOC stock (Mg ha^{-1}) was the highest in the FYM +100% NPK (73.0), and it was on par with FYM +50% NPK (72.9) > FYM (69.4) > NPK (63.3) > control (51.7) treatments (Fig. 6.3). In case of sorghum, the highest SOC stock (t ha^{-1}) of 68.5 was observed in the 25 kg N crop residue (CR) + 25 kg N (*Leucaena*) followed by that of 65.8 in the 25 kg N (CR) + 25 kg N (urea) > that in the 25 kg N (FYM) + 25 kg N (urea) (62.6) > 50 kg N (urea) (54.1) = 25 kg N (*Leucaena*) + 25 kg N (urea) (53.4) and the lowest (49.0) in the control (Srinivasarao et al. 2013b). However, maintenance of SOC is difficult in drylands as the extent of C loss is rapid due to high temperatures, and the fact that at least four times higher organic matter inputs are required in tropical regions than in temperate environments to maintain the SOC concentration. Hence, on-farm generation of organic matter with appropriate policy support needs to be promoted to maintain soil health and crop productivity (Srinivasarao et al. 2014b).

6.4.2 Drought Management

The risk involved in successful cultivation of crops depends on the nature of drought, its duration and frequency of occurrence within the season. These aberrations are expected to further increase in the coming years. Drought affects not only the food production at farm level but also the national economy and overall food security. Location-specific rainfed technologies are available to cope with different drought situations. Much of the research done in rainfed agriculture in India relates to conservation of soil and rainwater and to drought proofing. The key technologies for drought mitigation are in situ moisture conservation, rainwater harvesting and recycling, resilient crops and cropping systems including contingency crop plans, foliar sprays and integrated farming systems (Fig. 6.4) (Srinivasarao and Gopinath 2016).

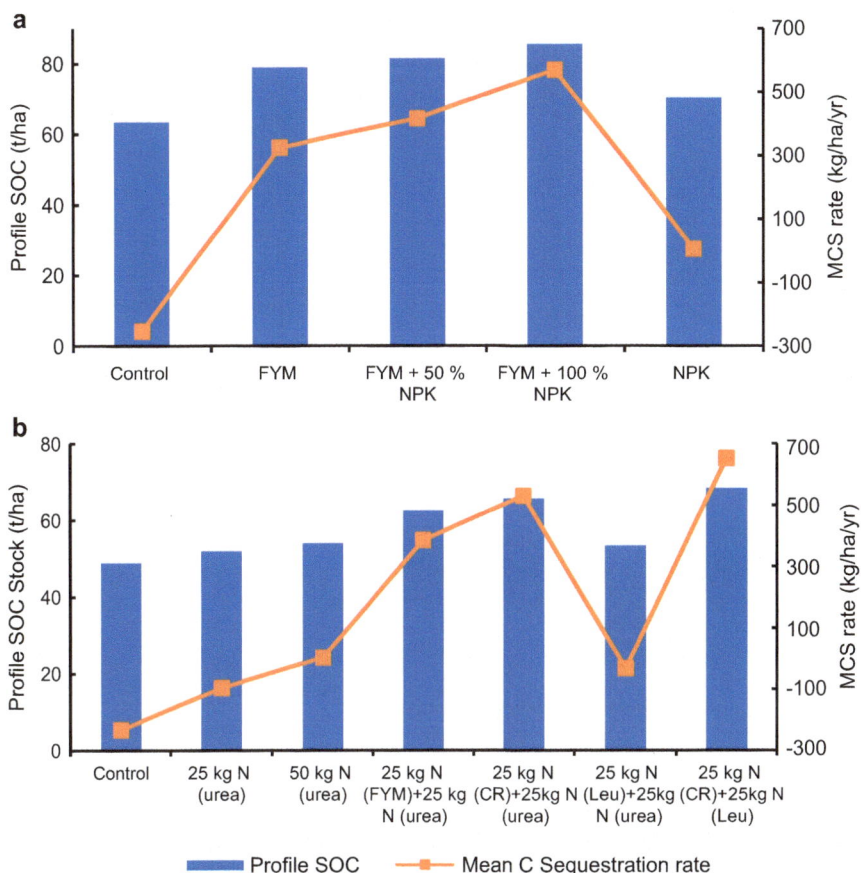

Fig. 6.3 Profile SOC and mean C sequestration rate as affected by (**a**) 13 years of groundnut-finger millet rotation (*left*) and (**b**) 22 years of sorghum cropping with differential manuring and fertilization under semiarid tropical conditions (Source: Srinivasarao et al. 2013b)

6.4.3 Site-specific Nutrient Management

Soils in most parts of India are deficient not only in NPK but also in secondary nutrients (S, Ca). Magnesium (Mg) deficiency is also prevalent in many rainfed areas. The data on soil analysis from farmers' fields in several districts of Andhra Pradesh, Karnataka, Tamil Nadu, Rajasthan, Madhya Pradesh and Gujarat states showed that almost all farms were low in SOC, low to moderate in available P, but generally sufficient in extractable K. However, there existed an extensive deficiency of sulphur (S), boron (B) and zinc (Zn) (Sahrawat et al. 2010). Rainfed crops suffer more from nutrient deficiency than from insufficient moisture, because of low rates of fertilizer use. Deficiencies of secondary nutrients vary greatly, mainly in soils under intensive cropping because of imbalanced fertilization resulting in negative nutrient budget or nutrient mining. Micronutrient deficiencies, particularly of Zn

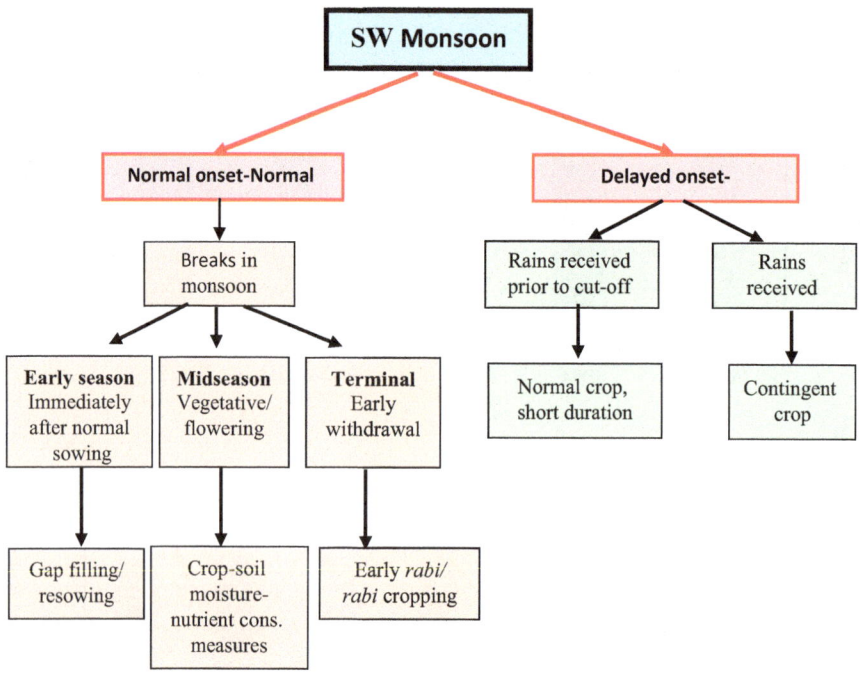

Fig. 6.4 Suggested drought contingency measures

and B, are among the emerging constraints to sustainable crop production in rainfed agriculture (Srinivasarao and Vittal 2007). Another study across diverse agroeco-logical regions highlighted the extent of Mg levels in major Indian soil types and recommended further attention on Mg nutrition in current intensive agriculture (Srinivasarao et al. 2015a) and micronutrients like B, Zn, Cu, Fe and Mn.

The correction of nutrient deficiencies can be achieved through site-specific nutrient management (SSNM). SSNM takes into account all nutrient deficiencies to ensure that crop demands are met, and soil fertility is improved, which in turn ensures higher nutrient use efficiency, crop productivity and economic returns (Dobermann 2004). The results of on-farm demonstrations across crops and soils in India showed that S application increased grain yield by 650 kg ha^{-1} (+24% over NPK) in cereals, 570 kg ha^{-1} (+32% over NPK) in oilseeds and 375 kg ha^{-1} (+20% over NPK) in pulses (Singh 2001).

In Andhra Pradesh and Telangana states of India, SSNM was demonstrated in eight districts to address nutrient deficiencies which exist within farmers' fields, with promising results (Table 6.1). Cotton yields increased in response to SSNM in Adilabad, Khammam and Warangal districts of Telangana by 17, 26 and 35%, respectively, compared with traditional farmer practices. Similarly, SSNM in groundnut increased the pod yields by 30% in Nalgonda district and 27% in Anantapur district, compared to farmers' practice. SSNM in castor, cowpea and maize improved crop yields by 51, 41 and 45%, respectively in Mahbubnagar, Kadapa and Ranga Reddy districts (Srinivasarao et al. 2010).

Table 6.1 Impacts of SSNM on yield of different crops

Name of the village & district	Crop	Yield (kg ha^{-1})		% yield increase
		SSNM	FP	
Seethagondi, Adilabad	Cotton	1151	1044	17
Dupahad, Nalgonda	Groundnut	1034	967	30
Thummalacheruvu, Khammam	Cotton	1162	1028	26
Zamistanpur, Mahbubnagar	Castor	634	536	51
Pampanur, Anantapur	Groundnut	1158	1101	27
B. Yerragudi, Kadapa	Cowpea	1101	965	41
Ibrahimpur, Ranga Reddy	Maize	2000	1862	45
Jafferguda, Warangal	Cotton	1260	1007	35

Source: Srinivasarao et al. (2010); *FP* Farmers' practice

6.4.4 Foliar Sprays

Nutrients not only help in better plant growth and development but also help to alleviate different kinds of abiotic stresses like drought. Leaf feeding is the use of foliar fertilizers to enhance the overall nutrient level in the plant and increase sugar production during times of stress. This form of foliar nutrition does not address any specific nutrient deficiency but supplies a small amount of all nutrients to keep leaf growth lush (Srinivasarao and Gopinath 2016). Several studies have indicated the beneficial effect of foliar sprays in different field crops (Table 6.2). The limited movement of nutrients from soil to plant root and shoot restrict the photosynthetic activities in leaves and pod filling. Under this situation, foliar application of 2% urea or DAP twice at flowering and pod filling stage can increase the seed yield of legume crops by up to 15% (Venkatesh and Basu 2011; Ali et al. 2012; Singh et al. 2014). Similarly in rice fallow pulses, foliar application of 2% DAP +0.5% ZnSO4 + 1% Fe twice (pre-flowering + flowering) had shown significant highest values of all yield attributes and yield in green gram and black gram (Ganapathy et al. 2008).

6.4.5 Rainwater Management

Out of the total annual precipitation (including snowfall) of around 4000 km^{3}in the country, the availability of surface water and replenishable groundwater is ~1869 km^3, comprising of 690 km^3 (37%) of surface water and 432 km^3 of groundwater (GOI 2002). Total annual national water use may exceed the utilizable water resource by 2050 or 2060, unless significant changes occur through increased water storage and efficient water management. Effective rainwater management is critical for drought mitigation and successful rainfed agriculture. It may not be possible to conserve all the rainwater in situ, in spite of adopting different soil and water conservation measures. The soil topographic features and climatic factors prevailing in most rainfed areas are highly conducive to generation of runoff. This inevitable runoff may be collected in small and medium reservoirs that can be utilized for providing supplemental/life-saving irrigation to the crop at critical growth stages.

Table 6.2 Effect of foliar spray of different chemicals on crops performance

Crop	Foliar spray	Performance
Maize	VAM-C	Yield increased by 16, 24, 33 and 15% in black
	50% SL @ 3.75 1 ha^{-1}	gram, maize, soybean and horse gram, respectively, compared to control (no foliar spray)
Cotton	$MgSO_4$ and $ZnSO_4$	Seed cotton yield increased by 300 kg ha^{-1} over
	Two sprays	farmers' practice
Toria	KNO_3 at 2% before flowering	Higher seed yield (738 kg ha^{-1}) with net return of Rs. 10171/ha
Wheat	Thiourea (1000 ppm)	Higher grain yield (2716 kg ha^{-1}) with net returns of INR 31566 ha^{-1}
	Seed soaking, foliar spray at maximum tillering and booting stage	
Maize	$ZnSO_4$ @ 0.5% + $FeSO_4$ @ 0.5%	Higher seed yield (1137 kg ha^{-1}) with net return of INR13833 ha^{-1}

Source: Srinivasarao and Gopinath (2016)

The importance of rainwater harvesting has increased in recent years due to the increased rainfall variability, depletion of groundwater levels and rising temperatures. Rainwater harvesting not only reduces runoff and soil loss but also facilitates groundwater recharge and prevents early sedimentation of the reservoirs.

Extensive studies on rainwater harvesting in dugout ponds have been undertaken both in alfisol and vertisol regions. These studies have shown that appropriate pond size varies with rainfall ranging from 200 m^3 to 3000 m^3. Even small ponds can be dug and each plastered as a cistern (50 m^3). For example, in alfisols of scarcity zone of Andhra Pradesh, a farm pond size of 10 m × 10 m with 2.5 m depth, side slopes of 1.5:1 with a storage capacity of 250 m^3 lined with soil and cement mixture (6:1 ratio) was sufficient for a catchment area of 5 ha (Srinivasarao et al. 2014a). Several studies conducted in different parts of the country have revealed the benefits accruing from supplemental irrigation to crops from stored rainwater during prolonged periods of dry spells (Table 6.3). The results of supplemental irrigation in medium-deep black soils at Bijapur indicated that seed yields with one life-saving irrigation could be enhanced by 32.5% in chickpea and 92.4% in pigeon pea (Guled et al. 2003). In another experiment at Solapur, one protective irrigation at flowering of chickpea enhanced the yield by 39.1% (Bangar et al. 2003). Similarly, trials conducted at farmers' fields in Vidarbha region of Maharashtra showed that pigeon pea yield increased by 66% with one supplemental irrigation. In chickpea, the yield increased by 167% with two supplemental irrigations (Taley 2012). At Rewa, pre-sowing irrigation to chickpea produced about 50% higher yield than rainfed crop (AICRPDA 2010).

Recycling of harvested water from the *Nadis* (small water harvesting structures) ensures double cropping (maize followed by chickpea) resulting in higher productivity of land and water. About 5.5 to 7.0 Mg ha^{-1} of chickpea green pods can be harvested with a net income of Rs. 25,000 ha^{-1} (Venkateswarlu et al. 2009). The post-monsoon crops suffer from the progressive increase of stress due to receding

Table 6.3 Response of pulses to supplemental irrigation

Location	Climate (MARF* mm)	Dominant soil type	Crop	Yield (kg ha^{-1}) Irrigated	Control	Yield increase (%)	Sources
Akola, Maharashtra	Semiarid hot moist (824)	Vertisols	Pigeon pea	1000	600	67	Taley (2012)
			Chickpea	1000	375	167	
Rewa, Madhya Pradesh	Subhumid hot dry (1088)	Vertisols	Chickpea	1905	1270	50	AICRPDA (2010)
Agra, Uttar Pradesh	Semiarid hot dry (665)	Inceptisols	Lentil	1353	1119	21	AICRPDA (2011)
Parbhani, Maharashtra	Semiarid hot moist (901)	Vertisols	Pigeon pea	748	435	72	AICRPDA-NICRA (2016)

*Mean annual rainfall

Table 6.4 Effect of supplemental irrigation on yield winter pulses

Treatment	Yield (Mg ha^{-1}) Chickpea	Lentil	Field pea	Rajmash
Rainfed	1.69	1.35	0.85	0.25
One irrigation	2.48	1.45	1.60	0.66
Two irrigations	2.65	1.40	1.72	1.13

Source: Pramanik (2009)

soil moisture status. Under these conditions, supplemental irrigation helps in overcoming moisture stress and achieves better crop yields (Table 6.4; Pramanik 2009).

In situ soil and water conservation (SWC) practices improve soil structure and soil porosity, increase infiltration and hydraulic conductivity and consequently increase soil water storage that helps crops to withstand moisture stress. These measures are more feasible and practical proposition under most situations and can be adopted by individual farmers with less draft and amenable even for a small holder. The suitability of a practice depends on the topography of the field, temporal and spatial distribution of rainfall, type of soil, crop, etc. Based on extensive research conducted at various locations in the country, several in situ moisture conservations practices have been recommended (Table 6.5) and some of these practices are being adopted in significant areas in the country (Srinivasarao and Gopinath 2016).

6.4.6   Watershed Management

Watershed management is a holistic approach towards optimizing the use of land, water and vegetation to alleviate drought, reduce floods, prevent soil erosion and improve water availability and increase fuel, fodder, fibre and agricultural production on a sustained basis. The watershed programme which is primarily a

Table 6.5 Location specific in-situ moisture conservations practices

Practice	Crops/cropping system	Remarks
Compartmental bunding	*Rabi* sorghum sunflower, safflower, chickpea, maize, pearl millet, cotton	The impact of the practice is more during suboptimal rainfall years. It also controls runoff. This practice is widely adopted in Bijapur, Bagalkot and Raichur districts of Northern Karnataka
Conservation furrow	Finger millet + pigeon pea (8:2), groundnut + pigeon pea (8:2), soybean + pigeon pea (4:2), cotton + soybean (1:1)	It enhances in situ moisture conservation; thus, the crops can overcome the effect of dry spells during vegetative and or reproductive stages of crops resulting in increased rainwater use efficiency, better performance of crops and additional net returns
Broad bed and furrow (BBF)	Soybean, groundnut	BBF system helps drain out excess water from black soils. Further, the rainwater conserved in the furrows helps in better performance of crop during long dry spells
Ridges and furrows	*Rabi* sorghum, pigeon pea + rice	The practice conserves 30–45% more moisture than farmers' practice, retains it for longer period (up to 60 days) and increases crop yield
Ridge planting	Pearl millet	Provides enough aeration and porosity to soil for enhanced root growth, safe disposal of excess rainwater and reduction in soil loss apart from moisture conservation during low rainfall period
Set furrow	Pearl millet-sunflower cropping, pigeon pea + groundnut (2:4)	Conserves more moisture and make it available for longer time to the crops. This helps to overcome the effect of drought
Inter-plot rainwater harvesting	Sunflower, sorghum, chickpea	This practice makes possible to take up two crops even in drought years mainly by allowing rainfall infiltration into the soil profile

Source: Adapted from Srinivasarao and Gopinath (2016)

land-based area development programme has contributed significantly towards conservation of soil and water. Creation of soil and water conservation structures and additional storage volume helped to enhance infiltration and contributed to the groundwater recharge at several locations of the country (Table 6.6; Samra 2004).

A meta-analysis of 311 watershed programmes revealed that watershed management resulted in better returns, equity and employment generation in the rainfall region of 700–1100 mm (Joshi et al., 2005). Based on the economic efficiency parameters, Joshi et al. (2005) estimated that the performance of micro-watersheds with an area up to 1250 ha was 42% less than that of large size (>1250 ha) watersheds. Livestock-pasture-based land use was found ideal for regions receiving rainfall up to 700 mm and crop horticulture for regions between 700–1100 mm of rainfall. The investment in watershed programme by the Government of India has contributed to realize a benefit-cost ratio of 2.14 (Table 6.7) indicating that the

Table 6.6 Effect of watershed management strategies on ground water recharge in different regions

Watershed	Surface storage capacity created (ha-m)	Observed rise in groundwater table (m)*
Chhajawa (Rajasthan)	20.0	2.0
Chinnatekur (Andhra Pradesh)	5.6	0.8
GR Halli (Karnataka)	4.0	1.5
Joladarasi (Karnataka)	4.0	0.2

Source: Samra (2004)

Table 6.7 Summary of benefits from the sample watershed studies

Indicator	Particulars[a]	Unit	No. of studies	Value
Efficiency	B/C ratio	Ratio	128	21.2
	IRR	%	40	6.5
Equity	Employment	Person (days ha^{-1} year^{-1})	39	6.7
Sustainability	Irrigated area	%	97	11.7
	Rate of runoff	%	36	6.8
	Soil loss	t ha^{-1} year^{-1}	51	39.3
	Cropping intensity	%	115	12.6

Source: Joshi et al. (2005), [a]*BC* benefit-cost, *IRR* internal rate of return

investment has yielded more than double the investment. Watershed programmes performed well with a mean B-C ratio of 2 which indicates that investment on watershed programmes is economically viable and substantially beneficial (Joshi et al. 2005). Collective action and community participation in watershed activities can be enhanced by ensuring tangible economic benefits for all the stakeholders through productivity enhancement, diversification of crops with high-value crops, income generating activities and by adopting holistic approach through convergence of activities and actors.

6.4.7 Biomass Energy and Waste Recycling

Crop residues are a precious resource, and effective management of residues, roots, stubbles and weed biomass can increase soil fertility through addition of SOM, recycling of plant nutrients and improvement of soil structure. Pathak (2006) estimated total available crop residues in India as 523.4 million tons/year and surplus as 127.3 million tons/year; residue generated and burnt along with its surplus is listed in Table 6.8. Crop residues of important rainfed crops such as cotton, castor, pigeon pea, maize, sunflower, etc. are mostly burnt on the field. For example, a survey of 100 households in Nalgonda district of Andhra Pradesh (Nandyalagudem) showed that about 1000 tons of cotton and 400 tons of pigeon pea residues are burned annually (Fig. 6.5). While many farmers realize the importance of crop

Table 6.8 Generation and surplus of crop residues (million tons year^{-1}) in India

State	Residue generated MNRE (2009)	Residue surplus	Residue burned IPCC Coefficient	Pathak et al. (2010)
Andhra Pradesh	43.89	6.96	5.73	2.73
Bihar	25.29	5.08	3.77	3.19
Chhattisgarh	11.25	2.12	1.84	0.83
Gujarat	28.73	8.9	6.69	3.81
Jharkhand	3.61	0.89	1.11	1.10
Karnataka	33.94	8.98	2.85	5.66
Madhya Pradesh	33.18	10.22	3.46	1.91
Maharashtra	46.45	14.67	6.27	7.41
Odisha	20.07	3.68	2.57	1.34
Rajasthan	29.32	8.52	3.58	1.78
Tamil Nadu	19.93	7.05	3.55	4.08
Uttarakhand	2.86	0.63	13.34	21.92
Uttar Pradesh	59.97	13.53	0.58	0.78
West Bengal	35.93	4.29	10.82	4.96
Arunachal Pradesh	0.4	0.07	0.06	0.04
Assam	11.43	2.34	1.42	0.73
Goa	0.57	0.14	0.08	0.04
Haryana	27.83	11.22	5.45	9.06
Himachal Pradesh	2.85	1.03	0.20	0.41
Jammu and Kashmir	1.59	0.28	0.35	0.89
Kerala	9.74	5.07	0.40	0.22
Manipur	0.9	0.11	0.14	0.07
Meghalaya	0.51	0.09	0.10	0.05
Mizoram	0.06	0.01	0.01	0.01
Nagaland	0.49	0.09	0.11	0.08
Punjab	50.75	24.83	8.94	19.62
Sikkim	0.15	0.02	0.01	0.01
Tripura	0.04	0.02	0.22	0.11
Rainfed (%)	78.60	67.82	79.07	66.26
India (Total)	501.76	140.84	83.66	92.81

residues and its recycling, decomposition of woody residues of cotton and pigeon pea is a major constraint. Thus, CRIDA has introduced a shredder to facilitate recycling of woody biomass and introduced rotavator for incorporating the crop residues in the soil (Fig. 6.6). In addition, labour availability and high cost have become constraints in adopting this technology.

Incorporation of crop residues has several positive impacts on physical, chemical and biological properties of soil through either incorporation or retention on the surface. Crop residues can also be composted and used as farmyard manure (FYM) and supplement to fertilizer. It increases hydraulic conductivity and reduces bulk density of soil by modifying soil structure and aggregate stability. Mulching with

Fig. 6.5 Burning of crop residues

Fig. 6.6 Crop residue recycling into soil through rotavator

plant residues raises the minimum soil temperature in winter due to reduction in upward heat flux from soil and decreases soil temperature during summer through the shading effect. Retention of crop residues on soil surface reduces runoff, decreases surface crust formation and enhances water infiltration. Thus, crop residue cover or surface mulching has tremendous positive impact on micro-site improvement under adverse environments which is a common feature of rainfed agriculture. These crop residues also contribute to improved soil organic matter, higher water retention, profile soil moisture storage and reduced evaporation losses, so that crops are protected during mid-season droughts.

6.5   Crop Improvement Strategies

Production of transgenic plant or crop expressing gene(s) which are functionally active in eliciting a stress countering action in plants in addition to other agronomic desirable traits as detailed in the above discussion is the goal of stress biotechnology in plants. Putting it in simple terms, the steps involved are identification of candidate genes, cloning them and incorporating them in varieties which already process high-quality agronomic traits. Although this sounds fairly simple, the complexity of each step is enormous. Combining the best genes in one plant is a long and difficult process. Identifying the candidate gene which in this case is genes with stress countering functionality is a far more difficult task than to actually transfer them into plants (Fig. 6.7).

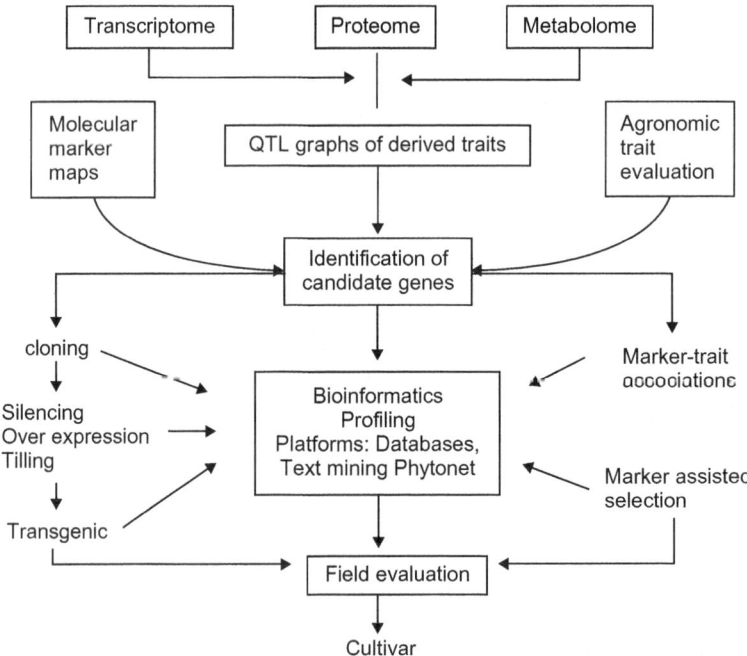

Fig. 6.7 Flow chart depicting sequence of events in the process of developing stress tolerant cultivar

6.5.1   Importance of Plant Traits

Plant traits which favour yield and also which have direct effect on tolerance mechanism are one of the important characteristics that need to be considered when breeding for tolerance to abiotic stresses by both conventional and molecular methods. The study of innate plant traits has yielded important autecological information on the differentiation of plants into so-called plant functional types (PFTs). Approaches in plant trait analysis can be used to assess the adaptation of plants to specific niches in their natural environment and the prevailing adaptation strategies. Plant traits of single species or those typical for larger entities play an important role as they largely determine how responsive or susceptible a plant or vegetation will be when exposed to a stress.

Considering that any improvement in stress tolerance must be a result of an underlying physiological change, it is surprising that direct selection for a physiological trait has not contributed more to yield progress in most food crops. Physiological changes are interpreted here in the broadest sense as any change to the growth, development, morphology, anatomy or physiology of a crop. Nevertheless, physiological changes such as flowering time and plant height have been important for yield progress, and molecular genetics regularly select for desirable expression of these traits to maintain adaptation and optimal yield. The traits associated with stress tolerance and yield increase will not be usually the same in most crops; hence, the molecular breeder is presented with a paradox of choices to select from, resulting in forced exclusion of certain traits which are mutually exclusive (Bohnert et al. 2006). Desirable traits are highly specific to the nature of stress experienced by the plant; the following is a brief discussion of general traits which stand apart from the stress-specific traits that one should keep in mind when breeding for multiple stress-tolerant crops.

6.5.2   Photosynthesis and Yield

Photosynthesis (PS) is the cornerstone of physiological process and the basis of dry matter production in plants. Photosynthetic rate is an important parameter characterizing the photosynthetic capacity of the photosynthetic apparatus. It also reflects the efficiency because it is a determinant of crop yield and light use efficiency. Photosynthesis is regulated through the control of leaf area and leaf senescence and the daily duration and extent of stomatal opening. These provide a crop with substantial flexibility. When the photosynthesis-yield relation is analysed, the partitioning of photosynthates to different organs, expressed as a partitioning coefficient or harvest index which is defined as the ratio of yield biomass to the total cumulative biomass at harvest, must be considered. The size of respiration loss is closely related to crop yield. Selection for low leaf respiration has led to yield increases in various crops (Barber 1998). For most crops, more than half of the economic yield is derived from photosynthesis after flowering. Therefore, photosynthesis at the reproductive stage is more directly related to yield size. The positive correlation between leaf photosynthesis and yield is observed mostly at this stage.

6.5.3   Water Use Efficiency and Transpiration Efficiency

Water use efficiency is defined as the dry matter production of a crop per unit of evapotranspiration (ET). Water relations and stomatal behaviour are important indices that reflect the ability of plants to economize essential requirements under prevailing climatic and edaphic conditions. In plants adapted to dry environments, anatomical and morphological changes at the leaf and whole plant levels prevent metabolic imbalance and help to improve water relations. Hence, it is paramount to improve water use efficiency (WUE) and tissue water status and a potentially increased growth and/or yield with no additional penalty in water consumption in order to obtain a high-yielding stress-resistant variety (Beale et al. 1999). Transpiration efficiency (TE) is a function of both environmental and plant attributes related to resistances to CO_2 fixation by leaves. Under some circumstances, the environment can have a significant influence on TE.

6.5.4   Osmotic Adjustment

Osmotic adjustment is a biochemical mechanism that helps plants acclimate to dry and saline conditions. Maintenance of cell turgor and plant growth requires sufficient increase in sap osmolality to compensate for external osmotic stress, the process of osmotic adjustment. Many drought-tolerant plants can regulate their solute potentials to compensate for transient or extended periods of water stress by making osmotic adjustments, which results in a net increase in the number of solute particles present in the plant cell (Buitink 2006). Osmotic adjustment occurs when the concentrations of solutes within a plant cell increases to maintain a positive turgor pressure within the cell. The cell actively accumulates solutes and as a result the solute potential drops, promoting the flow of water into the cell. Osmotic adjustment reduces the impact of stress on the growth and yield of crops.

6.5.5   Tools and Strategies

There has been a lack of systems approach in developing strategies for evolving plants tolerant to multiple stressed environments. Mechanisms accounting for genotypic differences in stress adaptation within a crop species are not thoroughly understood, and their genetic basis is still lesser. A number of factors have contributed to this lack of understanding. Firstly, much of the basic research in the area of plant stress focuses either on comparisons between species or the response of a single genotype to various stress treatments. Secondly, the predominance of an upstream focus in plant stress research has led to a greater emphasis on traits associated with survival under extreme stress than those associated with agronomic productivity under resource-limited conditions. Finally, a crop that experiences moisture deficit may simultaneously experience a number of additional abiotic and biotic stress factors that exacerbate drought stress (Feder and Walser 2005). This calls for a systems stress approach (SSA) which should form the basis of strategies in stress physiology and biochemistry dealing with several combinations of stresses. This area is a rapidly growing area in biology, due to progress in several fields. The most critical factor has been rapid progress in molecular biology, involving characterizing DNA sequence, gene expression profiles, protein-protein interactions, etc.

With the ever-increasing flow of biological data, serious attempts to understand stress systems as integrated systems is one of the important ways to generate multiple stress-tolerant genotypes. While many stress responses appear to be specific to that form of stress, some responses are general and confer tolerance to multiple stresses. The genes associated with these general stress responses may provide insight into biochemical networks underlying stress resistance and may prove as targets for stress resistance engineering in plants (Chen and Zhu 2004).

Molecular studies focused on the identification of genes that are activated by different forms of stresses and might thus contribute to tolerance. Recent progress in this area indicates that stress tolerance is a complex quantitative trait and no real diagnostic marker has been reported yet. In spite of the characterization of many stress-responsive genes, the functions of few genes have been established. These facts make it difficult to design a 'transgenic approach' to improve multiple stress tolerance. Research on biotechnological approaches for tolerance to abiotic stresses began within a decade of the molecular understanding of pathways induced in response to one or more of the abiotic stresses. In most of the cases, the transgenes expressed faithfully but only a limited level of tolerance was provided under stress conditions as compared to control plants (Denby and Gehring 2005).

6.6   Contingency Plans

Contingency crop planning is essentially aimed at stabilization of crop output in the situation of various weather aberrations including late onset of monsoon, mid-season and terminal droughts, cyclones, floods, heat wave, cold wave, etc. ICAR-CRIDA has developed district level agricultural contingency plans for 620 districts so far in collaboration with respective state agricultural universities, AICRPDA and All India Coordinated Research Project on Agrometeorology (AICRPAM). These plans are intended to benefit district authorities including line departments to overcome the adverse effect of weather aberrations particularly drought. The suggested drought contingency measures include change in crop/cropping system, crop management, soil nutrient and moisture conservation measures.

In view of the frequent weather aberrations impacting agricultural production round the year in some part of the country, the need was felt to implement contingency measures on real-time basis to minimize the losses in agriculture and allied sectors and to improve the efficiency of the production systems. Any contingency measure, either technology related (land, soil, water, crop) or institutional and policy based, which is implemented based on real- time weather pattern (including extreme events) in any crop growing season, is considered as real-time contingency planning (RTCP). If done timely and effectively, RTCP contributes to household and village food and fodder security (Srinivasarao et al. 2013a). Some of the methods/measures to be adopted as real-time contingency plan implementation during various weather aberrations are presented in Table 6.9.

The real-time contingency measures aim:

1. To establish a crop with optimum plant population during the delayed onset of monsoon
2. To ensure better performance of crops during seasonal drought (early/mid- and terminal drought) and extreme events, enhance performance, improve productivity and income
3. To minimize damage to horticultural crops/produce
4. To minimize physical damage to livestock, poultry and fisheries sector and ensure better performance
5. To ensure food security at village level
6. To enhance the adaptive capacity and livelihoods of the farmers

For example, during *kharif* season 2013, ridge/furrow or broad bed and furrow method implemented fields maintained almost normal soybean yields, while under farmers' practice (flatbed sowing), crop damage was almost 80–90% due to excessive rains which lead to water stagnation and poor drainage in deep black soil regions of Malwa, Madhya Pradesh. In Southern Karnataka, delay in onset of monsoon was managed with cultivation of medium-duration finger millet variety (GPU 28) with intercropping of pigeon pea and soil moisture conservation practice of conservation furrow. Mid-season droughts have been managed with foliar sprays, reduction in plant population and critical irrigation with harvested water (Srinivasarao et al. 2013a). Overall, the aberrations or extreme events can be managed with components of preparedness and real-time implementation of agriculture contingencies as observed in pilot models in about 34 villages across the country (Srinivasarao et al. 2015c).

6.7   Policies Related to Stress Management

The aspects that are to be addressed are increase in concessional credit to small and marginal farmers, for adoption of climate-resilient agricultural practices; evolution of a new verification system at field level to sensitize farmers through credit and input subsidies; establishment of efficient cooperatives and associations/groups to tackle critical needs of farmers like resource mobilization, custom hiring, marketing their outputs and efficient natural resource management; and evolution of a national policy on disaster management in agriculture since events like cyclones and floods devastate agriculture, horticulture and livestock. Much of the corporate social responsibilities' funds from leading private sector players in the country should be channelized to promote CRA in vulnerable regions. Channelizing investment in human resource development and capacity building through awareness and training among officials, extension workers and farmers is a kep policy initiative. Farmers should be given incentives for adoption of natural resource conservation practices

Table 6.9 RTCP measures for various climatic aberrations in arable crops

Climatic aberration	RTCP measures
Delayed onset of monsoon	Beyond the sowing window, choice of alternate crops or cultivars depends on the farming situation, soil, rainfall and cropping pattern in the location and extent of delay in the onset of monsoon
Early season drought	Re-sowing within a week to 10 days with subsequent rains for better plant stand when germination is less than 30%
	If the plant population is 50 to 75% of the optimum population, sow the same crop of shorter duration variety in the gaps either within or in between rows. If less than 50% optimum population, sow suitable contingent crop
	Thinning in small-seeded crops
	Interculture to break soil crust, remove weeds and create soil mulch
	Avoid top dressing of fertilizers till favourable soil moisture
	Opening conservation furrows at 10 to 15 m interval
	Pot watering along with gap filling when the crop stand is less than 75% in commercial crops like cotton
Mid-season drought	Blade harrowing in rows in field crops during dry spell helps in creating dust mulch and closing of cracks in black soils. After relief of dry spell, open conservation furrows at 1.2 m distance and spray with 2% urea particularly in pulse crops, castor (*Ricinus communis*) and sunflower (*Helianthus indica*). Apply additional 10 kg N/ha
	Supplemental/protective irrigation, if available
	Repeated interculture to remove weeds and create soil mulch
	Avoid top dressing of fertilizers until receipt of rains
	Surface mulching with crop residues
Terminal drought	Providing life-saving or supplemental irrigation, if available. Harvesting crop at physiological maturity with some realizable yield or harvest for fodder
	Prepare for winter (*rabi*) sowing in double-cropped areas
	Ratoon maize (*Zea mays*) or pearl millet (*Pennisetum glaucum*) or adopt relay crops as chickpea (*Cicer arietinum*), safflower (*Carthamus tinctorius*), *rabi* sorghum (*Sorghum bicolor*) and sunflower with minimum tillage after soybean (*Glycine max*) in medium to deep black soils in Maharashtra
Unseasonal heavy rainfall events	Re-sowing
	Providing surface drainage
	Application of hormones/nutrient sprays to prevent flower drop or promote quick flowering/fruiting and plant protection measures against pest/disease outbreaks with need based prophylactic/curative interventions
	At crop maturity stage, harvesting of produce and proper drying prevent seed germination and if untimely rains occur at vegetative stage, the contingency measures include:
	Draining out the excess water as early as possible
	Top dressing with fertilizers, after draining out excess water
	Gap filling either with available nursery or by splitting the tillers from the surviving hills in crops like rice (*Oryza sativa*)
	Ensure proper weed control, after draining out excess water
	Suitable plant protection measures in anticipation of pest and disease outbreaks
	Foliar spray with 1% KNO_3 or water soluble fertilizers like 19–19-19, 20–20-20, 21–21-21 at 1% to support nutrition
	Interculture at optimum soil moisture condition to loosen and aerate the soil and to control weeds
	Earthing up the crop for anchorage, etc.

(continued)

Table 6.9 (continued)

Climatic aberration	RTCP measures
Floods	In sand deposited crop fields/fallows, ameliorative measures include early removal or ploughing in of sand (depending on the extent of deposit) for facilitating sowing of succeeding crop
	Draining out of stagnant water and strengthening of field bunds, etc.
	Alternate crop plans for receding situations
	Community paddy nursery raising
	Grow flood tolerant paddy varieties like Swarna Sub-1
	Re-transplanting in damaged fields and transplanting new areas or direct seeding depending on seed availability
	Prevention of premature germination of submerged crop at maturity or of harvested produce by spray of salt solution
Heat wave	Light frequent irrigation
	Foliar spray with thiourea or KNO_3 at recommended dose
	Cultivation of short-duration varieties
	Early sowing of crops by adopting zero-till cultivation methods
Cold wave	Light frequent irrigation during evening hours
	Basin mulching
	Application of supplemental fertilizer dose
Frost	Change in planting time to avoid sensitive stages coinciding with frost period
	Thatching young plants
	Mulching ground cover to prevent loss of heat
Hailstorm	Use of anti-hail guns and anti-hail nets
	Spray fungicides to control secondary fungal infection
	Open trenches to drain out excess water from the field
	If crop is at maturity/harvesting stage, harvest and air-dry the produce
Cyclones	Providing field drainage, staking and propping of plantation crops
	Cleaning and drying of harvested field crops

Source: Adapted from Srinivasarao et al. (2016b)

and support to improve the existing indigenous technologies that are eco-friendly and sustainable in long-run. Encourage the role of non-governmental organizations, civil societies and public and philanthropic organizations to enhance adaptation preparedness among the local community. Forge international/regional partnerships to develop tools and technologies adaptable to suit local requirement through pooling financial and intellectual resources.

Introduce soil and water conservation practices on a large-scale to bring long-term sustainability of the systems. Farmers must be given incentives to adopt such practices which help to cope with abiotic stresses and contribute to climate-resilient agriculture. Building on priority infrastructure like markets and information gateways and creation of opportunities in the non-farm sector in vulnerable regions should help the farmers to diversify their incomes. Implementation of various measures to cope with abiotic stresses at field level needs well-structured institutional support for farmers with strong government policy and convergence among various institutions. The Ministry of Agriculture, Government of India, needs to facilitate the convergence process of various government schemes such as Mahatma Gandhi National Rural Employment Guarantee Act (MGNREGA), Mega Seed Project, National Food Security Mission (NFSM), National Horticulture Mission (NHM), Integrated Watershed Management Programme (IWMP), Soil health schemes, etc. for drought preparedness. The National Mission for Sustainable Agriculture (NMSA), one of the missions under the Prime Minister National Action Plan for Climate Change (NAPCC), may take a lead role in the implementation of adaptation measures to cope with abiotic stresses, by inclusion of this activity in state action plans (SAP) with a dedicated nodal institution /officers and budget provision (Srinivasarao and Gopinath 2016).

6.8   Conclusions and Future Prospects

To be able to feed the world in the coming 50–100 years, concerted efforts are required involving sound agricultural policies, well-planned research strategies and efficient delivery systems. Water is a critical natural resource and managing rainwater in situ or harvesting runoff water and its recycling is key to mitigate effects of moisture stress (drought) on crops. Location-specific needs of soil and water conservation measures *vis-a-vis* changing rainfall scenario will address water issues much better. Public-private partnerships (PPP) that create market links have proved successful in several watersheds sites and created win-win situations for all stakeholders involved, particularly in India. Therefore, it is necessary to formulate a coherent set of guidelines to enable governments and consortium partners to approach the private sector and begin fruitful collaborations in PPPs. The main emphasis in rainfed farming systems is to build the soil organic matter (SOM) for soil health restoration. Field burning of crop residues must be stopped, and constant efforts are needed to move these surplus residues into soil system either as manure or surface cover. There is an essential need for weather-based agro-advisory services (AAS) in farming activities for access to real-time weather information, timely agricultural operations, improved crop yields, reduced cost of cultivation, need-based changes in cropping patterns and finally improved livelihoods. There is an urgent need to revamp the weather/crop-based insurance programs. The risk in weather insurance is not only inherent in the location of reference weather station, but it is also a function of the design of the Weather Based Crop Insurance Scheme (WBCIS) product.

Evolving crop traits tolerant to multiple abiotic stresses is still in its infancy. It would be desirable that future work exploit the synergies of interfacing physiological and molecular/genetic research.

An integrated systems approach is essential in the study of complex quantitative traits which govern tolerance to multiple abiotic stresses. It is evident from the current work that a much clearer picture of abiotic stress signal transduction pathways is likely to emerge and more examples of genetic improvement for multiple tolerances by fine-tuning plant sensing and signalling systems are also likey to emerge which will help us understand the mechanism of abiotic stress tolerance in plants. The research must use the latest genomics resources combining new technologies in quantitative genetics, genomics and biomathematics with an eco-physiological understanding of the interactions between crop/plant genotypes and the growing environment. Most current research programmes lack this interdisciplinary approach. A comprehensive understanding of abiotic stress in crops across the full range of scenarios and regions would go a long way in preparing the nations for climate change. A multipronged strategy of using indigenous coping mechanisms, wider adoption of the existing technologies and/or concerted R&D efforts for evolving new technologies are needed for adaptation and mitigation. Policy incentives will play crucial role in adoption of climate-ready technologies in all sectors. The state agricultural universities and regional research centres will have to play major role in adaptation research which is more region and location specific, while national level efforts are required to come up with cost-effective mitigation options, new policy initiatives and global cooperation.

References

AICRPDA (2010) Annual report 2009–10, All India Coordinated Research Project for Dryland Agriculture, CRIDA, Hyderabad, 428 p
AICRPDA (2011) Annual report 2009–10, All India Coordinated Research Project for Dryland Agriculture, CRIDA, Hyderabad, 420 p
AICRPDA-NICRA (2016) Managing weather aberrations through real time contingency planning AICRPDA-NICRA Annual Report, 2015–16, All India Coordinated Research Project for Dryland Agriculture, ICAR-Central Research Institute for Dryland Agriculture, India, 226 p
Ali M, Kumar N, Ghosh PK (2012) Milestones on agronomic research in pulses in India. Indian J Agron 57:52–57
Arnholdt-schmidt (2005) Functional markers and a 'systemic strategy': Convergence between plant breeding, plant nutrition and molecular biology. Plant Physiol Biochem 43:817–820
Babu RC, Nguyen BD, Chamarerk V, Shanmugasundaram P, Chezhian P, Jeyaprakash P, Ganesh SK, Palchamy A, Sadasivam S, Sarkarung S, Wade LJ, Nguyen HT (2003) Genetic analysis of drought resistance in rice by molecular markers. Crop Sci 43:1457–1469
Bangar AR, Deshpande AN, Sthool VA, Bhanavase DB (2003) Farm pond – a boon to agriculture. Research Bulletin 403, Zonal Agricultural Research Station, Solapur, 37 p
Bapuji Rao B, SanthibhushanChowdary P, Sandeep VM, Pramod VP, Rao VUM (2015) Spatial analysis of the sensitivity of wheat yields to temperature in India. Agric Forest Meteorol 200:192–202
Barber J (1998) What limits the efficiency of photosynthesis and can there be beneficial improvements? In: Waterlow JC, Armstrong DG, Fowden L, Riley R (eds) Feeding a World population of more than eight billion people – a challenge to science. Oxford University Press, Cary, pp 112–123
Beale CV, Morison JIL, Long SP (1999) Water use efficiency of C-4 perennial grasses in a temperate climate. Agric Forest Meteoro l96:103–115
Bennett DC, Freeling M (1987) Flooding and the anaerobic stress response. In: Newmann DW, Wilson KG (eds) Models in plant physiology and biochemistry, vol III. CRC Press, Boca Raton, pp 79–82
Bohnert HJ, Qingqiu G, Pinghua I, Ma S (2006) Unraveling abiotic stress tolerance mechanisms – getting genomics going. Curr Opin Plant Biol 9:180–188
Boru G, Van Ginkel M, Krondtad WE, Boersma L (2001) Expression and inheritance of tolerance to waterlogging stress in wheat. Euphytica 117:91–98

Buitink J (2006) Transcriptome profiling uncovers metabolic and regulatory processes occurring during the transition from desiccation sensitive to desiccation-tolerant stages in *Medicago truncatula* seeds. Plant J 47:735–750

Ceccarelli S (1996) Adaptation to low/high input cultivation. Euphytica 92:203–214

Chen WQJ, Zhu T (2004) Networks of transcription factors with roles in environmental stress response. Trends Plant Sci 9:591–596

Denby K, Gehring C (2005) Engineering drought and salinity tolerance in plants: lessons from genome-wide expression profiling in Arabidopsis. Trends Biotechnol 23:547–552

Dobermann A (2004) Increasing productivity of intensive rice systems through site-specific nutrient management. Science Publishers and IRRI, Enfield, 410 p

Dunwell JM (2000) Transgenic approaches to crop improvement. J Exp Bot 51:487–496

Feder ME, Walser JC (2005) The biological limitations of transcriptomics in elucidating stress and stress responses. J Evol Biol 18:901–910

Firbank LG (2005) Striking a new balance between agricultural productivity and biodiversity. Ann Appl Biol 146:163–175

Flowers TJ, Yeo AR (1995) Breeding for salinity resistance in crop plants: where next? Aust J Plant Physiol 22:875–884

Ganapathy M, Baradhan G, Ramesh N (2008) Effect of foliar nutrition on reproductive efficiency and grain yield of rice fallow pulses. Legum Res 31(2):142–144

GOI (2002) National water policy, Ministry of water resources, Government of India, 10 p

Guled MB, Lingappa S, Itnal CJ, Shirahatti MS, Yaranal RS (2003) Resource conservation and management in rainfed agro-ecosystem- A compendium for research and development. Technical Bulletin 34, UAS, Dharwad, 268 p

Hall AE (2000) Crop responses to environment. CRC Press LLC, Boca Raton

Jonak C, Kiegerl S, Ligterink W, Barker PJ, Huskisson NS, Hirt H (2006) Stress signaling in plants: a mitogen-activated protein kinase pathway is activated by cold and drought. Proc Nat Acad Sci USA 93:11274–11279

Joshi PK, Jha AK, Wani SP, Joshi L, Shiyani RL (2005) Meta analysis to assess impact of watershed program and people's participation, Comprehensive assessment research report 8. Colombo, Sri Lanka: Comprehensive Assessment Secretariat, IWMI, pp 18

Kasuga M, Liu Q, Miura S, Yamaguchi-Shinozaki K, Shinozaki K (1999) Improving plant drought, salt and freezing tolerance by gene transfer of a single stress-inducible transcription factor. Nat Biotechnol 17:287–291

Kratsch HA, Wise RR (2000) The ultrastructure of chilling stress. Plant Cell Environ 23:337–350

Maji AK (2007) Assessment of degraded and wastelands of India. J Indian Soc Soil Sci 55:427–435

MNRE (2009) Ministry of New and Renewable Energy Resources www.mnre.gov.in/relatedlinks/

Moon H, Lee B, Choi G, Shin D, Prasad DT, Lee O, Kwak SS, Kim DH, Nam J, Bahk J, Hong JC, Lee SY, Cho MJ, Lim CO, Yun DJ (2003) NDP kinase 2 interacts with two oxidative stress-activated MAPKs to regulate cellular redox state and enhances multiple stress tolerance in transgenic plants. Proc Nat Acad Sci USA 100:358–363

Munns R (1993) Physiological processes limiting plant growth in saline soil: some dogmas and hypotheses. Plant Cell Environ 16:15–24

NRAA (2013) Contingency and compensatory agriculture plans for droughts and floods in India-2012, Position paper No. 6. National Rainfed Area Authority, NASC Complex, New Delhi, India, 87 p

Pai DS, Thapliyal V, Kokate PD (2004) Decadal variation in the heat and cold waves over India during 1971–2000. Mausam 55(2):281–292

Pathak BS (2006) Crop residue to energy. In: Chadha KL, Swaminathan MS (eds) Environment and agriculture. Malhotra Publishing House, New Delhi, pp 854–869

Pathak H, Bhatia A, Jain N, Aggarwal PK (2010) Greenhouse gas emission and mitigation in Indian agriculture – a review, In: Bijay-Singh (ed) ING Bulletins on Regional Assessment of Reactive Nitrogen, Bulletin No. 19, SCON-ING, New Delhi, pp 1–34

Pramanik SC (2009) Rainwater management techniques for successful production of pulses in rainfed areas. Indian Fmg 58(12):15–18

Raghavan K (1967) Climatology of severe cold waves in India. Indian J Meteorol Geophys 18(1):91–96

Reichmann SM (2002) The response of plants to metal toxicity: a review focusing on copper, manganese and zinc. Occasional Paper No. 14. Publisher: Australian Minerals and Energy Environment Foundation. Melbourne, Victoria, 3000, Australia

Richards RA (1996) Defining selection criteria to improve yield under drought. Plant Growth Regul 20:57–166

Rockstrom J, Hatibu N, Oweis T, Wani SP (2007) Managing water in rainfed agriculture. In: Molden D (ed) Water for food, water for life: a comprehensive assessment of water management in agriculture. Earth Scan/International Water Management Institute, London/Colombo, pp 315–348

Sahrawat KL, Wani SP, Parthasaradhi G, Murthy KVS (2010) Diagnosis of secondary and micronutrient deficiencies and their management in rainfed agroecosystems: Case study from Indian Semi-arid tropics. Commun Soil Sci Plant Anal 41(3):346–360

Samra JS (2004) Resource management options in multi-enterprise agriculture. In: Acharya CL, Gupta RK, Rao DLN, Subba Rao A (eds) Multi-enterprise systems for viable agriculture. NAAS, New Delhi, pp 19–39

Samra JS, Kaur P, Amrit Kaur M (2012) Spectral density analysis of the cold wave (2010–11 and 2011–12) and its impact on wheat productivity in Indian Punjab. Plenary Lecture in the Third International Agronomy Congress, New Delhi, November 27, 2012

Sikka AK, Bapuji Rao B, Rao VUM (2016) Agricultural disaster management and contingency planning to meet the challenges of extreme weather events. Mausam 67(1):155–168

Singh MV (2001) Evaluation of current micronutrient stocks in different agroecological zones of India for sustainable crop production. Fertil News 46(2):25–42

Singh AK, Singh V, Chauhan MP (2014) Effect of foliar application (nitrogen and phosphorus) on different agronomic and economic character in lentil (*Lens culinaris* M.) Res Environ Life Sci 7(1):65–67

Srinivasarao C, Gopinath KA (2016) Resilient rainfed technologies for drought mitigation and sustainable food security. Mausam 67(1):169–182

Srinivasarao C, Vittal KPR (2007) Emerging nutrient deficiencies in different soil types under rainfed production systems of India. Indian J Fertil 3:37–46

Srinivasarao C, Wani SP, Venkateswarlu B, Sahrawat KL, Kundu S, Dixit S, Gayatri Devi K, Rajesh C, Pardasaradhi G (2010) Productivity enhancement and improved livelihoods through participatory soil fertility management in tribal districts of Andhra Pradesh. Indian J Dryland Agr Res 25(2):23–32

Srinivasarao Ch, Ravindra Chary G, Mishra PK, Nagarjuna Kumar R, Maruthi Sankar GR, Venkateswarlu B, Sikka AK (2013a) Real time contingency planning: Initial experiences from AICRPDA, All India Coordinated Research Project for Dryland Agriculture (AICRPDA), Central Research Institute for Dryland Agrciulture (CRIDA), ICAR, Hyderabad – 500 059, India, 63 p

Srinivasarao C, Venkateswarlu B, Lal R, Singh AK, Kundu S (2013b) Sustainable management of soils of dryland ecosystems of India for enhancing agronomic productivity and sequestering carbon. Adv Agron 121:253–329

Srinivasarao Ch, Chary GR, Mishra PK, Subba Reddy G, Sankar GRM, Venkateswarlu B, Sikka AK (2014a) Rainfed farming – a compendium of doable technologies, All India Coordinated Research Project for Dryland Agriculture, CRIDA, Indian Council of Agricultural Research, Hyderabad, 194 p

Srinivasarao C, Lal R, Kundu S, MBB PB, Venkateswarlu B, Singh AK (2014b) Soil carbon sequestration in rainfed production systems in the semiarid tropics of India. Sci Total Environ 487:587–603

Srinivasarao C, Kundu S, Sharma KL, Reddy S, Pharande AL, Vijayasankarbabu M, Satish A, Singh RP, Singh SR, Ravindra Chary G, Osman M, Gopinath KA, Yasmin C (2015a) Magnesium balance in four permanent manurial experiments under rainfed agro-ecosystems of India. Crop Pasture Sci 66:1230–1240

Srinivasarao C, Lal R, Prasad JVNS, Gopinath KA, Singh R, Jakkula VS, Sahrawat KL, Venkateswarlu B, Sikka AK, Virmani SM (2015b) Potential and challenges of rainfed farming in India. Adv Agron 133:113–181

Srinivasarao Ch, Venkateswarlu B, Sikka AK, Prasad YG, Chary GR, Rao KV, Gopinath KA, Osman M, Ramana DBV, Maheswari M, Rao VUM (2015c) District agriculture contingency plans to address weather aberrations and for sustainable food security in India, ICAR-Central Research Institute for Dryland Agriculture, Natural Resource Management Division, Hyderabad, CRIDA-NICRA Bulletin 2015/3, 22 p

Srinivasarao Ch, Gopinath KA, Rama Rao CA, Raju BMK, Rejani R, Venkatesh G, Visha Kumari V (2016a) Dryland agriculture in South Asia: Experiences, challenges and opportunities. In: Farooq M, Siddique KHM (eds) Innovations in dryland agriculture, Springer International Publishing, pp 345–392

Srinivasarao C, Ravindra Chary G, Rani N, Baviskar VS (2016b) Real time implementation of agricultural contingency plans to cope with weather aberrations in Indian agriculture. Mausam 67(1):183–194

Taley SM (2012) Enhancing crop productivity in rainfed agriculture through rainwater conservation. In: Souvenir and abstracts of seminar on breaking yield barriers in major field crops, 6–7 January 2012, Akola Chapter of Indian Society of Agronomy, Akola, India, pp 83–92

Thiaw S, Hall AE (2004) Comparison of selection for either leaf-electrolyte-leakage or pod set in enhancing heat tolerance and grain yield of cowpea. Field Crop Res 86:239–253

Venkatesh MS, Basu PS (2011) Effect of foliar application of urea on growth, yield and quality of chickpea under rainfed conditions. J Food Legumes 24(2):110–112

Venkateswarlu B, Mishra PK, Ravindra Chary G, Maruthi Sankar GR, Subba Reddy G (2009) Rainfed farming – a compendium of improved technologies. All India Coordinated Research Project for Dryland Agriculture, CRIDA, Hyderabad, 132 p

Viswanathan C, Zhu JK (2004) Molecular perspectives on cross-talk and specificity in abiotic stress signaling in plants. J Exp Bot 55:225–236

Wendehenne D, Durner J, Klessig DF (2004) Nitric oxide: a new player in plant signalling and defence responses. Curr Opin Plant Biol 7:449–455

Heavy Metal Toxicities in Soils and Their Remediation

7

Arvind K. Shukla, Kulasekaran Ramesh, Ritu Nagdev, and Saumya Srivastava

Abstract

Pollution of soils with heavy metals from various sources has become a common feature across the globe due to increase in anthropogenic activities and industrial development and has attracted the attention of all stakeholders. In spite of the differential tolerance of plants to heavy metal toxicities, impairment in the productivity of most of the agricultural crops is steadfast throughout the globe. Biotransfer of these metals remains unabated from polluted sites and even through animal milk and dung. The remediation methods are broadly grouped into engineering, electrokinetics, and bioremediation. These have their own merits and demerits, but the bioremediation is quite effective and the current results are encouraging. Therefore, the sources of heavy metals to soils (including pathways), their effect on soils and plants, and few of the proven phytoremediation methods have been elaborated here.

7.1 Introduction

With rapid industrialization and urbanization, several environmental issues including soil pollution are attracting global attention. Industries do generate enormous quantities of wastes containing heavy metals, hydrocarbons, pesticides, and other toxic chemicals; those are usually dumped in the nearby open area/soil/water. Metallic element with relatively high density which is toxic even at suboptimal concentration is commonly referred to as "heavy metals" (Lenntech Water Treatment and Air Purification 2004), comprises a group of metals and metalloids with atomic

A.K. Shukla (✉) • K. Ramesh • R. Nagdev • S. Srivastava
ICAR-Indian Institute of Soil Science, Bhopal 462038, Madhya Pradesh, India
e-mail: arvindshukla2k3@yahoo.co.in; ramechek@gmail.com; ritunagdev@gmail.com; sonata906@gmail.com

© Springer Nature Singapore Pte Ltd. 2017
P.S. Minhas et al. (eds.), *Abiotic Stress Management for Resilient Agriculture*,
DOI 10.1007/978-981-10-5744-1_7

153

density greater than water or more than 4 g cm^{-3}, or more (Hawkes 1997). Heavy metals enter the soil through soil-applied agrochemicals, food processing waste, detergents, cosmetics and construction waste, etc. These can accumulate in soil and groundwater above toxic levels, thereby posing grave risk to human health. Heavy metals threatening the ecosystem are mainly lead (Pb), mercury (Hg), chromium (Cr), nickel (Ni), arsenic (As), cadmium (Cd), copper (Cu), and zinc (Zn). Soil contamination occurs due to addition of heavy metals through industrialization, urbanization and agricultural intentisification to a limited extent (Zhou 1995; Gowd et al. 2010; Luo et al. 2011). Extensive damage to soils and crops with these metals has been reported from several countries, viz., Australia (Markus and Mcbratney 1996), Hong Kong (Chen et al. 1997), Saudi Arabia (Al-Shayeb and Seaward 2001), Croatian capital of Zagreb (Romic and Romic 2003), Turkey (Aydinalp and Marinova 2003), Bolivia (Miller et al. 2004), Thailand (Zarcinas et al. 2004), Jordan (Al-Khashman 2007), India (Sharma et al. 2007; Chopra et al. 2009), Iran (Shakeri et al. 2009), Greece (Christoforidis and Stamatis 2009), Algeria (Mass et al. 2010), Nigeria (Ololade 2014), Nanxun County, Southeast China (Zhao et al. 2015), etc.

Soil acts as the sink for several heavy metals, and most of them do undergo neither microbial nor chemical degradation (Kirpichtchikova et al. 2006) unlike organic contaminants, and their concentrations persist in soils for long periods (Adriano 2003). Heavy metal-contaminated soil may pose risks to humans as well as the ecosystem through direct contact or transport through the food chain, drinking water (safety and marketability), phytotoxicity and/or decline in land usability for agricultural production, etc. (McLaughlin et al. 2000a, b; Ling et al. 2007).

Noticing heavy metal contamination in the soil is rather difficult since these are neither rich in color nor have odor. The impact of the pollution is a creeping poison as its damage to the soil environment could not be explicitly quantified in temporal means. Whenever the limit of the pollutants exceeds the maximum allowable limit for either crop/animal/soil biota, the repercussions become eminent. Therefore, Wood (1974) termed heavy metal contamination as chemical time bombs.

7.2   Sources and Forms of Heavy Metals

There are several sources for heavy metal contamination in soils (Fig. 7.1). The origin of heavy metal pollution can either be natural and/or anthropogenic sources besides other means. Though considered to be associated with industrialization, roadways and automobiles too ingest heavy metals to the soils. Zinc, copper, and lead released from road travel account for majority of the total metals in road runoff (Mishra and Shukla 2014). On the other hand, the loading could be due to excessive fertilizer, chemical usage, irrigation, atmospheric deposition, and pollution by waste materials. Organic manures also play role in their availability, e.g., the chelation of free metal ions regulates their availability and mobility in soils and/or water through the formation of metal-humate complexes with varying degrees of stability (Sanyal 2001; Sinha and Bhattacharya 2011). Heavy metals are present in the soil ecosystem as i) easily available to plants in dissolved form (in soil solution) and

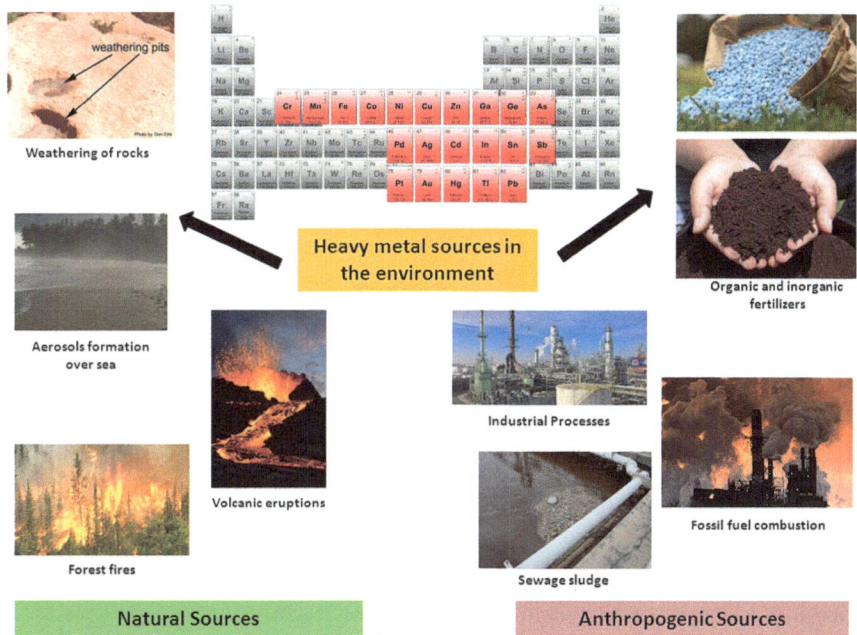

Fig. 7.1 Sources of heavy metals in soils

exchangeable form (in organic and inorganic components) and ii) long-term availability to plants and present in soil as lattices of soil minerals and as insoluble precipitates, e.g., Cd, a toxic nonessential metal, has become an environmental hazard to various forms of life on the earth. Generally, soil contamination occurs through either point source or diffuse pollution.

The rate of application of the contributors and its concentration are the chief factors for enhancing heavy metal contamination of agricultural soil besides the soil characteristics itself. Applications of soil amendments such as compost refuse as well as nitrate fertilizers too contribute to the heavy metal pollution, liming itself. Notwithstanding to this fact, sewage sludge has more pronounced effect than the above (Table 7.1).

Wastewater irrigation-based heavy metal contamination is a serious concern as it contaminates the agricultural produce. The inadvertent ingestion of contaminated water is also a potential pathway (Shukla and Tiwari 2013). As assessment made at Jabalpur, MP showed the extent of pollution due to use of sewage water in soil and plants from Jabalpur, Morena, Gwalior, Katni, Sagar, Dhar, Indore, and Dewas districts of Madhya Pradesh. The sludge samples were neutral to alkaline in reaction and rich in Zn, Cu, N, and K, with a tendency to accumulate more in the surface layer. The pollutant content was relatively higher in Gwalior as compared to Morena. Plants irrigated with the sewage either in Khan Nalla near Indore or Dewas factory area had higher content of heavy metal. Further, average contents of heavy metals in Katni were relatively higher in cauliflower as compared to brinjal. Although the

Table 7.1 Range of heavy metal concentrations (μg g^{-1}) in agricultural amendments

Agricultural amendments	Range of metals (μg g^{-1})					
	Cr	Ni	Cu	Zn	Cd	Pb
Sewage sludge	8.40–600	6–5300	50–8000	91–49,000	<1–3410	2–7000
Compost refuse	1.8–410	0.9–279	13–3580	82–5894	0.01–100	1.3–2240
Farmyard manure	1.1–55	2.1–30	2–172	15–556	0.1–0.8	0.4–27
Phosphate fertilizers	66–245	7–38	1–300	50–1450	0.1–190	4–1000
Nitrate fertilizers	3.2–19	7–34	–	1–42	0.05–8.5	2–120
Lime	10–15	10–20	2–125	10–450	0.04–0.1	20–1250
Pesticides	–	–	–	–	–	11–26

Modified from Ross (1994)

Table 7.2 DTP-extractable heavy metal contents (mg kg^{-1}) in sewage-polluted soils

Location	Cr	Cd	Ni	Pb
Uthamapalayam	0.04	0.15	0.09	2.27
Vedapatti	0.06	0.11	0.09	1.61
Palapatti	–	0.12	0.17	1.84
Vachinampalayam	–	0.11	0.04	3.21

Source: Stalin (2011)

contents of Zn and Pb were decreased, distribution of pollutant elements was irregular. By contrast, sewage-irrigated soil samples in the state of Tamil Nadu have shown insignificant buildup of heavy metals except Pb (Table 7.2) which was still below the toxic limit (Stalin 2011). The pathways for the entry of heavy metals to soils can broadly be divided into "solid wastes to soil pathway" and "agricultural supplies to soil pathway" which are discussed as under.

7.2.1   Solid Wastes to Soils Pathway

Heavy metals with varied complex composition arising from a variety of solid wastes are known to pollute the agroecosystem. Among all, mining for various purposes and industrial solid waste contamination are considered as the major culprits. Mining activities, particularly for extraction and manufacturing of metal products, result in pollutant generation released into the nearby agricultural soils (Parth et al. 2011). When the industrial wastes are piled, facilitation of sunlight, rain, and washing helps easy movement of heavy metals which later spread to the surrounding water and soils. Cr as metal pollutant exists as Cr^{3+} and Cr^{6+} in minerals in soils of chromite mining area of Kaliapani, Sukinda, Odisha, while the latter is highly mobile in soil and causes phytotoxicity in plants. More than 90% of soil samples collected from villages within 1 to 10 km radius from the chromite mines were polluted with Cr at an above toxic limit considering 0.05 ppm Cr^{+6} (Shukla and Tiwari 2013).

Dye factory effluent   **Gold Processing effluent**

Crops grown with sewage water   **Sewage water from city**

Fig. 7.2 Industrial effluents and soils irrigated with sewage water

Table 7.3 DTPA-extractable metal status in the soils (mg kg^{-1}) of industrial effluent irrigated areas

Nature of contaminated site	DTPA metal status (mg kg^{-1})			
	Cd	Ni	Fe	Pb
Dye	0.53	2.16	250	3.98
Gold processing	2.34	2.17	238	5.28
Sewage	1.90	6.69	216	15.01
Foundry	0.43	19.83	270	2.87
Electroplating	0.37	2.93	221	3.61
Textile	0.48	2.57	56	2.29
Painting	0.11	31.35	216	4.98
Casting	0.46	1.28	207	2.68

Source: Shukla and Behera (2012)

The soil irrigated with industrial effluents (Fig. 7.2), viz., casting, painting, and sewage effluents, had the highest metal status (Shukla and Behera 2012) and the content varied widely with the industry (Table 7.3). The highest DTPA-Ni was obtained in painting industrial effluent irrigated areas (31.35 mg kg^{-1}), Cd in gold processing effluent irrigated areas (2.34 mg kg^{-1}), and Pb in sewage effluent

Table 7.4 Regulatory guidelines for some heavy metals ($\mu g\ g^{-1}$) in agricultural soils

Standards	Cd	Cu	Pb	Zn	Mn	Ni	Cr
Indian standard (Awashthi 2000)	3–6	135–270	250–500	300–600	--	75–150	--
European union standards (EU 2002)	3	140	300	300	--	75	150

irrigated soils (15.01 mg kg^{-1}). Among the metals, the order of availability was Ni > Pb > Cd. Similarly the highest total Pb was obtained from gold processing and sewage effluent irrigated area (354 and 344 mg kg^{-1}), Cd in painting (11 mg kg^{-1}) and sewage-irrigated areas (9.81 mg kg^{-1}), and Ni in electroplating (317 mg kg^{-1}) which reflected in plant content too. Higher Ni in foundries (512 mg kg^{-1}) and electroplating (475 mg kg^{-1}), Pb in dye (498 mg kg^{-1}), textile (500 mg kg^{-1}), Cd in sewage irrigation (35.6 mg kg^{-1}), and sewage effluent irrigated areas (469 mg kg^{-1}) were also observed. The sewage-irrigated soils of Amritsar and Jalandhar were analyzed to contain 0.318 and 3.34 mg kg^{-1} DTPA-Cd and Pb, respectively (Shukla and Tiwari 2014) (Table 7.4).

7.2.2 Agricultural Supplies to Soil Pathway

Pollutants from agrochemical sources include fertilizers (Aydinalp and Marinova 2003), manures (Mullins et al. 1982), herbicides (Shrotriya et al. 1984), etc. which are directly applied on the soils for optimum crop productivity. Also the fertilizers and manures accidentally add arsenic, uranium, and vanadium through some phosphatic fertilizers to the soil. Besides, chromium, lead, mercury, and nickel are also added to the soil inadvertently. Small amounts of heavy metals are found in rock phosphate (Mortvedt 1995; Dissanayake and Chandrajith 2009; Lema et al. 2014). Animal manure is the main organic fertilizer that may entertain heavy metals. Heavy metals in biosolids exist in either inorganic form or organic complex form and could affect their chemical reactions in the soil. Repeated applications of these fertilizer sources aggravate their accumulation. Some countries have set tolerance limits on heavy metal in soils, and most of the fertilizer regulations relate Cd limits to P concentrations.

7.3 Basic Soil Chemistry and Risk of few Important Heavy Metals

The abundance of heavy metals at contaminated sites in the ascending order of Hg, Cu, Cd, Zn, As, Cr, and Pb (USEPA, 1996) is responsible for impairment in the crop production either due to their bioaccumulation or biomagnification when these enter the food chain from soil to higher animals besides groundwater contamination too. Notwithstanding to this fact, the fate and its transport to soil depend on its chemical form and speciation. After reaching the soil, these are adsorbed and redistributed into various forms with degree of toxicity (Shiowatana et al. 2001; Buekers 2007). The distribution of heavy metal in soil is controlled by (i) mineral dissolution and precipitation; (ii) adsorption, desorption, and ion exchange (iii); aqueous complexation; (iv) biological mobilization and immobilization; and (v) plant uptake (Levy et al. 1992).

7.3.1 Lead

The major forms of Pb are lead oxides, ionic lead, Pb(II), hydroxides, and lead-metal oxyanion complexes that are released into the soil environment. Under reduced soil conditions, sulfide and lead sulfide (PbS) are considered as the most stable solid forms. In general, the lead accumulation is confined to leafy vegetables and on the surface of root crops. Plants grown in lead contaminated soils do not accumulate excess amount of lead. However, direct eating of contaminated soil may led to Pb poisioning. Lead content in farm produce less than 300 ppm is generally considered as safe. The risk of lead poisoning through the food chain increases as the lead level rises above this concentration (Wuana and Okieimen 2011).

7.3.2 Chromium

Chromium (VI) is the chief form commonly distributed at contaminated sites, although Cr (III) oxidation state is not uncommon, depending on pH and redox conditions. Cr (VI) is more toxic and mobile than Cr (III). Under anaerobic conditions in presence of organic matter, sufide and ferrous ions, Cr (III) gets reduced to Cr (VI). The mobility of Cr is determined by the sorption characteristics of the soil. The leachability of Cr (VI) is proportional to soil pH. Cr is highly toxic for plants and causes various deleterious impacts on its growth and metabolism.

7.3.3 Arsenic

Arsenic could be seen in several oxidation states, viz., $-III$, 0, and III to V (Smith et al. 1995). Under aerobic conditions, As (V) dominates and under reducing conditions As (III) dominates. Under extreme reducing conditions, elemental arsenic and arsine and AsH_3 may be present. A nonessential and generally toxic to plants, arsenic, intercepts the root through inhibition of root extension and proliferation. Upon translocation to the aerial parts, it inhibits plant growth by arresting biomass accumulation and loss in fertility (Garg and Singla 2011).

7.3.4 Cadmium

Zn has the chemical similarity with Cd as a divalent Cd (II), and their intersubstitution may cause the malfunctioning of metabolic processes (Campbell 2006).

7.3.5   Mercury

Hg (highly toxic for plants) usually exists in mercuric (Hg^{2+}), mercurous (Hg_2^{2+}), elemental ($Hg°$), or alkylated form (methyl/ethyl mercury). Stability of Hg depends upon the redox potential and pH. Under oxidizing conditions, the mercurous and mercuric mercury are more stable, while organic and/or inorganic Hg may be reduced to elemental form under mild reducing conditions. This may then be converted to most toxic alkylated forms by biotic or abiotic processes. Sorption to sediments, soils, and humic materials is the important mechanism for the removal of Hg from solution (Wuana and Okieimen 2011).

7.4   Impact of Heavy Metals on Soils

7.4.1   Soil Microbial Activity

The environmental risk of heavy metal pollution on the functioning of soil microorganisms adjacent to large industrial complexes is an absolute necessity in the current industrial scenario. Generally poisoning and inactivation of enzyme systems in soil is considered as the heavy metal toxicity (Rai et al. 1981). For example, Cd concentration of $0.16\ \mu g\ g^{-1}$ protects 95% of soil invertebrates and 85% at $0.8\ \mu g\ g^{-1}$ (van Straalen and Denneman 1989). A decrease in the population of actinomycetes and bacteria was noticed by Hiroki (1992) in a heavy metal-contaminated (Cd, Zn, and Cu, 1.1–2.7, 234–571, and 310–751 mg kg^{-1} soil, respectively) paddy field. However, fungi remained unaffected. The degree of intolerance to heavy metals appears to be: fungi >bacteria > actinomycetes.

Although enzyme activity is related to soil property from biological organisms, some stimulatory effects on microbial populations were noted in a sandy soil but not in clay soil. However, irrespective of the soil, activities of urease and nitrate reductase were inhibited. As an exception, amidase activity was inhibited only at higher concentration (Hemida et al. 1997). Heavy metal stress in copper mining wasteland was exhibited in the form of microbial ecophysiological parameters, viz., ratio of microbial biomass C(Cmic)/organic C(Corg) and metabolic quotient (qCO_2) in red soil area, southern China (Liao et al. 2005). The microbial biomass C was negatively affected by the elevated metal levels (Wang et al. 2007). Denitrifying microbial community adapted to elevated levels of Pb by selecting for metal-resistant forms of nitrite reductases and Pb had marked impacts on the microbial community even at very low concentrations (Sobolev and Begonia 2008). However, the total bioactivity, richness, and microbial diversity decreased with concentration of heavy metal in Cd contaminated or uncontaminated soils from Hunan province of China (Xie et al. 2016).

7.4.2   Impact on the Plants

Anthropogenic activities are the main source of pollutants to soils, and the growth of plants growing on these soils gets impaired when the concentrations build up above the toxic limits (Chibuike and Obiora 2014). The capability of plants to accumulate essential metals equally enables them to acquire other nonessential metals (Djingova and Kuleff 2000). Plant growth is affected when the concentration exceeds optimal levels. High metal concentration had direct toxic effects, i.e., inhibition of activities of cytoplasmic enzymes and damage to cell structures through oxidative stress (Assche and Clijsters 1990; Jadia and Fulekar 2009); besides, soil microbial enzyme activities useful for plant metabolism may also be hampered. Indirectly the replacement of essential nutrients at cation exchange sites of plants may also occur (Taiz and Zeiger 2002). These toxic effects lead to a decline in plant growth and ultimately the mortality of plants occurs.

7.4.3   Heavy Metal Continuum in Soil-Water-Crop-Human System

Heavy metals in urban soils may enter into the human body either through skin absorption, inhalation of dust, etc. Soil, water, food, and blood samples from the Cherlapally (Uppal mandal), Patancheru and Ramachandrapuram (Ramachandrapuram mandal), and Munagala (Munagala mandal) in Ranga Reddy, Nalgonda, and Medak districts (surrounded by highly polluted industries), where farmers used industrial effluent for irrigation, were analyzed by Surendra-Babu et al. (2012) for heavy metals, viz., Pb, Cd, Cr, Co, and Ni (Table 7.5). Although irrigation water samples were within the safe limits except Co in Patancheru irrigated water, they may become toxic if they enter into food chain even at very low concentrations as they could bind certain proteins and essential elements, thereby rendering them in exhibiting normal function.

Table 7.5   Status of heavy metal contents in soil, water, plants, and human in polluted areas

Element	Water (mg kg^{-1})	Soil (mg kg^{-1})	Rice (mg kg^{-1})	Human beings (%)
Pb	0.294 (Tr.)	1.24(0.44)	15.83(3.17)	79 (100)
Cd	0.010(Tr.)	0.042(0.021)	1.92(0.31)	92(100)
Co	0.080(Tr.)	0.182(0.140)	5.21(0.12)	84(100)
Ni	0.200(Tr.)	0.502(0.226)	9.28(0.89)	72(100)
Ca	39.86 (29.03)	2.64(0.57)	14.39(1.44)	24 (92)
Fe	Traces (Tr.)	3.59(7.82)	26.81(22.75)	100(100)

Figures in parenthesis are for nonpolluted areas. Source: (Surendra-Babu et al. 2012)

Table 7.6 Heavy metal concentrations in blood serum, plasma, milk, and dung samples (n = 5)

Week	Part	Heavy metal concentration (mg kg⁻¹)			
		Pb	Cd	Ni	Cr
Background	Serum	0.014	ND	ND	ND
	Plasma	0.0216	ND	0.138	0.141
	Milk	0.003	0.029	ND	0.125
	Dung	32.6	1.36	5.72	0.684
2nd week	Serum	0.007	ND	ND	ND
	Plasma	0.552	0.092	1.532	0.962
	Milk	ND	0.05	9.30	9.35
	Dung	34.8	2.28	6.21	8.18
4th week	Serum	ND	0.044	ND	ND
	Plasma	1.211	0.154	0.930	0.994
	Milk	ND	0.05	4.34	7.95
	Dung	44.4	1.32	48.72	13.76

ND not detectable; Source: Stalin et al. (2014)

7.4.4 Soil-Animal-Milk-Human Continuum for Heavy Metals

The heavy metal contents (Cd, Pb, Cr, and Ni) were analyzed in blood, plasma, milk, and dung of Jersey cows when fed with Bajra-Napier grass grown in Ukkadam sewage farm of Coimbatore district (Stalin et al. 2014). The contents of heavy metals followed the order: dung > milk > blood (Table 7.6). In the milk samples, accumulation of heavy metals has shown an increasing trend from 10th day onward. Ni was absent up to 4th day and thereafter showed a buildup. The Cd content was nearly doubled after 10th day and remained so. The Cr content was initially low, thereafter a constant buildup with highest Cr (22.5 mg kg⁻¹) at 22nd day. In the serum samples, Ni and Cr could not be detected initially. Cd accumulation was observed during the 3rd and 4th week. Ni and Cr accumulated in the plasma at the end of the 4th week; Ni and Cr content of the plasma were seven times more than that of initial value. Similarly, Pb also increased in the plasma from the 2nd week itself. During the 7th day, the dung sample recorded the highest content of heavy metals and remained constant.

7.5 Remediation Measures

Remediation measures of contaminated soils are classified into in situ and ex situ treatment (USEPA 2007). In in situ, treatment of soil is at its original place, whereas in ex situ, the soil which is contaminated is moved, excavated, and removed from the site. Yao et al. (2012) have reviewed the remediation technologies (including physical, chemical, and biological remediation). Few proven remediation methods are discussed here under.

7.5.1   Engineering Remediation

Physical or chemical methods to manage heavy metal contamination of soils are referred to as engineering remediation. Soil washing, phytoremediation techniques, and the principles, advantages, and disadvantages of immobilization are available in Wuana and Okieimen (2011).

7.5.2   Physical Remediation

7.5.2.1 Replacement of Contaminated Soil, Soil Removal, and Soil Isolation

Complete replacement of contaminated soil with addition of large amount of non-contaminated soil or blending with the latter is referred as soil breeding. Through soil removal the contaminated soil is renewed with the clean soil if the area is very small. But, with additional engineering measures, soil isolation of the contaminated soil from the uncontaminated soil (Zheng et al. 2002) may also prove beneficial. This method needs huge cost and manpower.

7.5.2.2 Thermal Desorption of Soil

The thermal desorption utilizes the pollutant's volatility character by heating the contaminated soil through steam, microwave, and infrared radiation (e.g., Hg, As). These volatile metals are collected using vacuum pressure and are subsequently removed from the soil (Li et al. 2010). The traditional thermal desorption can be separated into high-temperature desorption according to the temperature ($320 \sim 560\ °C$) and low-temperature desorption ($90 \sim 320\ °C$). However, being laborious and costly, this method finds limited application.

7.5.3   Chemical Remediation

Chemical leaching, fixation, electrokinetic remediation, vitrification, etc. are referred to as chemical remediation. The process of vitrification involves heating of the soil at very high temperature, e.g., $1400–2000\ °C$, for the pollutant to get volatize or decompose. Important chemical remediation measures are as follows.

7.5.3.1 Soil Leaching/Chemical Leaching

Washing the contaminated soil with specific reagents to remove the heavy metal complex is the basis of soil leaching and soluble irons adsorbed on the solid phase particles. After separating from the soil, heavy metals are recycled from extracting solution. The various ionic forms of the heavy metals are transferred from soil to liquid phase and then recovered from the leachate. The leachate includes inorganic eluent, chelation agents, surfactant, etc. Chief chelating agents are ethylenediaminetetraacetic acid and citric acid in remediating the high permeability soil and tartaric acid to a little extent for average contamination (Wuana et al. 2010). Na_2 EDTA solutions were more effective than $Na_2S_2O_5$ as the former extracted lead over zinc and cadmium but to a limited extent on chromium. Cadmium and, especially, zinc removal by a 0.01 M Na$_2$ EDTA solution were improved considerably by inclusion of 0.1 M Na$_2$S$_2$O$_5$ (Abumaizar and Smith 1999).

7.5.3.2 Adsorption/Chemical Fixation

Addition of external reagents into the contaminated soil to form insoluble forms to decrease the migration of heavy metals to the environment is referred to as chemical fixation (Zhou et al. 2004) as their removal in polluted areas is very complex as they persist in soils for very long periods. It is based on the fact that almost all heavy metal ions can be fixed and adsorbed by clay mineral (bentonite, zeolite, etc.), a steel slag, furnace slag, etc. (Wang and Zhou 2004). For example, zeolites, the crystalline aluminosilicates (Ramesh et al. 2011), with large cation exchange capacity attract positive-charged ions and widely used for sequestration of cationic pollutants (Kumar et al. 2007). They offer absorption sites for small molecules and increase ion exchange sites in soils due to their porous structure. Zeolites can retain heavy metals in soil (Mühlbachová and Šimon 2003).

Clinoptilolite, a zeolite, is stable up to pH 2 (Ming and Mumpton 1989). Zeolite affinity to heavy metals has also been demonstrated by Tsadilas (2000). Alexander and Christos (2003) found that its sorption is up to 30 times more Pb than the soil. Based on the value of maximal sorption capacity of zeolite, addition of just 1% zeolite can retain up to 750 mg Pb kg^{-1} soil. It has high efficiency at range of 3–5 pH too (Alexander and Christos 2003). The potential of organo-zeolitic systems to revegetate metal polluted soils was demonstrated by Leggo et al. (2006).

7.5.3.3 Electrokinetic Remediation

Soil electrokinetic remediation (Kim et al. 2001) is a new economically effective technology utilizing various particle and fluid interactions governing the dynamics of an electrokinetic system. These interactions are a direct result of applied electrical potential across the system and are identified as electrokinetic phenomena. The physiochemical composition of clay particles in soils is the strong basis for utilizing electrokinetic remediation in soils. Use of clay particles could be an innovative solution for efficient removal of contaminants from soils to solve groundwater and soil pollution, as they have a net negative surface charge.

The principle is that the DC voltage is applied to form the electric field gradient on both sides of the electrolytic tank which contains the contaminated soil; contaminants in the soil are taken to the processing chamber to reduce the contamination, which is located at the two polar sides of electrolytic cell, through the way of electromigration, electric seepage, or electrophoresis. This method applies to low permeable soil (Hanson et al. 1992). Removal of heavy metals from clay and sandy soils was reviewed by Virkutyte et al. (2002).

7.5.4 Bioremediation

This type of remediation mainly includes phytoremediation and microbial remediation for heavy metal removal from soils. While phytoremediation is a form of bioremediation wherein plants are used to either sequester the environmental contaminants or convert them to harmless forms (Cunningham and Berti 1993; Raskin et al. 1994, 1997; Salt et al. 1995,1998), a predominant method followed widely (Alkorta and Garbisu 2001; Garbisu and Alkorta 1997; Garbisu et al. 2002; Tangahu et al. 2011; Moosavi and Seghatoleslami, 2013); bioremediation includes all methods and processes to biotransform a contaminated environment to uncontaminated status which primarily refers to use of microbes (Boopathy 2000) or microbial processes to degrade and transform environmental contaminants into harmless or less toxic forms (Garbisu and Alkorta 2003) either be ex situ or in situ. The physical removal of the contaminated material is referred to as ex situ, while treatment of the contaminated material in place is referred to as in situ. In situ bioremediation is one of the most attractive options, of which soil bioventing, the process of supplying oxygen to contaminated soil in hopes of stimulating microbial degradation of contaminants, is a promising technology (Hellekson 1999) wherein indigenous microbes are utilized to biodegrade organic constituents adsorbed to soils in the unsaturated zone.

7.5.4.1 Microbial Remediation

The microorganisms cannot degrade and destroy the heavy metals (Yao et al. 2012), but can affect their migration and transformation. Several mechanisms by which microbes remediate heavy metals include either individually or in combinations of electrostatic interactions, van der Waal forces, redox interactions, extracellular covalent bonding, and precipitation with their cell surfaces which have been reported in the literature. However their efficiency in remediation or the response depends on the concentration and availability of heavy metals (Goblenz et al. 1994). Ex and in situ bioremediation of refractory pollutants by specific microbes is possible (Iranzo et al. 2001), and is an efficient strategy due to its low cost and high efficiency (Rajendran et al. 2003). Sulfate-reducing bacteria were found to modify the form of Cd in sewage-irrigated soils (Jiang and Fan 2008). They can reduce the mobility and bioavailability of contaminants through various ways (Gang et al. 2010). However, acceptable solutions are not guaranteed. Microorganisms that use metals as terminal electron acceptors or reduce metals as a detoxification mechanism could aid in the removal of metals from contaminated environments (Garbisu and Alkorta 2003). *Bacillus pumilus* and *Pseudomonas aeruginosa* were identified as Pb-resistant bacteria (Chen et al. 2011). Highest efficiency of decabromodiphenyl ether removal and metal phytoextraction was obtained by using co-planting of *Sedum alfredii* with tall fescue inoculated with *Bacillus cereus* JP12 in co-contaminated soil at China. Bacterial inoculation increased plant biomass and decabromodiphenyl ether degradation. Soil microbial activity was promoted by planting tall fescue which enhanced degradation and mineralization of BDE-209. Soil microbial activity and community structure were altered during the remediation (Lu and Zhang 2014).

7.5.4.2 Rhizoremediation

Rhizoremediation, an emerging technology for large recalcitrant compounds, is a phytoremediation method involving plants and their rhizosphere microbes, either naturally or through introduction (Gerhardt et al. 2009). In this method, plant exudates stimulate the survival of bacteria, for an efficient degradation of pollutants involving root, their exudates, rhizospheric soil, and microbes. The spreading of the bacteria and its penetration to soil layers are facilitated by the plant root system. In order to improve the efficiency of phytoremediation or bioaugmentation, inoculation of pollutant-degrading bacteria on plant seed is suggested (Kuiper et al. 2004). Khan (2005) reviewed the role of rhizoremediation of heavy metals in soils. Generally, plant growth-promoting rhizobacteria, P-solubilizing bacteria, and arbuscular mycorrhizal fungi maintain soil fertility in conventional agriculture wherever agrochemicals are used very minimum. The insoluble glycoprotein, glomalin, produced by AM fungi could sequester trace elements and be considered for biostabilization. These phytoextraction strategies need more studies. They can be contaminant degraders (Gerhardt et al. 2009), besides promoting plant growth under stress conditions. The rhizoremediation process can be enhanced with the proper control of factors influencing plant growth as well as microbial activity in the rhizosphere environment (Tang et al. 2010). Plant- and microbe-mediated biotransformation of heavy metals into nontoxic forms and plants and their mechanisms are well known (Dixit et al. 2015). Some of the microorganisms useful in bioremediation of heavy metals are *Flavobacterium* spp., *Rhodococcus* spp., *Pseudomonas* spp., *Alcaligenes* spp., *Arthrobacter* and *Bacillus* spp., *Corynebacterium* spp., *Azotobacter* spp., *Nocardia* spp., *Mycobacterium* spp., *Methosinus* spp., etc. (Girma 2015).

7.5.4.3 Animal Remediation

Animal remediation is in accordance with the characterization of some lower animals by adsorbing, degrading, and migrating the heavy metals and thus minimizing their toxicity. For example, earthworm's activities can increase the availability of soil nutrients in soils which is a well-known fact. Hopkin (1989) has indicated that these earthworms have specificity and capacity to regulate metals in their bodies, but it could be species specific and can even reproduce in metal-contaminated soil (Spurgeon et al. 1994). *Aporrectodea caliginosa* and *Lumbricus rubellus* were found accumulating Zn, Cd, Pb, and Cu in their tissues (Dai et al. 2004). The earthworm species specific to a particular soil types and forms of metal have been reviewed by Nahmani et al. (2007) for specific metal uptake and accumulation. But very little attention has been paid to the impact of earthworms on soil metals either for metal mobility or availability (Sizmur and Hodson 2009). Recent developments in science have demonstrated the presence of metal-tolerant earthworms and change the fractional distribution of heavy metals in contaminated soil too, besides enhancement in the metal availability (Dandan et al. 2007). As they have the potential for bioaccumulation of metals in their chloragogenous tissues, they can serve as soil contamination indicator (Usmani and Kumar 2015). Dabke (2013) has found that earthworms *Eisenia fetida* (Annelida: Oligochaeta) removed heavy metals like chromium via bioaccumulation and also stimulate microbial remediation by increasing bacteria and improving soil aeration. Although *Eisenia fetida* can remove the chromium and cadmium metals, their effectiveness in removing cadmium is more

Fig. 7.3 Phytoremediation through growing marigold (**a**) and amaranthus (**b**)

than chromium (Aseman-Bashiz et al. 2014). In addition, Ni (Kaur and Hundal 2015) and lead (Kaur and Hundal 2015; Prashanth and Prabha 2016) were also taken up by earthworm. Sahu and Sharma (2016) could even find mercury uptake by earthworms *Eudrilus eugeniae* and Hyperiodrilus africanus (Ekperusi et al. 2016). The same could remediate diesel-contaminated soils also (Ekperusi and Aigbodion 2015).

7.5.4.4 Phytoremediation
Phytoremediation is a cost-effective, environment-friendly method (Datta and Darkar 2004), an emerging green technology that uses plants to degrade; stabilize from sediment, surface, groundwater, organics, and radionuclides; and remove soil contaminants, for example, *Pteris vittata*. Microbiota from the rhizosphere could enhance phytoremediation, besides the use of genetic engineering (Marques et al. 2009). Plants with rapid biomass gain with high metal uptake are needed (Khan et al. 2000). Graminaceous species and cultivars have a wide variation for Al tolerance. In cereal crops, the Al tolerance usually follows rice and rye (Secale cereale) > oat (*Avena sativa*) > wheat > barley (Bona et al. 1993). Plants may also be useful for metal decontamination of the polluted soils (Is et al. 2002). Phytoremediation has been grouped into phytoextraction (Lasat, 2002; Mahmood 2010; Sun et al. 2011; Pajević et al. 2016), phyto-degradation (Newman and Reynolds 2004), rhizofiltration (Yadav et al. 2011; Veselý et al. 2011), phytostabilization (China et al. 2014; Garaiyurrebaso et al. 2017), and phytovolatilization (Sakakibara et al. 2010). Among these, phytoextraction is generally most suitable for heavy metal contaminated soils like in the case of arsenic. Phytoextraction can be accomplished by using either tolerant, high-biomass plant species cottonwood (*Populus deltoides* Bartr.), cypress (*Taxodium distichum* L.), eucalyptus (*E. amplifolia* Naudin, *E. camaldulensis* Dehnh, and *E. grandis* Hill), and leucaena (*L. leucocephala* L.) or hyperaccumulator plant species. The latter (Fig. 7.3) has the advantage of producing more concentrated residue, facilitating the final disposal of the contaminant-rich biomass.

Relative Tolerance of Plants

Plants vary in their sensitivity/tolerance toward heavy metal toxicity. A sophisticated network of defense strategies is essential either to avoid or tolerate heavy metal intoxication. A screening experiment carried out at Tamil Nadu Agricultural University, Coimbatore (Shukla and Behera 2012), to identify the poor, moderate, and hyperaccumulators for the heavy metals indicated that among the food crops, highest fresh biomass yield was recorded with *Amaranthus* species while the lowest biomass for sunflower. The dominance of extractability was in the order of Pb > Ni > Zn > Cu > Cd in sewage-irrigated soil after harvest. The highest Pb availability was recorded in the soil grown with sorghum (16.46 mg kg^{-1}) while the lowest in amaranthus (7.38 mg kg^{-1}). Similarly the Cd availability was ranged from 0.34 to 0.706 mg kg^{-1} and highest extractability in cluster bean soils. With regard to Ni, the values varied from 3.47 to 8.71 mg kg^{-1}, and the highest extractability was noticed with radish. The soils grown with mustard recorded the higher Zn extractability (9.55 mg kg^{-1}) followed by amaranthus (9.35 mg kg^{-1}). Cu availability ranged between 3.50 and 7.70 mg kg^{-1}. The highest heavy metal contents such as Pb (498 mg kg^{-1}), Ni (442 mg kg^{-1}), Zn (516 mg kg^{-1}), and Cu (193 mg kg^{-1}) were recorded in mustard crop. The lowest tissue concentration of Pb was noted with brinjal (124 mg kg^{-1}), Cd and Ni in tomato (7.9 and 135 mg kg^{-1}, respectively), Zn in beans (74 mg kg^{-1}), and Cu in lablab (25.9 mg kg^{-1}). Shukla and Behera (2012) used the metal accumulation ratio using total soil metal status and plant tissue concentration to screen the crops for hyperaccumulation (Table 7.7). The crops having

Table 7.7 Metal accumulation ratio of food crops

Plant sp.	Metal accumulation ratio				
	Pb	Cd	Ni	Zn	Cu
Maize	1.81	0.88	1.88	1.34	1.09
Sunflower	0.77	0.66	1.04	0.32	0.63
Tomato	0.79	0.65	0.74	0.41	0.49
Okra	0.90	0.73	0.76	0.58	0.46
Amaranthus	2.31	0.74	2.14	1.00	1.09
Mustard	2.37	0.73	2.43	1.47	1.20
Spinach	0.88	0.85	1.61	0.87	0.85
Beans	0.66	0.96	0.80	0.21	0.48
Cluster bean	1.00	1.09	0.75	0.32	0.48
Cauliflower	0.63	1.03	0.77	0.41	0.64
Radish	0.67	2.21	0.78	0.21	0.55
Brinjal	0.59	1.14	0.83	0.26	0.52
Thandu keerai	1.83	1.00	2.18	0.71	1.09
Avarai	0.74	0.94	1.00	0.26	0.16
Sorghum	0.77	1.19	0.84	0.32	0.64
Agathi	0.71	1.07	0.77	0.34	0.47
Fodder cowpea	1.10	0.91	1.04	0.23	0.35

Source: Shukla and Behera (2012)

Table 7.8 Classification of hyperaccumulators and poor accumulators of heavy metals

Heavy metals	Hyperaccumulators	Poor accumulators
Pb	Mustard, amaranthus, maize, cluster bean, fodder cowpea	Brinjal, cauliflower, beans
Cd	Radish, cluster bean, cauliflower, Brinjal, Amaranthus, sorghum, *Sesbania*	Tomato, sunflower
Ni	Amaranthus, mustard, spinach, maize, cowpea, sunflower	Tomato, okra
Zn	Mustard, Amaranthus, maize	Radish, beans
Cu	Mustard, Amaranthus, maize	Lablab

Source: Shukla and Behera (2012)

a metal accumulation ratio of more than one were taken as hyperaccumulators and are included in Table 7.8.

Among the nonfood crops, the highest biomass yield was recorded with cockscomb (27.9 g pot^{-1}) followed by castor (22.9 g pot^{-1}), balsam (22.5 g pot^{-1}), and globe amaranth (22.1 g pot^{-1}) out of the eight crops tested. The poor biomass yield was registered with *Zinnia* (8.08 g pot^{-1}). The highest DTPA-extractable Pb (21.05 mg kg^{-1}), Ni (9.02 mg kg^{-1}), Zn (15.4 mg kg^{-1}), Cu (6.40 mg kg^{-1}), and Cd (0.926 mg kg^{-1}) were recorded in soil grown with castor crop. The lowest metal extractability was noticed with aster (Pb), cockscomb (Cd), globe amaranth (Ni), balsam (Zn), and *Zinnia* (Cu). The tissue heavy metal content varied with test crops, and the order of higher absorption was Pb > Zn > Ni > Cu > Cd. The highest tissue content of Pb and Cu was observed with castor, and the values varied from 151 to 440 mg kg^{-1} and 116 to 223 mg kg^{-1}, respectively. The absorption of Pb and Cu was the lowest in cockscomb and balsam. With regard to Cd, Ni, and Zn, the highest content was registered with marigold, and the lowest values were noted with cockscomb and aster. The values ranged from 10.4 to 13.3 mg Cd kg^{-1}, 141 to 300 mg Ni kg^{-1}, and 63 to 434 mg Zn kg^{-1}. To screen the nonfood crops for hyperaccumulation, the metal accumulation ratio was calculated by using total soil metal status and plant tissue concentration. For remediating the Pb-contaminated soils, castor and marigold were recommended by Shukla and Behera (2012). Among the 25 food and nonfood crops tested by them, mustard, amaranthus, maize in food crops and castor, and marigold under nonfood crops were found to possess higher hyperaccumulation potentials for remediating Pb-polluted soils. Since food crops cannot be used effectively, the nonfood crops were recommended to remediate the Pb-polluted soils.

Technologies for Phytoremediation

Amaranthus and marigold when cultivated in the Pb-contaminated sites indicate that their biomass yield, Pb availability, its absorption, and removal were significantly influenced by the levels of EDTA and organic manure addition. The order of higher biomass production was marigold > castor wild > castor hybrid > fodder cowpea > cluster bean > amaranthus. Among the organics, higher biomass yield was recorded with 5 t FYM ha^{-1} in amaranthus and cluster bean, while with fodder

Table 7.9 Effect of EDTA (mg kg^{-1}) and organics on the translocation coefficient of crops

EDTA	Amaranthus					Marigold				
	Nil	FYM	GLM	MI	Mean	Nil	FYM	GLM	MI	Mean
0	1.28	1.42	1.48	1.38	1.39	1.76	2.24	1.93	2.15	2.02
50	1.37	1.69	1.54	1.44	1.51	1.60	2.37	2.42	2.16	2.14
100	1.20	1.81	1.70	1.55	1.57	1.84	2.22	2.37	2.25	2.17
Mean	1.28	1.64	1.57	1.46		1.73	2.28	2.24	2.19	

Source: Shukla and Tiwari (2014)

cowpea, marigold, castor wild, and hybrid, addition of green leaf manure at 5 t ha^{-1} registered the highest biomass yield. The interaction effect was found nonsignificant (Shukla and Tiwari 2014).

Shukla and Tiwari (2014) have also found that increasing levels of EDTA increased the biomass production of both crops up to 50 mg kg^{-1} and showed a decline at 100 mg EDTA kg^{-1} of soil. Increasing levels of EDTA addition increased the extractability of Pb, and its availability in various pools and the order of higher availability was organically bound > exchangeable + adsorbed > water soluble Pb. Addition of 5 Mg FYM with 100 mg EDTA kg^{-1} recorded higher bioavailable fractions followed by green leaf manure (Table 7.9). Higher Pb removal and phytoextraction efficiency was noted with the addition of 5 mg FYM + 50 mg kg^{-1} EDTA for amaranthus and 5 t GLM + 100 mg kg^{-1} EDTA for marigold. However between the crops, marigold crop possesses higher TCF and BCF and highly efficient in removing more Pb from soil and thus can be recommended to decontaminate the Pb-polluted soils.

Application of organics/green leaf manures could remediate Pb pollution. Application of 100 mg EDTA kg^{-1} along with either 5 t FYM or green leaf manure ha^{-1} was found to be the best in increasing the availability of Pb and its absorption by crops. Increasing levels of EDTA addition increased the extractability of Pb, and the percent increase was marked in amaranthus (48.0%) followed by fodder cowpea (26.0%) > cluster bean (22%) and marigold (19.6%). EDTA and organics addition significantly increased the Pb content and its translocation from root to shoot.

7.6   Conclusions

Heavy metals refer to some significant elements of biological toxicity, including mercury (Hg), lead (Pb), chromium (Cr), cadmium (Cd), arsenic (As), etc., that enter soil as soil pollutants through various routes either knowingly or unknowingly due to increase in anthropogenic activities and industrialization. In recent years, rapid industrialization has increased the diversity of heavy metal deposition on the soil, resulting in serious environment deterioration. Crops grown in polluted soils get affected and pollutants are transported to edible plant parts gradually. Further, they migrate into the food chain by direct or indirect usage of respective crops. Heavy metals get recycled back to soil through various means like addition of crop residues and animal excreta etc. And thus the soil polluters follow the soil-plant-animal-human-soil continuum. Although there are several tolerant plants to various heavy metals at varied degrees, plants grown on these soils show an impairment in growth. Several available remediation methods, viz., bioremediation, microbial remediation, and electrokinetic remediation, have varying degrees of merits and demerits need to be selected based on the local conditions. Although bioremediation is an effective method of treating heavy metal polluted soils, the return of heavy metal to the soil cannot be ruled out.

References

Abumaizar RJ, Smith EH (1999) Heavy metal contaminants removal by soil washing. J Hazard Mater 70(1–2):71–86

Adriano DC (2003) Trace elements in terrestrial environments: biogeochemistry, bioavailability and risks of metals, 2nd edn. Springer, New York, pp 1–165

Alexander AP, Christos DT (2003) Lead (II) retention by Alfisol and clinoptilolite: cation balance and pH effect. Geoderma 115:303–312

Al-Khashman OA (2007) Determination of metal accumulation in deposited street dusts in Amman. Jordan. Environ Geochem Health 29(1):1–10

Alkorta I, Garbisu C (2001) Phytoremediation of organic contaminants in soils. Bioresour Technol 79(3):273–276

Al-Shayeb SM, Seaward MRD (2001) Heavy metal content of roadside soils along ring road in Riyadh (Saudi Arabia). Asian J Chem 13(2):407–423

Aseman-Bashiz E, Asgharnia H, Akbari H, Iranshahi L, Mostafaii GR (2014) Bioremediation of the soils contaminated with cadmium and chromium by the earthworm *Eisenia fetida*. Anuário do Instituto de Geociências - UFRJ 37(2):216–222

Assche F, Clijsters H (1990) Effects of metals on enzyme activity in plants. Plant Cell Environ 24:1–15

Awashthi SK (2000) Prevention of food adulteration act no 37 of 1954. Central and state rules as amended for 1999. Ashoka Law House, New Delhi

Aydinalp C, Marinova S (2003) Distribution and forms of heavy metals in some agricultural soils. Polish J Env Studies 12(5):629–633

Bona L, Wright RJ, Baligar VC, Matuz J (1993) Screening wheat and other small grains for acid soil tolerance. Landsc Urban Plan 27:175–178

Boopathy R (2000) Factors limiting bioremediation technologies. Bioresour Technol 74:63–67

Buekers J (2007) Fixation of cadmium, copper, nickel and zinc in soil: kinetics, mechanisms and its effect on metal bioavailability, Ph.D. thesis, Katholieke Universiteit Lueven, Dissertationes De Agricultura, Doctora atsprooef schrift nr. pp. 1–107

Campbell PGC (2006) Cadmium-a priority pollutant. Environ Chem 3(6):387–388

Chen TB, Wong JWC, Zhou HY, Wong MH (1997) Assessment of trace metal distribution and contamination in surface soils of Hong Kong. Environ Pollut 96(1):61–68

Chen B, Jia-nan L, Wang Z, Dong L, Jing-hua F, Juan-juan Q (2011) Remediation of Pb-resistant bacteria to Pb polluted soil. J. Environ Prot 2:130–141

Chibuike GU, Obiora SC (2014) Heavy metal polluted soils: effect on plants and bioremediation methods. Appl Environ Soil Sci. http://dx.Doi.Org/10.1155/2014/52708

China SP, Das M, Maiti SK (2014) Phytostabilization of mosaboni copper mine tailings: a green step towards waste management. Appl Ecol Environ Res 12(1):25–32

Chopra AK, Pathak C, Prasad G (2009) Scenario of heavy metal contamination in agricultural soil and its management. J App Natural Sci 1(1):99–108

Christoforidis A, Stamatis N (2009) Heavy metal contamination in street dust and roadside soil along the major national road in Kavala's region, Greece. Geoderma 151(3–4):257–263

Cunningham SD, Berti WR (1993) Remediation of contaminated soils with green plants: an overview. In-Vitro Cell Dev Biol 29:207–212

Dabke SV (2013) Vermi-remediation of heavy metal-contaminated soil. J Health Poll 4:4–10

Dai J, Becquerb T, Rouillerc JH, Reversata G, Bernhard-Reversata F, Nahmania J, Lavellea P (2004) Heavy metal accumulation by two earthworm species and its relationship to total and DTPA-extractable metals in soils. Soil Biol Biochem 36:91–98

Dandan W, Huixin L, Feng H, Xia W (2007) Role of earthworm-straw interactions on phytoremediation of cu contaminated soil by ryegrass. Acta Ecol Sin 27(4):1292–1299

Datta R, Darkar D (2004) Biotechnology in phytoremediation of metal contaminated soils. Proc Indian Nat Sci Acad B 70(1):99–108

Dissanayake CB, Chandrajith R (2009) Phosphate mineral fertilizers, trace metals and human health. J Nat Sci Found Sri Lanka 37(3):153–165

Dixit R, Wasiullah MD, Pandiyan K, Singh UB, Sahu A, Shukla R, Singh BP, Rai JP, Sharma PK, Lade H, Paul D (2015) Bioremediation of heavy metals from soil and aquatic environment: an overview of principles and criteria of fundamental processes. Sustain For 7:2189–2212

Djingova R and Kuleff I (2000) Instrumental techniques for trace analysis, In: Vernet JP (Ed), Trace elements: their distribution and effects in the environment, Elsevier, London

Ekperusi OA, Aigbodion IF (2015) Bioremediation of heavy metals and petroleum hydrocarbons in diesel contaminated soil with the earthworm: Eudrilus eugeniae. Ekperusi Aigbodion Springer Plus 4:540

Ekperusi OA, Aigbodion IF, Iloba BN, Okorefe S (2016) Assessment and bioremediation of heavy metals from crude oil contaminated soil by earthworms. Ethiopian J Environ Stud Manage 9(Suppl. 2):1036–1046

European Union (2002) Heavy metals in wastes, European commission on environment http://www.ec.europa.eu/environment/waste/studies/pdf/heavymetalsreport.pdf

Garaiyurrebaso O, Garbisu C, Blanco F, Lanzén A, Martín I, Epelde L, Becerril JM, Jechalke S, Smalla K, Grohmann E, Alkorta I (2017) Long-term effects of aided phytostabilisation on microbial communities of metal-contaminated mine soil. FEMS Microbiol Ecol 93(3):fiw252

Garbisu C, Alkorta I (1997) Bioremediation: principles and future. J Clean Tech Environ Toxicol Occup Med 6:1–16

Garbisu C, Alkorta I (2003) Basic concepts on heavy metal soil bioremediation. Eur J Mineral Proc Environ Protec 3(1):58–66

Garbisu C, Hernández-Allica J, Barrutia O, Alkorta I, Becerril JM (2002) Phytoremediation: a technology that uses green plants to remove contaminants from polluted areas. Rev Environ Health 17(3):173–188

Garg N, Singla P (2011) Arsenic toxicity in crop plants: physiological effects and tolerance mechanisms. Environ Chem Lett 9:303–321. doi:10.1007/s10311-011-0313-7

Gerhardt KE, Huang X, Glick BR, Greenberg BM (2009) Phytoremediation and rhizoremediation of organic soil contaminants: potential and challenges. Plant Sci 176(1):20–30

Girma G (2015) Microbial bioremediation of some heavy metals in soils: an updated review. Indian J Sci Res 6(1):147–161

Goblenz A, Wolf K, Bauda P (1994) The role of glutathione biosynthesis in heavy metal resistance in the fission yeast Schizosaccharomyces pombe. FEMS Microbiol Rev 14:303–308

Gowd SS, Reddy MR, Govil PK (2010) Assessment of heavy metal contamination in soils at Jajmau (Kanpur) and Unnao industrial areas of the ganga plain, Uttar Pradesh, India. J Hazard Mater 174:113–121

Hawkes JS (1997) Heavy metals. J Chem Educ 74:1369–1374

Hellekson D (1999) Bioventing principles, applications and potential. Restor Reclam Rev 5(2):1–9

Hemida SK, Omar SA, Abdel-Mallek AY (1997) Microbial populations and enzyme activity in soil treated with heavy metals. Water Air Soil Pollut 95(1):13–22

Hiroki M (1992) Effects of heavy metal contamination on soil microbial population. Soil Sci Pl Nutrit 38(1):141–147

Hopkin SP (1989) Ecophysiology of metals in terrestrial invertebrates. Elsevier, London

Iranzo M, Sainz-Pardo I, Boluda R, Sánchez J, Mormeneo S (2001) The use of microorganisms in environmental remediation. Ann Microbiol 51:135–143

Is L, Kim OK, Chang YY, Bae B, Kim HH, Baek KH (2002) Heavy metal concentrations and enzyme activities in soil from a contaminated Korean shooting range. J Biosci Bioeng 94(5):406–411

Jadia CD, Fulekar MH (2009) Phytoremediation of heavy metals: recent techniques. African J Biotech 8(6):921–928

Jiang W, Fan W (2008) Bioremediation of heavy metal–contaminated soils by sulfate-reducing bacteria. Ann N Y Acad Sci 1140:446–454

Kaur G, Hundal SS (2015) Bioremediation of heavy metal contaminated soil using earthworm *Eisenia fetida*. J Environ 04(02):25–29

Khan AG (2005) Role of soil microbes in the rhizospheres of plants growing on trace metal contaminated soils in phytoremediation. J Trace Elem Med Biol 18:355–364

Khan AG, Kuek C, Chaudhry TM, Khoo CS, Hayes WJ (2000) Role of plants, mycorrhizae and phytochelators in heavy metal contaminated land remediation. Chemosphere (41(1–2):197–207

Kim S, Moon S, Kim K (2001) Removal of heavy metals from soils using enhanced electrokinetic soil processing. Water Air Soil Pollut 125(1):259–272

Kirpichtchikova TA, Manceau A, Spadini L, Panfili F, Marcus MA, Jacquet T (2006) Speciation and solubility of heavy metals in contaminated soil using X-ray microfluorescence, EXAFS spectroscopy, chemical extraction, and thermodynamic modelling. Geochim Cosmochim Acta 70(9):2163–2190

Kuiper I, Lagendijk EL, Bloemberg GV, Lugtenberg BJJ (2004) Rhizoremediation: a beneficial plant-microbe interaction. Mol Plant-Microbe Inter 17(1):6–15

Kumar P, Jadhav PD, Rayalu SS, Devotta S (2007) Surface-modified zeolite–a for sequestration of arsenic and chromium anions. Curr Sci 92:512–517

Lasat MM (2002) Phytoextraction of toxic metals: a review of biological mechanisms. J Environ Qual 31:109–120

Leggo PJ, Ledesert B, Christie G (2006) The role of clinoptilolite in organo-zeolitic-soil systems used for phytoremediation. Sci Total Environ 363:1–10

Lema MW, Ijumba JN, Njau KN, Ndakidemi PA (2014) Environmental contamination by radionuclides and heavy metals through the application of phosphate rocks during farming and mathematical modeling of their impacts to the ecosystem. Int J Engg Res Gen Sci 2(4):852–863

Lenntech Water Treatment and Air Purification (2004) Water treatment. Lenntech, Rotterdamseweg, Netherlands (http://www.excelwater.com/thp/filters/Water-Purification. Inter J Adv Res. 2(6): 1043–1055

Levy DB, Barbarick KA, Siemer EG, Sommers LE (1992) Distribution and partitioning of trace metals in contaminated soils near Leadville, Colorado. J Environ Qual 21(2):185–195

Li J, Zhang GN, Li Y (2010) Review on the remediation technologies of POPs. Hebei Environl Sci:65–68

Liao M, Chen CL, Huang CY (2005) Effect of heavy metals on soil microbial activity and diversity in a reclaimed mining wasteland of red soil area. J Environ Sci (China) 17(5):832–837

Ling W, Shen Q, Gao Y, Gu X, Yang Z (2007) Use of bentonite to control the release of copper from contaminated soils. Aust J Soil Res 45(8):618–623

Lu M, Zhang Z (2014) Phytoremediation of soil co-contaminated with heavy metals and deca-BDE by co-planting of sedum alfredii with tall fescue associated with Bacillus Cereus JP12. Plant Soil 382:89–102

Luo C, Liu C, Wang Y, Liu X, Li F, Zhang G, Li X (2011) Heavy metal contamination in soils and vegetables near an e-waste processing site, South China. J Hazard Mater 186:481–490

Mahmood T (2010) Phytoextraction of heavy metals – the process and scope for remediation of contaminated soils. Soil Environ 29(2):91–109

Markus JA, Mcbratney AB (1996) An urban soil study: heavy metals in glebe, Australia. Aust J Soil Res 34(3):453–465

Marques APGC, Rangel AOSS, Castro PML (2009) Remediation of heavy metal contaminated soils: phytoremediation as a potentially promising clean-up technology. Critical Rev Environ Sci Tech 8:622–654

McLaughlin MJ, Zarcinas BA, Stevens DP, Cook N (2000a) Soil testing for heavy metals. Commun Soil Sci Plant Anal 31(11–14):1661–1700

McLaughlin MJ, Hamon RE, McLaren RG, Speir TW, Rogers SL (2000b) Review: a bioavailability-based rationale for controlling metal and metalloid contamination of agricultural land in Australia and New Zealand. Aust J Soil Res 38(6):1037–1086

Miller JR, Hudson-Edwards KA, Lechler PJ, Preston D, Macklin MG (2004) Heavy metal contamination of water, soil and produce within riverine communities of the Rio Pilcomayo basin, Bolivia. Sci Total Environ 320(2–3):189–209

Ming DW, Mumpton FA (1989) Zeolites in soils. In: Dixon JB, Weed SB (eds) Minerals in soil environments, 2nd edn. Soil Sci Soc Am, Madison, Wisconsin, pp 873–911

Mishra A, Shukla SK (2014) Heavy metal toxicity: a blind evil. J Forensic Res 5:e116. doi:10.4172/2157-7145.1000e116

Moosavi SG, Seghatoleslami MJ (2013) Phytoremediation: a review. Adv Agric Biol 1(1):5–11

Mortvedt JJ (1995) Heavy metal contaminants in inorganic and organic fertilizers. Fert Res 43(1):55–61

Mühlbachová G, Šimon T (2003) Effects of zeolite amendment on microbial biomass and respiratory activity in heavy metal contaminated soils. Plant Soil Environ 49:536–541

Mullins GL, Martens DC, Miller WP, Hallock DL (1982) Copper availability, form, and mobility in soils from three annual copper-enriched hog manure applications. J Environ Qual 11(2):316–320

Nahmani J, Hodson ME, Black S (2007) A review of studies performed to assess metal uptake by earthworms. Environ Pollut 145:402–424

Newman LA, Reynolds CM (2004) Phytodegradation of organic compounds. Curr Opin Biotechnol 15:225–230

Ololade IA (2014) An assessment of heavy-metal contamination in soils within auto-mechanic workshops using enrichment and contamination factors with geo-accumulation indexes. J Environ Prot 5:970–982

Pajević S, Borišev M, Nikolić N, Arsenov DD, Orlović S, Župunski M (2016) Phytoextraction of heavy metals by fast- growing trees: a review. In: A. Ansari et al. (eds.), Phytoremediation, Springer international publishing Switzerland 29–64

Prashanth VG, Prabha ML (2016) Bioremediation of contaminated lead soil by Eudrilus Eugeniae and synthesis of nanoparticles. Int J Medicine Res 1(2):31–34

Rai LC, Gaur JP, Kumar HD (1981) Phycology and heavy metal pollution. Biol Rev 56(2):99–151

Rajendran P, Muthukrishnan J, Gunasekharan P (2003) Microbes in heavy metal remediation. Indian J Exp Biol 41:935–944

Ramesh K, Reddy DD, Biswas AK, Subba-Rao A (2011) Zeolites and their potential uses in agriculture. Adv Agron 113:215–236

Raskin I, Kumar PBAN, Dushenkov S, Salt DE (1994) Bio concentration of heavy metals by plants. Curr Opin Biotechnol 5:285–290

Raskin I, Smith RD, Salt DE (1997) Phytoremediation of metals: using plants to remove pollutants from the environment. Curr Opin Biotechnol 8:221–226

Romic M, Romic D (2003) Heavy metals distribution in agricultural top soils in urban area. Environ Geol 43(7):795–805

Ross SM (1994) Toxic metals in soil–plant systems. Wiley, Chichester, p 469

Sahu P, Sharma S (2016) Mercury and lead accumulation by *Eudrilus eugeniae* in soils amended with Vermicompost. Biol Forum 8(1):565–569

Sakakibara M, Watanabe A, Inoue M, Sano S, Kaise T (2010) Phytoextraction and phytovolatilization of arsenic from As-contaminated soils by *Pteris vittata*, Proceedings of the Annual International Conference on Soils, Sediments, Water and Energy: Vol. 12 , Article 26

Salt DE, Blaylock M, Kumar PBAN, Dushenkov V, Ensley BD, Chet L, Raskin L (1995) Phytoremediation: a novel strategy for the removal of toxic metals from the environment using plants. Biotechnology 13:468–474

Salt DE, Smith RD, Raskin I (1998) Phytoremediation. Annu Rev Plant Physiol Plant Mol Biol 49:643–668

Sanyal SK (2001) Colloid chemical properties of soil humic substances– a relook. J Indian Soc Soil Sci 49:537–569

Shakeri IA, Moore F, Modabberi S (2009) Heavy metal contamination and distribution in the shiraz industrial complex zone soil, south shiraz, Iran. World Appl Sci J 6(3):413–425

Sharma RK, Agrawal M, Marshall F (2007) Heavy metal contamination of soil and vegetables in suburban areas of Varanasi, India. Ecotoxi Environ Safety 66(2):258–266

Shiowatana J, McLaren RG, Chanmekha N, Samphao A (2001) Fractionation of arsenic in soil by a continuous-flow sequential extraction method. J Environ Qual 30(6):1940–1949

Shrotriya N, Joshi JK, Mukhiya YK, Singh VP (1984) Toxicity assessment of selected heavy metals, herbicides and fertilizers in agriculture. Int J Environ Stu 22(3–4):245–248

Shukla AK, Behera SK (2012) Progress report 2007–10. All India Coordinated Research Project of Micro and Secondary Nutrients and Pollutant Elements in Soils and Plants, ICAR-IISS, Bhopal, p 102

Shukla AK, Tiwari PK (2013). Progress report 2011–2013. AICRP-MSN, pp 93–120

Shukla AK, Tiwari PK (2014) Progress report 2011–13. All India Coordinated Research Project of Micro and Secondary Nutrients and Pollutant Elements in Soils and Plants, ICAR-IISS, Bhopal, p 155

Sinha B, Bhattacharya K (2011) Retention and release isotherm in arsenic-humic/fulvic equilibrium study. Biol Fertil Soils 47:815–822

Sizmur T, Hodson ME (2009) Do earth worms impact metal mobility and availability in soil? – a review. Environ Pollut 157:1981–1989

Smith LA, Means JL, Chen A et al (1995) Remedial options for metals-contaminated sites. Lewis Publishers, Boca Raton

Sobolev D, Begonia MF (2008) Effects of heavy metal contamination upon soil microbes: lead-induced changes in general and denitrifying microbial communities as evidenced by molecular markers. Int J Environ Res Public Health 5(5):450–456

Spurgeon DJ, Hopkin SP, Jones DT (1994) Effects of cadmium, copper, lead and zinc on growth, reproduction and survival of the earthworm *Eisenia fetida* (Savigny): assessing the environmental impact of point-source metal contamination in terrestrial ecosystems. Environ Pollut 84:123–130

Stalin P (2011) Annual report 2010-11. AICRP-MSN Coimbatore Centre. pp. 5–6

Stalin P, Malathi P, Muthumanickam D (2014) Annual Report 2013–14, All India Coordinated Research Project of Micro and Secondary Nutrients and Pollutant Elements in Soils and Plants, Coimbatore Centre .1–.10

van Straalen NM, Denneman CAJ (1989) Ecotoxicological evaluation of soil quality criteria. Ecotox Environ Safety 18(3):241–251

Sun YB, Sun GH, ZhouQX XYM, Wang L, Liang XF, Sun Y, Qing X (2011) Induced-phytoextraction of heavy metals from contaminated soil irrigated by industrial wastewater with marvel of Peru (*Mirabilis jalapa* L.) Plant Soil Environ 57(8):364–371

Surendra-Babu P, Patnaik MC, Khadke KM (2012) Annual report 2011-12 all India Coordinated Research Project of Micro and Secondary nutrients and pollutant elements in soils and plants, Hyderabad center

Taiz L, Zeiger E (2002) Plant physiology. Sinauer Associates, Sunderland

Tang JC, Wang RG, Niu XW, Wang M, Chu HR, Zhou QX (2010) Characterization of the rhizoremediation of petroleum-contaminated soil: effect of different influencing factors. Biogeosciences 7:3961–3969

Tangahu BV, Abdullah SRS, Basri H, Idris M, Anuar N, Mukhlisin M (2011) A review on heavy metals (as, Pb and hg) uptake by plants through phytoremediation. Int J Chem Engg 2011:1–31

Tsadilas CD (2000) Effect of soil pH on the distribution of heavy metals among soil fractions. In: Iskandar I (Ed.), Environment restoration of metals contaminated soils. Lewis Publishers, pp.107–119

USEPA (1996) Report: recent developments for in situ treatment of metals contaminated soils, U.S. Environmental Protection Agency, Office of Solid Waste and Emergency Response. pp. 1–44

USEPA (2007) Treatment technologies for site cleanup: annual status report (12th edn), Technical report EPA 542-R-07-012, Solid Waste and Emergency Response [5203P], Washington DC, USA. pp.1-H2

Usmani Z, Kumar V (2015) Role of earthworms against metal contamination: a review. J Biodivers Environ Sci 6(1):414–427

Veselý T, Tlustoš P, Száková J (2011) The use of water lettuce (*Pistia stratiotes* L.) for rhizofiltration of a highly polluted solution by cadmium and lead. Int J Phytoremediation 13(9):859–872

Virkutyte J, Sillanpaa M, Latostenmaa P (2002) Electrokinetic soil remediation - critical overview. Sci Total Environ 289(1–3:97–121

Wang YP, Shi JY, Wang H, Lin Q, Chen XC, Chen YX (2007) The influence of soil heavy metals pollution on soil microbial biomass, enzyme activity, and community composition near a copper smelter. Ecotox Environ Safety 67(1):75–81

Wood JM (1974) Biological cycles for toxic elements in the environment. Science 183:1049–1052

Wuana RA, Okieimen FE (2011) Heavy metals in contaminated soils: a review of sources, chemistry, risks and best available strategies for remediation. ISRN Ecol 2011(402647):20–31. doi:10.5402/2011/402647

Wuana RA, Okieimen FE, Imborvungu JA (2010) Removal of heavy metals from a contaminated soil using chelating organic acids. Int J Environ Sci Technol 7(3):485–496

Xie Y, Fan J, Zhu W, Amombo E, Lou Y, Chen L, Fu J (2016) Effect of heavy metals pollution on soil microbial diversity and bermuda grass genetic variation. Front Plant Sci 7:755. doi:10.3389/fpls.2016.00755

Yadav BK, Siebel MA, vanBruggen JJA (2011) Rhizo filtration of a heavy metal (lead) containing wastewater using the wetland plant *Carex pendula*. Clean Soil Air Water 39(5):467–474

Yao Z, Li J, Xie H, Yu C (2012) Review on remediation technologies of soil contaminated by heavy metals. Procedia Environ Sci 16:722–729

Zarcinas BA, Pongsakul P, McLaughlin MJ, Cozens G (2004) Heavy metals in soils and crops in Southeast Asia 2. Thailand Environ Geochem Health 26(3):359–371

Zhao K, Fu W, Ye A, Zhang C (2015) Contamination and spatial variation of heavy metals in the soil-rice system in Nanxun county, southeastern China. Int J Environ Res Public Health 12:1577–1594

Zhou DM, Hao XZ, Xue Y (2004) Advances in remediation technologies of contaminated soils. Ecol Environ Sci 13(2):234–242

Current Trends in Salinity and Waterlogging Tolerance

8

Parbodh C. Sharma, Arvind Kumar, and T.V. Vineeth

Abstract

Soil salinity and waterlogging together impair crop production on at least one-fourth of the irrigated land worldwide and cause yield loses ranging from 15 to 80%. Much has been reported on plant accumulation to waterlogging and salinity in terms of physiological, biochemical and anatomical modifications. Genome-level profiling coupled with systematic genetic analysis is the need of the hour to understand the underlying mechanism regulating stress tolerance. The accumulation of organic osmolytes and proteins from the late embryogenesis abundant (LEA) superfamily adds on to maintenance of low intracellular osmotic potential of plants. A significant upregulation of several other pathways including calcium signalling, sulphur assimilation and ROS detoxification is associated with salinity stress response. Salinity stress induces widespread proteome modification in crop plants, and study of proteome has proved to be a very efficient approach to study plant salt stress tolerance. Approaches like metabolic fingerprinting, metabolite profiling and targeted analysis are gaining wide importance to investigate salinity stress tolerance of crops. Plants respond to low oxygen condition in three distinct stages. These are signal transfer (Stage a, 0–4 h), metabolic reprogramming (Stage b, 4–24 h) and morphological transformations brought about by the first two stages (Stage c, 24–48 h). Genomic studies have classified genes activated under waterlogging stress into three categories: low oxygen-sensing and cell signalling-involved genes, metabolic adjustment genes and genes that maintain plant internal microenvironments. Search for quantitative trait loci (QTLs) controlling vital traits linked to waterlogging tolerance has gained momentum in recent years. Integration of all the key omics approaches

P.C. Sharma (✉) • A. Kumar • T.V. Vineeth
Central Soil Salinity Research Institute, Karnal 132 001, Haryana, India
e-mail: pcsharma.knl@gmail.com; singh.ak92@gmail.com; vinee2705@gmail.com

© Springer Nature Singapore Pte Ltd. 2017
P.S. Minhas et al. (eds.), *Abiotic Stress Management for Resilient Agriculture*,
DOI 10.1007/978-981-10-5744-1_8

should be another vital target to research groups working across the globe to further progress towards the ultimate goal of developing salt tolerant crop varieties without compromising yield.

8.1 Introduction

Any external limitation (salinity, high temperature, water, biotic factors, etc.) that lowers plants' capacity to fix atmospheric carbon dioxide into triose phosphates is defined as stress (Grime 1977). Global agriculture is struck with the primary challenge to produce 70% more food for the burgeoning population in an era where productivity of crops is approaching towards stagnation. Moreover, total production is also decreasing rapidly because of the adverse effects of various environmental stresses. To limit this crop loss due to various abiotic and biotic stressors is a major area of research to feed the ever-increasing population (Shanker and Venkateswarlu 2011). Among the abiotic stressors, excess soluble salts, water deficit, waterlogging and heat negatively affect the growth and yield of staple food crops up to 70% (Ahmad et al. 2012).

Excess salts in the soil are one of the major constraints which impair crop production of at least one-fifth of the global irrigated land. The issue has been further aggravated by unscientific agricultural practices such as intensive irrigation (Zhu 2001). The negative effect of excess ions such as Na^+ and/or Cl^- on crop plants is called salt stress (Munns 2005). On the basis of type, characteristics and crop growth relationships in saline soils, two main types of soils have been coined by Szabolcs (1974). These are (i) saline soils, in which NaCl and Na_2SO_4 and SO_4^- of Ca^{2+} and Mg^{2+} are the major and partial contributors, respectively, and (ii) sodic soils – Na_2CO_3 is the major player in this category which is capable of alkaline hydrolysis.

Excessive soil moisture in rhizosphere or waterlogging is also another major abiotic stress in which diffusion of gases is reduced almost four orders of magnitude compared with that in the atmosphere. Waterlogging depleted the oxygen in the root zone either complete (anoxia) or partial (hypoxia) (Malik et al. 2002). Oxygen is rapidly depleted, whereas gases like CO_2 and ethylene accumulate rapidly approaching gradual anaerobiosis in hours or even in days. Measurements of intensity of waterlogging relate to the chemical changes which are associated with oxidation and reduction status of soil environment. With time of waterlogging, soil loses much or all of its O_2, and concentrations of other gases increase; certain microelements are reduced and increase in soil solution that showed phytotoxicity. In addition, energy deficiency or limitations faced by the plant which cause changes in root permeability are also important factors. In view of the changing climate, erratic distribution of rainfall may often lead to waterlogging. Soil physical properties, i.e. low hydraulic conductivity and crust on the soil surface or of a subsoil hardpan especially in alkaline soil, contribute to waterlogging event. The yield loss caused by waterlogging may range from 15 to 80%, varying with duration of the stress, soil types and tolerance of different species. The term 'waterlogging' is defined as a condition of the soils where excess water inhibits gas exchange of roots with the atmosphere. Waterlogging can be differentiated from 'flooding' because of the partial or complete submergence of the shoot in the case of flooding. Waterlogging may be defined as plant survival or relative growth rate, grain yield or biomass accumulation under waterlogging compared with non-waterlogged conditions (Setter and Waters 2003). Submergence blocks direct exchange of gases between the entire plant body and air resulting in decreased O_2 and CO_2 levels. Over and above complete submergence usually reduces the photosynthetic rate due to low concentrations of CO_2 and limited light access underwater (Colmer 2003).

8.2 Salt Stress

Build-up of excess salts over a period of time mediated by natural processes in the soil or groundwater is called as primary salinity. Weathering of parent materials which break down rocks and release soluble salts of various types and the deposition of oceanic salt carried in wind and rain are the two major contributors towards primary salinity. 'Cyclic salts' mainly constitute sodium chloride which is ocean salts carried into the land via wind. Human activities that change the hydrologic balance of the soil in terms of water applied (irrigation or rainfall) and water used by crops (transpiration) result in secondary salinization (Garg and Manchanda 2008). Land clearing and monocropping with annual crops combined with unscientific, intensive irrigation using saline ground water with poor drainage are the major causes of secondary salinity. As per the United Nations Food and Agriculture Organization (FAO) reports, more than 400 million hectares of the global area are affected by salinity (Koohafkan 2012). This trend of increase in salinization of proper agricultural land is projected to reduce the availability of cultivable land by 50% in 2050 (Rani et al. 2012). Removal of excess salts from the crop root zone by fresh water irrigation and followed by leaching or improving the tolerance ability of the crop plants to such elevated levels of salts are the two approaches to combat soil salinity (Plaut et al. 2013).

8.2.1 Crop Plants Under Saline Condition

Plant growth is affected by excess salts in two major ways. Firstly, the presence of excessive salts in the soil reduces the water potential of the soil solution and thereby impedes the water potential gradient which is highly inevitable for plants to take up water from the soil solution. This is called as the osmotic phase of salinity. Secondly, excess salts, if entered in to the sensitive plant tissues via the transpiration stream, may cause specific cellular distortions and growth reduction. This is called as the ionic phase of salinity (Greenway and Munns 1980). The adverse effects of salinity on crop plants can be summarized as follows:

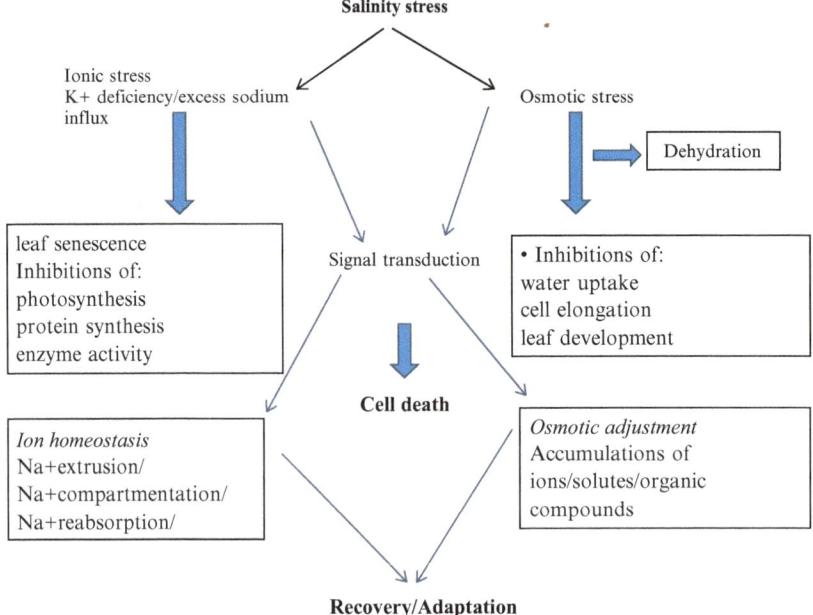

Fig. 8.1 Effects of high salinity on plants and the corresponding adaptive responses

- Excess salts create cellular osmotic imbalance and specific ionic effects which together hinder crop growth and development (Zhu 2002). Therefore, salinity affects almost all primary metabolic pathways of plant leading to its demise.
- Salinity mediated cellular ionic imbalance leads to altered cellular K^+/Na^+ ratio. The high sodium content in external soil solution negatively impacts intracellular K^+ influx.
- The build-up of cellular Na^+ and Cl^- ions can be ultimately detrimental to the cell. The excess Na^+ dissipates the membrane potential and therefore facilitates the uptake of Cl^- down the gradient.
- Extremely high Na^+ (> 100 mM) inhibits the activity of many enzymes, affects cell division and expansion and specifically damages the cellular membrane network paving way for extreme growth reduction.
- Increased Na^+ ions negatively affect net carbon assimilation and may produce various reactive oxygen species due to altered cellular energetic.
- Alteration in the cellular K^+ content negatively affects activity of enzymes requiring K^+ as cofactor and disturbs the cellular osmotic balance and may also affect stomatal functions.
- High salinity can be detrimental to sensitive plant tissues like leaves because of the specific ionic effects. It may cause salt injury, and excess salts tend to accumulate preferentially in the older leaves, thereby protecting the younger leaves from desiccation (Munns et al. 2006).
- High salinity has been reported to negatively affect cortical microtubule organization and helical growth in *Arabidopsis* (Shoji et al. 2006). Figure 8.1 depicts the overall effect of salinity on plants and corresponding adaptive behaviour.

8.2.2   Adverse Effects of Waterlogging or Submergence Stress

Waterlogging triggers a number of different stress responses in plants. Whereas oxygen concentrations in well-drained, porous soil are nearly equal to atmospheric concentrations (20.6% oxygen, 20.6 kPa), the diffusion coefficient of oxygen in water is four times lower than that in air. When flooding occurs, soil gases are replaced with water thereby reduced the entry of oxygen into the soil and making it difficult for roots and other parts to carry out respiration. Free water is essential for the growth of all higher plants. However, excess water in the rhizosphere of plants is detrimental or lethal when it forms a barrier between soil and free transfer of gases, such as oxygen (O_2) and carbon dioxide (CO_2) (Drew 1997), with the effect of meagre supply of oxygen being the most significant. Symptoms of waterlogging stress in young plants may also be acquainted with hydroponic plants lacking in proper aeration (Trought and Drew 1980). Fully immersed plants may also experience shading stress (Sarkar and Panda 2009). The waterlogging symptoms in plant traits are given in Table 8.1.

8.2.3   Elemental Toxicities Under Waterlogging

The decreased redox potential incepted by flooding stress increases the accumulation and solubility of many toxic metals, including manganese, iron, sulphur and phosphorus (Jackson and Drew 1984). Besides elemental toxicities in root tips, enhanced concentration of secondary metabolites such as phenolics and volatile fatty acids could be injurious in the low-pH rhizosphere (Shabala 2011). Waterlogging further reduced soil pH by higher concentration of CO_2 and volatile organic acids accumulated in the root zone (Greenway et al. 2006). Under waterlogged soils ethylene plays as metabolite toxicity which supresses the root expansion; however, re-entrance of oxygen in recovery phase and ethanol in anoxic cells transformed into acetaldehyde that may cause cell injuries (Bailey-Serres and Voesenek 2008). Waterlogging is an inherent phenomenon of heavy textured sodic soils, and it was demonstrated that under waterlogged conditions, the accumulation of Na^+ and Cl^- in shoots is greater than salt stress condition because energy (ATP) crisis caused by waterlogging-induced anoxic stress attenuates the exclusion of Na^+ and Cl^- (Teakle et al. 2010).

8.2.4   Salt Tolerance Strategies in Plants

The specific cellular mechanism causing minimum loss of yield in a saline soil compared to the maximum yield in a normal soil is called salinity tolerance. Resistance strategy to excess salts can be mainly grouped into (1) avoidance and (2) tolerance (cellular and molecular approaches) mechanisms (Fu et al. 2011; Vinocur and Altman 2005). The halophytes which can tolerate high amounts of salt display an efficient avoidance mechanism whereby which it separates salts from sensitive

Table 8.1 Effect of waterlogging on morphological and physiological traits

Traits/Effects	References
Morphological traits	
Chlorosis of lower leaves	Van Ginkel et al. (1992)
Early senescence of lower leaves	Dong et al. (1983) and Dong and Yu (1984)
Decreased plant height	Sharma and Swarup (1989) and Wu et al. (1992)
Delayed ear emergence	Sharma and Swarup (1989)
Reduced root and shoot growth	Huang and Johnson (1995)
Lower number of spike-bearing tillers	Sharma and Swarup (1989) and Wu et al. (1992)
Fewer grains per spikelet and reduced kernel weight	Van Ginkel et al. (1992)
Histological traits	
Reduced diameter of metaxylem and protoxylem vessels of the nodal roots	Huang et al. (1994)
Enhanced formation of aerenchyma cells in the cortical tissue of both seminal and nodal roots	Huang et al. (1994) and Boru (1996)
Leakage of cell electrolytes	Wang et al. (1996)
Physiological traits	
Reduced uptake of N, P, K^+, Ca, Mg and Zn while increasing Na^+, Fe and Mn absorption under alkaline soil conditions	Sharma and Swarup (1989) and Stieger and Feller (1994b)
Reduced root respiration	Wu et al. (1992)
Lowered rates of plant photosynthesis, stomatal conductance and transpiration	Dong and Yu (1984)
Inhibited transport of sugars from the shoots to the roots	Waters et al. (1991a, b)
Inhibited biosynthesis of new tissue	Attwell et al. (1985)
Spikelets fertility linked to lower transpiration and hence low uptake of boron (and other nutrients)	Rawson and Subedi (1996)
Ethylene production increases and acts as a trigger (not promoter) of accelerated wheat plant senescence	Dong et al. (1983)
Exogenous cytokinins delayed degradation of chlorophyll and other biochemical processes	Dong and Yu (1984)
Enhancement of ACC (1-aminocyclopropane-1-carboxylic acid), its precursor and ethylene in older leaves	Dong et al. (1986)
Less nitrogen concentrates and accumulates in the upper leaves	McDonald and Gardner (1987)
Nitrogen remobilization from lower leaves is accelerated on flooded soils and explains their chlorosis	Stieger and Feller (1994b)
Reduced rooting depth and increased root porosity	Yu et al. (1969)

Adopted from Samad et al. (2001)

plant parts. The avoidance mechanisms are mostly a result of whole plant level responses which try to keep the plant away from excess salts (Kumar Parida and Jha 2010). Some tactics of salt avoidance in halophytes are as follows:

- Exclusion: This is the primary avoidance mechanism by which excess ions are pumped out of the root cell.
- Secretion: Certain crop species possess specialized cells called as salt glands and specialized hairs through which they secrete excess salts.
- Shedding: Most crop species preferentially accumulate salts in the older leaves and shed them off to protect the actively growing younger leaves from the detrimental effects of salt.
- Succulence: This is an adaptive behaviour of certain species which possess thick leaves with large mesophyll cells and lesser surface area which minimizes water loss.
- Stomatal response: This is seen in certain halophytes where, in order to limit the intake of Na^+, their guard cell uses K^+ instead of Na^+ (Roy et al. 2014).

These avoidance mechanisms, in general, cannot be manipulated on a practical scale to achieve improved salt resistance. The mechanism behind avoidance is in turn dependent on certain other mechanism at the cellular level (Vinocur and Altman 2005). However, tolerance mechanism, which is the ability of crop plants to modify their cellular ambience in such a way as to grow and develop under saline condition, is amenable to manipulations. Improvement in salinity tolerance of crop plants has been obtained by classical breeding, comprising of large-scale screening of land races and germplasm (Roy and Sengupta 2014), transgressive segregation (Shahbaz and Ashraf 2013) and triple test cross (Sadat Noori and Sokhansanj 2004). Genetic engineering is another tool which has been applied in the development of salt-tolerant crop species by introgression of genes/QTLs controlling salt stress signalling at the cellular level (Hossain et al. 2007; Roy et al. 2014; Vinocur and Altman 2005). Omics tools like genomics, transcriptomics, proteomics, metabolomics, etc. are another aspect of advanced molecular biology which aids to identify and functionally characterize the salt tolerance components and mechanisms at the molecular level (Inan et al. 2004), and these are briefed below.

8.2.5 Genomics Under Salt Stress

Plants respond to abiotic stresses at the cellular level and whole plant level on a synergistic basis. The primary signal related to any abiotic stress is first sensed by the receptors (histidine kinases, F-box proteins, G-protein-coupled receptors), which leads to a spike in the production of numerous secondary signal molecules, such as Ca^{2+}, inositol phosphates, ROS and abscisic acid. These molecules together with the residual stress signal activate a set of transcription factors inside the nucleus to upregulate or downregulate multiple stress-responsive genes, the products of which ultimately lead to improved salt stress tolerance (Fig. 8.2). Tolerant crop species may contain certain unique differentially expressing stress-responsive genes as compared to sensitive species. Genome-level profiling, together with systematic genetic analysis, has a huge potential in bringing out novel candidate genes/QTLs along with key signalling networks that control salt stress tolerance.

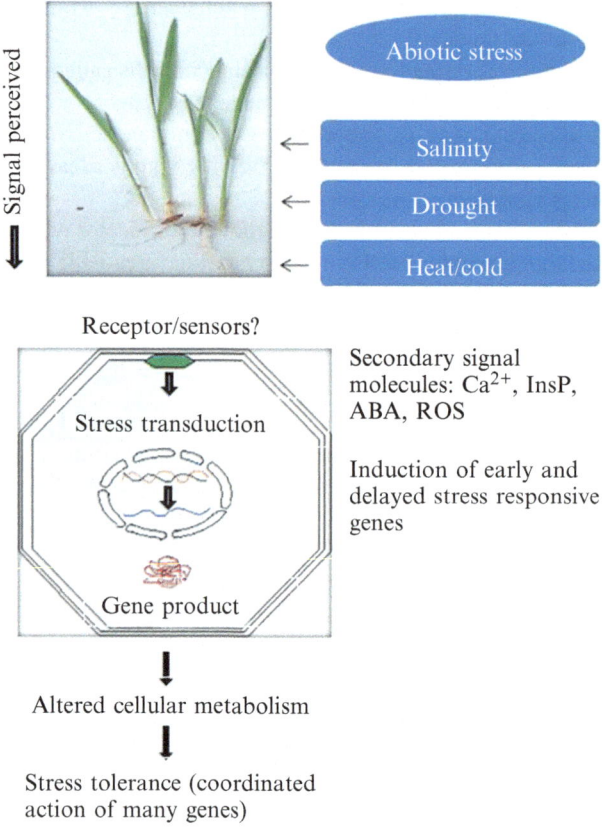

Fig. 8.2 Generic pathway for plant response to stress

8.2.6 Role of *NHX* and *SOS* in Salinity Tolerance

Plants maintain a low cellular Na^+/K^+ ratio by virtue of two major antiporters, i.e. Na^+/H^+ exchanger 1 (NHX1) located on the vacuolar membrane (Blumwald and Poole 1985) and cell membrane-localized salt overly sensitive 1 (SOS1) (Yamaguchi et al. 2013). A genetic screen identified three distinct salt overly sensitive *SOS* mutant loci in *Arabidopsis thaliana*, which make plants extremely sensitive to elevated levels of Na^+ and Li^+ but not to the general osmotic stress (Zhu et al. 1998). Initial research groups working on *SOS* mutants found retarded growth of these mutants under low K^+ conditions, which led them to assume that *SOS* mutant loci are vital components for K^+ acquisition and signal transduction during salinity stress (Wu et al. 1996; Zhu et al. 1998). Allelic tests confirmed that *SOS* mutants delineate three SOS loci, i.e. *SOS1*, *SOS2* and *SOS3* (Martínez-Atienza et al. 2007). Mutants of *SOS1* exhibited deficiency in the high-affinity K^+ uptake in roots, which suggested a major role of *SOS1* locus in high-affinity K^+ absorption into roots (Wu et al. 1996). Interestingly, however, the *SOS1* gene was later found to encode $Na^+/$

H⁺ antiporter, located at the cell membrane which pumps Na⁺ out of the cell (Shi et al. 2000). Apart from its role as a Na⁺ transporter, *SOS1* possesses a large cytosolic domain including Na⁺ sensor.

SOS2 and *SOS3* loci were found to encode a protein kinase and a Ca^{2+}-binding protein, respectively (Halfter et al. 2000). They were later classified into larger protein families of calcineurin B-like proteins (CBL) and CBL-interacting protein kinases (CIPK), and therefore *SOS2* and *SOS3* are also known as *CIPK24* and *CBL4*, respectively (Martínez-Atienza et al. 2007). The complete model depicting the functional salt overly sensitive pathway was later proposed in which the N-myristoylated SOS3 forms a dimer with SOS2, and this complex gets targeted to the plasma membrane where it phosphorylates and activates the Na⁺ efflux protein SOS1 (Qui et al. 2002). Calcineurin has been reported to play a pivotal role in the regulation of Na⁺ and K⁺ transport in yeast. Calcineurin B mutants displayed increased sensitivity to growth inhibition by Na⁺ and Li⁺ stresses (Mendoza et al. 1994). Apart from its regulation by *SOS2*, *SOS1* may also be regulated by *SOS4*. *SOS4* is involved in the formation of pyridoxal-5-phosphate. *SOS1* protein consists of a putative binding sequence for this cofactor, and thus this cofactor may act as a ligand for *SOS1* protein (Shi and Zhu 2002). Quan et al. (2007) reported SOS3-like calcium-binding protein8 (SCaBP8) that together with SOS3 activates *SOS2*. Lin et al. (2009) further reported that *SOS2* also phosphorylates and activates downstream SCaBP8 but not SOS3. Recently, *SOS4* and *SOS5* have also been characterized. *SOS4* encodes a pyridoxal (PL) kinase involved in the biosynthesis of pyridoxal-5-phosphate (PLP). *SOS5* has been reported to play a major role in cell wall formation and maintenance of its wholeness which are critical aspects of normal growth and cellular expansion under salt stress (Mahajan et al. 2008). Overall, it can be concluded that each and every component of this versatile pathway interacts with other branching components and may cross talk to maintain ionic homeostasis and thereby attain improved salt tolerance.

Enhanced salt tolerance has been reported in *Arabidopsis*, tomato and rice plants which constitutively overexpressed *NHX1* gene (Zhang et al. 2001). These *NHX1*overexpressing plants showed increased vacuolar K⁺ content and increased translocation of K⁺ from root to shoot (Bassil et al. 2011). NHX-type proteins are highly important for efficient localization of cellular K⁺ in vacuoles and are key candidate players involved in maintenance of cellular pH homeostasis (Barragan et al. 2012). Moreover, tomato *LeNHX3* maps to a quantitative trait locus (QTL) related to leaf Na⁺ accumulation (Villalta et al. 2008). Apart from its primary role as antiporters, NHX proteins play a much wider role as osmoregulators, cell growth regulators, vesicular trafficking, protein processing and overall plant development (Leidi et al. 2010; Krebs et al. 2010).

8.2.7 *HKT* Genes for Salinity Tolerance and Root to Shoot Na⁺ Partitioning

The first identified *HKT* gene was that of wheat (*Triticum aestivum*) (*TaHKT2;1*) which mediates Na⁺/K⁺ cation transport (Rubio et al. 1995). Further, this led to the identification and characterization of multiple *HKT* genes from different crop species (Horie et al. 2009). *AtHKT1;1* and its rice ortholog *OsHKT1;5* played a major role in pumping out excess Na⁺ from the xylem sap into the parenchyma cells thereby keeping Na⁺ away from the sensitive leaves (Davenport et al. 2007). In such a kind of HKT-mediated xylem, Na⁺ unloading is highly imperative for improved salt tolerance in many crops (James et al. 2006; Byrt et al. 2007). Analysis of QTLs controlling salt tolerance led to the identification of another strong salt tolerance QTL named *Nax1* in wheat (Munns 2003). This locus has been mapped to a region that encodes HKT transporter (*TaHKT1;4*) which plays a major role in removing excess Na⁺ from xylem into surrounding cells thereby protecting the more sensitive leaf blade from this dangerous ion (Huang et al. 2006).

8.2.8   Genes Contributing to Osmotic or Protective Function

Salinity mediated accumulation of compatible solutes like proline, N containing compounds like glycine betaine, polyols, polyamines and compounds belonging to the late embryogenesis abundant (LEA) superfamily plays a major role in lowering the intracellular water potential of plants and thereby regaining the lost water potential gradient (Verslues et al. 2006). Proline is one among the major amino acids whose level increases several folds under salt stress, whereas its catabolism is stimulated during recovery from stress (Sharma and Verslues 2010). Proline has been proposed to play a different role during recovery phase, in terms of signalling agent, regulator of cell proliferation, cell death and inducer of stress recovery-related genes (Szabados and Savoure 2010). Mutant plants lacking the *P5CS1* gene coding for pyrroline-5-carboxylate synthase enzyme involved in proline biosynthesis have been reported to be salt hypersensitive (Szekely et al. 2008). Apart from its primary role in osmotic adjustment, proline has been found to play key roles as molecular chaperone, redox buffer and ROS scavenger and in protection of membranes and associated proteins during stress conditions (Verbruggen and Hermans 2008). Glycine betaine is another N-containing compound which has been reported to act as an organic osmolytes in several crop species. Crop plants accumulating glycine betaine under stress have thrown light on its mechanism of action by protecting the cellular membranous network and associated macromolecules under stress (Guinn et al. 2011). Glycine betaine has also been reported to act as a key regulator of antioxidant enzyme activities (Chen and Murata 2011), whereas its direct role in ROS quenching is yet to be conclusively proven. Table 8.2 depicts the list of genes encoding various compatible solutes whose overexpression improved salt tolerance in different plant species.

8.2.9   Genes that Control Cell and Tissue Growth Rates

Genes controlling cell and tissue growth rates are a part of a larger signalling network comprising of a relay of units starting from the sensor, intermediary kinases

Table 8.2 Osmolyte(s) and their corresponding gene(s) whose overexpression imparts salt stress tolerance

Solute type	Natural conct. (mM)	Overexpression studies that resulted into increased salt tolerance				
		Solute(s)	Gene(s)	Species	Conct. (mM)	Reference
N-containing solutes	1–50	Proline, glycine, betaine	*P5CS mod*, *codA*, *P5CS*	Tobacco, rice	5–60	Nounjana et al. (2012)
Trehalose	0.005–0.10	Trehalose	*otsA, otsB*, *AtTPS1*, *AtTPS2*	Rice	0.3	Baea et al. (2005)
Straight chain polyhydric	1–50	Mannitol, sorbitol	*mt1D*, *S6PDH*, *OemaT1*	Wheat, tobacco, persimmon	2–60	Abebe et al. (2003)
Cyclic polyhydric alcohols	1–200	Ononitol	*imt1*	Tobacco	35	Conde et al. (2007)

Adapted from Parihar et al. (2015)

and phosphatases, hormones, transcription factors and an array of secondary signalling molecules. Among these players, transcription factors are vital in connecting the sensory units to the ultimate tolerance responses. Major families of transcription factors including basic helix-loop-helix (bHLH), basic leucine zipper (bZIP), WRKY, APETALA2/Ethylene Response Factor (AP2/ERF) and NAC have been reported to exhibit differential expression pattern under salt stress condition (Golldack et al. 2011; Yang et al. 2009; Jiang and Deyholos 2009; Kasuga et al. 1999; Cui et al. 2013; Jiang et al. 2009; Tran et al. 2004). These transcription factors specifically control the expression levels of various downstream target genes that ultimately affect the tolerance level to salt stress. Modulations in transcriptome have been reported to occur as early as 3 h after imposition of salt stress (Geng et al. 2013). Ethylene has been recently proven to enhance salt tolerance in *Arabidopsis* by maintaining a lower Na^+/K^+ ratio under stress in shoot tissues (Jiang et al. 2013). Elevated ethylene in *eto1* mutants lacking ethylene overproducer1 displayed oxidative respiratory burst leading to ROS production in the root stellar tissues. This elevated ROS levels, in turn, reduced the Na^+ influx in roots and reduced Na^+ xylem loading and thereby maintained a lower cellular Na^+/K^+ ratio under stress conditions (Jiang et al. 2013).

8.2.10 Transcriptomic Approach Under Salt Stress

In the holistic study of the entire mRNA transcripts that collectively represent the spatial and temporal gene expression of a cell, tissue of a plant species under a specific biological situation is defined as transcriptomics (Thompson and Goggin 2006).

Sharma et al. (2015) reported transcriptome sequencing of *B. juncea* var. CS-52 grown under control and imposed salt stress conditions. Differential gene expression analysis under imposed salt stress revealed numerous genes which may be specifically linked to various signalling pathways contributing to salt tolerance. Differential expression analysis of major salt-responsive genes in *Brassica juncea* and *Brassica nigra* (salt tolerant and susceptible, respectively) showed constitutive expression pattern of majority of the genes in *Brassica juncea* under control conditions. Specific upregulation of cellular metabolic pathways including osmolytes biosynthesis, sulphur assimilation, calcium signalling and ROS scavenging has been reported to be correlated with response to high salt stress. Efficient energy conservation strategy was quite evident in the salt-tolerant *Brassica* species which downregulated the expression of genes involved in photosynthesis, carbohydrate metabolism and other anabolic pathways and upregulated genes involved in stress response such as ABA biosynthetic enzyme coding gene and 9-cis-epoxy-carotenoid dioxygenase 4 *(NCED4)*.

Glutathione is another very important antioxidant molecule which acts as a redox sensor and plays a major role in maintaining lower levels of ROS via the glutathione-ascorbate cycle (Gill and Tuteja 2010). Numerous antioxidant compounds and enzymes constitute the glutathione-ascorbate cycle. Superoxide dismutase (SOD) is one component which catalyzes dismutation of lethal superoxide radical into hydrogen peroxide (H_2O_2). This H_2O_2 is further scavenged and reduced to water by ascorbate peroxidase with the concomitant oxidation of ascorbate (ASC) to monodehydroascorbic acid (MDHA). Monodehydroascorbate reductase enzyme converts the MDHA back in to ascorbate either directly or via a two-step process mediated by dehydroascorbate reductase. DHAR in turn uses reduced glutathione (GSH) which is regenerated from its oxidized form GSSG by the action of glutathione reductase (GR), leading to removal of ROS. Sharma et al. (2015) showed significant upregulation of various ROS scavenging genes like *SOD, GPX, APX, DHAR and MDHAR* under imposed salt stress though their initial transcript levels were also much higher in tolerant species as compared to the sensitive one. They also examined induction of proline biosynthesizing gene delta1-pyrroline-5-carboxylate synthase (*P5CS*) under salt stress conditions. Many researchers have previously suggested constitutive expression of salt-responsive genes in tolerant cultivars as a critical aspect of improved salt stress tolerance (Taji et al. 2004).

Recently, two proteins, a calcium-binding protein, namely, RSA1 (short root in salt medium 1) and a bHLH transcription factor (RITF1), have been reported to be a part of calcium-sensing metabolic pathway playing a major role in tolerance to salt stress in *Arabidopsis* (Guan et al. 2013). These two proteins interact with each other and form the RSA1-RITF1 complex which controls expression of major genes related to ROS scavenging and maintenance of cellular ionic homeostasis under salt stress (Guan et al. 2013). RITF1 was grouped along with peak 1% genes exhibiting more than 100-fold induction under salt stress by Sharma et al. (2015) giving clear indications that this bHLH transcription factor plays a major role in regulating downstream gene expression under imposed salt stress and may serve as a molecular marker of stress tolerance. *RITF1* also controls the expression of *SOS1* suggesting a connection in dual calcium-signalling pathways.

Overexpression of the transcription factor MYb 44 in a phosphorylation-dependent manner has been reported to confer salt stress tolerance in many crop species. ZAT7 is another zinc finger protein reported from *Arabidopsis* which regulated the induction of many genes responsible for improved salt tolerance (Ciftci-Yilmaz et al. 2007). OsNAC063 is another classical example of a transcription factor that was highly induced in rice roots exposed to excess salts (Yokotani et al. 2009). *Arabidopsis* transgenic overexpressing ONAC063 displayed improved tolerance to high salt concentration and osmotic pressure suggesting a possible role of ONAC063 in the large salt stress signalling network. Another bHLH transcription factor from wild rice (OrbHLH2) has also recently been proven to enhance osmotic and ionic stress tolerance in *Arabidopsis* (Zhou et al. 2009).

Mitogen-associated protein kinases (MAPK) is another group of genes which have been associated with abiotic and biotic stress tolerance in animals and plants since time immemorial (Kyriakis and Avruch 2012; Rodriguez et al. 2010; Samajova et al. 2013). Single, double, triple and further downstream MPKs have been reported to be a part of the MPK cascade where in which a phosphorelay mechanism activates downstream MPK which in turn activates the target MPK. The mode of action of MPKs has been conclusively proven to be via phosphorylation of serine or threonine residues of target proteins. Majority of the target proteins of MAP kinases possesses a kinase interaction motif $(R/K^+-x (2-6)-I/LxI/L)$ which has been shown to assist substrate binding (Schweighofer et al. 2007). *Arabidopsis* has been reported to have 20 MAPKs and 10 MAPKKs, out of which MKK4/MKK5 and their target proteins, MPK3/MPK6 in particular, are strongly linked with salt stress signalling. Recently, Campo et al. (2014) showed that overexpression of *OsCPK4* improved salinity tolerance in rice by reducing peroxidation of the cellular membraneous network.

8.2.11 Proteomics and Salt Stress

Comparison of proteome between a salt-tolerant and susceptible cultivar reveals the differentially expressing proteins in terms of protein structure and activity. Salt stress upregulates specific signalling pathways leading to modulations in gene expression and ultimately leads to alter relative abundance and activity of various proteins. The differentially expressing proteins bring about profound changes in the cellular metabolism which can be broadly classified as short-term and long-term stress acclimatization responses. Short-term responses include those which are aimed at reducing the direct effect of salt stress such as lowering the water potential and maintenance of cellular ion homeostasis. Long-term adaptations include structural modifications at cellular level resulting in altered crop growth and development. A mere study of differential abundance of proteins in tolerant variety does not give much idea about protein function under salt stress, and therefore functional analysis of differentially expressing proteins should be done as a part of validation

Table 8.3 Major protein functional groups revealing differential protein abundance under salinity

Protein functional group	Increased protein	Decreased protein
Signalling	Annexin; calmodulin; OsRPK1; calreticulin MAPK; large GTP-binding protein, β subunit of heterotrimeric G protein; OsRac2; 14–3-3	Ras GTPase, 14–3-3
Gene expression regulation, cell growth and division	NAC-α, HB1B, OSAP1, RNA helicase	NAC-α, poly(A)-binding protein
Photosynthesis	23 kDa (PsbP), ferredoxin-NADPH reductase, OEE2, RuBisCO activase, TPI, GAPDH, Glucose-6-P dehydrogenase	LHC, PC, OEE1, OEE2, RuBisCO LSU and SSU, RuBisCO activase, carbonic anhydrase, GAPDH, SBP, PGK, PRK, TK
Protein biosynthesis and degradation	eIF, eEF, eIF5A3; ribosomal proteins L12, L31, S29; proteasome 20S, 26S; GS	eIF-4E2, eEF, L10, L12, S3a, S12, L29, S5, GS
Protein folding	DnaK chaperone, HSP70, small HSP, RuBisCO-binding protein subunit β (CPN60-β)	HSP90
Respiratory pathway and sucrose matabolism	FBP aldolase, GAPDH, TPI, ENO, succinyl-CoA ligase β subunit, MDH (NAD), cytochrome c oxidase subunit 6b-1, ATP synthase CF1β,δ,ε ADK, NDPK1, ADH	ATP synthase CF1α,β-3, mtATP β-3

Adapted from Kosova et al. (2013)

of comparative proteome. Functional analysis can be done by studying the subcellular localization, interacting partners, post-translational modifications and influence of phenotype on the gene coding a particular protein. Salt stress-modulated proteins can be functionally categorized into various classes: energy metabolism, signalling, transporter (photosynthesis, respiration), primary metabolism, hormonal metabolism, stress countering proteins (ROS scavengers, PR proteins), cytoskeleton-linked proteins, secondary metabolism, etc. (Kosová et al. 2013) (Table 8.3).

Salinity stress induces widespread proteome alteration in crop plants, and analysis of proteome is a very efficient tool to understand plant responses to salt stress (Sobhanian et al. 2011; Tahir et al. 2013). Zhang et al. (2012) reported a healthy database comprising of more than 2000 proteins specifically responding to salt stress, spread across 34 plant species, which has greatly improved our understanding of salt stress tolerance mechanism. The basic tool applied in proteomic studies is still two-dimensional polyacrylamide gel electrophoresis (2-DE) which has been prolifically used in studies related to abiotic stress tolerance in crops (Ma et al. 2012). Here, we focus mainly on the major protein functional groups affected by salt stress.

Photosynthesis is one among the key physiological processes affected by salinity (Munns et al. 2006; Chaves et al. 2009). Proteomic analysis by Jia et al. (2015)

displayed a specific set (12) of photosynthetic proteins that were salt stress responsive, which involves six proteins related to light reactions and six enzymes involved in carbon assimilation reactions. Surprisingly, majority of the proteins related to light reaction, including the oxygen-evolving complex, D1 protein of photosystem II, subunit VII of photosystem I, Chl a/b-binding protein and Lhcb6 protein, were induced under salt stress condition. However, the major proteins of the Calvin cycle, including the chloroplast-coded large subunit of ribulose bisphosphate carboxylase oxygenase (RuBisCO), sedoheptulose-1,7-bisphosphatase and phosphoribulokinase, was drastically downregulated under salt stress condition (Jia et al. 2015). Altogether it can be hypothesized that the reduction in net photosynthetic rate under salt stress may be primarily attributed to functional distortions in the Calvin cycle. RuBisCO activase (RCA) protein has also been reported to be significantly accumulated in *Brassica* species 48 h after imposition of salt stress (Jia et al. 2015). RCA is primarily involved in the ATP-dependent activation of RuBisCO (Ashraf and Harris 2013). RCA has been reported to play a vital role in the maintenance of net photosynthetic rate at low atmospheric CO_2 levels because of the decrease in stomatal conductance under salt stress (de Abreu et al. 2014). Therefore, induction of RCA may have contributed to overall salt stress tolerance in *Brassica* species (Jia et al. 2015).

Salinity stress can negatively affect carbon assimilation process indirectly by lowering chlorophyll content (Tang et al. 2014). However, the exact effect of salt stress on the biosynthetic pathway of chlorophyll still remains unclear. Reduced activities of major enzymes involved in the chlorophyll biosynthetic pathway have been attributed to be the major cause of impaired chlorophyll biosynthesis in rice (Turan and Tripathy 2015). Glutamate-1-semialdehyde aminotransferase (GSA-AT) is a major enzyme of the chlorophyll biosynthetic pathway (Grimm 1990), and antisense construct of this gene displayed moderate to severe reduction in chlorophyll (Hartel et al. 1997; Tsang et al. 2003). Jia et al. (2015) found that the transcript abundance of GSA-AT decreased after 24 h of imposed salt stress. However, the protein level recovered to normal levels after 48 and 72 h of salt treatment which may be the major reason behind the improved salt tolerance of *Brassica* plants.

Salt stress has been reported to specifically activate a set of Ca-binding proteins including cell membrane protein annexin (Pang et al. 2010) and calmodulin (Cheng et al. 2009). Calcium-binding proteins have been reported to mediate salt stress signalling in an ABA-dependent manner (Lee et al. 2004a, b), and very high transcript abundance was observed in the shoot and root tissues of *Salicornia europaea* (Wang et al. 2009) and tobacco (Manaa et al. 2011) under salt stress condition. Guanosine triphosphate-binding proteins are another group of proteins which have been reported to show differential abundance under salt stress condition. These proteins come under the Rab family and play a major role in endocytosis and vesicle trafficking (Wang et al. 2008b; Pang et al. 2010).

8.2.12 Metabolomics

Metabolomics is one among the most modern tools for critically analyzing plant responses to salt stress (Lu et al. 2013). Metabolites are the end products of diverse cellular pathways which adequately represent the interaction between the biological unit and its ambient environment. Metabolomics has tremendously added to our knowledge regarding salinity mediated changes in metabolites (Urano et al. 2009). Presently, a multi-combinatorial approach is employed bringing transcriptomics, proteomics and metabolomics on one single platform (Wu et al. 2013). Primary tools presently being employed in plant metabolomics are metabolite profiling, metabolic fingerprinting and targeted analysis. Currently, metabolomics is gaining wide popularity and is being used with high precision to analyze abiotic stress responses including salt stress tolerance (Alvarez et al. 2008; Renberg et al. 2010). Metabolomic studies in rice have so far analysed the physiology of seed germination, differential abundance of metabolites between wild type and mutants, changes in metabolome across developmental stages and the differences in nature of metabolites across different varieties (Shu et al. 2008).

Plants respond to any kind of abiotic stress by virtue of transient, short-term or sustained metabolic changes. For instance, upon exposure to abiotic stresses like salinity, drought or cold, there occurs a sudden alteration in the carbohydrate metabolism, whereas the osmotic adjustment and resultant spike in osmolytes occur after several days of exposure to the stress (Lugan et al. 2010). Certain metabolic changes like sugar alcohol and amino acid levels are similar to all types of abiotic or biotic stresses, but there are unique metabolomic alterations to any particular stress. Interestingly, proline build-up has been reported to occur upon drought, salinity and cold stress but not upon high temperature stress (Lugan et al. 2010). Another typical metabolomic shift specifically observed under salinity stress is the decline in TCA cycle intermediates and organic acids (Gong et al. 2005), which generally increase under water deficit or heat stress (Urano et al. 2009). The LEA family proteins which are hydrophilic in nature and have an osmoprotective role have been reported to be elevated under salt stress (Kosová et al. 2010). Specific salt stress inducible LEA proteins are the LEA3 proteins in Indica rice varieties like Pokkali and Nona Bokra and TAS14 in tomato (Moons et al. 1995). Recently, Wu et al. (2013) showed the salt stress-mediated spike of several osmolytes in barley leaves including glycine betaine, proline, alanine, sugar alcohols, inositol, raffinose, glucose, etc. which provide salinity tolerance.

8.2.13 Chromatin Modifications and Epigenetics in Salt Tolerance

Chromatin modifications, broadly called as epigenetic changes, have been reported to play a major role in plant acclimatization towards different kinds of abiotic and biotic stresses (Jiang et al. 2012). Numerous studies have reported that such kind of epigenetic changes is involved in the tolerance responses of crop plants to salt stress in the same generation as the stress occurs. Epigenetic histone modification has been cited to be the specific reason for enhanced salt stress tolerance in *Arabidopsis* seedlings subjected to stress recovery and stress cycle. Histone modification affected the expression of various transcription factors and thereby modulated downstream target gene expression and showed less Na⁺ ion accumulation in its shoot tissues (Sani et al. 2013). Another study proved hypersensitivity to salinity stress because of failure in cytosine demethylation at a putative small RNA target site of the *AtHKT1*;1 promoter, leading to reduced gene expression (Baek et al. 2011). Previous research groups have clearly shown distinct differences in the expression pattern of methylases and other chromatin modifier genes among diverse rice genotypes and tissues under salinity stress (Karan et al. 2012). Hence, demethylation of such genes in the root tissues of rice may be an active epigenetic response. But, till date, epigenetic inheritance of salinity tolerance from one generation to another has not been proven conclusively (Chao et al. 2013).

8.2.14 Marker Assisted Selection as a Promising Breeding Tool

One of the major constraints in development of salt-tolerant crop varieties through conventional breeding is that it takes a long time. The traditional method of outcrossing among genetically diverse germplasm and selection based on their performance in the field is too laborious and lengthy process. So the search for an alternative methodology has been in place for a long time. Marker-assisted selection (MAS), employing various markers SSR, SCAR, ISSR, SNP, etc. to map QTLs governing traits associated with salt tolerance, has been quite streamlined (Ashraf and Foolad 2013). The prerequisite for MAS is tight linkage between a useful trait and a genetic marker with polymorphic alleles between parental lines (Ashraf and Foolad 2013). Hence, the indispensable basis for efficient breeding with MAS is a thorough understanding of useful traits and its variability within the concerned plant species (Ashraf and Foolad 2013). One classical example is Saltol, a QTL identified in rice that is responsible for more than 65% of the phenotypic variation in Na⁺ ion uptake under salt stress (Ashraf and Foolad 2013). So this technology indeed has the strong potential of assisting in the process of development of salt-tolerant varieties. A major constraint in this process is the appropriate selection of markers linked with favourable traits in the context of salinity tolerance. Genetic engineering and recombinant DNA technology have widely been criticized as single gene modification technology, which seldom works efficiently in improving salt tolerance which is governed by multiple genes and is affected by environment. On the contrary, it has been proposed by previous research groups that improving salinity tolerance practicable by manipulating only one or a few major constituents of the regulatory gene network instead of engineering numerous molecular mechanisms (Golldack et al. 2011). Gene pyramiding is another approach widely employed to stack several beneficial genes in to a favourable background to ultimately combine useful traits and improve tolerance without yield reduction. However, Mendelian segregation of traits limits this approach and hence makes breeding more and more complicated. Genome editing is gaining wide popularity in the area of abiotic stress tolerance, whereby engineered CRISPR-Cas (clustered regularly intersperced short palindromic repeats-associated proteins) and zinc finger nucleases are used to target specific genes and either knock them down or overexpress them (Ainley et al. 2013; Belhaj et al. 2013).

8.3 Waterlogging

Waterlogging tolerance may be defined as the higher survivability or better plant growth under waterlogged conditions proportionate to well-drain conditions. In crop improvement programmes, it is defined as the subsistence of relatively high grain yields under waterlogged conditions relative to non-waterlogged conditions (Setter and Waters 2003). Flooding-tolerant plant acquired some special morphological traits or metabolic adaptations in response to deficient oxygen to survive or to sustain their growth (Table 8.4). Anatomical and biochemical are the two major adaptations to anoxia or hypoxia. In the former case, the structure of the tissue is modified, to create spaces for gas diffusion and to minimize oxygen demand. The mechanisms of waterlogging or hypoxia tolerance include (i) the maintenance of high internal aeration through constitutive aerenchyma (Armstrong et al. 1994) and (ii) metabolic adaptation under hypoxia with the substantial storage of carbohydrates for ethanolic fermentation (Brandle 1991).

Table 8.4 Adaptive traits for waterlogging tolerance

Adaptive traits	Correlated response	References
Phenology	Seed/seedling vigour	Gardner and Flood (1993)
	Long season	McDonald and Gardner (1987) and Gardner and Flood (1993)
	Dormancy (seeds or whole plant tissues)	Setter (2000)
	Slow growth	McDonald et al.(2001a)
Morphology and anatomy	Nodal/adventitious root development	Huang et al. (1994, 1997) and Malik et al.(2002)
	Survival of seminal roots	Barrett-Lennard et al. (1988) and Watkin et al. (1998)
	Aerenchyma	Watkin et al.(1998) and McDonald et al.(2001a,b)
	Increased root porosity/intercellular spaces	Huang et al. (1997), Malik et al. (2001) and McDonald et al. (2001a, b).
	Increased suberin/lignin; barriers to radial O_2 loss	Jackson and Drew (1984), Watkin et al.(1998) and McDonald et al.(2001b)
	Root membrane integrity	Sangen et al. (1996)

(continued)

Table 8.4 (continued)

Adaptive traits	Correlated response	References
Nutrition and nutrient toxicities	Root length and depth	Huang et al.(1997) and McDonald et al. (2001a, b)
	Cell function for nutrient uptake, incl. K^+/Na^+ selectivity	Akhtar et al. (1994) and Huang et al. (1995)
	Leaf chlorosis	Drew and Sisworo (1977, 1979)
	Microelement tolerance – Fe^{2+}	Huang et al. (1995)
	Microelement tolerance – Mn^{2+}	Huang et al. (1995) and Ding and Musgrave (1995)
Root metabolism: respiration, anaerobic catabolism and anoxia tolerance	Reduced respiration	Huang and Johnson 1995
	Anaerobic catabolism	Thomson et al. (1989) and Waters et al. (1991b)
	High carbohydrate concentrations	Barrett-Lennard et al. (1988), Albrecht et al. (1993) and Huang and Johnson (1995)
	Anoxia tolerance	Waters et al.(1991a, b) and Greenway et al. (1992)
Recovery and prevention of post anoxic damage	Recovery ability	Barrett-Lennard et al. (1988), Albrecht et al. (1993) and Malik et al. (2002)
	Antioxidants and antioxidative enzymes, e.g. SOD, catalase, glutathione reductase.	Albrecht and Wiedenroth (1994), Wang et al. (1996) and Biemelt et al. (1998)

Adopted form Setter and Waters (2003)

8.4 Hypoxia and Anoxia

Plant or cellular oxygen status can be defined as normoxia (normal O_2 levels), hypoxia (reduced O_2 levels) or anoxia (lacking O_2). In hypoxia condition oxygen in pore space of soil becomes reduced to a point below optimum level. The reduced O_2 limits ATP production by oxidative phosphorylation and a larger proportion of ATP yield through glycolysis. It is the common phenomenon of excessive soil moisture stress in waterlogged soils and occurs during short-term and long-term flooding of water when the roots are submerged, but the shoot remains in the atmosphere (Sairam et al. 2008, Morard and Silvestre 1996). Anoxia is the usual form of water stress in the soil that is caused by long-term flooding or waterlogging in which complete lack of oxygen under rhizosphere. In case of anoxia, ATP is produced only by way of glycolysis. Cells exhibit low ATP contents, diminished protein synthesis and impaired division and elongation. However if anoxic conditions persist, many plant cells die. The supply of oxygen to root cells is influenced by several factors, including soil porosity, water content, temperature, root density and aerobic microorganisms. Oxygen concentrations in root tissues also vary according to root depth, root thickness, the volume of intercellular gaseous spaces and cellular metabolic activity.

Over the longer term, acclimation to hypoxic or anoxic conditions can take the form of developmental responses involving modifications in growth behaviour, morphology and anatomy. Injury caused by anoxia or hypoxia is a consequence of acidification of cytoplasm, which in turn results in greatly diminished protein synthesis, mitochondrial degradation, inhibition of cell division and elongation, disrupted ion transport and root meristem cell death. Flooding-tolerant plants are able to avoid cytoplasmic acidosis and continue to make ATP during short-term inundation by stimulating ethanolic fermentation. To survive under waterlogging stress, plants must generate sufficient ATP, regenerate NADP+ and NAD+ and avoid accumulation of toxic metabolites. Periods of oxygen deficit can trigger developmental responses that promote acclimation to hypoxic or anoxic conditions.

8.4.1 Morphological Adaptations for Waterlogging Tolerance

Morphological adaptations to hypoxic environment include the characteristic development of aerenchyma in the root parenchyma that transports oxygen down to submerged tissues. This aerenchyma development, which can reach up to 55% of the cross-sectional area of roots, is often enhanced in response to flooding. They reduce the diffusion barrier and enhance gas exchange between the cells (Steffens et al. 2011). Another common strategy is maintaining surface gas films or over leaves, maintained by hydrophobic leaf hairs. These thin extensive layers facilitate gas diffusion (both of O_2 in the dark and CO_2 in the light) to the submerged leaf or root by enlarging the effective gas-water interface beyond that just at the surface of stomata or lenticels (Petersen et al. 2009). The various responses to submergence such as enhanced shoot growth as found in flooding-tolerant species such as rice have been recognised as the low oxygen escape syndrome (LOES). Three different stages of response were articulated by the plants under low oxygen condition (Dennis et al. 2000). These are signal transduction (Stage a, 0–4 h), metabolic adjustment (Stage b, 4–24 h) and morphological changes induced by the first two stages (Stage c, 24–48 h) (Steffens et al. 2011). The first one is the determinative stages, which switch over the plants from normal metabolism pattern to low oxygen-responsive pattern, which contributes significantly to the survival of seedlings (Liu et al. 2012b). Survival under waterlogged conditions is largely attributable to the ability to improve gas exchange between plants and environment as well as gas transport from aerial parts to hypogeal organs (Changdee et al. 2009). In Stage c, formation of aerenchyma (gas-filled spaces) in the root, stems and in other shoot organs is observed (Liu et al. 2012b). Aerenchyma is developed constitutively in many wetland plants as well as rice, which is essential for plants to tolerate frequent flooding, and is regarded as an efficient morphological adaptation to alleviate low oxygen stress. Another major adaptive trait of plants under waterlogging stress is conformation of new adventitious roots (Mano and Omori 2007).

8.4.2   Aerenchyma Formation and Adventitious Roots Development

The major anatomical adaptation to waterlogging is the formation of aerenchyma. It is formed constitutively in wetland species, whereas in others it is a result of hypoxia condition of waterlogging. Three major pathways of aerenchyma formation are known in plants. Type 1 is lysigenous aerenchyma which is developed when previously formed cells die within a tissue to create a gas space where cortex cells undergo PCD, resulted from exhaustion of sugars during waterlogging or submergence, and it is found in rice, wheat, barley and corn (Bailey-Serres and Voesenek 2008). Type 2 is schizogenous aerenchyma, which develops by splitting of the common cell wall previously connected without cell death when intercellular gas spaces form within a tissue. Lysigenous aerenchyma formation is initiated by ethylene formed in hypoxic conditions. Ethylene promotes the formation of aerenchyma by accumulating in plant organs during waterlogging or submergence due to the reduced diffusion rate (Steffens et al. 2011). Type 3 is expansigenous aerenchyma (Bailey-Serres and Voesenek 2008) or secondary aerenchyma (Shimamura et al. 2003), a white spongy tissue filled with large gas spaces. It is formed from living cell division or enlargement without cell separation or death. It is located in adventitious roots, root nodules, stems, hypocotyls and taproots under waterlogged conditions (Shimamura et al. 2003). Aerenchyma formation involves multiple signal transduction pathways, in which Ca^{2+}, protein phosphorylation and G-protein are crucial signal components (He et al. 1996). Aerenchyma formation in corn and rice is the death of cells in the mid-cortex of the root. However, one major difference is the need for the cell walls of the dying cells to be removed, and this is achieved by the induction and release of cell wall-degrading enzymes. Adventitious roots can facilitate internal O_2 transport and a barrier to the radial O_2 loss in the subapical regions of roots (Colmer 2003). It is a physical response mainly resulting from secondary cell wall deposits in outer hypodermal layer, which is a dominant mechanism of reducing ROL compared with the respiratory activity alternations in the hypodermal/epidermal layers (Garthwaite et al. 2008). Although wheat, barley and *H. marinum* all form adventitious roots containing aerenchyma, wheat and barley are much less tolerant to waterlogging than *H. marinum*. (Garthwaite et al. 2006).

8.4.3   Metabolic Adaptation

The metabolic adaptations to oxygen deficiency include anaerobic respiration, maintenance of carbohydrate supply for anaerobic respiration, avoidance of cytoplasmic acidification and development of antioxidative defence system. Metabolic adaptations to waterlogging and anaerobiosis include the transformation of metabolic process towards the anaerobic generation of ATP by glycolysis and fermentation of pyruvate to regenerate NADP while producing ethanol. Although ethanol is toxic, it readily diffuses out of cells, while the alternative fermentation of pyruvate to the less toxic lactate is limited by its effect on cytoplasmic acidification. Immediate responses to waterlogging or anoxia in mesophytes include the shifting of metabolism and substantial decreases in root/stem hydraulic conductance resulting induction of leaf epinasty and stomatal closure in some species. Although several hormones have been implicated, the most important component of the long-distance signalling under waterlogging is 1-aminocyclopropane-1-carboxylic acid (ACC), which is transported from the roots in the xylem and oxidized to ethylene (C_2H_4) in the leaves by ACC oxidase, which itself increases in response to flooding (Jackson, 2002). The enhanced export ACC from the roots results an inhibition of ACC oxidase by low O_2 concentrations in the roots. The ethylene released is directly involved in stimulation of cell expansion in the adaxial petiole cells leading to epinastic curvature and in overall inhibition of leaf growth. Changes in gene expression rapidly created under anoxia were created by waterlogging. Preliminary, constitutive protein synthesis interrupted and transit polypeptides are produced. Subsequently, a range of new anaerobic proteins are transcribed. Most of these are metabolic enzymes involved in establishing anaerobic metabolism as, in the absence of oxygen, the citric acid cycle and oxidative phosphorylation cannot function. Three pathways of anaerobic metabolism have been described in plants. Induction of pyruvate decarboxylase and alcohol dehydrogenase results in alcoholic fermentation, with the production of ethanol and carbon dioxide. Induction of lactate dehydrogenase results in lactic acid fermentation, with the production of lactate. The third pathway involves the production of alanine from glutamate and pyruvate. Whenever cytosolic pH of the tissue decreased, transition from lactate to alcoholic fermentation occurs. Alanine fermentation is common in some (e.g. barley), but not all roots in waterlogging and alanine aminotransferase are induced by hypoxia in these species.

8.4.4 Anaerobic Respiration

Plant cells produce energy in presence of oxygen through aerobic respiration which includes glycolysis, TCA cycle and oxidative phosphorylation. However in absence of oxygen, Krebs cycle and oxidative phosphorylation are blocked, and cells inevitably undergo anaerobic respiration to fulfil the demand for energy (Davies 1980). The efficiency of energy production by glycolysis and fermentation is much lower than that of aerobic respiration. Besides, the end products of glycolytic and fermentative pathway, such as ethanol, lactic acid and carbon dioxide, pose an additional hazard to the cell. It is well reported that the maintenance of an active glycolysis and an induction of fermentative metabolism are adaptive mechanisms for plant tolerance to anoxia (Sairam et al. 2008). Anaerobic respiration includes glycolysis and fermentation. Generation of energy under anaerobic condition is largely achieved through glycolysis. For the continued operation of glycolytic pathway, the regeneration of NAD+, a cofactor from NADH, is essential (Drew 1997). Large amount of pyruvate produced in glycolysis as an end product and converted to alternative products to recycle NADH to NAD+. Ethanolic fermentation or lactate fermentation is the most important process by which NADH can be recycled to NAD+ during oxygen deficiency (Ricard et al. 1994). In long-term waterlogged environment, anoxia is always headed by hypoxia (Setter and Waters 2003). Hypoxia accelerates the induction of glycolytic and fermentative enzymes, for example, aldolase, enolase, ADH and PDC (Germain et al. 1997, Albrecht et al. 2004). This induction can improve or at least sustain the glycolytic rate in anoxic plants contributing higher tolerance to anoxia. In wheat, increased activities of ADH and PDC (Johnson et al. 1989) have been found in response to hypoxia resulting higher ethanol production contributing greater tolerance to anoxia (Waters et al. 1991b). An accumulation of lactate promotes acidification of the cytoplasm of anoxia sensitive plants, such as maize, wheat and barley (Menegus et al. 1989, 1991). However, the enhanced lactate transport out of the roots into the surrounding medium may help to avoid cytoplasmic acidification (Xia and Saglio 1992). Moreover, lowered cytoplasmic pH leads to the activation of PDC and inhibition of LDH (Davies 1980) resulting a shift from lactate fermentation to ethanolic fermentation.

8.4.5 Role of Soluble Sugars

Plants increase their osmotic potential leading to enhanced stress resistance by accumulating osmolytes such as proline, glycine betaine and non-reducing sugars and polyols. Soluble sugars also contribute to the regulation of ROS signalling during abiotic stresses (Seki et al. 2007). Soluble sugars are involved in the metabolism and protection of both ROS-producing and ROS-scavenging pathways, such as mitochondrial respiration, photosynthesis and oxidative-pentosephosphate pathway (Couée et al. 2006). Under hypoxia or anoxia energy metabolism pathway, shift from aerobic to anaerobic mode through which energy requirements of the tissue is greatly restricted as very few 2ATPs are generated per molecule of glucose in anaerobic respiration. However 36 ATP molecules can be produced for each glucose molecule in aerobic respiration. Sufficient energy for sustaining the metabolism in roots in hypoxic or anoxic condition is derived from anaerobic metabolic pathway by which plant can survive under waterlogging stress (Jackson and Drew 1984). Therefore, maintenance of adequate fermentable sugars in hypoxic or anoxic roots is one of the important adaptive mechanisms to waterlogging (Sairam et al. 2009). The amount of root sugar reserve and activity of sucrose-hydrolysing enzymes are important determinants for waterlogging tolerance of crop plants (Sairam et al. 2009). The roots of comparatively tolerant genotypes contain greater sugar content (total, reducing and non-reducing sugar) than in susceptible genotypes. Moreover, waterlogging induces and increases the content of reducing sugar through increased activity of sucrose synthase (SS) in tolerant genotypes. The tolerant genotypes show increased expression of mRNA for sucrose synthase, while susceptible genotypes show very little expression under waterlogged condition (Sairam et al. 2009). Accessibility of adequate sugar in the roots with increased activity of sucrose synthase for providing reducing sugars for anaerobic respiration is one of the important mechanisms of waterlogging tolerance. The higher concentration of soluble carbohydrate in roots and shoots of wheat has been reported when the crop is exposed to long-term oxygen deficit (Albrecht et al. 1993). The ratio of the root to shoot sugar increases for waterlogging-tolerant wheat genotypes under hypoxia (Huang and Johnson 1995). The relatively large amount of sugars transported to root facilitates the energy supply for root respiration and ion uptake (Huang 1997).

8.4.6 Antioxidant Defence System for Waterlogging Stress

Waterlogging also stimulates oxidative stress through increasing reactive oxygen species, such as superoxide anion ($O_2 \bullet -$), singlet oxygen ($^1O_2$), hydrogen peroxide (H_2O_2) and hydroxyl radical ($OH \bullet$) and perhydroxyl radical ($O_2H \bullet$) (Arbona et al. 2008). When a plant is faced to hypoxia/anoxia conditions, different types of reactive oxygen species (ROS) are produced which can damage or kill cells. All ROS are highly reactive and can destroy lipids, nucleic acids and proteins; however, some ROS such as $O_2 \bullet -$ and H_2O_2 are also function as signals in response to a number of abiotic stresses. Plants dispose of excessive ROS through the use of antioxidant defence systems (enzymatic and nonenzymatic) present in several subcellular compartments of the cell. High concentrations of major antioxidant enzymes are very important for plants subjected to different intensities of waterlogging. Waterlogging stress increased superoxide dismutase (SOD), catalase (CAT) and ascorbate peroxidase (APX) activities depending on the plant genotypes, for illustration stress-sensitive cultivars showed lower activities than tolerant cultivars (Arbona et al. 2008). Among the different antioxidant enzymes, SODs are the first enzymes for detoxification of ROS, dismuting $O_2^{\bullet -}$ to H_2O_2, and CAT and APX subsequently convert H_2O_2 to H_2O. Glutathione peroxidase (GPX) consumes GSH to convert H_2O_2 to GSSG and H_2O. Other enzymes such as glutathione reductase (GR), dehydroascorbate reductase (DHAR) and monodehydroascorbate reductase (MDAR) are antioxidant enzymes of the ascorbate-glutathione cycle, but they do not detoxify ROS directly. Regulation of the concentrations of antioxidants and antioxidant enzymes (such as SOD, CAT, APX, MDHAR, DHAR, GR and GPX) constitutes an important mechanism for avoiding oxidative stress. Antioxidant enzymes activity increased during waterlogging in pigeon pea that scavenge not only the post-hypoxic ROS build-up, but the system also detoxifies the cellular system of ROS. Antioxidant enzyme activities in waterlogging-tolerant genotype ICP 301 was comparatively greater than susceptible genotype Pusa 207 which could be one of the important factors for higher tolerance to waterlogging (Kumutha et al. 2009). The more effective expression of antioxidant mechanisms in the tolerant cultivar also contributed to its lower level of lipid peroxidation (Yin et al. 2009). Bin et al. (2010) suggested that in maize seedlings, increased POX, APX, GR, CAT and SOD activities led to an efficient H_2O_2 scavenging system and enhanced protection against oxidative stress caused by waterlogging. However, the activities were higher in waterlogging-tolerant genotypes. Sairam et al. (2011) observed that under waterlogging, the activity of three antioxidative enzymes (SOD, APX and GR) showed a continuous increase up to 8 days of waterlogging in waterlogging-tolerant mung bean genotypes; in susceptible genotypes the increase in the activity of these enzymes was observed only following 2–4 days of waterlogging. In all subsequent stages, there was a decline in activity of all three enzymes compared to the control and plants waterlogged for 2–4 days. Waterlogging pretreatments before anthesis can effectively improve the tolerance of wheat to waterlogging occurring during the generative growth by maintaining relatively higher activities of ROS-scavenging than the non-hardening treatment (Li et al. 2011a). The major nonenzymatic antioxidants in plants include ascorbate, reduced glutathione (GSH), α-tocopherol and carotenoids; polyamines and flavonoids also provide some protection from free radical injury.

8.4.7 Gene Expression Pattern Under Waterlogging Stress

According to differential gene expression patterns under oxygen deficiency, these genes are divided into three groups: low oxygen-sensing and cell signalling-involved genes, metabolic adjustment genes and genes that maintain plant internal microenvironments. Low oxygen-sensing and cell signalling-involved genes induce the synthesis of mitochondria alternative oxidase (AOX) (Klok et al. 2002) and stimulate ROS-related RopGAP4 (Rop GTPase activating protein4) (Baxter-Burrell et al. 2002). Some genes can also encode important protein involved in Ca^{2+} signalling such as calmodulin (Snedden and Fromm 2001) and CAP (calmodulin-associated peptide) (Subbaiah et al. 2000). Metabolic adjustment genes regulate plant switching from aerobic respiration to adaptive anaerobic fermentation by increasing anaerobic proteins (ANPs). Genes that maintain plant internal microenvironments activate plant glutamate decarboxylases (GADs) interacting with calmodulin, which lead to pH regulation (Klok et al. 2002; Zou et al. 2011). RAP2.6 L gene is one of the 145 genes, which belong to the AP2/ERF (APETALA2/ethylene responsive element-binding factor) superfamily in *Arabidopsis*. The crucial function of AP2/ERF transcription factor family is to regulate plant growth and respond to abiotic stresses (Liu et al. 2012a; Sun et al. 2010a). It is known that RAP2.6 L is induced by ABA and its overexpression enhances plant resistance to jasmonic acid (JA), salicylic acid (SA), ABA and ET in *Arabidopsis* (Sowmya et al. 2011). It was confirmed that overexpression *of RAP2.6 L* delayed plant premature senescence and reduced oxidant damage, water loss and membrane leakage under waterlogging stress. Moreover, *RAP2.6 L* overexpression resulted in a significant increase of transcripts of ROS-involved enzyme genes, ABA-signalling and biosynthesis genes and waterlogging-responsive genes and a decrease of transcripts of *ABI1* (ABA insensitive1), suggesting that *RAP2.6 L* plays a role in *ABI1*-mediated ABA-signalling pathway (Liu et al. 2012a).

Nonsymbiotic hemoglobins (nsHbs) are also crucial for a large number of various cellular processes in plants, such as the adaptation to hypoxic stress. To further identify its function, a maize *nsHb* gene (*ZmHb*) and its 50 flanking genes were cloned and characterized in one maize inbred line. It was revealed that transcript of *ZmHb* was mainly active at regions from plant root tips to vascular tissues. Through ectopic expression in transgenic tobacco, it was shown that *ZmHb* was involved in root tolerance to multiple stresses with similar regulation patterns (Zhao et al. 2008).

Aside from the alterations in abundance of transcript and translation, waterlogging adaptive changes in post-transcriptional and post-translational regulation are also crucial. In maize, nine prolyl 4-hydroxylases (*P4H*) genes were identified, and the effects of alternative splicing (AS) on the expression of these genes were studied. Five *P4H* genes were alternatively spliced into at least 19 transcripts. *zmP4H* genes displayed different expression patterns under waterlogging due to AS. The difference of AS events was also found in the same genes between different inbred lines and at different growth stages, which further proved that these transcripts were specifically dynamic during waterlogging stress. For *zmP4H8* transcripts modified by AS, both *zmP4H8–4* and *zmP4H8–5* were induced under waterlogging. In contrast, both *zmP4H2* and *zmP4H2–1* were repressed, despite zmP4H2/zmP4H2–1 ratio was reduced slightly under waterlogging (Zou et al. 2011). MicroRNAs (miRNAs) have been identified as important post-transcriptional regulators of gene expression (Bartel 2004, 2009). miRNAs and their downstream targets were identified as important factors in plant responses to many abiotic and biotic stresses. By quantitative gene expression assay of 24 candidate mature miRNAs in 3 inbred *Zea mays* lines, 5 of the miRNAs (miR159, miR164, miR167, miR393, miR408 and miR528) were found to play a key regulatory role in root development and energy saving under short-term waterlogging.

Further study by computational approaches constructed a miRNA-mediated gene regulatory and biochemical network. These were hypothesized to participate in three signal transduction pathways: ethylene and ABA-dependent signalling pathway, auxin-signalling pathway and cupredoxin-mediated oxidative stress responding pathway. By comparing differential expression patterns in three inbred lines, it was found that miRNAs were more active in signal transduction at early stage. The signal of aerenchyma formation and lateral root development was suppressed in the tolerant line, while the signal of rapid energy consumption for lateral root development was enhanced in the sensitive line (Liu et al. 2012b). N-end rule pathway is an important post-translational regulative pattern for plants responsive to anoxia or hypoxia conditions (Gibbs et al. 2011; Licausi et al. 2011).

8.4.8   Genetics and Gene Identification for Waterlogging Tolerance

Considerable effort has been put in identifying quantitative trait loci controlling the waterlogging tolerance. In most of the illustrations, waterlogging tolerance-related traits were used as indications of the tolerance. Quantitative trait loci (QTLs) for components of grain yield such as grains per spike, spikes per plant, kernel weight and spike length were also positioned of waterlogging tolerance in maize, rice, barley and soybean (Table 8.5). A total of 32 QTLs associated with waterlogging tolerance were positioned all over the chromosomes using the ITMI population W7984/Opata85. Positive associations were detected for plant height index, root length index, survival rate, germination rate index, leaf chlorophyll content index, root dry weight index, shoot dry weight index and total dry weight index in the population.

Table 8.5 QTLs identified for flooding/waterlogging tolerance in different crops

Phenotypic traits	Crops	Population Type and size	Parents of population	Chromosome	References
Root aerenchyma formation	Barley	Dh, 92	Franklin × TX9425	2H	Li et al. (2008)
	Maize	F2, 195	B73 × teosinte (Z. luxurians)	2, 5, 9, 10	Mano et al. (2008)
	Maize	BC4F1, 123	Mi29 × teosinte	1	Mano & Omori (2009)
Adventitious root formation	Maize	F2, 94	B64 × teosinte (Z. mays sp. Huehuetenangensis)	4, 5, 8	Mano et al. (2005)
	Rice	Ril, 96	IR1552 (O. indica) × Azucena (O. japonica)	3, 4, 9	Zheng et al. (2003)
Flooding tolerance index	Soybean	Ril, 156	Misuzudaizu × Moshidou Gong 503	LG C2	Githiri et al. (2006)
Waterlogging score	Barley	Dh, 177	Yerong × Franklin	2H, 3H, 4H	Zhou (2011)
		Dh, 188	TX9425 × Naso Nijo	2H, 4H, 5H, 7H	Xu et al. (2012)
		Dh, 172	Franklin × YYXT	2H, 3H, 4H, 6H	Zhou et al. (2012a, b)
	Soybean	Ril, 103	A5403 × Archer	LG A1, LG F, LG N	Cornelious et al. (2005)
		Ril, 67	P9641× Archer	LG A1, LG F, LG N	
Waterlogging tolerance index	Soybean	Ril, 175	Su88-M21 × NJRISX	LG L2	Sun et al. (2010b)
Internode elongation	Rice	F2, 94	C9285 (O. indica) × T65	12	Hattori et al. (2007), (2008)
			W0120 (O. rufipogon) × T65		
Leaf chlorosis	Barley	Dh, 92	Franklin × TX 9425	1H, 2H, 3H, 7H	Li et al. (2008)
		Dh, 177	Yerong × Franklin	1H, 2H, 3H, 4H, 5H, 7H	Li et al. (2008)
Seedling vigour	Rice	Ril, 282	Lemont (O. japonica) × Teqing (O. indica)	1, 3, 5, 10	Zhou et al. (2007)
Plant biomass reduction	Barley	Dh, 92	Franklin × TX9425	4H	Li et al. (2008)
		Dh, 177	Yerong × Franklin	4H	

(continued)

Table 8.5 (continued)

Phenotypic traits	Crops	Population Type and size	Parents of population	Chromosome	References
Germination rate	Rice	BC2F2, 423	Khao Hlan on (*O. japonica*) × IR64	1, 3, 9	Angaji et al. (2010)
	Rice	F2:3466	IR72 × Madabaruc	1, 2, 9, 12	Septiningsih et al. (2012)
	Rice	Ril, 81	Kinmaze (*O. japonica*) × DV85 (*O. indica*)	1, 2, 5, 7	Jiang et al. (2004)
	Soybean	Ril, 151	Nannong 1138-2 × Kefeng No. 1	LGA1, GD1a, LG G	Wang et al. (2008a)
Grains per spike	Barley	Dh,156	Yerong × Franklin	2H, 5H, 7H	Xue et al. (2010)
Grain yield				2H, 7H	
Kernel weight				2H	
Plant height		Dh,177	Yerong × Franklin	6H	
	Maize	F2, 228	HZ32 × K12	1, 4, 7, 10	Qiu et al. (2007)
Plant survival	Barley	Dh, 177	Yerong × Franklin	2H, 5H	Li et al. (2008)
Root dry weight	Maize	F2, 228	HZ32 × K12	1, 6, 9	Qiu et al. (2007)
Root length				7	
Spike length	Barley	Dh, 156	Yerong × Franklin	2H, 3H	Xue et al. (2010)
Spikes per plant				2H, 6H, 7H	
Total dry weight	Maize	F2, 228	HZ32 × K12	3, 4, 9	Qiu et al. (2007)

One of the QTLs for GRI positioned on 7A explained 23.92% of phenotypic variation. These results suggested that synthetic hexaploid wheat W7984 is an important genetic resource for waterlogging tolerance in wheat (Yu et al. 2014). Four major QTLs (*Sft1, Sft2, Sft3* and *Sft4*) were identified in soybean for normal seedling rate and germination rate under waterlogging stress. The QTL-*Sft2* responsible for seed coat pigmentation showed the largest effect on seed-flooding tolerance. However *Sft1, Sft3* and *Sft4* were controlling for seed coat colour and seed weight independently (Sayama et al. 2009). One single QTL was identified from tolerant genotype for grain yield and improved plant growth which shows 95% higher yield in tolerant line and 16% increase in height. Further this QTL was validated in another population of soybean (Van Toai et al. 2001, Reyna et al. 2003). Qiu et al. (2007) map a major QTL on chromosome 9 for relative dry weight of shoot, and QTL was positioned near a known anaerobic response gene in maize. Seven QTLs for waterlogging tolerance identified in barley DH population. All of these seven QTLs were located on different chromosomes. The QTL-*Qwt4–1* on 4H contributed to reduction in leaf chlorosis and increased plant biomass under waterlogging stress (Wenzl et al. 2006).

8.4.9   Evaluation and Development of Waterlogging-Tolerant Genotypes

Evaluation of genotypes for waterlogging tolerance is difficult as it varied upon timing, duration and severity of stress, and in certain extent waterlogging tolerance was also found correlated with varietal geographic distribution, for example, varieties from East Asia were more tolerant than those from Western Asia (Setter and Waters 2003). This circumstance can be explained by the natural or artificial selection for tolerance traits related with the climate variations. As wheat genotypes respond to waterlogging stress differently under varying environmental or experimental conditions. Therefore it is reasonable to combine several adaptive traits to evaluate the levels of waterlogging tolerance under specific environment condition (Pang et al. 2004; Zhou et al. 2012a, b). Chlorophyll fluorescence Fv/Fm parameter may be promising and indomitable for quick screening of lucerne genotypes for waterlogging tolerance (Smethurst and Shabala 2003). Recovering ability of root system from transient waterlogging or hypoxia is important for waterlogging tolerance. Different tolerance mechanisms are identified when exposed to intermittent waterlogging. These include adventitious root development, aerenchyma formation, least reduction of photosynthetic rate, chlorophyll content and fluorescence, and rapid recovery after drainage seems to be important survival traits for waterlogging tolerance (Pang et al. 2004). On the contrary, the measurement of net O_2 and ion fluxes from the root surface and net CO_2 assimilation are applicable for screening a small amount of lines. It is not practical for breeders to use these indices to screen thousands of lines during a limited time period. In consequence, identification of molecular markers associated with waterlogging tolerance and marker-assisted selection could be appreciable methods for waterlogging screening (Zhou et al. 2012a, b). Efforts made over the last four decades at the Central Soil Salinity Research Institute, Karnal, India, have demonstrated that development of salt-/waterlogging-tolerant wheat varieties is the key to successful biological approach of land reclamation. In this regard, good success in transferring and deployment of waterlogging-tolerant genes in suitable salt-tolerant background with the help of conventional approaches has been achieved. Three genetic stocks, i.e. KRL 35, KRL 99 and KRL 3–4, have been registered with NBPGR, New Delhi. The genetic stock KRL 35 (registered with NBPGR, New Delhi, in 2004, IC No. 408332, Registration No. 4013) has also been rated as the genotype most tolerant to waterlogging and B toxicity. KRL 99 (registered with NBPGR, New Delhi, in 2007, IC No. 546936, Registration No. 7046) has been shown excellent performance under waterlogging for 15 days after 22 days of sowing at high sodicity (pH 2:9.5). Genetic stock KRL 3–4 (registered with NBPGR, New Delhi, in 2009, IC No. 408331, Registration No. 9087) developed for salinity, sodicity and waterlogging tolerance. It has red colour grains, high

Table 8.6 Percent leaf chlorosis and agronomic characteristics of synthetic hexaploid wheats (*Triticum turgidum/Aegilops tauschii*) tolerant to waterlogging

Reg. no.	Designation	Parentage	Chlorosis (%)	Grain wt. (g 1000^{-1})	Anthesis (d)	Height (cm)
GP-639	CIGM86.953-SH19	Dverd 2/A. tauschii (221)	6.3	49.3	100	84
GP-640	CIGM90.863-SH64	Botno/A. tauschii (617)	6.7	44.7	104	102
GP-641	CIGM89.567-SH54	Ceta/A. tauschii (895)	8.8	37.6	104	93
GP-642	CIGM92.1723-SH82	68.111/Rgb-U/ Ward/3/A. tauschii (454)	9.2	42.4	98	92
Ckeck 1	CM80232	Ducula (BW tolerant check)	13.7	34.6	92	81
Ckeck 2	CM33027	Seri M82 (BW sensitive check)	76.8	26.4	92	73
		LSD (0.05)	3.5	4.1	3	7

Adapted from Villareal et al. (2001)

level of tolerance to sodicity, salinity and waterlogged sodic conditions due to lower uptake of sodium under salinity and sodicity. Excellent performance under water-logging for 20 days after 22 days of sowing has also been recorded for KRL 3–4. Villareal et al. (2001) developed four synthetic hexaploid wheat germplasm lines from durum wheat (*Triticum turgidum* L.) crossed by *Aegilops tauschii* Coss. These germplasms showed waterlogging tolerance under field conditions. Protocols for the *T. turgidum/A. tauschii* hybridization, embryo rescue and colchicine treatment were involved in the synthetic hexaploid wheat production. The four germplasm lines were found tolerant to waterlogging stress screened out on the basis of percent leaf chlorosis, and agronomic data across 3 years of field tests are presented in Table 8.6. Advanced breeding lines were subjected to waterlogging treatment for 7 weeks, commencing at 15 days after emergence to allow uniform crop emergence until about boot stage. During the duration of waterlogging treatment, continuous standing water was maintained within a range of 3 to 8 cm. Percentage leaf chloro-sis taken after 7 weeks of excess water stress was used as the primary criteria for waterlogging tolerance.

An important example of specific adaptation to a submergence tolerance is in rice, where a single major gene is the critical element in an adaptive trait. On mil-lions of hectares, rice is grown by poor farmers, particularly in eastern India and Bangladesh, and is subject to flash flooding that completely submerges plants. A highly tolerant Indian landrace, FR13A, was used as a donor for the trait in genetic analyses that identified a single major quantitative trait locus (QTL), designated *sub1*, which controlled 60–70% of phenotypic variation for the trait in the screening system. Submergence 1 (sub1) locus is located on chromosome 9, which is associ-ated with submergence tolerance (Xu and Mackill, 1996). Sub1 was further mapped to a 0.16 cM region (Xu et al. 2000) and fine mapped to 0.075 cM on chromosome 9 (Xu et al. 2006). Sub1 locus dissected into three major genes, sub1a, sub1b and sub1c, through positional cloning and reveals sub1a has two allelic forms, in which sub1a-1 was associated with tolerant lines and sub1a-2 with intolerant lines (Xu et al. 2006). Sub1a-1 allele has been introgressed into Swarna, BR11 and IR 64 varieties using marker-assisted backcrossing. These are locally adapted and widely grown rice cultivars in Asia which were released as 'Swarna-Sub1' in India, Indonesia and Bangladesh, 'BR11-Sub1' in Bangladesh, and 'IR64-Sub1' in the Philippines and Indonesia (Bailey-Serres et al. 2010). These cultivars had significant impact on rice growing in Asia, for example, Swarna-Sub1 submerged for 7 to 14 days across 128 villages of Odisha offers a 45% increase in yields over the popular cultivar 'Swarna' with no yield penalty (Dar et al. 2013).

8.5   Conclusion and Future Perspectives

Past research groups have advanced great leaps and bounds in understanding specific stress signalling events comprising of the sensors, secondary messengers, transcription factors, target genes and final stress response. However, challenges and research gaps are still in plenty for upcoming enthusiastic researchers to fill in. Salinity is still one among the major threats to global agriculture, rendering more and more agricultural lands as fallow. Salt stress affects all major primary plant physiological processes like photosynthesis, nutrient uptake, assimilation, respiration, hormonal metabolism, etc. Another indirect effect of excess salts in the root zone/sensitive plant parts is the increased production of ROS which damages the cellular membraneous network and associated macromolecules like proteins, lipids and nucleic acids. Therefore, it is highly imperative to critically understand key candidate players in the salt stress signalling pathway and modify/manipulate them to achieve better tolerance without significant compromise in yield. Integration of all the key omics approaches should be another vital target to research groups working across the globe to further progress towards the ultimate goal of developing salt-tolerant crop varieties without compromising yield (Fig. 8.3).

Tolerance to waterlogged conditions is affected by many external factors like soil properties, water level, developmental stage of the crop, ambient temperature, etc. Another important aspect is the relative time span of waterlogging treatment, which is highly critical in determining the extent of differences among tolerant, medium tolerant and susceptible genotypes. Waterlogging tolerance is a quantitative trait governed by multiple QTLs and has low heritability. Low heritability and difficulties in evaluating for waterlogging tolerance due to external factors make MAS highly imperative in selecting useful traits linked to tolerance. Multiple useful alleles can be combined in to popular backgrounds using MAS. Accurate locations of the QTL and linked markers are very important aspects of efficient marker-assisted selection. Stringent and accurate phenotyping are the most important factor that determines the accuracy of the mapped QTL, and therefore very reliable screening facility along with efficient selection indices is highly imperative. Further

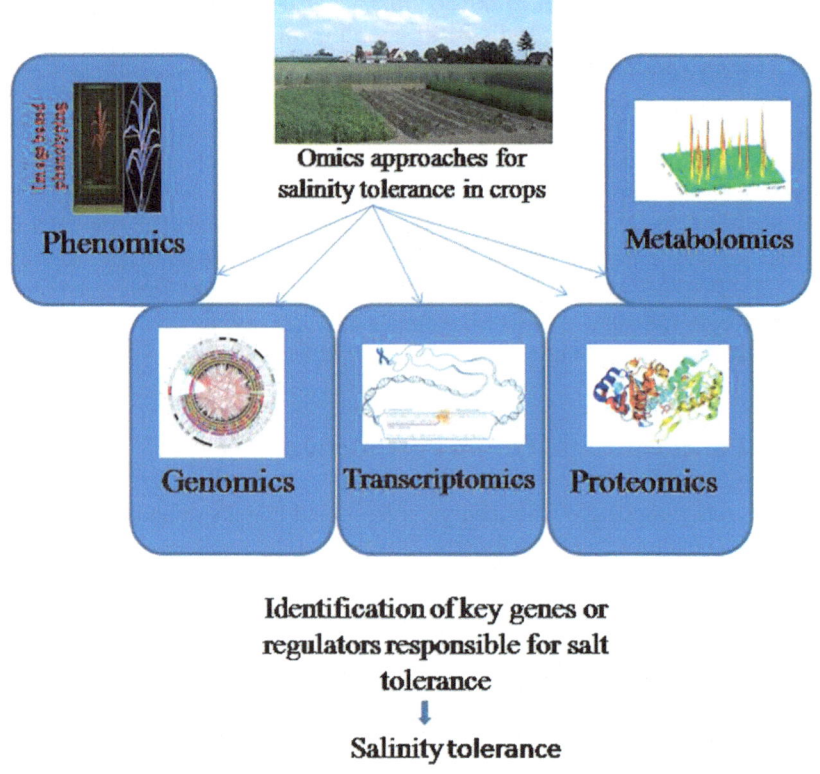

Fig. 8.3 'Omics-based'approaches and their integration to develop salinity tolerant crops. Flow of information from various 'omics-based' platforms such as phenomics, genomics, transcriptomics, proteomics and metabolomic needs to be integrated to understand complex traits such as salinity tolerance. Ultimately, the key genes/regulators responsible for salt tolerance need identification and validation using the tools of functional genomics

validation of already identified QTLs should also be done before introgressing these into popular backgrounds.

Future prospects include several aspects like regulatory elements that control the expression of major genes in the salt stress signalling pathway, synergistic effect of combining the cellular transport processes, targeted mutagenesis using genome editing technology, salt stress inducible promoter identification and analysis, etc. A judicious blend of molecular breeding using marker-assisted selection and modern biotechnological tools like genome editing is the need of the hour to develop salt-tolerant crop varieties without significant yield reduction in a short span of time.

References

Abebe T, Guenzi AC, Martin B, Cushman JC (2003) Tolerance of mannitol-accumulating trans-genic wheat to water stress and salinity. Plant Physiol 131:1748–1755

Ahmad P, Hakeem KR, Kumar A, Ashraf M, Akram NA (2012) Salt-induced changes in photosynthetic activity and oxidative defense system of three cultivars of mustard (*Brassica juncea* L.) Afr J Biotechnol 11:2694–2703

Ainley WM, Sastry-Dent L, Welter ME, Murray MG, Zeitler B, Amora R, Corbin DR, Miles RR, Arnold NL, Strange TL, Simpson MA, Cao Z, Carroll C, Pawelczak KS, Blue R, West K, Rowland LM, Perkins D, Samuel P, Dewes CM, Shen L, Sriram S, Evans SL, Rebar EJ, Zhang L, Gregory PD, Urnov FD, Webb SR, Petolino JF (2013) Trait stacking via targeted genome editing. Plant Biotechnol J 11:1126–1134

Akhtar J, Gorham J, Qureshi RH (1994) Combined effect of salinity and hypoxia in wheat (*Triticum aestivum* L.) and wheat-Thinopyrum amphiploids. Plant Soil 166:47–54

Albrecht G, Kammerer S, Pranznik W, Wiedenroth EM (1993) Fructan content of wheat seedlings (*Triticum aestivum* L.) under hypoxia and following reaeration. New Phytol 123:471–476

Albrecht G, Wiedenroth EM (1994) Protection against activated oxygen following re-aeration of hypoxically pretreated wheat roots. The response of the glutathione system. J Exp Bot 45:449–455

Albrecht G, Mustroph A, Fox TC (2004) Sugar and fructan accumulation during metabolic adjustment between respiration and fermentation under low oxygen conditions in wheat roots. Plant Physiol 120:93–101

Alvarez S, Marsh EL, Schroeder SG, Schachtman DP (2008) Metabolomic and proteomic changes in the xylem sap of maize under drought. Plant Cell Environ 31:325–340

Angaji SA, Septiningsih EM, Mackill DJ, Ismail AM (2010) QTLs associated with tolerance of flooding during germination in rice (Oryza sativa L.) Euphytica 172:159–168

Arbona V, Hossain Z, Lo'pez Climent MF, Pe'rez Clemente RM, Go'mez Cadenas A (2008) Antioxidant enzymatic activity is linked to waterlogging stress tolerance in citrus. Physiol Plant 132(452):466

Armstrong W, Brändle R, Jackson MB (1994) Mechanisms of flood tolerance in plants. Acta Bot Neerl 43:307–358

Ashraf M, Foolad MR (2013) Crop breeding for salt tolerance in the era of molecular markers and marker-assisted selection. Plant Breed 132:10–20

Ashraf M, Harris PJC (2013) Photosynthesis under stressful environments: an overview. Photosynthetica 51(2):163–190

Attwell BJ, Greenway H, Barrett Lennard EG (1985) Root function and adaptive responses in conditions of oxygen deficiency. In: Muirhead WA, Humphreys E (ed.). Root zone limitations to crop production on clay soils: symposium of the Australian Society of Soil Science Inc., Riverina Branch. Melbourne, Vic., Australia. Commonwealth Scientific and Industrial Research Organization. pp. 65–75

Baea H, Herman E, Bailey B, Bae HJ, Sicher R (2005) Exogenous trehalose alters *Arabidopsis* transcripts involved in cell wall modification, abiotic stress, nitrogen metabolism, and plant defense. Physiol Plant 125:114–126

Baek D, Jiang J, Chung JS, Wang B, Chen J, Xin Z, Shi H (2011) Regulated AtHKT1 gene expression by a distal enhancer element and DNA methylation in the promoter plays an important role in salt tolerance. Plant Cell Physiol 52:149–161

Bailey-Serres J, Voesenek L (2008) Flooding stress: acclimations and genetic diversity. Annu Rev Plant Biol 59:313–339

Bailey-Serres J, Fukao T, Ronald P, Ismail A, Heuer S, Mackill D (2010) Submergence tolerant rice: SUB1's journey from landrace to modern cultivar. Rice 3:138–147

Barragan V, Leidi EO, Andrés Z, Rubio L, De Luca A, Fernández JA, Cubero B, Pardo JM (2012) Ion exchangers NHX1 and NHX2 mediate active potassium uptake into vacuoles to regulate cell turgor and stomatal function in *Arabidopsis*. Plant Cell 24:1127–1142

Barrett-Lennard EG, Leighton P, Buwalda F, Gibbs J, Armstrong W, Thomson CJ, Greenway H (1988) Effects of growing wheat in hypoxic nutrient solutions and of subsequent transfer to aerated solutions I. Growth and carbohydrate status of shoots and roots. Aust J Pl Physiol 15:585–598

Bartel DP (2004) MicroRNAs: genomics, biogenesis, mechanism, and function. Cell 116:281–297

Bassil E, Tajima H, Liang YC et al (2011) The *Arabidopsis* Na$^+$/H$^+$ antiporters NHX1 and NHX2 control vacuolar pH and K$^+$ homeostasis to regulate growth, flower development and reproduction. Plant Cell 23:3482–3497

Baxter-Burrell A, Yang ZB, Springer PS, Bailey-Serres J (2002) RopGAP4-dependent Rop GTPase rheostat control of *Arabidopsis* oxygen deprivation tolerance. Science 296:2026–2028

Belhaj K, Chaparro GA, Kamoun S, Nekrasov V (2013) Plant genome editing made easy: targeted mutagenesis in model and crop plants using the CRISPR/Cas system. Plant Methods 9:39

Biemelt S, Keetman U, Albrecht G (1998) Re-aeration following hypoxia or anoxia leads to activation of the antioxidative defence system in roots of wheat seedlings. Plant Physiol 116:651–658

Bin T, Xu SZ, Zou XL, Zheng YL, Qiu FZ (2010) Changes of antioxidative enzymes and lipid peroxidation in leaves and roots of waterlogging-tolerant and waterlogging-sensitive maize genotypes at seedling stage. Agric Sci China 9:651–661

Blumwald E, Poole RJ (1985) Na/H antiport in isolated tonoplast vesicles from storage tissue of *Beta vulgaris*. Plant Physiol 78:163–167

Boru G (1996) Expression and inheritance of tolerance to waterlogging stresses in wheat (*Triticum aestivum* L.), Ph.D Thesis, Oregon State University, Corvallis, OR. 88 pp.

Brandle RA (1991) Flooding resistance of rhizomatous amphibious plants. In: Jackson MD, Davies DD, Lambers H (eds) Plant life under oxygen deprivation. Academic Publisher, The Hague, pp 35–46

Byrt CS, Platten JD, Spielmeyer W, James RA, Lagudah ES, Dennis ES, Tester M, Munns R (2007) HKT1;5-like cation transporters linked to Na$^+$ exclusion loci in wheat, Nax2 and Kna1. Plant Physiol 143:1918–1928

Campo S, Baldrich P, Messeguer J, Lalanne E, Coca M, Segundo BS (2014) Overexpression of a calcium-dependent protein kinase confers salt and drought tolerance in rice by preventing membrane lipid peroxidation. Plant Physiol 165:688–704

Changdee T, Polthanee A, Akkasaeng C, Morita S (2009) Effect of different waterlogging regimes on growth, some yield and roots development parameters in three fiber crops (*Hibiscus cannabinus* L., *Hibiscus sabdariffa* L. and *Corchorus olitorius* L.) Asian J Plant Sci 8:515–525

Chao DY, Dilkes B, Luo H, Douglas H, Yakubova E, Lahner B, Salt DE (2013) Polyploids exhibit higher potassium uptake and salinity tolerance in *Arabidopsis*. Science 341:658–659

Chaves MM, Flexas J, Pinheiro C (2009) Photosynthesis under drought and salt stress: regulation mechanisms from whole plant to cell. Ann Bot 103(4):551–560

Chen TH, Murata N (2011) Glycine betaine protects plants against abiotic stress: mechanisms and biotechnological applications. Plant Cell Environ 34:1–20

Cheng Y, Qi Y, Zhu Q, Chen X, Wang N, Zhao X, Chen H, Cui X, Xu L, Zhang W (2009) New changes in the plasma-membrane-associated proteome of rice roots under salt stress. Proteomics 9:3100–3114

Ciftci-Yilmaz S, Morsy MR, Song L, Coutu A, Krizek BA, Lewis MW, Warren D, Cushman J, Connolly EL, Mittler R (2007) The EAR motif of the Cys2/His2-typezinc finger protein Zat7 plays a key role in the defense response of *Arabidopsis* to salinity stress. J Biol Chem 282:9260–9268

Colmer TD (2003) Aerenchyma and an inducible barrier to radial oxygen loss facilitate root aeration in upland, paddy and deep-water rice (*Oryza sativa* L.) Ann Bot 91:301–309

Conde C, Silva P, Agasse A, Lemoine R, Delrot S, Tavares R, Geròs H (2007) Utilization and transport of mannitol in Olea europaea and implications for salt stress tolerance. Plant Cell Physiol 48:42–53

Cornelious B, Chen P, Chen Y, Leon ND, Shannon JG, Wang D (2005) Identification of QTLs underlying water-logging tolerance in soybean. Mol Breed 16:103–112

Cui MH, Yoo KS, Hyoung S, Nguyen HT, Kim YY, Kim HJ, Ok SH, Yoo SD, Shin JS (2013) An *Arabidopsis* R2R3-MYB transcription factor, AtMYB20, negatively regulates type 2C serine/threonine protein phosphatases to enhance salt tolerance. FEBS Letter 587:1773–1778

Dar MH, de Janvry A, Emerick K, Raitzer D, Sadoulet E (2013) Flood tolerant rice reduces yield variability and raises expected yield, differentially benefitting socially disadvantaged groups. Sci Rep 3:3315–3324

Davenport RJ, Muñoz-Mayor A, Jha D, Essah PA, Rus A, Tester M (2007) The Na⁺ transporter AtHKT1;1 controls retrieval of Na⁺ from the xylem in *Arabidopsis*. Plant Cell Environ 30:497–507

Davies DD (1980) Anaerobic metabolism and the production of organic acids. In: Davies DD (ed) The biochemistry of plants, vol 2. Academic Press, New York, pp 581–611

de Abreu CE, Araujo Gdos S, Monteiro-Moreira AC, Costa JH, Leite Hde B, Moreno FB et al (2014) Proteomic analysis of salt stress and recovery in leaves of *Vigna unguiculata* cultivars differing in salt tolerance. Plant Cell Rep 33(8):1289–1306

Dennis ES, Dolferus R, Ellis M, Rahman M, Wu Y, Hoeren FU, Grover A, Ismond KP, Good AG, Peacock WJ (2000) Molecular strategies for improving waterlogging tolerance in plants. J Exp Bot 51(342):89–97

Ding N, Musgrave ME (1995) Relationship between mineral coating on roots and yield performance of wheat under waterlogging stress. J Exp Bot 46:939–945

Dong JG, Yu SW (1984) Effect of cytokinin on senescence and ethylene production in waterlogged wheat plants. Acta Phytophysiol Sinica 10:55–62

Dong JG, Yu SW, Yu SW (1983) Effect of increased ethylene production during different periods on the resistance of wheat plants to waterlogging. Acta Phytophysiol Sinica 9:383–389

Dong JG, Yu SW, Li ZG (1986) Changes in ethylene production in relation to 1-amino cyclopropane-à1-carboxylic acid and its malonyl conjugate in waterlogged wheat plants. Acta Bot Sin 28:396–403

Drew MC (1997) Oxygen deficiency and root metabolism: injury and acclimation under hypoxia and anoxia. Annu Rev Plant Physiol Plant Mol Biol 48:223–250

Drew MC, Sisworo EJ (1977) Early effects of flooding on nitrogen deficiency and leaf chlorosis in barley. New Phytol 79:567–571

Drew MC, Sisworo EJ (1979) The development of waterlogging damage in young barley plants in relation to plant nutrient status and changes in soil properties. New Phytol 82:301–314

Fu XZ, Ullah Khan E, Hu SS, Fan QJ, Liu JH (2011) Overexpression of the betaine aldehyde dehydrogenase gene from Atriplex hortensis enhances salt tolerance in the transgenic trifoliate orange (*Poncirus trifoliata* L. Raf.) Environ Exp Bot 74:106–113

Gardner WK, Flood RG (1993) Less waterlogging damage with long season wheats. Cereal Res Commun 21:337–343

Garg N, Manchanda G (2008) Salinity and its effects on the functional biology of legumes. Acta Physiol Plant 30:595–618

Garthwaite AJ, Steudle E, Colmer TD (2006) Water uptake by roots of *Hordeum marinum*: formation of a barrier to radial O₂ loss does not affect root hydraulic conductivity. J Exp Bot 57:655–664

Garthwaite AJ, Armstrong W, Colmer TD (2008) Assessment of O₂ diffusivity across the barrier to radial O₂loss in adventitious roots of *Hordeum marinum*. New Phytol 179:405–416

Geng Y, Wu R, Wee CW, Xie F, Wei X, Chan PM, Tham C, Duan L, Dinneny JR (2013) A spatio-àtemporal understanding of growth regulation during the salt stress response in *Arabidopsis*. Plant Cell 25:2132–2154

Germain V, Ricard B, Raymond P, Saglio PH (1997) The role of sugars, hexokinase and sucrose synthase in the determination of hypoxically induced tolerance to anoxia in tomato roots. Plant Physiol 114:167–175

Gibbs DJ, Lee SC, Isa NM, Gramuglia S, Fukao T, Bassel GW, Correia CS, Corbineau F, Theodoulou FL, Bailey-Serres J, Holdsworth MJ (2011) Homeostatic response to hypoxia is regulated by the N-end rule pathway in plants. Nature 479:415–418

Gill SS, Tuteja N (2010) Reactive oxygen species and antioxidant machinery in abiotic stress tolerance in crop plants. Plant Physiol Biochem 48:909–930

Githiri SM, Watanabe S, Harada K, Takahashi R (2006) QTL analysis of flooding tolerance in soybean at an early vegetative growth stage. Plant Breed 125:613–618

Golldack D, Lüking I, Yang O (2011) Plant tolerance to drought and salinity: stress regulating transcription factors and their functional significance in the cellular transcriptional network. Plant Cell Rep 30:1383–1391

Gong Q, Li P, Ma S, Rupassara SI, Bohnert HJ (2005) Salinity stress adaptation competence in the extremophile Thellungiella halophila in comparison with its relative *Arabidopsis thaliana*. Plant J 44:826–839

Greenway H, Munns R (1980) Mechanisms of salt tolerance in non-halophytes. Annu. Rev Plant Physiol 31:149–190

Greenway H, Waters I, Newsome J (1992) Effects of anoxia on uptake and loss of solutes in roots of wheat. Aust J Plant Physiol 19:233–247

Greenway H, Armstrong W, Colmer TD (2006) Conditions leading to high CO_2 (>5 kPa) in water-logged flooded soils and possible effects on root growth and metabolism. Ann Bot 98: 9–32

Grime JP (1977) Evidence for the existence of three primary strategies in plants and its relevance to ecological and evolutionary theory. Am Nat 111:1169–1194

Grimm B (1990) Primary structure of a key enzyme in plant tetrapyrrole synthesis: glutamate 1-semialdehyde aminotransferase. Proc Natl Acad Sci U S A 87(11):4169–4173

Guan Q, Wu J, Yue X, Zhang Y, Zhu J (2013) A nuclear calcium-sensing pathway is critical for gene regulation and salt stress tolerance in *Arabidopsis*. PLoS Genet 9:e1003755

Guinn EJ, Pegram LM, Capp MW, Pollock MN, Record MT Jr (2011) Quantifying why urea is a protein denaturant, whereas glycine betaine is a protein stabilizer. Proc Natl Acad Sci U S A 108:16932–16937

Halfter U, Ishitani M, Zhu JK (2000) The Arabidopsis SOS2 protein kinase physically interacts with and is activated by the calcium-binding protein SOS3. Proc Natl Acad Sci 97:3735–3740

Hartel H, Kruse E, Grimm B (1997) Restriction of chlorophyll synthesis due to expression of glu-tamate 1-semialdehyde aminotransferase antisense RNA does not reduce the light-harvesting antenna size in tobacco. Plant Physiol 113(4):1113–1124

Hattori Y, Miura K, Asano K, Yamamoto E, Mori H, Kitano H, Matsuoka M, Ashikari M (2007) A major QTL confers rapid internode elongation in response to water rise in deepwater rice. Breed Sci 57:305–314

Hattori Y, Nagai K, Mori H, Kitano H, Matsuoka M, Ashikari M (2008) Mapping of three QTLs that regulate internode elongation in deepwater rice. Breed Sci 58:39–46

He C, Morgan PW, Drew MC (1996) Transduction of an ethylene signal is required for cell death and lysis in the root cortex of maize during aerenchyma formation induced by hypoxia. Plant Physiol 112:463–472

Horie T, Hauser F, Schroeder JI (2009) HKT transporter-mediated salinity resistance mechanisms in *Arabidopsis* and monocot crop plants. Trends Plant Sci 14:660–668

Hossain Z, Mandal AKA, Datta SK, Biswas AK (2007) Development of NaCl tolerant line in *Chrysanthemum morifolium* Ramat through shoot organogenesis of selected callus line. J Biotechnol 129:658–667

Huang B (1997) Mechanisms of plant resistance to waterlogging. In: Basra AS, Basra RK (eds) Mechanisms of environmental stress resistance in plants. Harwood Academic Publishers, Amsterdam, pp 59–82

Huang B, Johnson JW (1995) Root respiration and carbohydrate status of two wheat genotypes in response to hypoxia. Ann Bot 75(4):427–432

Huang B, Johnson JW, Scott NeSmith D, Bridges DC (1994) Growth, physiology, and anatomical response of two wheat genotypes to waterlogging and nutrient supply. J Exp Bot 54:193–202

Huang B, Johnson JW, Scott NeSmith D, Bridges DC (1995) Nutrient accumulation and distribu-tion of wheat genotypes in response to waterlogging and nutrient supply. Plant Soil 173:47–54

Huang B, Johnson JW, Box JE, NeSmith S (1997) Root characteristics and physiological activities of wheat in response to hypoxia and ethylene. Crop Sci 37:812–818

Huang S, Spielmeyer W, Lagudah ES, James RA, Platten JD, Dennis ES, Munns R (2006) A sodium transporter (HKT7) is a candidate for Nax1, a gene for salt tolerance in durum wheat. Plant Physiol 142:1718–1727

Inan G, Zhang Q, Li P, Wang Z, Cao Z, Zhang H, Zhang C, Quist TM, Goodwin SM, Zhu J, Shi H, Damsz B, Charbaji T, Gong Q, Ma S, Fredricksen M, Galbraith DW, Jenks MA, Rhodes D, Hasegawa PM, Bohnert HJ, Joly RJ, Bressan RA, Zhu JK (2004) Salt cress. A halophyte and cryophyte *Arabidopsis* relative model system and its applicability to molecular genetic analyses of growth and development of extremophiles. Plant Physiol 135:1718–1737

Jackson MB, Drew MC (1984) Effects of flooding on growth and metabolism of herbaceous plants. In: Kozlowski TT (ed) Flooding and plant growth. Academic Press, New York, pp 47–128

James RA, Davenport RJ, Munns R (2006) Physiological characterization of two genes for Na$^+$ exclusion in durum wheat, Nax1 and Nax2. Plant Physiol 142:1537–1547

Jia H, Shao M, He Y, Guan R, Chu P, Jiang H (2015) Proteome dynamics and physiological responses to short-term salt stress in Brassica napus leaves. PLoS One 10(12):e0144808

Jiang Y, Deyholos MK (2009) Functional characterization of Arabidopsis NaCl-inducible WRKY25 and WRKY33 transcription factors in abiotic stresses. Plant Mol Biol 69:91–105

Jiang L, Hou M, Wang C, Wan J (2004) Quantitative trait loci and epistatic analysis of seed anoxia germinability in rice (Oryza Sativa). Rice Sci 11:238–244

Jiang Y, Yang B, Deyholos MK (2009) Functional characterization of the *Arabidopsis bHLH92* transcription factor in abiotic stress. Mol Gen Genomics 282:503–516

Jiang C, Belfield EJ, Mithani A, Visscher A, Ragoussis J, Mott R, Smith JA, Harberd NP (2012) ROS-mediated vascular homeostatic control of root-to-shoot soil Na delivery in *Arabidopsis*. EMBO J 31:4359–4370

Jiang C, Belfield EJ, Cao Y, Smith JA, Harberd NP (2013) An Arabidopsis soil-salinity-tolerance mutation confers ethylene-mediated enhancement of sodium/potassium homeostasis. Plant Cell 25:3535–3552

Johnson JR, Crobb BG, Drew MC (1989) Hypoxic induction of anoxia tolerance in root tips of Zea mays. Plant Physiol 91:837–841

Karan R, DeLeon T, Biradar H, Subudhi PK (2012) Salt stress induced variation in DNA methylation pattern and its influence on gene expression in contrasting rice genotypes. PLoS One 7:e40203

Kasuga M, Liu Q, Miura S, Yamaguchi-Shinozaki K, Shinozaki K (1999) Improving plant drought, salt, and freezing tolerance by gene transfer of a single stress-inducible transcription factor. Nat Biotechnol 17:287–291

Klok EJ, Wilson IW, Wilson D, Chapman SC, Ewing RM, Somerville SC, Peacock WJ, Dolferus R, Dennis ES (2002) Expression profile analysis of the low-oxygen response in *Arabidopsis* root cultures. Plant Cell 14:2481–2494

Koohafkan P (2012) Water and cereals in drylands. Organization of the United Nations and Earthscan, The Food and Agriculture

Kosová K, Vítámvás P, Prášil IT (2010) The role of dehydrins in plant stress response. In: Pessarakli M (ed) Handbook of plant and crop stress. CRC Press/Taylor and Francis, Boca Raton, pp 239–285

Kosová K, Prášil IT, Vítámvás P (2013) Protein contribution to plant salinity response and tolerance acquisition a review. Int J Mol Sci 14:6757–6789

Krebs M, Beyhl D, Görlich E, Al-Rasheid KA, Marten I, Stierhof YD, Hedrich R, Schumacher K (2010) Arabidopsis V-ATPase activity at the tonoplast is required for efficient nutrient storage but not for sodium accumulation. Proc Natl Acad Sci U S A 107:3251–3256

Kumar Parida A, Jha B (2010) Salt tolerance mechanisms in mangroves: a review. Trees 24:199–217

Kyriakis JM, Avruch J (2012) Mammalian MAPK signal transduction pathways activated by stress and inflammation: a 10-year update. Physiol Rev 92:689–737

Lee YS, Park SR, Park HJ, Kwon YW (2004a) Salt stress magnitude can be quantified by integrating salinity with respect to duration. Proceedings of 4th international crop Science congress, Brisbane, Aust 26 Sept-1 Oct pp 1–5

Lee SM, Lee EJ, Yang EJ, Lee JE, Park AR, Song WH, Park OK (2004b) Proteomic identification of annexins, calcium-dependent membrane binding proteins that mediate osmotic stress and abscisic acid signal transduction in Arabidopsis. Plant Cell 16:1378–1391

Leidi EO, Barragan V, Rubio L, El-Hamdaoui A, Ruiz MT, Cubero B, Fernandez JA, Bressan RA, Hasegawa PM, Quintero FJ, Pardo JM (2010) The AtNHX1 exchanger mediates potassium compartmentation in vacuoles of transgenic tomato. Plant J 61:495–506

Li HB, Vaillancourt R, Mendham NJ, Zhou MX (2008) Comparative mapping of quantitative trait loci associated with waterlogging tolerance in barley (*Hordeum vulgare* L.) BMC Genomics 9:401

Li C, Jiang D, Wollenweber B, Li Y, Dai T, Cao W (2011) Waterlogging pretreatment during vegetative growth improves tolerance to waterlogging after anthesis in wheat. Plant Sci 180:672–678

Licausi F, Kosmacz M, Weits DA, Giuntoli B, Giorgi FM, Voesenek LACJ, Perata P, Dongen JT (2011) Oxygen sensing in plants is mediated by an N-end rule pathway for protein destabilization. Nature 479:419–422

Lin H, Yang Y, Quan R, Mendoza I, Wu Y, Du W, Zhao S, Schumaker KS, Pardo JM, Guoa Y (2009) Phosphorylation of SOS3-like calcium binding protein8 by SOS2 protein kinase stabilizes their protein complex and regulates salt tolerance in Arabidopsis. Plant Cell 21:1607–1619

Liu P, Sun F, Gao R, Dong H (2012a) RAP2.6L overexpression delays waterlogging induced premature senescence by increasing stomatal closure more than antioxidant enzyme activity. Plant Mol Biol 79:609–622

Liu Z, Kumari S, Zhang L, Zheng Y, Ware D (2012b) Characterization of miRNAs in response to short term waterlogging in three inbred lines of Zea mays. PLoS One 7:e0039786

Lu Y, Lam H, Pi E, Zhan Q, Tsai S, Wang C, Kwan Y, Ngai S (2013) Comparative metabolomics in Glycine max and Glycine Soja under salt stress to reveal the phenotypes of their offspring. J Agric Food Chem 61:8711–8721

Lugan R, Niogret MF, Leport L, Guegan JP, Larher FR, Savoure A et al (2010) Metabolome and water homeostasis analysis of *Thellungiella salsuginea* suggests that dehydration tolerance is a key response to osmotic stress in this halophyte. Plant J 64:215–229

Ma H, Song L, Shu Y, Wang S, Niu J, Wang Z et al (2012) Comparative proteomic analysis of seedling leaves of different salt tolerant soybean genotypes. J Proteome 75(5):1529–1546

Mahajan S, Pandey GK, Tuteja N (2008) Calcium and salt stress signaling in plants: shedding light on SOS pathway. Arch Biochem Biophys 471:146–158

Malik AI, Colmer TD, Lambers H, Schortemeyer M (2001) Changes in physiological and morphological traits of roots and shoots of wheat in response to different depths of waterlogging. Aust J Pl Physiol 28:1121–1131

Malik AI, Colmer TD, Lambers H, Setter TL, Schortemeyer M (2002) Short-term waterlogging has long-term effects on the growth and physiology of wheat. New Phytol 153:225–236

Manaa A, Ahmed HB, Valot B, Bouchet JP, Aschi-Smiti S, Causse M, Faurobert M (2011) Salt and genotype impact on plant physiology and root proteome variations in tomato. J Exp Bot 62:2797–2813

Mano Y, Muraki M, Fujimori M, Takamizo T, Kindiger B (2005) Identification of QTL controlling adventitious root formation during flooding conditions in teosinte (Zea mays ssp. huehuetenangensis) seedlings. Euphytica 142:33–42

Mano Y, Omori F (2007) Breeding for flooding tolerant maize using "teosinte" as a germplasm resource. Plant Roots 1:17–21

Mano Y, Omori F (2009) High-density linkage map around the root aerenchyma locus Qaer1.06 in the backcross populations of maize Mi29 * teosinte "Zea nicaraguensis". Breed Sci 59:427–433

Mano Y, Omori F, Kindiger B, Takahashi H (2008) A linkage map of maize × teosinte Zea luxurians and identification of QTLs controlling root aerenchyma formation. Mol Breed 21:327–337

Martínez-Atienza J, Jiang X, Garciadeblas B, Mendoza I, Zhu JK, Pardo JM, Quintero FJ (2007) Conservation of the salt overly sensitive pathway in rice. Plant Physiol 143:1001–1012

McDonald GK, Gardner UK (1987) Effect of waterlogging on the grain yield response of wheat to sowing date in southwestern Victoria. Aust J Exp Agric 27:661–670

McDonald MP, Galwey NW, Colmer TD (2001a) Waterlogging tolerance in the tribe *Triticeae*: the adventitious roots of Critesion marinum have a relatively high porosity and a barrier to radial oxygen loss. Plant Cell Environ 24:585–596

McDonald MP, Galwey NW, Ellneskog-Staam P, Colmer TD (2001b) Evaluation of Lophopyrum elongatum as a source of genetic diversity to increase waterlogging tolerance of hexaploid wheat (*Triticum aestivum*). New Phytol 151:369–380

Mendoza I, Rubio F, Rodriguez-Navarro A, Pardo JM (1994) The protein phosphatase calcineurin is essential for NaCl tolerance of Saccharomyces cerevisiae. J Biol Chem 269:8792–8796

Menegus F, Cattaruzza L, Chersi A, Fronza G (1989) Differences in the anaerobic lactate- succinate production and in the changes of cell sap pH for plants with high and low resistance to anoxia. Plant Physiol 90:29–32

Menegus F, Cattaruzza L, Mattana M, Beffagna N, Ragg E (1991) Response to anoxia in rice and wheat seedlings. Changes in the pH of intracellular compartments, glucose-6-phosphate level, and metabolic rate. Plant Physiol 95:760–767

Moons A, Bauw G, Prinsen E, van Montagu M, van Der Straeten D (1995) Molecular and physiological responses to abscisic acid and salts in roots of salt-sensitive and salt-tolerant Indica rice varieties. Plant Physiol 107:177–186

Morard E, Silvestre J (1996) Plant injury due to oxygen deficiency in the root environment of soilless culture: a review. Plant Soil 184:243–254

Munns R (2003) Genetic control of sodium exclusion in durum wheat. Aust J Agric Res 54:627–635

Munns R (2005) Genes and salt tolerance: bringing them together. New Phytol 167:645–663

Munns R, James RA, Lauchli A (2006) Approaches to increasing the salt tolerance of wheat and other cereals. J Exp Bot 57:1025–1043

Nounjana N, Nghiab PT, Theerakulpisuta P (2012) Exogenous proline and trehalose promote recovery of rice seedlings from salt-stress and differentially modulate antioxidant enzymes and expression of related genes. J Plant Physiol 169:596–604

Pang J, Zhou M, Mendham N, Shabala S (2004) Growth and physiological responses of six barley genotypes to waterlogging and subsequent recovery. Aust J Agric Res 55:895–906

Pang Q, Chen S, Dai S, Wang Y, Chen Y, Yan X (2010) Comparative proteomics of salt tolerance in *Arabidopsis thaliana* and *Thellungiella halophila*. J Proteome Res 9:2584–2599

Plaut Z, Edelstein M, Ben Hur M (2013) Overcoming salinity barriers to crop production using traditional methods. Crit. Rev. Plant Sci 32:250–291

Qiu F, Zheng Y, Zhang Z, Xu S (2007) Mapping of QTL associated with waterlogging tolerance during the seedling stage in maize. Ann Bot 99:1067–1081

Quan R, Lin H, Mendoza I, Zhang Y, Cao W, Yang Y, Shang M, Chen S, Pardo JM, Guo Y (2007) SCaBP8/CBL10, a putative calcium sensor, interacts with the protein kinase SOS2 to protect Arabidopsis shoots from salt stress. Plant Cell 19:1415–1431

Qui QS, Guo Y, Dietrich MA, Schumaker KS, Zhu JK (2002) Regulation of SOS1, a plasma membrane Na$^+$/H$^+$ exchanger in Arabidopsis thaliana, by SOS2 and SOS3. P Natl Acad Sci 99:8436–8441

Rani CR, Reema C, Alka S, Singh PK (2012) Salt tolerance of Sorghum bicolor cultivars during germination and seedling growth. Res J Recent Sci 1(3):1–10

Rawson HM, Subedi KD (1996) Hypothesis for why sterility occurs in wheat in Asia. In: Rawson HM (ed) Sterility in wheat in subtropical Asia: extent, causes and solutions, ACIAR Proceedings no, vol 72, pp 132–134

Renberg L, Johansson AI, Shutova T, Stenlund H, Aksmann A, Raven JA, Gardeström P, Moritz T, Samuelsson G (2010) A metabolomic approach to study major metabolite changes during acclimation to limiting CO_2 in *Chlamydomonas reinhardtii*. Plant Physiol 154:187–196

Reyna N, Cornelious B, Shannon JG, Sneller CH (2003) Evaluation of a QTL for waterlogging tolerance in southern soybean germplasm. Crop Sci 43:2077–2082

Ricard B, Couee I, Raymond P, Saglio PH, Saint-Ges V, Pradet A (1994) Plant metabolism under hypoxia and anoxia. Plant Physiol Biochem 32:1–10

Rodriguez MC, Petersen M, Mundy J (2010) Mitogen-activated protein kinase signaling in plants. Annu Rev Plant Biol 61:621–649

Roy C, Sengupta DN (2014) Effect of short term NaCl stress on cultivars of S. lycopersicum: a comparative biochemical approach. J. Stress Physiol Biochem 10(1):59–81

Rubio F, Gassmann W, Schroeder JI (1995) Sodium-driven potassium uptake by the plant potassium transporter HKT1 and mutations conferring salt tolerance. Science 270:1660–1663

Sadat Noori SA, Sokhansanj A (2004) Triple test cross analysis for genetic components of salinity tolerance in spring wheat. Aust J Sci 15(1):13–19

Sairam RK, Kumutha D, Ezhilmathi K, Deshmukh PS, Srivastava GC (2008) Physiology and biochemistry of waterlogging tolerance in plants. Biol Plant 52:401–412

Sairam RK, Kumutha D, Chinnusamy V, Meena RC (2009) Waterlogging-induced increase in sugar mobilization, fermentation, and related gene expression in the roots of mung bean (*Vigna radiata*). J Plant Physiol 166:602–616

Sairam RK, Dharmar K, Lekshmy S, Chinnusam V (2011) Expression of antioxidant defense genes in mung bean (*Vigna radiata* L.) roots under water-logging is associated with hypoxia tolerance. Acta Physiol Plant 33:735–744

Samad A, Meisner CA, Saifuzzaman M, van Ginkel M (2001) Waterlogging tolerance. In: Reynolds MP, Ortiz-Monasterio JI, McNab A (eds) Application of physiology in wheat breeding. CIMMYT, Mexico, pp 136–144

Samajova O, Plihal O, Al-Yousif M, Hirt H, Samaj J (2013) Improvement of stress tolerance in plants by genetic manipulation of mitogen activated protein kinases. Biotechnol Adv 31:118–128

Sangen W, He L, Li Z, Zeng J, Chai Y, Hou L (1996) A comparative study of the resistance of barley and wheat to waterlogging. Acta Agron Sin 22(2):228–232

Sani E, Herzyk P, Perrella G, Colot V, Amtmann A (2013) Hyperosmotic priming of *Arabidopsis* seedlings establishes a long-term somatic memory accompanied by specific changes of the epigenome. Genome Biol 14:R59

Sarkar RK, Panda D (2009) Distinction and characterisation of submergence tolerant and sensitive rice cultivars, probed by the fluorescence OJIP rise kinetics. Funct Plant Biol 36:222–233

Sayama T, Nakazaki T, Ishikawa G, Yagasaki K, Yamada N, Hirota N, Hirata K, Yoshikawa T, Saito H, Teraishi M, Okumoto Y, Tsukiyama T, Tanisaka T (2009) QTL analysis of seed flooding tolerance in soybean (*Glycine max* [L.] Merr.) Plant Sci 176:514–521

Schweighofer A, Kazanaviciute V, Scheikl E, Teige M, Doczi R, Hirt H, Schwanninger M, Merijn K, Schuurink R, Mauch F, Buchala A, Cardinale F, Meskienea I (2007) The PP2C-type phosphatase AP2C1, which negatively regulates MPK4 and MPK6, modulates innate immunity, jasmonic acid, and ethylene levels in Arabidopsis. Plant Cell 19:2213–2224

Septiningsih EM, Sanchez DL, Singh N, Sendon PMD, Pamplona AM, Heuer S, Mackill DJ (2012) Identifying novel QTLs for submergence tolerance in rice cultivars IR72 and Madabaru. Theor Appl Genet 124:867–874

Setter TL (2000) Farming systems for waterlogging prone sandplain soils of the south coast final report of GRDC project no DAW292. Department of Agriculture, Western Australia. 68 pp

Setter TL, Waters I (2003) Review of prospects for germplasm improvement for waterlogging tolerance in wheat, barley and oats. Plant Soil 253:1–34

Shabala S (2011) Physiological and cellular aspects of phytotoxicity tolerance in plants: the role of membrane transporters and implications for crop breeding for waterlogging tolerance. New Phytol 190:289–298

Shahbaz M, Ashraf M (2013) Improving salinity tolerance in cereals. Crit Rev Plant Sci 32:237–249

Shanker AK, Venkateswarlu B (2011) Abiotic stress in plants-mechanisms and adaptations. In: TechJaneza Trdine 9, 51000 Rijeka, Croatia

Sharma DP, Swarup A (1989) Effect of nutrient composition of wheat in alkaline soils. J Agric Sci (UK) 112:191–197

Sharma S, Verslues PE (2010) Mechanisms independent of abscisic acid (ABA) or proline feedback have a predominant role in transcriptional regulation of proline metabolism during low water potential and stress recovery. Plant Cell Environ 33:1838–1851

Sharma R, Mishra M, Gupta B, Parsania C, Singla-Pareek SL, Pareek A (2015) De novo assembly and characterization of stress transcriptome in a salinity-tolerant variety CS 52 of *Brassica juncea*. PLoS One 10(5):e0126783

Shi H, Zhu JK (2002) SOS4, a pyridoxal kinase gene, is required for root hair development in Arabidopsis. Plant Physiol 129:585–593

Shi H, Ishitani M, Kim C, Zhu JK (2000) The Arabidopsis thaliana salt tolerance gene SOS1 encodes a putative Na^+/H^+ antiporter. Proc Natl Acad Sci U S A 97:6896–6901

Shimamura S, Mochizuki T, Nada Y, Fukuyama M (2003) Formation and function of secondary aerenchyma in hypocotyl, roots and nodules of soybean (*Glycine max*) under flooded conditions. Plant Soil 251:351–359

Shoji T, Suzuki K, Abe T, Kaneko Y, Shi H, Zhu JK, Rus A, Hasegawa PM, Hashimoto T (2006) Salt stress affects cortical microtubule organization and helical growth in *Arabidopsis*. Plant Cell Physiol 47:1158–1168

Smethurst CF, Shabala S (2003) Screening methods for waterlogging tolerance in lucerne: comparative analysis of waterlogging effects on chlorophyll fluorescence, photosynthesis, biomass and chlorophyll content. Funct Plant Biol 30:335–343

Snedden WA, Fromm H (2001) Calmodulin as a versatile calcium signal transducer in plants. New Phytol 151:35–66

Sobhanian H, Aghaei K, Komatsu S (2011) Changes in the plant proteome resulting from salt stress: toward the creation of salt-tolerant crops? J. Proteomics 74(8):1323–1337

Sowmya K, Shiv V, Rahman MH, Kav NNV (2011) Functional characterization of four APETALA2- family genes (RAP2.6, RAP2.6L, DREB19 and DREB26) in Arabidopsis. Plant Mol Biol 75:107–127

Steffens B, Geske T, Sauter M (2011) Aerenchyma formation in the rice stem and its promotion by H_2O_2. New Phytol 190:369–378

Stieger PA, Feller U (1994b) Senescence and protein re-mobilization in leaves of maturing wheat planes grown on waterlogged soil. Plant Soil 166(2):173–179

Subbaiah CC, Kollipara KP, Sachs MM (2000) A Ca^{2+}-dependent cysteine protease is associated with anoxia-induced root tip death in maize. J Exp Bot 51:721–730

Sun F, Liu P, Xu J, Dong H (2010a) Mutation in RAP2.6L, a transactivator of the ERF transcription factor family, enhances Arabidopsis resistance to Pseudomonas syringae. Physiol Mol Plant Pathol 74:295–302

Sun H, Zhao T, Gai J (2010b) Inheritance and QTL mapping of waterlogging tolerance at seedling stage of soybean. Acta Agron Sin 36:590–595

Szabados L, Savoure A (2010) Proline: a multifunctional amino acid. Trends Plant Sci 15:89–97

Szabolcs I (1974) Salt affected soils in Europe. Martinus Nijhoff, The Hague, p 63

Szekely G et al (2008) Duplicated P5CS genes of *Arabidopsis* play distinct roles in stress regulation and developmental control of proline biosynthesis. Plant J 53:11–28

Tahir I, Sabir M, Iqbal M (2013) Unravelling salt stress in plants through proteomics. In. Ahmad P (ed) Salt stress in plants: signalling, omics and adaptations, Springer, pp 47–61

Taji T, Seki M, Satou M, Sakurai T, Kobayashi M, Ishiyama K, Narusaka Y, Narusaka M, Zhu JK, Shinozaki K (2004) Comparative genomics in salt tolerance between *Arabidopsis* and *Arabidopsis*-related halophyte salt cress using Arabidopsis microarray. Plant Physiol 35:1697–1709

Tang X, Mu X, Shao H, Wang H, Brestic M (2014) Global plant-responding mechanisms to salt stress: physiological and molecular levels and implications in biotechnology. Crit Rev Biotechnol:84–91. doi:10.3109/07388551.2014.889080

Teakle NL, Amtmann A, Real D, Colmer TD (2010) Lotus tenuis tolerates combined salinity and waterlogging: maintaining O_2 transport to roots and expression of an NHX1-like gene contribute to regulation of Na^+ transport. Physiol Plant 139:358–374

Thompson GA, Goggin FL (2006) Transcriptomics and functional genomics of plant defence induction by phloem-feeding insects. J Exp Bot 57:755–766

Thomson CJ, Atwell BJ, Greenway H (1989) Response of wheat seedlings to low O_2 concentrations in nutrient solution I. Growth, O_2 uptake and synthesis of fermentative end-products by root segments. J Exp Bot 40:985–991

Tran LS, Nakashima K, Sakuma Y, Simpson SD, Fujita Y, Maruyama K, Fujita M, Seki M, Shinozaki K, Shinozak KY (2004) Isolation and functional analysis of *Arabidopsis* stressäinducible NAC transcription factors that bind to a drought-responsive cis-element in the early responsive to dehydration stress 1 promoter. Plant Cell 16:2481–2498

Trought MCT, Drew MC (1980) The development of waterlogging damage in young wheat plants in anaerobic solution cultures. J Exp Bot 31:1573–1585

Tsang EW, Yang J, Chang Q, Nowak G, Kolenovsky A, McGregor DI et al (2003) Chlorophyll reduction in the seed of Brassica napus with a glutamate 1-semialdehyde aminotransferase antisense gene. Plant Mol Biol 51(2):191–201

Turan S, Tripathy BC (2015) Salt-stress induced modulation of chlorophyll biosynthesis during de-etiolation of rice seedlings. Physiol Plant 153:477–491

Urano K, Maruyama K, Ogata Y, Morishita Y, Takeda M, Sakurai N et al (2009) Characterization of the ABA-regulated global responses to dehydration in *Arabidopsis* by metabolomics. Plant J 57:1065–1078

Van Ginkel M, Rajaram S, Thijssen M (1992) Waterlogging in wheat: Germplasm evaluation and methodology development. In: Tanner DG, Mwangi W (eds) Seventh regional wheat workshop for Eastern, Central and Southern Africa. Nakuru, Kenya: CIMMYT. pp. 115–124

Van Toai TT, Martin SK, Chase K, Boru G, Schnipke V, Schmitthenner AF, Lark KG (2001) Identification of a QTL associated with tolerance of soybean to soil waterlogging. Crop Sci 41:1247–1252

Verbruggen N, Hermans C (2008) Proline accumulation in plants: a review. Amino Acids 35:753–759

Verslues PE, Agarwal M, Agarwal KS, Zhu J, Zhu JK (2006) Methods and concepts in quantifying resistance to drought, salt and freezing, abiotic stresses that affect plant water status. Plant J 45:523–539

Villalta I, Reina-Sánchez A, Bolarín MC, Cuartero J, Belver A, Venema K, Carbonell EA, Asins MJ (2008) Genetic analysis of Na+ and K+ concentrations in leaf and stem as physiological components of salt tolerance in tomato. Theor Appl Genet 116:869–880

Villareal RL, Sayre K, Banuelos O, Mujeeb-Kazi A (2001) Registration of four synthetic hexaploid wheat (*Triticum turgidum/Aegilops tauschii*) germplasm lines tolerant to waterlogging. Crop Sci 41:274

Vinocur B, Altman A (2005) Recent advances in engineering plant tolerance to abiotic stress: achievements and limitations. Curr Opin Biotechnol 16:123–132

Wang S, Liren H, Zhengwei L (1996) A comparative study on the resistance of barley and wheat to waterlogging. Acta Agron Sin 22:228–232

Wang F, Zhao T, Yu D, Chen S, Gai J (2008a) Inheritance and QTL analysis of submergence tolerance at seedling stage in soybean Glycine max (L.) Merr Acta AgronSin 34:748–753

Wang MC, Peng ZY, Li CL, Li F, Liu C, Xia GM (2008b) Proteomic analysis on a high salt tolerance introgression strain of *Triticum aestivum/Thinopyrum ponticum*. Proteomics 8:1470–1489

Wang X, Fan P, Song H, Chen X, Li X, Li Y (2009) Comparative proteomic analysis of differentially expressed protein in shoots of *Salicornia europaea*, under different salinity. J Proteome Res 8:3331–3345

Waters I, Kuiper PJC, Watkin E, Greenway H (1991a) Effects of anoxia on wheat seedlings. I. Interaction between anoxia and other environmental factors. J Exp Bot 42:1427–1435

Waters I, Morrell S, Greenway H, Colmer TD (1991b) Effects of anoxia on wheat seedlings. II. Effects of O₂ supply prior to anoxia on tolerance to anoxia, alcoholic fermentation and sugars. J Exp Bot 42:1437–1447

Watkin ELJ, Thomson CJ, Greenway H (1998) Root development in two wheat cultivars and one triticale cultivar grown in stagnant agar and aerated nutrient solution. Ann Bot 81:349–354

Wenzl P, Li H, Carling J, Zhou M, Raman H, Paul E, Hearnden P, Maier C, Xia L, Caig V, Ovesná J, Cakir M, Poulsen D, Wang J, Raman R, Smith KP, Muehlbauer GJ, Chalmers KJ, Kleinhofs A, Huttner E, Kilian A (2006) A high-density consensus map of barley linking DArT markers to SSR, RFLP and STS loci and agricultural traits. BMC Genomics 7:206–227

Wu JG, Liu SF, Li FR, Zhou JR (1992) Study on the effect of wet injury on growth and physiology winter wheat. Acta Agric Uni Hen 26:31–37

Wu SJ, Ding L, Zhu JK (1996) SOS1, a genetic locus essential for salt tolerance and potassium acquisition. Plant Cell 8:617–627

Wu D, Cai S, Chen M, Ye L, Chen Z, Zhang H, Dai F, Wu F, Zhang G (2013) Tissue metabolic responses to salt stress in wild and cultivated barley. PLoS One 8:e55431

Xia JH, Saglio PH (1992) Lactic acid efflux as a mechanism of hypoxic acclimation of maize root tips to anoxia. Plant Physiol 100:40–46

Xu K, Mackill DJ (1996) A major locus for submergence tolerance mapped on rice chromosome 9. Mol Breed 2:219–224

Xu K, Xu X, Ronald PC, Mackill DJ (2000) A high-resolution linkage map in the vicinity of the rice submergence tolerance locus Sub1. Mol Gen Genet 263:681

Xu K, Xu X, Fukao T, Canlas P, Maghirang-Rodriguez R, Heuer S, Ismail AM, Bailey-Serres J, Ronald PC, Mackill DJ (2006) Sub1A is an ethylene responsive-factor-like gene that confers submergence tolerance to rice. Nature 442:705–708

Xu R, Wang J, Li C, Johnson P, Lu C, Zhou M (2012) A single locus is responsible for salinity tolerance in a Chinese landrace barley (Hordeum vulgare L.) PLoS One 7:e43079

Xue D, Zhou MX, Zhang XQ, Chen S, Wei K, Zeng FR, Mao Y, Wu FB, Zhang GP (2010) Identification of QTLs for yield and yield components of barley under different growth conditions. J Zhejiang Univ Sci B 11:169–176

Yamaguchi T et al (2013) Sodium transport system in plant cells. Front Plant Sci 4:410

Yang O, Yang O, Popova OV, Süthoff U, Lüking I, Dietz KJ, Golldack D (2009) The Arabidopsis basic leucine zipper transcription factor AtbZIP24 regulates complex transcriptional networks involved in abiotic stress resistance. Gene 436:45–55

Yin D, Chen S, Chen F, Guan Z, Fang W (2009) Morphological and physiological responses of two chrysanthemum cultivars differing in their tolerance to waterlogging. Environ Exp Bot 67:87–93

Yokotani N, Ichikawa T, Kondou Y, Matsui M, Hirochika H, Iwabuchi M, Oda K (2009) Tolerance to various environmental stresses conferred by the salt-responsive rice gene ONAC063 in transgenic Arabidopsis. Planta 229:1065–1075

Yu PT, Stolzy LH, Letey J (1969) Survival of plants under prolonged flooded conditions. Agron J 61:844–847

Yu M, Mao SL, Chen GY, Liu YX, Li W, Wei YM, Liu CJ, Zheng YL (2014) QTLs for waterlogging tolerance at germination and seedling stages in population of recombinant inbred lines derived from a cross between synthetic and cultivated wheat genotypes. J Integr Agric 13:31–39

Zhang HX, Hodson JN, Williams JP, Blumwald E (2001) Engineering salt tolerant Brassica plants: characterization of yield and seed oil quality in transgenic plants with increased vacuolar sodium accumulation. Proc Natl Acad Sci 98:12832–12836

Zhang YM, Liu ZH, Wen ZY et al (2012) The vacuolar Na^+/H^+ antiport gene TaNHX2 confers salt tolerance on transgenic alfalfa (Medicago sativa). Funct Plant Biol 39:708–716

Zhao L, Gu R, Gao P, Wang G (2008) A nonsymbiotic hemoglobin gene from maize, ZmHb, is involved in response to submergence, high-salt and osmotic stresses. Plant Cell Tissue Organ Cult 95:227–237

Zheng BS, Yang L, Zhang WP, Mao CZ, Wu YR, Yi KK, Liu FY, Wu P (2003) Mapping QTLs and candidate genes for rice root traits under different water-supply conditions and comparative analysis across three populations. Theor Appl Genet 107:1505–1515

Zhou M (2011) Accurate phenotyping reveals better QTL for waterlogging tolerance in barley. Plant Breed 130:203–208

Zhou MX, Li HB, Mendham NJ (2007) Combining ability of waterlogging tolerance in barley. (Hordeum vulgare L.) Crop Sci 47:278–284

Zhou J, Li F, Wang J, Ma Y, Chong K, Xu Y (2009) Basic helix-loop-helix transcription factor from wild rice (OrbHLH2) improves tolerance to salt and osmotic stress in Arabidopsis. J Plant Physiol 166:1296–1306

Zhou H, Zhao J, Yang Y, Chen C, Liu Y, Jin X, Chen L, Li X, Deng XW, Schumaker KS, Guoc Y (2012a) UBIQUITIN-SPECIFIC PROTEASE16 modulates salt tolerance in Arabidopsis by regulating Na^+/H^+ antiport activity and serine hydroxymethyltransferase stability. Plant Cell 24:5106–5122

Zhou M, Johnson P, Zhou G, Lic C, Lance R (2012b) Quantitative trait loci for waterlogging tolerance in a barley cross of Franklin*YuYaoXiangTian Erleng and the relationship between waterlogging and salinity tolerance. Crop Sci 52:2082–2088

Zhu JK (2001) Plant salt tolerance. Trends Plant Sci 6:66–71

Zhu JK (2002) Salt and drought stress signal transduction in plants. Annu Rev Plant Biol 53:247–273

Zhu JK, Liu J, Xiong L (1998) Genetic analysis of salt tolerance in *Arabidopsis*: evidence for a critical role of potassium nutrition. Plant Cell 10:1181–1191

Zou XL, Jiang YY, Zheng YL, Zhang MD, Zhang ZX (2011) Prolyl 4-hydroxylase genes are subjected to alternative splicing in roots of maize seedlings under waterlogging. Ann Bot 108:1323–1335

Impacts and Management of Temperature and Water Stress in Crop Plants

Kiruba Shankari Arun-Chinnappa, Lanka Ranawake, and Saman Seneweera

Abstract

Plant growth and development are affected by various abiotic stresses like drought, submergence, salinity and high and low temperature. These abiotic stresses cause average yield losses of greater than 50% in a majority of crop plants. Food production needs to be doubled by 2050 to meet the growing demands of an increasing global population. Significant damage is being caused to crops, especially through temperature and water stress associated with climate change. High- and low-temperature stresses affect crop productivity by adverse biochemical changes in plants. Similarly, drought and water stress also affect the crop's performance throughout the growing season. Understanding plant responses and molecular and physiological changes occurring during these stresses is necessary to improve current cultivars and release new cultivars with enhanced resistance to such stresses. An overview of the impacts of high- and low-temperature stress, drought and submergence in plant growth and development and the physiological and molecular responses of plants is discussed. Strategies adopted by plants to overcome these stresses through avoidance and tolerance mechanisms are also briefly discussed.

K.S. Arun-Chinnappa • S. Seneweera (✉)
Centre for Crop Health, University of Southern Queensland,
Toowoomba, QLD, 4350, Australia
e-mail: Saman.Seneweera@usq.edu.au

L. Ranawake
Faculty of Agriculture, University of Ruhuna, Ruhuna, Sri Lanka

© Springer Nature Singapore Pte Ltd. 2017
P.S. Minhas et al. (eds.), *Abiotic Stress Management for Resilient Agriculture*,
DOI 10.1007/978-981-10-5744-1_9

9.1   Introduction

Plants are sessile organisms and have developed the ability to survive at extreme environmental conditions that affect their development. Adverse environmental factors can limit crop production as much as 70% (Boyer 1982). Abiotic stresses like high/low temperatures, freezing, drought and salinity cause major losses to crop productivity worldwide. These stresses in crops are of major concern since these limit plant growth, metabolism and thus productivity. Average yield loss of greater than 50% is accounted for due to these abiotic stresses in a majority of crop plants (Bray et al. 2000). Recent changes in climate have aggravated abiotic stresses in plants, and two major factors threatening crop productivity are temperature and water. Increasing concentrations of greenhouse gases like CO_2, methane, chlorofluorocarbons and nitrous oxide are gradually increasing the average ambient global temperature (Wahid et al. 2007). As per IPCC (Intergovernmental Panel of Climatic Change), average global temperature may rise 1 and 3 °C above current temperature by 2025 and 2100, respectively (Jones et al. 1999). Increase in temperature causes heat stress and very low temperatures cause chilling and freezing stress in plants. Plants undergo extensive reprogramming at cellular and molecular level to tolerate low temperatures, and this affects plant growth and productivity (Conroy et al. 1994; Yadav 2010). These extreme conditions affect growth and developmental processes of plants by altering the morphology, physiology and biochemistry of plants at the cellular and whole plant level (Seneweera et al. 2005). However, some plants have adopted survival mechanisms through avoidance and tolerance to heat and cold stresses.

Only 2% of fresh water is available for consumption (Shiklomanov 1993). The availability of natural water resources for day-to-day consumption and agricultural purposes varies with the location and the season of the year. Seasonal changes, as well as unexpected changes, in water availability create problems for sustainability in agriculture. Excess and limited availability of water causes submergence and drought, respectively. According to UNCCD, 12 million hectares in the world are lost due to drought or desertification annually. Further, excess water affects 15 million hectares or more lowland rice-growing areas in South and Southeast Asia. Both limited and excess water cause stress to plants and limit their productivity. Knowledge of the physiological and molecular responses of plants to these stresses and their tolerance mechanisms is essential for improving crop resistance and thereby increasing productivity (Fernando et al. 2016). An overview of plant response to temperature and water stresses in terms of physiological and molecular aspects, strategies adopted by plants to alleviate these stresses and biotechnological approaches to overcome such stresses is presented in this chapter.

9.2   High- and Low-Temperature Stress

Temperature above and below the optimum hinders the growth and development of crop plant and is considered as heat and cold stress, respectively (Kotak et al. 2007). Heat stress can be lethal and lead to major crop losses. Severe losses due to heat waves were observed in the USA during 1980 and 1988 and caused a total loss of 55 and 71 billion dollars, respectively (Lobell et al. 2011; Mittler et al. 2012). Heat stress affects all crop plants; in wheat, grain yield loss of 3–5% was observed when wheat was grown at one-degree increase in average temperature above 15 °C under controlled environments (Gibson and Paulsen 1999). Rice, a staple food of Asian countries, is also highly susceptible to high-temperature stress. About 3 million hectares of rice cultivation was affected by heat stress in China in 2003 causing a total loss of 5.18 million tons of yield (Peng et al. 2004). There has been a continuous increase of 0.13 °C in global average temperature per decade since 1950. An increase of 0.2 °C per decade for the next two to three decades is expected globally (Lobell et al., 2011). In Australia, by 2030, the annual mean temperature will be increased by 0.2–4 °C (Zheng et al. 2012). Low-temperature stress (LTS) also affects plant growth and crop production. Substantial loss due to LTS accounts for $2 billion each year. In some cases, cold stress may not cause yield losses; however, it may result in reduced quality. One of the major losses due to LTS was accounted for in 1995 during early frost events in the USA costing $1 billion loss in corn and soybean crop cultivation (Sanghera et al. 2011). In Australia, frost events cause major losses in wheat cultivation especially when it strikes the crops at the reproductive stage (Frederiks et al. 2012).

9.2.1   Growth Effects

Heat stress and low-temperature stress affect seed germination, photosynthesis, reproductive development and yield. It also results in oxidative stress (Hasanuzzaman et al. 2013a). In rice, seed germination is influenced by high- and low-temperature stress. Mild heat stress caused due to a slight increase in temperature causes delayed germination. However, extreme temperature stress (heat and cold) might inhibit the seed germination completely, and this is called as thermo-inhibition (Fig. 9.1; Takahashi 1961). At plant level, especially in crop plants like rice and wheat, heat stress during reproductive stages is more detrimental than during the vegetative stage (Wollenweber et al. 2003; Xie et al. 2009). Rise in temperature by 1–2 °C more than the optimum affects grain filling in crops by shortening the period of filling and so affects the yield in cereals (Hasanuzzaman et al. 2013a). In wheat, optimum temperature for anthesis and grain filling is about 12–22 °C, and any increase in temperature above this affects grain filling and grain size (Farooq et al., 2011). In rice, HTS during grain filling causes spikelet sterility and also reduces the duration of the grain-filling phase resulting in yield reduction (Fig. 9.1; Xie et al. 2009).

Fig. 9.1 Effects of temperature and water stress in plants. High (hot) and low (cold) temperature and excess (submergence) and scarcity of water (drought) are stresses that affect crop productivity and yield. Cellular and physiological effects of these stresses are summarised in this picture

Low-temperature stress (LTS) can be classified into two categories, freezing and chilling stress. Chilling stress is caused by exposing plants to low temperatures (0–15 °C) that injures the plant but does not form ice crystals, whereas freezing stress is caused by exposing the plants to temperatures below 0 °C that causes injury by ice crystal formation (Fig. 9.1). LTS also affects plants at different growth stages leading to reduction in growth and yield (Hasanuzzaman et al. 2013b). LTS mainly affects early stages of plant growth especially the seedling stage. Injuries caused by

LTS during seedling stages are discoloration and reduction in seedling growth, whitening and yellowing of leaf and failure to grow after transplantation (Buriro et al. 2011; Hasanuzzaman et al. 2013b).

Frost is a major low-temperature stress especially in Australia that hinders barley, wheat and legumes like field pea, fava bean, lentil and chickpea cultivation. Damage caused to agricultural crops mentioned above, post-head emergence or pod emergence, is economically significant (Frederiks et al. 2012). Radiant frost is witnessed when crops are grown in spring with optimum day temperature but experience rapid fall in temperature during nights. Consequently, rapid radiation occurs resulting in a super cooling effect that drops the plant canopy and actual plant temperatures to as low as −5 °C. This causes whole plant freezing resulting in plant death (Frederiks 2010).

9.2.2   Physiological Aspects

High-temperature stress (HTS) affects the plants at different growth stages resulting in physiological impairment and thus reduced growth. Photosynthesis is one of the physiological processes affected greatly by heat stress (Fig. 9.1). In wheat crops, excessive heat damages chlorophyll by altering the structural organisation of thylakoids thereby affecting its functionality and also by reducing the chlorophyll content (Xu et al. 1995). Similar damages of heat stress to the photosynthetic apparatus and process are observed in rice (Xie et al. 2009).

Excessive heat also decreases the photosynthetic pigments, which in turn affects photosynthesis (Cao et al. 2009). Further, excess heat induces water loss from the leaf, negatively affecting leaf water potential resulting in closure of stomata. This is one of the major phenomena affecting photosynthesis under heat stress. In *Vitis vinifera*, about 15–30% of photosynthesis reduction during heat stress was caused by stomatal closure (Fig. 9.1; Greer and Weedon 2012). HTS affects the reproductive health of crops; exposure to high temperatures during flowering affects its fertility due to impaired pollen and ovary development. HTS causes water loss in cells, which impairs cell size resulting in reduced growth in crop plants. At extreme HTS conditions, plants undergo programmed death of cells and tissues as a result of water loss and denaturation of proteins and enzymes (Hasanuzzaman et al. 2010).

LTS also causes damages in crop plants similar to HTS. LTS causes ice crystal formation in the apoplasm of cells in the leaves which results in cellular dehydration. This in turn results in stomatal closure and affects the cellular homeostasis in plants (Hasanuzzaman et al. 2013b). Cellular dehydration in roots resulting from LTS leads to osmotic stress caused by water imbalance in the cells (Chinnusamy et al. 2007; Hasanuzzaman et al. 2013b). LTS also affects photosynthesis by damaging the organisation of chlorophyll and thylakoids similar to heat stress (Fig. 9.1; Hasanuzzaman et al. 2013b). Stability and structure of the plasma membrane are affected due to cellular dehydration and chemical instability of its lipid components (Yadav 2010).

9.2.3 Plant Responses and Tolerance

HTS for short and long periods can affect morphology and physiology and induce biochemical changes in plants depending on the developmental stage, leading to reduced growth and yield. Crop plants subjected to high temperatures over long periods of time exhibit symptoms like scorching of leaves and twigs; sunburns on leaves, branches and stems; leaf senescence and abscission; shoot and root growth inhibition; fruit discoloration and damage; and reduced yield (Wahid et al. 2007). High temperature can lead to shortening of the life cycle of plants. Mild changes in temperature may induce changes in membrane properties and activate calcium channels. Calcium signalling leads to alteration in plant metabolism resulting in acclimation to mild heat stress. However, if plants are subjected to high-temperature stress for prolonged periods, they produce ROS (reactive oxygen species) inducing oxidative stress throughout the plant (Mittler et al. 2012). Harmful ROS like singlet oxygen (O^{2-}), superoxide ($\cdot O2$), hydroxyl (OH^{-1}) and hydrogen peroxide (H_2O_2) are produced. These react with proteins, DNA and all the components of the cell disturbing cellular homeostasis (Asada 2006; Moller et al. 2007). Heat stress-induced oxidative stress is very damaging and causes protein degradation and membrane instability (Hasanuzzaman et al. 2013a). Excess accumulation of ROS in the plants induces programmed cell death and damages the whole plant (Qi et al. 2010).

Survival mechanisms of plant to HTS can be classified into avoidance and tolerance. Avoidance of HTS can be short term by altering the leaf orientation and membrane lipid composition or transpirational cooling, or it can be long term with morphological and phenological adaptations (Hasanuzzaman et al. 2013a). Plants also adopt escape mechanisms to HTS by reducing their life cycle through early maturation; however, this causes yield reduction (Rodríguez et al. 2005; Hasanuzzaman et al. 2013a). Ability of plants to grow, survive and produce economic yield at HTS is termed as tolerance. Tolerance of plants may also be classified as short and long term. Short-term tolerance mechanisms include morphological and anatomical changes. At molecular level, HTS induces heat stress transcription factors which in turn induce production and accumulation of HSPs (heat shock proteins) and other heat-induced transcripts as molecular chaperones to protect plant metabolism (Mittler et al. 2012). Plants produce antioxidants to counteract the damaging effects of excess ROS accumulated due to HTS. Various enzymatic and non enzymatic ROS scavengers are synthesised, and these detoxify the plant system and prevent programmed cell death.

Temperature that causes cold stress varies from one plant to another. Optimal temperature for one plant might cause cold stress to others. However, whenever a plant experiences cold stress, it responds by showing various phenotypic and physiological symptoms. Phenotypic symptoms include stunted plant growth, bushy appearance, reduced leaf expansion, chlorosis of leaves, wilting and even necrosis at severe conditions. LTS at the time of reproductive development leads to sterility in plants (Hasanuzzaman et al. 2013b). In fruit crops, extreme cold stress induces excessive fruit drop and fruit cracking (Yang et al. 2010). At the cellular level, cold stress induces ice crystal formation resulting in dehydration of the cell. Ice crystals are lethal to cells since it damages the cell membranes. LTS also induces oxidative stress by excess ROS accumulation leading to programmed cell death (Hasanuzzaman et al. 2013b).

Plants adapt to cold stress by certain avoidance and tolerance mechanisms. One of the major effects of cold stress in plants is membrane disintegration. Cold stress in plants reduces the fluidity of membranes resulting in rigid membranes. This membrane rigidification enhances cold acclimation in plants by inducing cold-responsive genes. Similar to HTS, calcium signalling and antioxidants play an important role in tolerance to LTS in plants. Further, plants tolerate or recover from cold stress through repair mechanisms, restructuring the plasma membrane and accelerating osmolite synthesis (Yadav 2010; Hasanuzzaman et al. 2013b).

9.2.4   Management Strategies

Employing a combination of genetic improvement and cultural practices can alleviate impacts of heat and cold stress in plants. Cultivars that endure and tolerate these stresses without any economic loss in yield can be developed through genetic improvement. Cultural practices include adjusting and modifying planting time and planting density of plants to avoid these stress conditions (Wahid et al. 2007; Hasanuzzaman et al. 2013b). Other practices include external application of protectants like osmo-protectants, phytohormones, signalling molecules and trace elements that can protect during adverse conditions especially during HTS (Hasanuzzaman et al. 2013a).

Cultivars that are tolerant to high- and low-temperature stresses can be identified through proper screening methods and selection criteria through conventional breeding techniques in fields. However, other stress factors like pathogens and pests could be a hindrance in identifying resistant cultivars in the field. Glasshouse screening could be used as an alternative for field screening; nonetheless, certain factors like low genetic variation and unreliable selection criteria make it difficult to identify heat- or cold-tolerant varieties (Wahid et al. 2007; Hasanuzzaman et al. 2013b). Quantitative trait loci (QTL) associated with heat stress-tolerant traits have been identified in wheat, rice and maize (Talukder et al. 2014; Yang et al. 2002; Frova and Sari-Gorla 1994; Collins et al. 2008). These traits were associated with the reproductive stage of growth in plants (Collins et al. 2008). QTLs associated with resistance to chilling at seedling stage have been identified in crop plants like maize, sorghum, rice and tomato (Presterl et al. 2007; Lou et al. 2007; Goodstal et al. 2005; Knoll and Ejeta 2008). Loci associated with freezing tolerance at vegetative phase of plant growth have been identified across *Triticeae* species and also in *Arabidopsis* model plant (Alonso-Blanco et al. 2005; Båga et al. 2007; Collins et al. 2008). Very little is known about the QTLs for cold tolerance associated with the reproductive phase of crop plants (Collins et al. 2008).

Recent advances in biotechnology approaches have paved the way to develop high- and low -temperature-tolerant crops through genetic modification. Transgenic crops expressing heat shock proteins and heat shock transcription factors have increased heat tolerance in many crops like maize, rice, carrots and tobacco (Hasanuzzaman et al. 2013a). Manipulating the expression of small non-coding RNAs especially microRNAs could also increase heat tolerance and alleviate heat sensitivity in crop plants. Genetically modified crops expressing antifreeze proteins from winter flounder fish in wheat have shown increased frost tolerance characteristics. Antifreeze proteins bind to growing ice crystals in the cell and inhibit its growth increasing the tolerance of plants to freezing (Khanna and Daggard 2006).

9.3   Drought and Submergence Stress

Both water scarcity and excess water affect plant growth, development and production. Severity and duration of each stress determine the total production loss in agriculture. According to UNCCD, 12 million hectares in the world are lost due to drought or desertification annually. This land extent reduces 20 million tons of annual grain production for consumption (http://www.unccd.int). IPCC reported that global temperature would be raised by an average of 1.1–6.4 °C by the end of the twenty-first century which would increase the area under extreme arid lands by 1–30%. On the other hand, under the climate change context, annual rainfall patterns have changed, and excess water affects 15 million hectares of lowland rice-growing areas in South and Southeast Asia.

9.3.1   Physiological Aspects

Drought reduces water potential of the leaves by changing the cell's turgor pressure. Less water in the plant induces ABA production that leads to stomatal closure. This reduces CO_2 assimilation affecting the photosynthetic machinery in thylakoids of the mesophyll cells, which are the source of fluorescence signals. Cell peroxidation damage occurs in the drought-stressed plants with the accumulation of reactive oxygen species, namely, singlet oxygen (O^{2-}), superoxide ($\bullet O_2$), hydroxyl (OH^{-1}) and hydrogen peroxide (H_2O_2). Further, there is a reduction in the electron transport rate that creates oxidative stress reducing the photosynthetic capacity of the plants under drought stress. Changes in metabolites such as sugars, acids and carotenoids under drought stress directly affect the quality of the plant's end product (Ripoll et al. 2016).

Unpredicted waterlogging caused by heavy rains followed by poor drainage is a serious problem in a changing climate scenario. Under waterlogging conditions, soil gets saturated, and under submergence stress, the plant is partially or completely covered by the water body. In both instances, soil becomes anaerobic, and under anaerobic conditions, soil produces toxic components and emits different gases to the environment. These conditions accelerate plant death.

Diffusion of gases such as O_2, CO_2 and ethylene is reduced under submergence stress. Hypoxia and anoxia conditions under submergence stress are critical for plant growth (Fig. 9.1). Hypoxia is the low diffusion of O_2 to the roots under water through the plant canopy, which is above the water level. Anoxia is the total absence of O_2 to the root system due to complete submergence of the whole canopy (Lone et al. 2016). Reduction of O_2 diffusion under water hampers aerobic respiration. Anaerobic respiration produces the required energy for the plants at anoxia once aerobic respiration is arrested (Pradhan 2016). Fermentation enzymes are required for anaerobic respiration, and it produces different metabolites that maintains the stability of pH in the cell contents.

When photosynthesis is affected by less gas diffusion and poor light penetration, unavailability of soluble sugars decreases the activity of the glycolytic pathway. When the plant is able to survive utilising stored food reservoirs or using anaerobic respiration till the water recedes, the plant must be regrown in an aerated environment. The strong reactive oxygen species (ROS) are produced in plants under aerated environment upon de-submergence. Antioxidant mechanisms are required for the survival of the plant in the presence of ROS (Xu et al. 2010).

9.3.2 Plant Response and Tolerance

Plant response to water stress is a complex mechanism controlled by gene expression followed by biochemical metabolism through varied physiological processes. Leaf area, plant height, biomass and dry matter production are the morphological growth indices affected by drought stress. These indices and physiological traits such as photosynthetic efficiency, maximum quantum yield, photochemical and non-photochemical quenching, gas exchange measurements, stomatal conductance, relative water content and malondialdehyde (MDA) content are considered in the evaluation of drought stress of the plants (Sairam et al. 1998; Yousfi et al. 2016). Plants use different mechanisms like acceleration of life cycle; development of water absorption and retention mechanisms or organs within the plant; osmotic, metabolic and morphological adjustments; and genetic modifications to avoid, tolerate or overcome drought stress.

Under submergence stress, plants change their architecture depending on the type of flood. Complete submergence stress is the condition where plants are completely covered by floodwater for a considerable period of time. With the quiescence strategy, plant growth is arrested to conserve energy so it can be utilised once the water recedes. This strategy is important for tolerance under complete submergence stress. If the plant remains submerged for a long period, it will die upon complete depletion of food reserves. Elongation growth is a strategy used by plants to keep the canopy above the water level and is associated with tolerance under flash flooding where the water level increases suddenly but is not retained for a longer period to cover the plant canopy completely. However, if all the food resources were depleted before the canopy emerges from the water body, it would cause the death of the plants.

There is a natural multifaceted alteration in plant anatomy and metabolism to function under low O_2. The arrangement of soft tissues to provide space and a continuous gas channel for facilitation of an internal O_2 pathway from the canopy to the root system is called aerenchyma (Bailey-Serres and Voesenek 2008). Ethylene stimulates aerenchyma formation and root growth under submergence stress. Metabolic adjustment under submergence stress is driven by various plant hormones such as ethylene, gibberellin and abscisic acid. Finally, factors such as ATP production under limited O_2 diffusion, management of cytosolic pH due to the accumulation of various metabolites, effects of the reactive oxygen species produced under anaerobic conditions, the persistence of the antioxidant mechanism and aerenchyma development decide the fate of the plant under submergence stress.

9.3.3 Management Strategies

Evolution of plants under drought conditions has created drought-tolerant plants. Selection of such plant types for cultivation was practised in the early days. However, low heritability, polygenic nature of the trait, epistasis and genotype-by-environment interactions were the limitations of this type of selection. Later, selected plant materials were bred with elite lines for the production of better cultivars, but the complexity of drought tolerance mechanism hampers the breeding for drought tolerance.

QTL analysis reveals the segments of chromosomes determining the variation of the considered agronomic or physiological trait. Tightly linked markers to the prominent QTLs are used in marker-assisted selection in agriculture. The accuracy of QTL mapping and marker density of the genetic linkage map determines marker-assisted QTL efficiency (Cattivelli et al. 2008). The polygenic nature of drought tolerance, impact of a single chromosomal location and environment interaction of QTL determine the overall success of the breeding program. Finding QTLs related to yield and stress tolerance is much more promising in marker-assisted selection since the tolerance trait and its relationship with the yield would not be strong.

Many drought-tolerant genes have been identified in various studies. On the other hand, the complexity of the drought-tolerant transcriptome has also been revealed. The mechanism of drought tolerance has been studied further by leveraging the information from *Arabidopsis* a model plant to other crops by genome synteny. Metabolic engineering or overproduction of desired metabolites for the tolerance of the stress is another approach in improving plants for stress tolerance. Plant gene transformation techniques are important to introduce tolerant/resistant genes into genomes of elite breeding lines. Inoculation of plant growth-promoting bacteria is another attempt to overcome drought stress.

Natural variation in submergence tolerance can be seen within crop species including their wild, weed or landraces. Finding new genetic materials with substantial tolerance to submergence is needed to broaden gene pools of modern crops. Submergence tolerance of a crop species varies with the developmental stage, flooding intensity and duration. Finding tolerant materials according to stress level is a must in the development of submergence tolerance in plants. A strong submergence-tolerant QTL in rice, *Sub1*, is presently utilised as a marker for assisted breeding and gene transformation (Neeraja et al. 2007). *Sub1* suppresses plant elongation under submergence stress. Finding such QTLs or genes with different tolerant mechanisms is needed to fill the gaps in submergence tolerance at different growth stages under varying stress conditions in crops.

9.4 Conclusion

Abiotic stresses due to temperature and water are being amplified as a consequence of climate change. To overcome this problem and meet the global demand for food, tolerant crops must be selected or developed through appropriate technologies. Physiological aspects, mechanisms of tolerance and management strategies for better crop production must be thoroughly studied to effectively use such technologies.

References

Alonso-Blanco C, Gomez-Mena C, Llorente F, Koornneef M, Salinas J, Martínez-Zapater JM (2005) Genetic and molecular analyses of natural variation indicate CBF2 as a candidate gene for underlying a freezing tolerance quantitative trait locus in *Arabidopsis*. Plant Physiol 139:1304–1312

Asada K (2006) Production and scavenging of reactive oxygen species in chloroplasts and their functions. Plant Physiol 141:391–396

Båga M, Chodaparambil SV, Limin AE, Pecar M, Fowler DB, Chibbar RN (2007) Identification of quantitative trait loci and associated candidate genes for low-temperature tolerance in cold-hardy winter wheat. Funct Integr Genomics 7:53–68

Bailey-Serres J, Voesenek LACJ (2008) Flooding stress: acclimations and genetic diversity. Annu Rev Plant Biol 59:313–339

Boyer JS (1982) Plant productivity and environment. Science 218(4571):443–448

Bray EA, Bailey-Serres J, Weretilnyk E (2000) Responses to abiotic stresses. In: Gruissem W, Buchannan B, Jones R (eds) Biochemistry and molecular biology of plants. American Society of Plant Physiologists, Rockville, pp 1158–1249

Buriro M, Oad FC, Keerio MI, Tunio S, Gandahi AW et al (2011) Wheat seed germination under the influence of temperature regimes. Sarhad J Agric 27:539–543

Cao YY, Duan H, Yang LN, Wang ZQ, Liu LJ, Yang JC (2009) Effect of high temperature during heading and early filling on grain yield and physiological characteristics in indica rice. Acta Agron Sin 35:512–521

Cattivelli L, Rizza F, Badeck FW, Mazzucotelli E, Mastrangelo AM, Francia E et al (2008) Drought tolerance improvement in crop plants: an integrated view from breeding to genomics. Field Crop Res 105:1–14

Chinnusamy V, Zhu J, Zhu JK (2007) Cold stress regulation of gene expression in plants. Trends Plant Sci 12:444–451

Collins NC, Tardieu F, Tuberosa R (2008) Quantitative trait loci and crop performance under abiotic stress: where do we stand? Plant Physiol 147:469–486

Conroy J, Seneweera S, Basra A, Rogers G, Nissen-Wooller B (1994) Influence of rising atmospheric CO_2 concentrations and temperature on growth, yield and grain quality of cereal crops. Func Plant Biol 21:741–758

Farooq M, Bramley H, Palta JO, Siddique KHM (2011) Heat stress in wheat during reproductive and grain filling phases. Crit Rev Plant Sci 30:1–17

Fernando N, Manalil S, Florentine S, Chauhan B, Seneweera S (2016) Glyphosate resistance of C_3 and C_4 weeds under rising atmospheric CO_2. Front Plant Sci 7:910. https://doi.org/10.3389/fpls.2016.00910

Frederiks TM (2010) Frost resistance in cereals after ear emergence. PhD Thesis, University of Southern Queensland

Frederiks TM, Christopher JT, Harvey GL, Sutherland MW, Borrell AK (2012) Current and emerging screening methods to identify post-head-emergence frost adaptation in wheat and barley. J Exp Bot 63:5405–5416

Frova C, Sari-Gorla M (1994) Quantitative trait loci (QTLs) for pollen thermo-tolerance detected in maize. Mol Gen Genet 245:424–430

Gibson LR, Paulsen GM (1999) Yield components of wheat grown under high temperature stress during reproductive growth. Crop Sci 39:1841–1846

Goodstal FJ, Kohler GR, Randall LB, Bloom AJ, St Clair DA (2005) A major QTL introgressed from wild *Lycopersicon hirsutum* confers chilling tolerance to cultivated tomato (*Lycopersicon esculentum*). Theor Appl Genet 111:898–905

Greer DH, Weedon MM (2012) Modelling photosynthetic responses to temperature of grapevine (Vitis vinifera cv. Semillon) leaves on vines grown in a hot climate. Plant Cell Environ 35:1050–1064

Hasanuzzaman M, Hossain MA, Fujita M (2010) Physiological and biochemical mechanisms of nitric oxide induced abiotic stress tolerance in plants. Am J Plant Physiol 5:295–324

Hasanuzzaman M, Nahar K, Alam MM, Roychowdhury R, Fujita M (2013a) Physiological, biochemical, and molecular mechanisms of heat stress tolerance in plants. Int J Mol Sci 14:9643–9684

Hasanuzzaman M, Nahar K, Fujita M (2013b) Extreme temperature responses, oxidative stress and antioxidant defense in plants. In: Vahdati K, Leslie C (eds) Abiotic stress – plant responses and applications in agriculture. InTech, Rijeka, pp 169–205

Intergovernmental panel on climate change (IPCC), Climate Change 2007: Working Group I: The Physical Science Basis, https://www.ipcc.ch/publications_and_data/ar4/wg1/en/spmsspm-projections-of.html. Retrieved on 18th July 2016

Jones PD, New M, Parker DE, Martin S, Rigor IG (1999) Surface air temperature and its changes over the past 150 years. Rev Geophys 37:173–199

Khanna HK, Daggard GE (2006) Targeted expression of redesigned and codon optimised synthetic gene leads to recrystallization inhibition and reduced electrolyte leakage in spring wheat at sub-zero temperatures. Plant Cell Rep 25:1336–1346

Knoll J, Ejeta G (2008) Marker-assisted selection for early-season cold tolerance in sorghum: QTL validation across populations and environments. Theor Appl Genet 116:541–553

Kotak S, Larkindale J, Lee U, Von Koskull-Doring P, Vierling E, Scharf KD (2007) Complexity of the heat stress response in plants. Curr Opin Plant Biol 10:310–316

Lobell DB, Schlenker W, Costa-Roberts J (2011) Climate trends and global crop production since 1980. Science 333:616–620

Lone AA, Khan MH, Dar ZA, Wani SH (2016) Breeding strategies for improving growth and yield under waterlogging conditions in maize: a review. Maydica 61:1–11

Lou QJ, Chen L, Sun ZX, Xing YZ, Li J, Xu XY et al (2007) A major QTL associated with cold tolerance at seedling stage in rice (*Oryza sativa* L.) Euphytica 158:87–94

Mittler R, Finka A, Goloubinoff P (2012) How do plants feel the heat? Trends Biochem Sci 37:118–125

Moller IM, Jensen PE, Hansson A (2007) Oxidative modifications to cellular components in plants. Annu Rev Plant Biol 58:459–481

Neeraja CN, Maghirang-Rodriguez R, Pamplona A, Heuer S, Collard BCY, Septiningsih EM et al (2007) A marker-assisted backcross approach for developing submergence-tolerant rice cultivars. Theor Appl Gene 115:767–776

Peng S, Huang J, Sheehy JE, Laza RC, Visperas RM, Zhong X, Centeno GS, Khush GS, Cassman KG (2004) Rice yields decline with higher night temperature from global warming. Proc Nat Acad Sci USA 101:9971–9975

Pradhan B, Sarkar M, Kundagrami S (2016) Role of slow elongation of stem and induction of alcohol dehydrogenase (Adh) enzyme for increment of survival under submergence condition in Rice (Oryza sativa L.) Imperial J Interdisc Res 2(7):486–492

Presterl T, Ouzunova M, Schmidt W, Möller EM, Röber FK, Knaak C et al (2007) Quantitative trait loci for early plant vigour of maize grown in chilly environments. Theor Appl Genet 114:1059–1070

Qi Y, Wang H, Zou Y, Liu C, Liu Y, Wang Y et al (2010) Over-expression of mitochondrial heat shock protein 70 suppresses programmed cell death in rice. FEBS Lett 585:231–239

Ripoll J, Urban L, Brunel B, Bertin N (2016) Water deficit effects on tomato quality depend on fruit developmental stage and genotype. J Plant Physiol 190:26–35

Rodríguez M, Canales E, Borrás-Hidalgo O (2005) Molecular aspects of abiotic stress in plants. Biotechnol Appl 22:1–10

Sairam RK, Deshmukh PS, Saxena DC (1998) Role of antioxidant systems in wheat genotypes tolerance to water stress. Biol Plant 41(3):387–394

Sanghera GS, Wani SH, Hussain W, Singh NB (2011) Engineering cold stress tolerance in crop plants. Curr Genomics 12:30–43

Seneweera S, Makino A, Mae T, Basra AS (2005) Response of rice to p(CO2) enrichment: the relationship between photosynthesis and nitrogen metabolism. J Crop Improve 13:31–53

Shiklomanov I (1993) World fresh water resources, in: Peter H Gleick (editor), water in crisis: a guide to the World's fresh water resources. Oxford University Press, New York

Takahashi N (1961) The relation of water absorption to germination of rice seed. Sci Rep Res Inst Tohoku Univ D 12:61–69

Talukder S, Babar M, Vijayalakshmi K, Poland J, Prasad P, Bowden R et al (2014) Mapping QTL for the traits associated with heat tolerance in wheat (Triticum aestivum L.) BMC Genet 15:97

Wahid A, Gelani S, Ashraf M, Foolad MR (2007) Heat tolerance in plants: an overview. Environ Exp Bot 61:199–223

Wollenweber B, Porter JR, Schellberg J (2003) Lack of interaction between extreme high-temperature events at vegetative and reproductive growth stages in wheat. J Agron Crop Sci 189:142–150

Xie XJ, Li BB, Li YX, Shen SH (2009) High temperature harm at flowering in Yangtze River basin in recent 55 years. Jiangsu J Agric Sci 25:28–32

Xu Q, Paulsen AQ, Guikema JA, Paulsen GM (1995) Functional and ultrastructural injury to photosynthesis in wheat by high temperature during maturation. Environ Exp Bot 35:43–54

Xu Z, Zhou G, Shimizu H (2010) Plant responses to drought and rewatering. Plant Signal Behav 5:649–654

Yadav SK (2010) Cold stress tolerance mechanisms in plants. A review. Agron Sustain Dev 30:515–527

Yang J, Sears RG, Gill BS, Paulsen GM (2002) Quantitative and molecular characterization of heat tolerance in hexaploid wheat. Euphytica 126:275–282

Yang Y-H, Zhu X-C, Wang S-C, Hu G-B, Hee H, Huang X-M (2010) Developmental problems in over-winter offseason longan fruit. I: effect of temperature. Sci Hortic 126:351–358

Yousfi S, Marquez AJ, Betti M, Araus JL, Serret MD (2016) Gene expression and physiological responses to salinity and water stress of contrasting durum wheat genotypes. J Integr Plant Biol 58:48–66

Zheng B, Chenu K, Fernanda Dreccer M, Chapman SC (2012) Breeding for the future: what are the potential impacts of future frost and heat events on sowing and flowering time requirements for Australian bread wheat (Triticum aestivum) varieties? Glob Chang Biol 18:2899–2914

Plant Bioregulators: A Stress Mitigation Strategy for Resilient Agriculture

10

Ratna Kumar Pasala, Paramjit Singh Minhas,
and Goraksha C. Wakchaure

Abstract

Plant growth and productivity are stimulated by plant bio-regulators (PBRs), which are biochemical compounds that are applied even in small quantities at appropriate plant growth stages. To enhance the productivity, particularly, PBRs are being extensively used in agriculture and horticultural crops. Though nutrient allocation and source-sink transitions have their central role in plant growth and development and, most of the PBRs stimulate redox signaling under abiotic stress conditions. Since climate change and degrading natural resources are projected to amplify the stresses, particularly soil moisture deficit, high temperature, and soil salinity, PBRs are likely to play a crucial role in plant growth regulation under these stress conditions. However, critical evaluation should consider the utility of PBRs to enhance crop productivity under stresses induced by abiotic factors. Research efforts so far have centered on the crop and agroecosystem specificity, optimal doses, and schedule of their application for optimizing crop yields under stress conditions. These efforts are being complemented by investigations on genes and gene regulatory network at molecular level to tailor crop plants for climate resilience. In addition to complying with regulation governing the use of biochemicals, case studies related to crop yield in major cereals are

R.K. Pasala (✉) • P.S. Minhas
National Institute of Abiotic Stress Management, Indian Council for Agricultural Research, Baramati, Maharashtra, India
e-mail: pratnakumar@gmail.com; minhas_54@yahoo.co.in

G.C. Wakchaure
ICAR-National Institute of Abiotic Stress Management, Baramati, Pune 413115, Maharashtra, India
e-mail: goraksha.wakchaure@gmail.com

© Springer Nature Singapore Pte Ltd. 2017
P.S. Minhas et al. (eds.), *Abiotic Stress Management for Resilient Agriculture*,
DOI 10.1007/978-981-10-5744-1_10

235

also reported. In this chapter, prospects and pathways of PBRs are described as an emerging stress-alleviating technology for crop production in harsh agroeco-systems, specifically those featured by drought, heat, and salinity stress.

10.1 Introduction

Understanding the adaptive responses of plants is of paramount importance particularly when grown under harsh environments. A series of experiments conducted in the past have clearly established that plant growth regulators and hormones play a vital role in determining growth, development, and productivity of crops. This fundamental knowledge served as basis for a number of biochemicals as plant growth promoters in crop production. This tends to be the integral part of modern agriculture, which is often affected by adverse environmental factors such as drought, high temperature, and salt stress. In addition, recent surge in low external input sustainable agriculture (LEISA) demands enhanced intervention of modern plant biology that can integrate knowledge of plant responses to environmental factors at molecular, cellular, whole plant, as well as cropping system levels.

Through LEISA, the plant stress tolerance can be improved with an exogenous use of stress-alleviating chemicals (Wahid and Shabbir 2005; Wahid et al. 2007; Farooq et al. 2009). Plant bio-regulators (PBRs) are powerful tools for maximizing yield and quality and increasing net income to farmers. Control of vegetative vigor, stimulation of flowering, regulation of crop load, reduction of fruit drop, and delay or stimulation of fruit maturity and ripening are best examples of regulation with the exogenous applications. Novel PBRs with possible benefits for fruit growers are continually being made available by industries. In addition, research is in progress to find new uses for bio-regulator products that have been made available for specific uses.

Every aspect of plant growth and development is controlled by plant hormones, and these serve as key integrators of exogenous (environmental) and endogenous (developmental) cues. The classes of phytohormones are auxins, gibberellins (GA), cytokinins, ethylene, abscisic acid (ABA), brassinosteroids (BS), salicylic acid (SA), and jasmonates, with strigolactones representing a relatively new addition (Stamm et al. 2011). Stress may induce common responses such as enhancement of plant hormones. For instance, wounding can induce increased production of ethylene, auxin, and ABA. Since many kinds of stresses including water, salt, and temperatures induce ABA synthesis, ABA is considered as a plant stress hormone. It regulates several important aspects of plant growth and development. Other than plant hormones, exogenous PBRs such as inorganic and organic chemicals and booster's application will also have strong impact on plant adaptation to abiotic stress either independently or synergistically with one another (Srivastava et al. 2016). Many PBRs influence with the signaling pathways of one another thereby promoting plant tolerance to abiotic stress. Various researchers established the inter-linking of redox signaling pathway among different PBRs to accomplish the goal of tolerance (Srivastava et al. 2016). The coordination of the responses triggered by the multiple stimuli is controlled by a network of intricate signal transduction pathways. This network has many signaling components and a small number of highly interconnected components which are central for the functioning of the network (Hetherington and Woodward 2003). For instance, nitric oxide (NO) mediates the action of stomatal closure of ABA (Garcia-Mata and Lamattina2007). PBRs also promote the uptake and metabolism of nutrient elements as well. Among the stress-alleviating compounds, thiourea is one of the important molecules with two functional groups; "thiol" is important to oxidative stress response and "imino" partly fulfills the nitrogen requirement. Apart from thiol compounds, other chemicals including plant growth regulators – 6-benzyladenine (BA), prohexadione-calcium (Pro-Ca), N-2-chloro-4-pyridinyl-N-phenyl- urea (CPPU), 2,4-dichlorophenoxy acetic acid (2,4-D), 3,5,6-trichloro-2-pyridyloxyacetic acid (3,5,6-TPA), amino ethoxy vinyl glycine (AVG), dithiothreitol (DTT), potassium nitrate (KNO_3), thiourea (TU), salicylic acid (SA), and silicon (Si) products – were used, to a small extent, to screen for their ability to increase agricultural yield without compromising quality.

Stressful environments interrupt the balance between generation and utilization of reactive oxygen species (ROS) leading to toxicity by enhancing production of ROS such as superoxide radicals (O_2^-), hydrogen peroxide (H_2O_2), hydroxyl radicals (OH^-), etc., in plants thereby creating a state of oxidative stress in them (Panda et al. 2003a, b). This increased ROS level in plants causes oxidative damage to biomolecules such as lipids, proteins, and nucleic acids, thus altering the redox homeostasis (Smirnoff 1993; Gille and Singler 1995; Srivastava et al. 2016). When applied exogenously at suitable concentrations, PBRs enhance the efficiency of antioxidant system, upregulate osmolytes, and enhance the expression of stress-responsive genes in plants (Srivastava et al. 2016; Knorzer et al. 1999) (Fig. 10.1). PBRs are also proposed to sustain agriculture by integrating redox signaling as a possible unifying mechanism (Srivastava et al. 2016).

Several PBRs such as TU, SA, Si, NO, H_2O_2, hydrogen-rich water, BS and polyamines (PA), etc., have been tested for enhancing the plant stress tolerance as well as the crop yield (Jisha et al. 2013; Srivastava et al. 2016). Most of these are applied either for seed priming or as foliar spray. Seed priming is a pre-sowing treatment that partially hydrates seeds without allowing the radicle emergence. The seed priming may be induced alone or in combination with the foliar application of PBRs at the time or early flowering or grain-filling stage. The mechanism of seed priming mediated action is not well understood, but it may act in two ways. First, seed priming sets in motion germination-related activities (e.g., respiration, endosperm weakening, gene transcription and translation, etc.) that facilitate the transition of quiescent dry seeds into germinating state and lead to improved germination potential. Secondly, priming imposes abiotic stress on seeds that represses radicle protrusion but stimulates stress responses, e.g., accumulation of late embryogenesis abundant (LEAs), potentially inducing cross-tolerance (Chen and Arora 2013). These two strategies together also impose a "memory" in seeds, which can be recruited upon a subsequent stress exposure and may mediate a greater stress tolerance in the subsequent generation (Pastor et al. 2013).

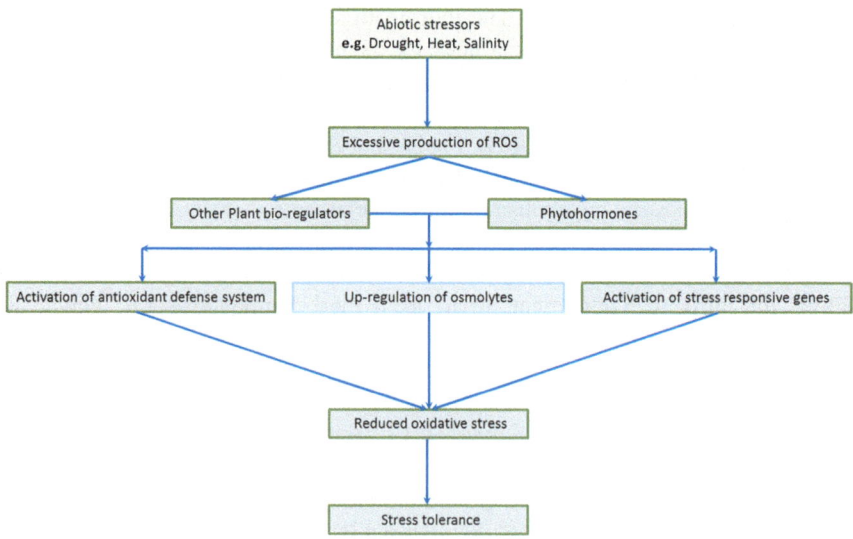

Fig. 10.1 Schematic representation of abiotic stress tolerance mechanisms induced by plant bio-regulators and phytohormones; impact on antioxidative system, upregulation of osmolytes, and gene activation (Source: Ratnakumar et al. 2016a, b)

10.2 Plant Bio-regulators (PBRs)

PBRs include various plant hormones (Farooq et al. 2015; Ratnakumar et al. 2015; Srivastava et al. 2016) which affect plant's ability to respond to its environment (Table 10.1 and 10.2). These interact with specific target tissues to cause physiological responses, such as growth and development. Each response is often the result of two or more hormones acting together. Among them, many hormones can be synthesized in the laboratory, therefore increasing the quantity of hormones available for commercial applications. For instance, GA is widely used as PBRs. Likewise, SA is another important plant hormone that regulates a wide range of metabolic and physiological responses in plants such as seed germination, seedling establishment, cell growth, respiration, stomatal closure, senescence-associated gene expression, responses to abiotic stresses, basal thermo-tolerance, nodulation in legumes, and fruit yield. Hence, the PBRs enhance plant growth and yield (Vlot et al. 2009; Rivas-San Vicente and Plasencia 2011) and for particular stress alleviation. However, judicious use of SA dose is required as the excess concentration proves toxic for the plant growth (Donovan et al. 2013). Exogenous application of cytokinin, preferably kinetin, to the foliage of plants has also been shown to increase crop productivity. The application of low concentrations of potassium together with cytokinin provides synergistic effect (US Patent 2012). Similarly, early application of trehalose has also shown to enhance the health and vigor of plant resulting in better production of sugar content (US Patent 2013). The application of plant sterols and steroid

Table 10.1 Drought, heat stress, and salinity induced plant tolerance by plant bio-regulator (PBRs), viz., SA (salicylic acid), TU (thiourea), and PGPR (plant growth-promoting rhizobacteria)

Stress	Plant	PBRs	Studied parameters	Response	References
Drought	Wheat	SA	Content of ascorbate and glutathione	+	Kang et al. (2013)
	Wheat	SA	Moisture content, dry matter accumulation, carboxylase activity of Rubisco, SOD, and total chlorophyll	+	Singh and Usha (2003)
	Barley	SA	ABA content in leaves	+	Bandurska and Stroinsky (2005)
	Wheat	SA	Stomatal regulation, maintaining leaf chlorophyll content, increasing water-use efficiency, and stimulating root growth	+	Anosheh et al. (2012)
	Salvia officinalis	SA	Remobilization of stored food	+	Abreu and Munne-Bosch (2008)
	Blackgram	PGPR	Production of IAA, cytokinins, antioxidants, and ACC deaminase	+	Figueiredo et al. (2008)
	Wheat	SA	Chlorophyll pigments and chlorophyll a/b ratio	−	Moharekar et al. (2003)
	Wheat	TU	Individual grain weight	+	Sahu and Singh (1995)
Heat	*Arabidopsis*	SA	Oxidative stress	+	Alonso-Ramı'rez et al. (2009)
	Mustard	SA	H_2O_2 content and CAT activity	+ and −	Dat et al. (1998a, b)
	Agrostis stolonifera	SA	POX activity and CAT activity	+ and −	Larkindale and Huang (2004)
	Gram	SA	Protein and proline content induction of stress enzymes, viz., POX and APX	+	Chakraborty and Tongden (2005)
	Wheat	SA	Proline content	+	Khan et al. (2013)
Salt	Wheat	SA	IAA and cytokinin levels, proline content, ABA accumulation, SOD and POX activity	+	Shakirova et al. (2003)
	Wheat	SA	Osmotic potential, shoot and root dry mass, K^+/Na^+ ratio, and photosynthetic pigment content	+	Kaydan et al. (2007)

(continued)

Table 10.1 (continued)

Stress	Plant	PBRs	Studied parameters	Response	References
	Barley	SA	Photosynthetic rate, membrane stability	+	El Tayeb (2005)
	Mustard	SA	Growth, photosynthetic parameters and activities of enzymes (nitrate reductase, carbonic anhydrase, CAT, POX, and SOD), proline content	+	Yusuf et al. (2008)
	Mung bean	SA	Antioxidant system	+	Khan et al. (2014)
	Tomato	SA	Activation of aldose reductase and ascorbate peroxidase, accumulation of osmolytes such as proline	+	Tari et al. (2002, 2004) and Szepesi et al. (2005)

The indication marks "+" represents the positive and "−" represents negative responses
Source: Ratnakumar et al. (2016a, b)

hormones such as brassinosteroids (BS) is essential for plant growth, reproduction, and responses to various abiotic and biotic stresses. The use of these BS (Vriet et al. 2012) has also been proposed as a promising strategy for crop improvement.

10.3 Biotic Agents

Apart from chemical-based PBRs, various biotic agents (Table 10.2) have also been used for the same purpose. Nagaraju et al. (2012) have used *Trichoderma harzianum* isolates to enhance the plant growth and resistance of sunflower toward downy mildew disease. Many bacteria not only promote the growth of plants but also protect the plants against various abiotic and biotic stress agents including flooding, drought, salts, metals, organic contaminants, wilting, and pathogens (Glick 2014).

10.4 Effects of PBRs on Crop Growth

Salicylic acid (SA) and other salicylates, thiourea and other thiol compounds, and KNO_3 and other nitrites are known to affect various physiological and biochemical activities of plants and may play a key role in regulating plant growth and productivity (Arberg 1981). There are conflicting evidences on the role of SA in flowering. In SA-deficient *Arabidopsis*, initiation of flowering failed when irradiated with UV-C, and substantial flowering occurred when grown under non-stress condition than wild-type plants (Shulaev et al. 1995). SIZ1, a SUMO E3 ligase, negatively

Table 10.2 Chemical-based PBRs and biotic agents, their mode of action in crop plants and orchards

PBRs	Type of Plant	Mode of action	References
Chemical-based PBRs			
TU, SA, Si, NO, H₂O₂, hydrogen-rich water, brassinosteroids, polyamines	Crop plants and crop yield	Improves plant stress tolerance	Srivastava et al. (2016)
Chlormequat chloride; ethephon and mepiquat chloride	*Cereals, cotton crop*	*Crop growth, development, and maturity of various crops*	Jisha et al. (2013) and Srivastava et al. (2016)
Naphthaleneacetic acid (NAA)	Apple and several other crops	Enhance flower	Srivastava et al. (2016)
Potassium (K)	*Plant cell*	*Osmosis regulation and photosynthesis*	Arteca (1995) and Wakchaure et al.(2016a)
Thiourea (TU)	Wheat	Grain yield	Sahu et al. (2006), Wakchure et al.(2016) and Ratnakumar et al. (2016a, b)
Prohexadione-calcium (P-ca),	Several fruit tree species, wheat	Inhibitor of gibberellic acid (GA)	Wakchure et al. (2016)
Benzyladenine (BA)	Nursery trees	Improved lateral branching	
Trehalose		Enhance the health and vigor of plant production of sugar content	US patent (2013)
Brassinosteroids (BS)		Hypocotyl cell elongation involves a microtubule regulatory protein, microtubule destabilizing protein40	Wang et al. (2012)

(continued)

Table 10.2 (continued)

PBRs	Type of Plant	Mode of action	References
SA and SNP (nitric oxide source)		Promoted Fe uptake, translocation and activation, modulated the balance of mineral elements, and protect Fe deficiency	Kong et al. (2014)
Biotic agents			
Pseudomonas asplenii	*Phragmites australis* seeds	Improved germination and protected the plants	Bashan et al. (2008)
PGPRs (plant growth-promoting *rhizobacteria*)	Induce production of auxin or inhibit ethylene synthesis or mineralization of nutrients		Steenhoudt and Vander-Leyden (2000)
Trichoderma harzianum	Sunflower	Plant growth and resistance, downy mildew	Nagaraju et al. (2012)
Pseudomonas fluorescens		Promote plant growth	Glick (2014)
Biotic agents + chemical-based PBRs			
Bradyrhizobiumdiazoefficiens + (SA + OSA)	Soybean	Seed yield enhancement under drought	Ratnakumar et al. (2017)

regulates flowering via an SA-dependent pathway (Jin et al. 2008). It was reported that the dry matter accumulation was enhanced in *Brassica juncea*, with SA foliar application, while higher concentrations of SA had an inhibitory effect. In sugarcane GA_3 increased internodal elongation, while glyphosate, CEPA, and other regulators increase the deposition of sucrose; diquat and 2-chloroethylphosphonic acid (CEPA) inhibit flowering, and paraquat desiccates leaves just prior to harvest to facilitate leaf removal or burning (Kossuth 1984). GA increased the rate of cell division and stimulation of vegetative growth (Arteca 1995). The leaf area was found to increase significantly after KNO_3 application probably due to promoting role of potassium (K) in plant growth. Generally, the essential element K has a great regulatory role within plant cells and organs such as activating on enzymes, osmosis regulation and photosynthesis, and loading and unloading of sugars in phloem (osmosis regulation and photosynthesis). DCPTA delay the natural senescence of mature leaves thereby contributing to enlarged leaf canopies and improved carbon assimilation per unit leaf area.

Flowering is another important parameter that is directly related to yield and productivity of plants. SA application induces flowering in a number of plants (Cleland and Ajama 1974). However, the rigorous mechanism of flower-inducing property of SA is not been explored. Kumar et al. (2000) studied the cumulative effect of SA in combination with GA, kinetin, NAA, ethral, and chloro chloro chloride (CCC) and found a synergistic effect between SA and GA on flowering as compared to other combinations of hormones. Likewise, ethylene modulates plant growth and development under normal conditions and is also a key feature in the response of plants to a wide range of stresses. Ethylene is synthesized in plants in response to various stresses typically in response to the presence of metals, organic and inorganic chemicals, cold or heat stress, drought or flood, ultraviolet light, insect and nematode damage, phytopathogens (both fungi and bacteria), and mechanical wounding (Glick 2014). A small fraction ethylene synthesis, which consumes the existing pool of 1-aminocyclopropane-1-carboxylate (ACC, the precursor of ethylene) in stressed plant tissues, is believed to be responsible for initiating the transcription of genes that encode plant defensive/protective proteins (Robison et al. 2001). However, higher level of ethylene production following synthesis by the plant of additional ACC in response to a stress is generally injurious to plant growth and usually involved in initiating senescence, chlorosis, and leaf abscission (Glick 2014). There has been a major increase in the utilization of BRs in agricultural applications as a mean to boost crop productivity and stress tolerance (Choudhary et al. 2012). Interactions between BRs and ABA regulate the expression of many genes that govern several biological processes.

Interaction between soil bacterium may be negative, positive, or neutral, and sometimes the effect of soil bacterium changes according to the conditions of soil. For instance, when the availability of chemical nitrogen fertilizer is abundant in soil, a nitrogen-fixing bacterium is useless for the plants. Similarly, many bacteria are useful for the plants only under environmental stress conditions and are unlikely to have any beneficial effect under optimal conditions (Glick et al. 2007). It is important to mention here that not all strains of a particular bacterial genus or species promote plant growth due to varied genetic makeup and metabolic capabilities. Some strains of *Pseudomonas fluorescens*, for instance, may actively promote plant growth, while other strains of this species have no measurable effect on plants (Glick 2014). Plant growth-promoting bacteria may facilitate plant growth and development either directly or indirectly. The indirect effect on promotion of plant growth occurs by preventing harmful effects of plant pathogens, and direct role of the PGBRs involves either acquisition of mineral nutrients or modulation of plant growth by alteration of phytohormones such as auxin, cytokinin, and ethylene (Glick 2012). It has been found that most of the plant growth-promoting rhizobacteria isolated from Graminaceae grasses growing in a meadow polluted with heavy metals exhibited ACC deaminase activity, which resulted in plant growth promotion (Dell'Amico et al. 2005). Some of the abiotic stresses whose effects can be ameliorated in this way include temperature extremes, flooding, drought, metals and metalloids, hypoxia, salt, and organic contaminants (Glick 2012). Foliar spray and rhizosphere irrigation with purple phototrophic bacteria (PPB) on a herb *Stevia rebaudiana* native to certain regions of South America increased the growth and also yield of stevioside (ST), one of the steviol glycosides, which is 300 times sweeter than cane sugar and non-calorific, and have specific immunomodulatory activities (Wu et al. 2013).

10.4.1 Germination

SA is known to improve germination, plant growth, rate of transpiration, stomatal regulation, photosynthesis, ion uptake, and transport in plants (Metwally et al. 2003; Khodary 2004; He et al. 2010; Khan et al. 2014, 2015). Low doses of SA have potential to improve seed germination and seedling establishment under different abiotic stress conditions (Rajjou et al. 2006; Alonso-Ramirez et al. 2009b). Lee et al. (2010) reported the role of SA in seed germination under salinity. They found that SA promotes germination under high salinity by modulating antioxidant activity in *Arabidopsis*. Further, they suggested that SA was not essential for germination under normal growth conditions, but under saline conditions, it promotes seed germination by reducing oxidative damage. Similarly, Hanieh et al. (2013) determined the effect of pre-sowing treatment of SA on seed germination of sweet pepper and found that pre-sowing seed treatment of SA caused better germination percentage and faster growth rate. SA improves seed germination by promoting the synthesis of proteins that are essential for germination and the mobilization or degradation of seed proteins accumulated during seed maturation (Rajjou et al. 2006). BR-related mutants (*det2-1* and *bri1-1*) showed increased sensitivity to the inhibitory effects of ABA during seed germination in comparison with wild-type plants (Steber and McCourt 2001). However, primary root and hypocotyl elongation assays in the *sax1* mutant revealed hypersensitive responses to ABA, as well as auxin and ethylene (Ephirtikhine et al. 1999). In *Phragmites australis*, inoculation of seeds with *Pseudomonas asplenii* improved germination and protected the plants from growth inhibition (Bashan et al. 2008).

10.4.2 Root Growth

Prolific root system is important for stress tolerance and improved water uptake particularly under abiotic stress conditions in order to harvest better crop yield (Farooq et al. 2009; Flowers 2004). Normally, sensitive genotypes have poor prolific roots than susceptible genotypes. Corresponding to this, a pronounced increment in root growth of sensitive wheat varieties to salt and high-temperature stress was reported with the application of TU and that strongly correlate to grain yield (Sahu et al. 2006). Thus, in view of TU being water soluble, readily absorbable in the tissues that enhance the plant, it has ability to ameliorate stress effects under field conditions.

Superiority of synthetic growth hormone NAA over other bio-regulators attributed its unique role in delaying senescence process, hastening root and shoot growth, and setting more fruits (Wein et al. 1989). Keithly et al. (1990) reported that photosynthate partitioning in DCPTA-treated (foliar application) sugar beet plants appeared to be balanced between the demands of plant growth and taproot sucrose accumulation supply. Increased taproot weight of 30-μM-DCPTA-treated taproots resulted in an 81% increase in sucrose yield.

SA increased the level of cell division within the apical meristem of seedling roots of wheat plants (Sakhabutdinova et al. 2003). Similarly, applications of 10–8 and 10–6 M SA increased root length (San-Miguel et al. 2003). The exogenous application of SA to plants results in an interference with the ion transportation and absorption in the membranes of root cells (Harper and Balke 1981). In soybean plants treated with 10 nM, 100 μM, and up to 10 mM SA, root growth increased up to 45% (Shakirova et al. 2003).

BSs were unable to antagonize the ET effects on hypocotyl growth in the etiolated fer-2 mutant, indicating FER-dependent BS effects on ET-induced growth responses (Deslauriers and Larsen 2010; Cheung and Wu 2011). BR-mediated hypocotyl cell elongation involves a microtubule regulatory protein, microtubule destabilizing protien 40 (MDP40) (Wang et al. 2012). In some plants, the treatment of seeds or cuttings with non-pathogenic bacteria, such as *Agrobacterium*, *Alcaligenes*, *Bacillus*, *Pseudomonas*, *Streptomyces*, etc., induces root formation (Esitken et al. 2003). PGPRs (plant growth-promoting rhizobacteria) might induce the production of auxin or inhibit ethylene synthesis or mineralization of nutrients (Steenhoudt and Vander-Leyden 2000). Desbrosses et al. (2009) investigated the PGPR-*Arabidopsis* interaction to establish the signaling pathways involved in controlling plant development and observed an ethylene-independent and auxin-independent mechanism, regulating the elongation of root hair in *Arabidopsis*.

10.4.3 Nutrient Uptake, Mobilization, and Translocation

Application of NAA (20 ppm) maximized the seed (1.76 Mg ha^{-1}) and straw (4.7 Mg ha^{-1}) yield of fenugreek and was superior to other bio-regulators like

GA. The uptake of N, P, and K was the maximum with 20 ppm NAA. Hormone application increased physiological and metabolic activities of the plant as a result more uptake of nutrients by plants from the soil. The favorable effect of plant growth regulators in enhancing the yields and nutrient uptake was also reported in cotton and green gram, respectively. Nitric oxide (NO) or its donor has been shown in reports by Zhang et al. 2012 and Zottini et al. 2007. SA and SNP (nitric oxide source) promoted Fe uptake (Table 10.2), translocation, and activation, modulated the balance of mineral elements, and protected Fe deficiency-induced oxidative stress (Kong et al. 2014). Little information is available on the role of BRs under nutrient-deficient conditions. Wang et al. (2012) reported that BRs play a negative role in regulating iron (Fe) deficiency-induced ferric reductase (FRO), expressions of CsFRO1 (transcripts encoding FRO) and CsIRT1 (Fe transporter), as well as Fe translocation from roots to shoots.

In wheat, the mobilization flow rate of sucrose to grain determines the rate at which carbohydrate can accumulate in the ear. Sahu and Singh (1995) observed that improved productivity of wheat under soil and foliar treatments of thiourea (TU) was mainly due to enhanced grain weight. Metabolic transport of sucrose to grains via effects on phloem loading was enhanced with application of TU. Mobilization of dry matter (reserves) from leaves to grains increased in wheat with TU spray at tiller stage. Improvement in harvest index under TU treatments lends further credence to the role of TU in improving dry matter partitioning to grains (Sahu and Solanki 1991). Reduced grain yield with 20 kg ha^{-1} soil-applied TU compared with 10 kg ha^{-1} treatment might be attributed to inhibitory effects perhaps on phloem transport of sucrose. The role of TU as a thiol compound and growth-regulating chemical in improving productivity of field crops such as wheat and brassica was evaluated by Srivastava et al. (2009), Sahu et al. (2005), and NIASM (2013) (Table 10.1). Inoculation of rhizobacteria increased uptake of nutrient elements like Ca, K, Fe, Cu, Mn, and Zn by plants through stimulation of proton pump ATPase (Mantelin and Touraine 2004; Kumar et al. 2014). In crop plants, mineral element uptake also increased in combined inoculants of *Bacillus* and *Microbacterium* (Karlidag et al. 2009). PGPRs might increase the nutrient uptake of plants through organic acid production and decreasing the soil pH in rhizosphere.

10.4.4 PBRs Induced Plant Adaptation and Mitigation

10.4.4.1 Drought

Exposure of plants to water stress leads to serious physiological and biochemical dysfunctions including reduction in turgor, growth, photosynthetic rate, stomatal conductance, and damages of cellular components (Janda et al. 2003). SA has significant role in controlling abiotic stresses including drought and salinity stress. SA has high potential for improving stress tolerance in agriculturally important crops. Its utility however depends on various factors like concentration of SA applied, mode of application, and stage of plant growth. At low concentrations, SA has been found to alleviate abiotic stress, and at higher concentrations, it induces oxidative

stress (Rivas-San Vicente and Plasencia 2011; Miura and Tada 2014). Higher tolerance to drought stress was also observed in the plants raised from the grains soaked
in aqueous solution of acetyl salicylic acid (Hamada and Hakimi 2001). It has been
suggested that SA-induced drought tolerance is associated with an enhanced antioxidant system (Horvath et al. 2007; Mutlu et al. 2009; Zhou et al. 2009). However, few
studies have shown that abiotic tolerance induced by SA could be related to the
altered expression of the genes encoding osmotin, pathogenesis-related proteins, and
heat shock proteins (Ding et al. 2002; Kim and Delaney 2002; Clarke et al. 2004). To
explore the molecular mechanisms involved, an effort was made by Kang et al.
(2013).Treatment with SA increased drought tolerance of common bean (*Phaseolus
vulgaris*) and tomato (*Solanum lycopersicum*) plants (Senaratna et al. 2000).
Exogenously applied SA has also been reported to modulate activities of intracellular antioxidant enzymes SOD and POD and increase plant tolerance to environmental stresses (Sakhabutdinova et al. 2004; Senaratna et al. 2000). Application of SA
alleviated adverse effects of drought stress in wheat plants by increasing improving
stomatal regulation, maintaining leaf chlorophyll content, increasing water-use efficiency, and stimulating root growth (Anosheh et al. 2012). Leaf senescence is a
highly regulated physiological process, allowing the remobilization of stored food
from the older leaves to the rest of the plant, during stressful conditions, and SA
involved in the promotion of drought-induced leaf senescence in *Salvia officinalis*
plants (Abreu and Munne-Bosch 2008). Singh and Usha (2003) revealed that the
wheat seedlings subjected to drought stress when treated with SA generally exhibited higher moisture content and also higher dry matter accumulation, ABA accumulation, and carboxylase activity of Rubisco, SOD, and total chlorophyll content
compared to the untreated plants (Bandurska and Stroinsky 2005; Ghasempour et al.
2001). Improvement in grain weight with soil and foliar treatments of thiourea that
increased productivity of wheat (Sahu and Singh 1995) is showed.

SA potentially generates a wide array of metabolic responses and also affects the
photosynthetic parameters and plant water relations. However, a reduction in chlorophyll content was observed in plants pretreated with SA (Anandhi and Ramanujam
1997; Pancheva et al. 1996), synthesis of carotenoids, and xanthophylls (Moharekar
et al. 2003), enhanced the rate of deep oxidation in wheat, transpiration rate
decreased significantly in *Phaseolus vulgaris* and *Commelina communis*, and
induced the closure of stomata (Larque-Saavedra 1978, 1979) in contrast to
enhanced photosynthesis and growth of soybean (C_3 plant) and corn (C_4 plant)
(Khan et al. 2003). SA also has the capacity of osmotic adjustment by maintaining
low MDA contents and decreased Na/K ratio in leaves (Fayez and Bazaid 2014).
Brassinosteroids enhance tolerance to drought and cold stress by modulations of the
expression of drought- and cold-stress marker genes (Kagale et al. 2007). However,
the exact biochemical link between the Brassinosteroids-signal cascade and stress
tolerance remains a mystery.

Soil and foliar treatments of thiourea (TU) increased the number of ears and
grains/ear, indicating an improved storage capacity (Werdan et al. 1975). Because
of its cytokine-like activity, TU might have also delayed leaf senescence (Halmann
1990). In maize, foliar spray of TU increased both canopy photosynthesis and

photosynthetically active leaf surface during grain filling (Sahu et al. 1993). Chloroplasts isolated from mature leaves of 30-μM-DCPTA-treated plants, as compared with that of controls, showed a 23% increase in the total soluble protein to chlorophyll ratio. This parallels to an observed increase in activated ribulose 1, 5-bisphosphate carboxylase/oxygenase (Rubisco) activity in vitro per unit chlorophyll. The Rubisco activity increased 87% per dm^2 leaf area of 30-μM-DCPTA-treated plants (Keithly et al. 1990). Increased Rubisco activity largely accounted for increase in net photosynthesis in DCPTA-treated plants.

Tolerance to drought stress is enhanced in PGPR-inoculated plants (Figueiredo et al. 2008). Plants achieve this tolerance either due to the production of IAA, cytokinins, antioxidants, or ACC deaminase. Reports are also available regarding the role of PGPR in conferring resistance to water stress in plants such as tomatoes and peppers under water-deficient conditions (Aroca and Ruiz-Lozano 2009). More efforts are needed to investigate the mechanistic approach of PGPR in eliciting tolerance to different stresses. This would improve our understanding of induced systemic tolerance to water stress in modern agriculture.

10.4.4.2 Salinity

High salinity induces serious metabolic perturbations in plants, as it generates ROS which disturb the cellular redox system in favor of oxidized forms, thereby creating an oxidative stress that may damage DNA, inactivate enzymes, and cause lipid peroxidation (Smirnoff 1993) (Table 10.1). The results of Srivastava et al. (2009) recommend that TU treatment maintains the integrity and functioning of mitochondria in seeds as well as seedlings exposed to salinity. Thus, TU has the potential to be used as an effective bio-regulator to impart salinity tolerance under field conditions and might prove to be of high economic importance by opening new avenues for both basic and applied research. However, most of the literature indicates that exogenous application of SA to the stressed plants can potentially alleviate the toxic effects, generated by salinity. An enhanced tolerance against salinity stress was observed in wheat seedlings (Hamada and Hakimi 2001) and tomato plants (Tari et al. 2002, 2004; Szepesi et al. 2005; Rai 2002). Wheat seedlings accumulated large amounts of proline under salinity stress (Shakirova et al. 2003). It is widely reported that the pretreatment with SA resulted in the accumulation of ABA which induces the synthesis of a wide range of anti-stress proteins and lowered the level of active oxygen species and activities of SOD and POX in the roots of young wheat seedlings (Shakirova et al. 2003; Sakhabutdinova et al. 2004). Exogenous application of SA enhanced the photosynthetic rate, maintained the stability of membranes (El Tayeb 2005) alleviated by exogenous application of SA in *Arabidopsis* seedlings (Borsani et al. 2001; Kaydan et al. (2007), and observed that pre-sowing soaking treatment of seeds with SA positively affected the osmotic potential, shoot and root dry mass, K^+/Na^+ ratio, and contents of photosynthetic pigments (chlorophyll a and b and carotenoids) in wheat seedlings, under both saline and nonsaline conditions. Various antioxidant enzymes (CAT, POX, and SOD) were increased with a concomitant increase in proline content (Yusuf et al. 2008; Ismail 2013, Khan et al. 2014). In both abscisic acid (ABA) levels, increase in both drought and high-salinity conditions, along with gene expression, changes (Zeller et al. 2009). Intensive cross talk of ABA with different signaling pathways produces abundance of proteins and secondary messengers that act as regulators or modulators of ABA responses (Christmann et al. 2006). Both ABA and ethylene responses integrate at the level of DELLA proteins under salt stress and under conditions of high salinity and vegetative growth of plants occurs through activation of ABA and ethylene signaling (Achard et al. 2006). The DELLA-mediated growth control and plant survival under high salt conditions was shown to be due to an increased accumulation of enzymes that detoxify reactive oxygen species (ROS) (Achard et al. 2008). BS crosstalk with SA regulates plant responses to abiotic stress. Exogenously applied BRs could not confer salt stress tolerance in the SA-insensitive npr1-1 mutant in *Arabidopsis* (Divi et al. 2010). This supports the idea that BR-induced salt tolerance in *Arabidopsis* partially depends on NPR1, a master regulator of the SA-mediated defense signaling pathway (Divi et al. 2010). In *Brassica juncea*, salt stress tolerance increased with exogenous applications of BS and SA. The combined application of BS and SA was most effective in alleviating the salt stress when compared with their individual treatments (Hayat et al. 2012).

Other PGBRs which enhance salt tolerance in plants include mycorrhizal associations of plants. Different mechanisms are employed by arbuscular mycorrhizal fungi to enhance salt tolerance of host plants (Azcon and El-Altrash 1997; Giri and Mukerji 2004; Sheng et al. 2009), inhibit high uptake of Na and Cl (Daei et al. 2009), improve water uptake (Ruiz-Lozano and Azcon 2000), and accumulate proline and polyamines (Evelin et al. 2009; Ibrahim et al. 2011) and antioxidant defense system (SOD and CAT) (Farooq et al. 2015, Wu et al. 2010) (Fig. 10.2). Arbuscular

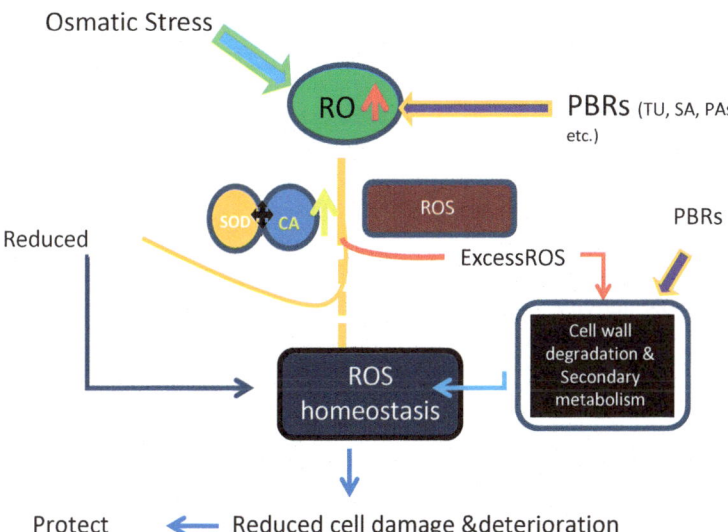

Fig. 10.2 Schematic representation of ROS scavenging induced by plant bio-regulators; impact on antioxidative system and reduced cell damage and deterioration (Source: Ratnakumar et al. 2016a, b)

mycorrhizal association also improves osmotic adjustment, which helps in mainte-nance of the leaf turgor pressure, which in turn affects photosynthesis, transpiration, stomatal conductance, and water-use efficiency (Juniper and Abbott 1993). Under saline condition, mycorrhizal inoculation of three different arbuscular mycorrhizal fungi, *Glomus mosseae*, *G. deserticola*, and *Gigaspora gergaria*, significantly increased growth responses, nutrient contents, acid and alkaline phosphatases, and proline and total soluble protein of wheat plants compared to non-mycorrhizal ones (Abdel-Fattah and Asrar 2012). Foliar application of *Moringa* (*Moringa oleifera*) leaf extract (MLE; 30 times diluted), benzyl amino purine (BAP; 50 mg L-1), and hydrogen peroxide (H_2O_2; 120 µM) at tillering, jointing, booting, and heading growth stages decreased the shoot Na+ and Cl- contents, with simultaneous increase in shoot K^+ contents (Yasmeen et al. 2013).

10.4.4.3 Heat Stress

Deviation from optimum temperature results in serious perturbations in plant growth and development (Mittler 2002). Both salinity and high temperatures have a common facet of oxidative damage (Wahid et al. 2007). Looking at the structure of thiourea, both "imino" and "thiol" functional groups and with foliar spray, amino group provides a ready source of nitrogen, and thiol has a great role in alle-viating oxidative stress damage on the physiologically more important mesophyll tissue which results that foliar applied thiourea was effective in improving salt (6–11%) and high-temperature tolerance (4–10%) of wheat varieties. However, SA plays a key role in providing tolerance against temperature stress. SA stimulates alternative respiratory pathway in mitochondria by inducing expression of alterna-tive oxidase, the terminal electron acceptor of the alternative respiratory pathway, and releases unused potential energy as heat (Vlot et al. 2009). SA has been found to provide protection against heat stress in plants (Karlidag et al. 2009; Khan et al. 2013). Exogenous application of SA also partially reverses the inhibitory effect of oxidative (0.5 mM paraquat) and heat stress (50 °C for 3 h) on seed germination (Alonso-Ramirez et al. 2009a, b). Jasmonic acid also acts with SA, resulting in basal thermo-tolerance in *Arabidopsisthaliana* (Clarke et al. 2009). A foliar spray of SA conferred heat tolerance to mustard (Dat et al. 1998a, b), in potato plantlets (Lopez-Delgado et al. 1998), and in plants of *Agrostis stolonifera* (Larkindale and Huang 2004). These authors further reported that the pretreatment with SA had no effect on POX activity, whereas the CAT activity was shown to decline as com-pared to control. Moreover, the treatment enhanced the activity of enzyme ascor-bate peroxidase. Contrary to this, an enhanced activity of CAT and SOD was observed in heat-stressed plants of *Poa pratensis*, after the treatment with SA (He et al. 2005). The application of SA in *Cicer arietinum* reduced heat stress and induction of various stress enzymes, viz., POX and APX (Chakraborty and Tongden 2005; Table 10.1).

10.4.4.4 Transferring Knowledge to Farmer's Field

The use of PBRs is a unique facet of biotechnology and a new approach of manipu-lating plant biochemistry for enhancing productivity and quality. Although, there is a wide range of PBRs that are commonly used to improve plant growth, development, defense, and productivity, the molecular mechanisms of their effects still remain to be fully elucidated. In addition, their commercialization depends upon several factors such as their stability under the field conditions, inertness, cost-effectiveness, ease of application, and versatility toward different stresses. From the practical point of view, both endogenous and exogenous substances can be regarded as PBRs if they exert an influence on the growth and development of the plant, in low concentrations, and without having any biocidal or nutritive action. Since the environmental conditions before, during, and after application of PBR may influence its effect on a given plant, it is critical to develop specific protocols for each PBR before it can widely be commercialized. In fact, changing environmental conditions in the field causing fluctuations of multiple parameters simultaneously such as temperature, humidity makes it difficult to utilize the results obtained from controlled experiments into practical applications for growers. In addition, plants face multiple stresses under farmer's field conditions, and combinations of PGRs may be needed to achieve significant impact on growth and productivity. Another aspect which needs to be thoroughly explored is the economic feasibility of commercializing the PBRs from the point of view of agrochemical industry and the customers/users which are the smallholder farmers.

10.4.4.5   Integrating PBRs with Other Management Activities

Under stressful conditions, crop management is a key component for optimizing inputs and reducing yield losses. Plant growth regulators play a vital role in determining growth, development, and productivity of crops. Advances in ABA signaling science have already allowed refined development of crop management techniques for reducing irrigation input (Wilkinson et al. 2012). Ethylenephysiology is accessible to manipulation via crop management, for example, applications of 1-MCP (1-methylcyclopropene). Liquid foliar applications of "Invinsa" (Agro-Fresh Inc., USA), which produces gaseous 1-MCP, can act on a range of ethylene-associated process. Crop management using plant growth-promoting rhizobacteria (PGPRs), as either seed treatments or soil additions, which reduce plant ethylene accumulation under stress by metabolizing the ethylene precursor at the root-soil interface, also increases yields of legumes in the field (Belimov et al. 2007). This provides scope for genetic or management modulation of water use and carbon gain in field crops.

10.4.5   PBRs for Enhancement of Water Productivity and Yield: Case Studies

10.4.5.1   Wheat

The use of plant bio-regulators has been shown to mitigate the impacts of water stress and benefit crops underwater scarce conditions. Therefore, studies were reported the impact of irrigation regimes and plant bio-regulators (PBRs) on grain yield and water productivity of spring wheat (*Triticum aestivum* L) (Wakchaure et al. 2016a; Ratnakumar et al. 2016a, b). PBR's application through exogenous sprays included 10 mM thiourea (TU), 10 uM salicylic acid (SA), 1.5 g L^{-1} potassium nitrate (KNO$_3$), 25 ppm gibberellic acid (GA$_3$), and 8 ppm ortho-silicic acid (OSA) at crown root initiation (CRI), flag leaf, and seed milking stages and control (no PBR). Seven irrigation levels were generated through a line source sprinkler system (LSS), viz., application of irrigation water (IW) equaling 1.0, 0.85, 0.70, 0.55, 0.40, 0.25, and 0.10 times the CPE (cumulative open pan evaporation). The maximum yield obtained with PBRs varied between 4.11 and 4.46 Mg ha^{-1} at IW/ CPE 0.85 against 4.09 Mg ha^{-1} without PBR, while the yield decline equaled 0.35– 0.42 Mg ha^{-1} for every 0.1 IW/CPE for PBRs against 0.43 Mg ha^{-1} without PBR. The overall improvement in grain yield and total biomass with PBRs ranged between 5.9–20.6% and 4.8–15.3%, respectively. Specifically, TU and SA showed a major role under medium (IW/CPE 0.40–0.69) and severe (0.10–0.39) stress conditions in terms of maintenance of leaf water content, modulating the stomatal opening and better water usage, and thereby improved yield by 0.41–0.88 Mg ha^{-1}. The maximum water productivity ranged between 1.20 and 1.35 kg m^{-3} with different PBRs, while it was 1.18 kg m^{-3} without PBR, and the latter could be achieved with 19–56% lesser irrigation water with PBRs. Overall conclusions are that the effects of deficit irrigation could be substantially enhanced in terms of grain yield and water productivity when used conjunctively with PBRs like TU and SA. Thus, for integrating PBRs with supplemental irrigation, large-scale testing is required for defining their economic spray schedules under water scarcity conditions.

10.4.5.2 Cultivar Responses

Supplemental irrigation, drought-tolerant cultivars, and the use of plant bioregulators are now being proposed as key strategies to unlock the yield potential and stabilize the yield of wheat grown in rainfed areas. Studies were conducted using line source sprinklers (LSS) to determine the interactive effects of quantities of supplemental irrigation and exogenous foliar sprays of 10 mM thiourea (TU) and 32 ppm of orthosilicic acid (OSA) on wheat (*Triticum aestivum* L.) varieties, viz., HD2189, LOK1, NIAW 301, NIAW 34, and PBW 550 (Ratnakumar et al. 2016b). The irrigation quantities were fully irrigated (I1:31.8 cm), mild (I2: 28.9 cm), medium (I3, 25.9 cm and I4, 22.7 cm), and severe (I5, 19.9 cm and I6, 17.2 cm) water stress conditions. Wheat cultivars varied in their response to water deficits and those popular with farmers, e.g., NIAW 301 showed higher water productivity under deficit irrigation. Response to TU, OSA also varied across water regimes and was higher under moderate to severe stress. Foliar application of TU and OSA at root crown initiation, flag leaf, and grain-filling stages improved yield by 6–9% at fully irrigated; 18–19% at medium stress; 12–17% at severe water stress conditions and water productivity by 0.12–0.10 Mg ha^{-1} with TU, 0.13–0.09 Mg ha^{-1} with OSA at fully irrigated and medium stress; and 0.11–0.03 Mg ha^{-1} with TU, 0.12–0.03 Mg ha^{-1} with OSA at severe stress conditions. TU and OSA induced efficient use of water through increased relative water content, modulating canopy temperatures, and enhanced total soluble sugars and sink partition that are essential for enhanced water productivity under deficit irrigation. The overall interpretation is that varieties like NIAW 301, NIAW 34, and LOK1 with higher water productivity under medium and severe water stress intensities, though having comparatively low potential yields, should be preferred, and exogenous application of PBRs like TU and OSA could further enhance wheat productivity.

10.4.5.3   Sorghum

Studies were also conducted to evaluate the interactive effects of plant bio-regulators (PBRs) and supplemental irrigation on growth and grain yield of sorghum (Wakchaure et al. 2016b). Exogenous application of PBRs included 10 uM salicylic acid (SA), 100 mg L^{-1} sodium benzoate (SB), 500 ppm thiourea (TU), 1.5% potassium nitrate (KNO3) at seedling elongation (20DAS), reproductive (50 DAS), and panicle emergence (75 DAS) stages and control (no spray of PBR). Line source sprinkler system (LSS) was used to apply variable quantities of irrigation water (IW), i.e., equaling 0.95, 0.80, 0.65, 0.50, 0.35, 0.20, and 0.05 times the CPE (cumulative open pan evaporation). The maximum grain yield (3.60–3.88 Mg ha^{-1}) was obtained at IW/CPE 0.80 and declined at 0.43–0.49 Mg ha^{-1} for every 0.1 IW/CPE for PBRs – and the corresponding values were 3.49 and 0.53 Mg ha, 1 without PBR. The application of PBRs mitigated water stress and improved gain yield by 6.8–18.5%. SA was more effective under moderate (IW/CPE 0.79–0.50), while SB and TU were better under severe water deficits (IW/CPE 0.49–0.05). PBRs maintained higher leaf water content; lower canopy temperature modulated the stomatal opening and ultimately the source – sink relations – thereby improving the yield and water productivity under deficit irrigation. The maxima of water productivity varied between 1.16 and 1.41 kg m^{-3} with PBRs – while it was 1.12 kg m^{-3} without PBR, and the latter could be achieved with 25.2–49.7% lesser irrigation water (IW) with PBRs. It is concluded that PBRs like SB and TU present viable option for improving sorghum yield and water productivity under the conditions of deficit irrigation.

10.4.5.4   Soybean

The interactive effect of bio-regulators and irrigated water applied through (LSS) on soybean (J-335) was also studied during the *kharif* season. Treatment combinations were (i) spraying of six bio-regulators, namely, thiourea (10 mM), salicylic acid (10 μM), silixol (160 ppm), KNO$_3$ (1.5%), GA3 (25 ppm), and water (control) at flowering and seed-filling stages. The highest (2.15 Mg ha^{-1}) yield was obtained with application of salicylic acid. The yield was decreased with reduction in amount of water received from the sprinkler. The yield was reduced to 91.6, 83.8, 78.4, 70.3, and 62.4% when applied water was 29.0, 25.9, 22.7, 19.9, and 17.2 cm, respectively.

10.5   Conclusions

Ample research investigations have been carried out to estimate efficacy of the PBRs under drought, saline, and high-temperature conditions in crops. Many of such studies depended on pot culture or single location or season experiment. Often

R.K. Pasala et al.

scientific leads were not carried forward for large-scale evaluation and validation in field conditions, possibly for several reasons including biosafety measures that may have adverse impact on human health, soil quality, and eco-friendly usage. Further, this could also be attributed to lack of information on effect of residues in grain and soil that can impact on human health and ecosystem of which soil microbes especially on rhizobacteria are the useful and vulnerable component. Hence, there is a need for non-residual deliverable product, and cost-effective PBR-based formulation and technology that can address the alleviation of abiotic stresses.

References

Abdel-Fattah GM, Asrar AWA (2012) Arbuscular mycorrhizal fungal application to improve growth and tolerance of wheat (*Triticum aestivum* L.) plants grown in saline soil. Acta Physiol Plant 34(1):267–277
Abreu ME, Munne-Bosch S (2008) Salicylic acid may be involved in the regulation of drought-induced leaf senescence in perennials: a case study in field-grown *Salvia officinalis* L. plants. Environ Exp Bot 64(2):105–112
Achard P, Cheng H, De-Grauwe L, Decat J, Schoutteten H, Morits T, Van-der-Straeten D, Peng J, Hareberd NP (2006) Integration of plant responses to environmentally activated phytohormonal signals. Science 311(5757):91–94
Achard P, Renou JP, Berthome R, Harbred NP, Genschik P (2008) Plant DELLAs restrain growth and promote survival of adversity by reducing the levels of reactive oxygen species. Curr Biol 18(9):656–660
Alonso-Ramirez A, Rodriguez D, Reyes D, Jimenez JA, Nicolas G, Lopez-Climent M, Gomez-Cadenas A, Nicolas C (2009a) Cross-talk between gibberellins and salicylic acid in early stress responses in *Arabidopsis thaliana* seeds. Plant Signal Behav 4(8):750–751
Alonso-Ramirez A, Rodriguez D, Reyes D, Jimenez JA, Nicolas G, Lopez-Climent M, Gomez-Cadenas A, Nicolas C (2009b) Evidence for a role of gibberellins in salicylic acid-modulated early plant responses to abiotic stress in *Arabidopsis* seeds. Plant Physiol 150(3):1335–1344
Anandhi S, Ramanujam MP (1997) Effect of salicylic acid on black gram (*Vigna mungo*) cultivars. Indian J Plant Physiol 2:138–141
Anosheh HP, Emam Y, Ashraf M, Foolad MR (2012) Exogenous application of salicylic acid and chlormequat chloride alleviates negative effects of drought stress in wheat. Adv Stud Biol 4(11):501–520
Arberg B (1981) Plant growth regulators. Mono-substituted benzoic acid. Swed Agric Res 11:93–105
Aroca R, Ruiz-Lozano JM (2009) Induction of plant tolerance to semi-arid environments by beneficial soil microorganisms–a review. In: Climate change, Intercropping, Pest control and beneficial microorganisms. Springer, Dordrecht, pp 121–135
Azcon R, El-Altrash F (1997) Influence of arbuscular mycorrhizae and phosphorus fertilization on growth, nodulation and N2 fixation (15N) in *Medicago sativa* at four salinity levels. Biol Fertil Soils 24(1):81–86
Bandurska H, Stroinsky A (2005) The effect of salicylic acid on barley response to water deficit. Acta Physiol Plant 27(3b):379–386
Bashan Y, Puente ME, De-Bashan LE, Hernandez JP (2008) Environmental uses of plant growth-promoting bacteria. Plant Microbe Int:69–93
Belimov AA, Dodd IC, Safronova VI, Hontzeas N, Davies WJ (2007) Pseudomonas Brassica cearum strain Am3 containing 1-aminocyclopropane-1-carboxylate deaminase can show both pathogenic and growth-promoting properties in its interaction with tomato. J Exp Bot 58:1485–1495

Borsani O, Valpuesta V, Botella MA (2001) Evidence for a role of salicylic acid in the oxidative damage generated by NaCl and osmotic stress in Arabidopsis seedlings. Plant Physiol 126:1024–1030

Chakraborty U, Tongden C (2005) Evaluation of heat acclimation and salicylic acid treatments as potent inducers of thermos tolerance in *Cicer arietinum* L. Curr Sci 89:384–389

Chen K, Arora R (2013) Priming memory invokes seed stress-tolerance. Environ Exp Bot 94(2013):33–45

Choudhary SP, Yu JQ, Yamaguchi K, Shinozaky K, Tran LSP (2012) Benefits of brassinosteroid cross talk. Trends Plant Sci 17(10):594–605

Christmann A, Moes D, Himmelbach A, Yang Y, Tang Y, Grill E (2006) Integration of abscisic acid signaling into plant responses. Plant Biol 8(3):314–325

Clarke SM, Mur LA, Wood JE, Scott IM (2004) Salicylic acid dependent signaling promotes basal thermo-tolerance but is not essential for acquired thermos tolerance in Arabidopsis thaliana. Plant J 38(3):432–447

Clarke SM, Cristescu SM, Miersch O, Harren FJ, Wasternack C, Mur LA (2009) Jasmonates act with salicylic acid to confer basal thermo-tolerance in *Arabidopsis thaliana*. New Phytol 182(1):175–187

Cleland CF, Ajama A (1974) Identification of the flower-inducing factor isolated from aphid honeydew as being salicylic acid. Plant Physiol 54:904–906

Daei G, Ardekani MR, Rejali F, Teimuri S, Miransari M (2009) Alleviation of salinity stress on wheat yield, yield components, and nutrient uptake using arbuscular mycorrhizal fungi under field conditions. J Plant Physiol 166(6):617–625

Dat JF, Foyer CH, Scott IM (1998a) Changes in salicylic acid and antioxidants during induced thermo-tolerance in mustard seedlings. Plant Physiol 118:1455–1461

Dat JF, Lopez-Delgado H, Foyer CH, Scott IM (1998b) Parallel changes in H_2O_2 and catalase during thermo-tolerance induced by salicylic acid or heat acclimation in mustard seedlings. Plant Physiol 116:1351–1357

Dell'amico E, Cavalca L, Andreoni V (2005) Analysis of rhizobacterial communities in perennial *Graminaceae* from polluted water meadow soil, and screening of metal-resistant, potentially plant growth-promoting bacteria. FEMS Microbiol Ecol 52(2):153–162

Desbrosses G, Contesto C, Varoquaux F, Galland M, Touraine B (2009) PGPR-Arabidopsis interactions is a useful system to study signaling pathways involved in plant developmental control. Plant Signal Behav 4(4):319–321

Deslauriers SD, Larsen PB (2010) Feronia is a key modulator of brassinosteroid and ethylene responsiveness in Arabidopsis hypocotyls. Mol Plant 3(3):626–640

Ding CK, Wang C, Gross KC, Smith DL (2002) Jasmonate and salicylate induce the expression of pathogenesis-related-protein genes and increase resistance to chilling injury in tomato fruit. Planta 214(6):895–901

Divi UK, Rahman T, Krishna P (2010) Brassinosteroid-mediated stress tolerance in Arabidopsis shows interactions with abscisic acid, ethylene and salicylic acid pathways. BMC Plant Biol 10(1):151–162

Donovan MP, Nabity PD, Delucia EH (2013) Salicylic acid-mediated reductions in yield in *Nicotiana attenuate* challenged by aphid herbivory. Arthropod Plant Interact 7:45–52

Ephirtikhine G, Fellner M, Vannini C, Lapous D, Barbier-Brugoo H (1999) The sax1 dwarf mutant of Arabidopsis thaliana shows altered sensitivity of growth responses to abscisic acid, auxin, gibberellins and ethylene and is partially rescued by exogenous brassinosteroid. Plant J 18(3):303–314

Esitken A, Karlidag H, Ercisli S, Turan M, Sahin F (2003) The effect of spraying a growth promoting bacterium on the yield, growth and nutrient element composition of leaves of apricot (*Prunus armeniaca* L. cv. Hacihaliloglu). Crop Pasture Sci 54(4):377–380

Evelin H, Kapoor R, Giri B (2009) Arbuscular mycorrhizal fungi in alleviation of salt stress: a review. Ann Bot 104(7):1263–1280

Farooq M, Wahid A, Kobayashi N, Fujita D, Basra SMA (2009) Plant drought stress: effects, mechanisms and management. Agron Sustain Dev 28:185–212

Farooq MA, Saqib ZA, Akhtar J, Bakhat HF, Ratna-Kumar P, Dietz K (2015) Protective role of silicon (Si) against combined stress of salinity and boron (B) toxicity by improving antioxidant enzymes activity in rice. Silicon, 1-5

Fayez KA, Bazaid SA (2014) Improving drought and salinity tolerance in barley by application of salicylic acid and potassium nitrate. J Saudi Soc Agric Sci 13(1):45–55

Figueiredo MV, Burity HA, Martinez CR, Chanway CP (2008) Alleviation of drought stress in the common bean (*Phaseolus vulgaris* L.) by co-inoculation with *Paenibacillus polymyxa* and *Rhizobium tropici*. Appl Soil Ecol 40(1):182–188

Ghasempour HR, Anderson EM, Gaff DF (2001) Effects of growth substances on the protoplasmic drought tolerance of leaf cells of the resurrection grass, *Sporobolus stapfianus*. Aust J Plant Physiol 28:1115–1120

Gille G, Singler K (1995) Oxidative stress in living cells. Folia Microbiol 2:131–152

Giri B, Mukerji KG (2004) Mycorrhizal inoculant alleviates salt stress in *Sesbania aegyptiaca* and *Sesbania grandiflora* under field conditions: evidence for reduced sodium and improved magnesium uptake. Mycorrhiza 14(5):307–312

Glick BR (2014) Bacteria with ACC deaminase can promote plant growth and help to feed the world. Microbiol Res 169(1):30–39

Glick BR, Todorovic B, Czarny J, Cheng Z, Duan J, Mcconey B (2007) Promotion of plant growth by bacterial ACC Deaminase. Crit Rev Plant Sci 26:227–242

Hamada AM, Hakimi AMAA (2001) Salicylic acid versus salinity drought induced stress on wheat seedlings. Rostl Vyr 47:444–450

Hanieh A, Mojtaba D, Zabiollah Z, Vahid A (2013) Effect of pre-sowing salicylic acid seed treatment on seed germination and growth of greenhouse sweet pepper plants. Indian J Sci Technol 6:11

Hayat S, Maheshwari P, Wani AS, Irfan M, Alyemeni MN, Ahmad A (2012) Comparative effect of 28 homo-brassinolide and salicylic acid in the amelioration of NaCl stress in *Brassica juncea* L. Plant Physiol Biochem 53:61–68

Hetherington AM, Woodward FI (2003) The role of stomata in sensing and driving environmental change. Nature 424(6951):901–908

Horvath E, Szalai G, Janda T (2007) Induction of abiotic stress tolerance by salicylic acid signaling. J Plant Growth Regul 26(3):290–300

Ibrahim AH, Abdel-Fattah GM, Eman FM, Abd-El-Aziz MH, Shohr AE (2011) Arbuscular mycorrhizal fungi and spermine alleviate the adverse effects of salinity stress on electrolyte leakage and productivity of wheat plants. Phyton 51(2):261–268

Janda TG, Szalai K, Rios-Gonzalez O, Veise P (2003) Comparative study of frost tolerance and antioxidative activity in cereals. Plant Sci 164:301–306

Jin JB, Jin YH, Lee J, Miura K, Yoo CY, Kim WY, Van OM, Hyun Y, Somers DE, Lee I, Yun DJ, Bressan RA, Hasegawa PM (2008) The SUMO E3 ligase, AtSIZ1, regulates flowering by controlling a salicylic acid-mediated floral promotion pathway and through effects on FLC chromatin structure. Plant J 53(3):530–540

Jisha KC, Vijaykumari K, Puthur JT (2013) Seed priming for abiotic stress tolerance: an overview. Acta Physiol Plant 35:1381–1396

Kagale S, Divi UK, Krochko JE, Keller WA, Krishna P (2007) Brassinosteroid confers tolerance in *Arabidopsis thaliana* and *Brassica napus* to a range of abiotic stresses. Planta 225(2):353–364

Kang GZ, Li GZ, Liu GQ, Xu W, Peng XQ, Wang CY, Zhu YJ, Guo TC (2013) Exogenous salicylic acid enhances wheat drought tolerance by influence on the expression of genes related to ascorbate-glutathione cycle. Biol Plant 57(4):718–724

Karlidag H, Yildirim E, Turan M (2009) Salicylic acid ameliorates the adverse effect of salt stress on strawberry. Sci Agric 66(2):180–187

Khan W, Prithiviraj B, Smith DL (2003) Photosynthetic responses of corn and soybean to foliar application of salicylates. J Plant Physiol 160(5):485–492

Khan MIR, Iqbal N, Masood A, Per TS, Khan NA (2013) Salicylic acid alleviates adverse effects of heat stress on photosynthesis through changes in proline production and ethylene formation. Plant Signal Behav 8:e26374

Khan MIR, Asgher M, Khan NA (2014) Alleviation of salt-induced photosynthesis and growth inhibition by salicylic acid involves glycine betaine and ethylene in mungbean (*Vigna radiata* L.) Plant Physiol Biochem 80:67–74

Khan MIR, Fatma M, Per TS, Anjum NA, Khan NA (2015) Salicylic acid-induced abiotic stress tolerance and underlying mechanisms in plants. Front Plant Sci 6:462–472

Kim HS, Delaney TP (2002) Arabidopsis SON1 is an F-box protein that regulates a novel induced defense response independent of both salicylic acid and systemic acquired resistance. Plant Cell 14(7):1469–1482

Knorzer OC, Lederer B, DurnerJ BP (1999) Antioxidative defense activation in soybean cells. Physiol Plant 107:294–302

Kong J, Dong Y, Xu L, Liu S, Bai X (2014) Effects of foliar application of salicylic acid and nitric oxide in alleviating iron deficiency induced chlorosis of *Arachis hypogaea* L. Bot Stud 55(9)

Kumar P, Lakshmi NJ, Mani VP (2000) Interactive effects of salicylic acid and phytohormones on photosynthesis and grain yield of soybean (*Glycine max* L. Merrill). Physiol Mol Biol Plants 6:179–186

Kumar K, Madhuri K, Murugan V, Sakthivel K, Aanantharaj A, Singh AK, Roy SD (2014) Growth enhancement in vegetable crops by multifunctional resident plant growth promoting rhizobacteria under tropical island ecosystem. Afr J Microbiol Res 8(25):2436–2448

Larkindale J, Huang B (2004) Thermo tolerance and antioxidant systems in *Agrostis stolonifera*: involvement of salicylic acid, abscisic acid, calcium, hydrogen peroxide, and ethylene. J Plant Physiol 161:405–413

Larque-Saavedra A (1978) The anti-transpirant effect of acetyl salicylic acid on *Phaseolus vulgaris* L. Physiol Plant 43(12):6–128

Larque-Saavedra A (1979) Stomatal closure in response to acetylsalicylic acid treatment.Z Pflanzenphysiol 93(4):371–375

Lee S, Kim SG, Park CM (2010) Salicylic acid promotes seed germination under high salinity by modulating antioxidant activity in Arabidopsis. New Phytol 188(2):626–637

Mantelin S, Touraine B (2004) Plant growth-promoting bacteria and nitrate availability: impacts on root development and nitrate uptake. J Exp Bot 55(294):27–34

Metwally A, Finkemeier I, GeorgiM DKJ (2003) Salicylic acid alleviates the cadmium toxicity in barley seedlings. Plant Physiol 132(1):272–281

Mittler R (2002) Oxidative stress, antioxidants and stress tolerance. Trends Plant Sci 9:405–410

Miura K, Tada Y (2014) Regulation of water, salinity, and cold stress responses by salicylic acid. Front Plant Sci 23:5–14

Moharekar ST, Lokhande SD, Hara T, Tanaka R, Tanaka A, Chavan PD (2003) Effect of salicylic acid on chlorophyll and carotenoid contents of wheat and moong seedlings. Photosynthetica 41:315–317

Mutlu S, Atici Ö, Nalbantoglu B (2009) Effects of salicylic acid and salinity on apoplastic antioxidant enzymes in two wheat cultivars differing in salt tolerance. Biol Plant 53(2):334–338

Nagaraju A, Sudisha J, Murhty SM, Ito S (2012) Seed priming with *Trichoderma harzianum* isolates enhances plant growth and induces resistance against *Plasmopara halstedii*, an incident of sunflower downy mildew disease. Aus Plant Pathol 41:609–620

NIASM (2013) Crop water productions using line source sprinklers – reponses to bioregulators. Annual Reprot, NIASM, Baramati, pp. 27–29

Pancheva TV, Popova LP, Uzunova AM (1996) Effect of salicylic acid on growth and photosynthesis in barley plants. J Plant Physiol 149:57–63

Panda SK, Chaudhary I, Khan MH (2003a) Heavy metal induced lipid peroxidation and affects antioxidants in wheat leaves. Biol Plant 46:289–294

Panda SK, Sinha LB, Khan MH (2003b) Does aluminum phytotoxicity induce oxidative stress in green gram (*Vigna radiate*). Bulg J Plant Physiol 29:77–86

Pastor V, Luan E, Mauch-Mani B, Ton J, Flors V (2013) Primed plants do not forget. Environ Exp Bot 94:46–56

Rajjou L, Belghazi M, Huguet R, Robin C, Moreau A, Job C, Job D (2006) Proteomic investigation of the effect of salicylic acid on Arabidopsis seed germination and establishment of early defense mechanisms. Plant Physiol 141(3):910–923

Ratnakumar P, Singh Y, Kumar PS, Bhagat KP, Nangare DD, Taware PB, Minhas PS (2015) Ray of hope after hailstorm: a delightful boon by bioregulator – ortho silicic acid. ICAR Newsletter, pp 2–3

Ratnakumar P, Deokate PP, Rane J, Jain N, Kumar V, VandenBerhe D, Minhas PS (2016a) Effect of stabilized silicic acid foliar sprays on wheat under water stress conditions. J Func Environ Bot 6:17–19

Ratnakumar P, Minhas PS, Wakchaure GC, Chaudary RL, Deokate PP (2016b) Yield and water production functions of wheat (*Triticum aestivum* L.) cultivars and response to exogenous application of thiourea and ortho-silicic acid. Int J Agric Environ Res 2(6):1228–1250

Ratnakumar P, Govindasamy V, Minhas PS (2017) Yield enhancement through bio-fertilizer priming and effect of bio-regulators in soybean under soil moisture stress. Inter-drought-V Conference proceeding book. (In press)

Rivas-San Vicente M, Plasencia J (2011) Salicylic acid beyond defence: its role in plant growth and development. J Exp Bot 62(10):3321–3338

Robison MM, Shah S, Tamot B, Pauls KP, Moffatt BA, Glick BR (2001) Reduced symptoms of Verticillium wilt in transgenic tomato expressing a bacterial ACC deaminase. Mol Plant Pathol 2(3):135–145

Ruiz-Lozano JM, Azcon R (2000) Symbiotic efficiency and infectivity of an autochthonous arbuscular mycorrhizal Glomus sp. from saline soils and *Glomus deserticola* under salinity. Mycorrhiza 10(3):137–143

Sahu MP, Singh D (1995) Role of thiourea in improving productivity of wheat. J Plant Growth Regul 14:169–173

Sahu MP, Solanki NS (1991) Role of sulfhydryl compounds in improving dry matter partitioning and grain production of maize. J Agron Crop Sci 167:356–359

Sahu MP, Solanki NS, Dashora LN (1993) Effects of thiourea, thiamine and ascorbic acid on growth and yield of maize (*Zea mays* L.) J Agron Crop Sci 171:65–69

Sahu MP, Kumawat SM, D'Souza SF, Ramaswamy NK, Singh G (2005) Sulphydryl bioregulator technology for increasing mustard production. Research Bulletin RAU-BARC, Bikaner, pp. 52

Sahu MP, Kumawat SM, Ramaswamy NK, D'Souza SF (2006) Sulphydryl bioregulator technology for increasing wheat productivity. Research bulletin RAU-BARC:1–56

Sakhabutdinova AR, Fatkhudinova DR, Shakirova FM (2004) Effect of salicylic acid on the activity of antioxidant enzymes in wheat under conditions of salination. Appl Biochem Microbiol 40(5):501–505

Senaratna T, Touchell D, Bunn E, Dixon K (2000) Acetyl salicylic acid (aspirin) and salicylic acid induce multiple stress tolerance in bean and tomato plants. Plant Growth Regul 30(2):157–161

Shakirova FM, Sakhabutdinova AR, Bezrukova MV, Fatkhudinova RA, Fathudinova DR (2003) Changes in hormonal status of wheat seedlings induced by salicylic acid and salinity. Plant Sci 164:317–322

Sheng M, Tang M, Chen H, Yang B, Zhang F, Huang Y (2009) Influence of arbuscular mycorrhizae on the root system of maize plants under salt stress. Can J Microbiol 55(7):879–886

Shulaev V, Leon J, Raskin I (1995) Is salicylic acid a translocated signal of systemic acquired resistance in tobacco? Plant Cell 7(10):1691–1701

Singh B, Usha K (2003) Salicylic acid induced physiological and biochemical changes in wheat seedlings under water stress. Plant Growth Regul 39:137–141

Smirnoff N (1993) The role of active oxygen in response of plants to water deficit and desiccation. New Phytol 125:27–58

Srivastava AK, Ramaswamy NK, Mukhopadhayaya R, Jincy MGC, D'Souza SF (2009) Thiourea modulates the expression and activity profile of mtATPase under salinity stress in seed of *Brassica juncea*. Ann Bot 103:403–410

Srivastava AK, Ratnakumar P, Minhas PS, Suprasanna P (2016) Plant bioregulators for sustainable agriculture; integrating redox signaling as a possible unifying mechanism. Adv Agron 137:238–278

Stamm P, Ramamoorthy R, Kumar PP (2011) Feeding the extra billions: strategies to improve crops and enhance future food security. Plant Biotechnol Rep 5(2):107–120

Steenhoudt O, Vander-Leyden J (2000) Azospirillum, a free-living nitrogen-fixing bacterium closely associated with grasses: genetic, biochemical and ecological aspects. FEMS Microbiol Rev 24(4):487–506

Tari I, Csiszar J, Szalai G, Horvath F, Pecsvaradi A, Kiss GY, Szepesi A, Szabo M, Erdei L (2002) Acclimation of tomato plants to salinity stress after a salicylic acid pre-treatment. Acta Biol Szegediensis 46:55–56

Tari I, Simon LM, Deer KA, Csiszar J, Bajkan SZ, Kiss GY, Szepesi A (2004) Influence of salicylic acid on salt stress acclimation of tomato plants: oxidative stress responses and osmotic adaptation. Acta Physiol Plant 26S:237–241

Vlot AC, Dempsey DMA, Klessig DF (2009) Salicylic acid, a multifaceted hormone to combat disease. Annu Rev Phytopathol 47:177–206

Vriet C, Russinova E, Reuzeau C (2012) Boosting crop yields with plant steroids. Plant Cell 24:842–857

Wahid A, Shabbir A (2005) Induction of heat stress tolerance in barley seedlings by pre-sowing seed treatment with glycine betaine. Plant Growth Regul 46:133–141

Wahid A, Parveen M, GelaniS BSMA (2007) Pretreatment of seeds with H2O2 improves salt tolerance of wheat seedling by alleviation of oxidative damage and expression of stress proteins. J Plant Physiol 164:283–294

Wakchaure GC, Minhas PS, Ratnakumar P, Chaudhary RL (2016a) Optimizing supplemental irrigation for wheat (*Triticum aestivum* L.) and the impact of plant bio-regulators in a semi-arid region of Deccan plateau in India. Agric Water Manag 172:9–17

Wakchaure GC, Minhas PS, Ratnakumar P, Chaudhary RL (2016b) Effect of plant bioregulators on growth, yield and water production functions of sorghum [*Sorghum bicolor* (L.) Moench]. Agric Water Manag 177:138–145

Wakchaure GC, Minahas PS, Ratanakumar P, Chaudhary RL (2016c) Plant bioregulators for raising crop and water productivity. ICAR News 22(4):1–2

Wang ZY, Bai MY, Oh E, Zhu JY (2012) Brassinosteroid signaling network and regulation of photo morphogenesis. Annu Rev Genet 46:701–724

Werdan K, Heldt HW, Milovancev M (1975) The role of pH in the regulation of carbon fixation in the chloroplast stroma: studies on CO_2 fixation in the light and dark. Biochem Biophys Acta 396:276–282

Wilkinson SA, Kudoyarova GR, Veselov DM, Arkhipova TN, Davies WJ (2012) Plant hormone interactions: innovative targets for crop breeding and management. J Exp Bot 63:3499–3509

Wu QS, Zou YN, He XH (2010) Contributions of arbuscular mycorrhizal fungi to growth, photosynthesis, root morphology and ionic balance of citrus seedlings under salt stress. Acta Physiol Plant 32(2):297–304

Yasmeen A, Basra SMA, Farooq M, Urrehman H, Hussain N (2013) Exogenous application of moringa leaf extract modulates the antioxidant enzyme system to improve wheat performance under saline conditions. Plant Growth Regul 69(3):225–233

Yusuf M, Hasan SA, Ali B, Hayat S, Fariduddin Q, Ahmad A (2008) Effect of salicylic acid on salinity induced changes in *Brassica juncea*. J Integr Plant Biol 50(8):1–4

Zeller G, Henz SR, Widmer CK, Sachsenberg T, Ratsch G, Weighel D, Laubinger S (2009) Stress-induced changes in the *Arabidopsis thaliana* transcriptome analyzed using whole-genome tiling arrays. Plant J 58(6):1068–1082

Zhou ZS, Guo K, Elbaz AA, Yang ZM (2009) Salicylic acid alleviates mercury toxicity by preventing oxidative stress in roots of *Medicago sativa*. Environ Exp Bot 65(1):27–34

Zottini M, Costa A, De-Michele R, Ruzzene M, Carimi F, Schiavo FL (2007) Salicylic acid activates nitric oxide synthesis in Arabidopsis. J Exp Bot 58(6):1397–1405

M.P. Sahu

Abstract

Abiotic stresses, viz. drought, high temperature and salinity, are the major constraints in enhancing agricultural productivity. The predicted climate change is likely to further aggravate these environmental stresses. There has been unprecedented increase in the frequency and intensity of extreme events such as drought, heavy rainfall, flooding and high temperatures. Though sporadic in nature, these extreme events are expected to expand covering many regions. Hence for sustainable agriculture under these conditions, it is necessary to have crop plants endowed with mechanisms to cope with environmental stresses. Plants cope with these stresses and thrive under harsh environment through sophisticated and efficient mechanisms to re-establish and maintain redox homeostasis. These mechanisms involve efficient stress sensing and signaling processes. In addition, stresses lead to a vital reorientation of transport and partitioning of assimilate for plant growth and productivity under environmental stresses. Thiourea, a sulphydryl compound, has been observed to improve assimilate partitioning, photosynthetic efficiency and crop productivity. Since thiourea has redox regulatory properties imparted by –SH group, it can influence thiol-disulphide cycle through effects on plant thioredoxin system, which has a key role in plant tolerance to abiotic stresses. In several field studies, thiourea has been observed to mitigate drought, salinity and heat stress, and it enhances the crop yields when applied through foliar sprays. Thus, it acts as an effective bioregulater by regulation of cell metabolic activities and by restoring cellular redox homeostasis in crop plants under stressful environments. Hence, thiourea has considerable potential for achieving food security under the changing scenario of climate change.

M.P. Sahu (✉)
Swami Keshwanand Rajasthan Agricultural University, Bikaner 334006, Rajasthan, India
e-mail: mpsdean@gmail.com

© Springer Nature Singapore Pte Ltd. 2017
P.S. Minhas et al. (eds.), *Abiotic Stress Management for Resilient Agriculture*,
DOI 10.1007/978-981-10-5744-1_11

11.1   Introduction

More than 50% yield losses in agricultural crops are attributed to abiotic stresses (Wang et al. 2007). Among them, the losses due to high temperature, salinity, drought and low temperature can be as high as 40, 20, 17 and 15%, respectively (Ashraf et al. 2008). Climate change can aggravate these environmental stresses and can substantially reduce global crop yields as reported in the case of maize (3.8%) and wheat (5.5%) (Lobell et al. 2011). This can be partially attributed to steep decrease in crop productivity when temperatures exceed critical physiological thresholds (Battisti and Naylor 2009; Wheeler et al. 2000). According to the Intergovernmental Panel on Climate Change (IPCC), climate change affects crop production in several regions of the world, with negative effects more common than positive, and developing countries are highly vulnerable to further negative impacts (IPCC 2014). Increases in the frequency and intensity of extreme events such as drought, heavy rainfall, flooding and high maximum temperatures are already occurring and expected to accelerate in many regions (Porter et al. 2014). Despite these challenges, the very low agricultural productivity of food-insecure countries presents a great opportunity (Brown and Funk 2008). There is wide geographic variation in crop productivity, even across regions that experience similar climates (Godfray et al. 2010). Even under irrigated conditions of Southeast Asia where the average maximum climate-adjusted rice yields are 8.5 Mg ha^{-1}, the actually achieved yield is just 60% of this (Cassman 1999). Similar yield gaps exist in rain-fed wheat in central Asia and rainfed cereals in Argentina and Brazil.

11.2   Overcoming Productivity Constraints

The crop productivity can be improved by three approaches that involve photosynthesis, the key process responsible for plant growth and development. These include enhancing rate of canopy photosynthesis, efficiency of photosynthetic carbon fixation and translocation of photosynthate for economic yield (Geiger and Giaquinta 1982). All these three factors are closely linked to harvestable yield (Gifford et al. 1984). Thus, while attempting crop yield enhancement, it is critical to assess if economic yield can be enhanced through improved efficiency in use of photosynthate available to the plant or by improving net photosynthate supply to meet the demand (Wardlaw 1980). The decision can be difficult due to dynamic nature of these processes and interlinks between them as the demand for assimilates can directly influence net assimilation rate and translocation patterns (Wardlaw 1968). In cereals most of the carbohydrate in the grain emanates from post-anthesis photosynthesis (Barnell 1936; Carr and Wardlaw 1965; Lupton 1966).

Either the photosynthesis or the grain is often considered as limitations to yield in cereals, while the assimilate translocation from photosynthesizing part to grain was initially neglected (Evans and Wardlaw 1976). The translocation system has the transport and the partitioning of assimilates as its major components, and both of them have undergone changes during the evolution of cereals. Further, the vascular system which facilitates transport is unlikely to be a major limitation to yield in cereals. Instead, the limitation in the diffusion of assimilates into the phloem at the source (leaves) as well as from the phloem of inflorescence into the grains can be a more significant limitation (Evans and Wardlaw 1976; Gifford and Evans 1981).

In general, plants are well adapted to diverse environmental conditions. However, their growth and productivity is adversely affected by stresses imposed by salinity, drought, cold or heat, which are getting amplified due to the ongoing climate changes. Plants cope with these stresses by maintaining ion and cellular homeostasis (Conde et al. 2011). Cellular homeostasis is maintained by sophisticated mechanisms for signaling and sensing the stress, minimizing the toxic elements generated in the cell, compatible solute and osmoprotectant accumulation and a reorientation of transport and compartmentation of solutes.

11.3 The Available Yield Optimization Technology

Optimum crop productivity under environmental stress depends upon a delicate balance between photosynthesis and photosynthate partitioning when a particular stress leads to the physiological dysfunction in crops. In terms of carbon economy, the route to increased crop yield lies between carbon source (photosynthesis) and the carbon sink (economic product); hence both these components need to be improved simultaneously. High-yielding varieties and good management practices can maximize the interception of solar radiation by crop canopy. They can also help in prioritizing excessive investment of assimilates into reproductive parts rather than vegetative parts. Improving crop photosynthesis rate prior to grain filling by means of environmental amelioration can be the most effective strategy for increasing yield (by increasing the number of grains competent to fill). Increasing sink size (grain number per ha) by chemical or breeding methods can be another approach to achieve higher yields of seed crops. This can be achieved through reducing abscission or abortion of flowers and young seeds without compromising the ability to fill these extra sinks (Gifford et al. 1984).

While in irrigated agriculture increased sink size (grain number per ha) has been brought about by dwarf stature of the Mexican wheat and other cereals, in rainfed lands, this has not been achieved successfully. Therefore, it is now widely realized that the Green Revolution has bypassed the drylands, more particularly the arid lands in India. Low crop productivity and sometimes crop failure due to environmental stresses are the common occurrences in arid lands. Therefore, molecular biology-based plant breeding or molecular agronomy can alone break the yield barrier of crops in arid environments.

11.4 Role of Assimilate Transport and Partitioning

The ability of plants to accumulate assimilated carbon is a function of its photosynthetic capacity and the pattern of carbon allocation among its parts, both of which appear to be genetically determined (Gifford et al. 1984). In nature this capacity also depends on the plant's ability to maintain dry matter accumulation under stresses imposed by a variety of environmental factors (Boyer 1982). These stresses do not allow plants to realize their genetic potential for productivity. This is evident from the fact that agricultural yields are only 12–30% of record yields for a variety of crops under field conditions. Under water stress, for example, direct responses such as stomata closure initially may inhibit photosynthesis. Subsequent responses, mediated by altered gene expressions, can result in changes in allocation of newly fixed or stored carbon, synthesis of osmotic agents and expansion of root, which ultimately enhances uptake of water, reopens stomata and restores the previous rate of photosynthesis. Phloem translocation provides communication from sources to sinks, while xylem transport can complete the loop, carrying materials such as growth regulators from sinks to sources (Geiger and Servaites 1991).

In arid areas, despite constraints imposed by lack of water and high temperatures, the crop plants mostly survive in average rainfall years. But the yields are seldom high even though the biomass production is of a satisfactory level. The major limitation is inefficient assimilate partitioning in crop plants at grain filling because of high aridity and radiation load. Most probably the sucrose transport proteins are rendered weak and may get even inactivated due to the formation of free radicals like superoxides under environmental stress conditions. The situation becomes still worse when rainfall is below average and the crops mostly fail, producing only some biomass and almost no yield.

The allocation and partitioning of assimilated carbon provide resources for acclimation to environmental stress. Allocation comprises the processes determining the biochemical fate of carbon that has become available for distribution, or partitioning, among plant parts. Without other specifications, allocation refers to the initial determination of recently assimilated carbon. As a result of both biochemical conversions and compartmentation, newly assimilated carbon may be used immediately or set aside as a reserve to be mobilized and used later. Mobilization of reserves that were previously allocated to storage can also supply carbon needed for stress responses. Plants have mechanisms for appropriate allocation of current photosynthates for internal translocation and for storage in plant tissues. These mechanisms enable plants to maintain a steady supply of carbon for development in addition to restoring and maintaining homeostasis under environmental stress (Geiger and Servaites 1991).

11.5 Role of Sulphydryl Compounds in Assimilate Transport and Partitioning

The mechanism of assimilate transport and partitioning in crop plants is still poorly understood. However, several evidences suggest that the loading of sucrose into phloem vessels may be a rate-limiting step in assimilate transport. Since the translocation of ^{14}C–sucrose in leaf discs of sugar beet was inhibited by sulphydryl blocker, para-chloromercuribenzoic sulphonic acid (PCMBS), it was concluded by Giaquinta (1976) that sulphydryl groups are necessary for the activity of sucrose transport protein. These compounds may not enhance the stover and biological yields in pearl millet but can enhance harvest index and hence grain yield as evident from responses to foliar sprays of sulphydryl compounds such as mercaptoethanol and thiourea (Sharma 1988). These observations were later confirmed with maize where mercaptoethanol, mercaptoethylamine and thiourea enhanced the grain yield by 18.1, 29.2 and 34.1%, respectively (Sahu and Solanki 1991). This was mainly due to the respective increase in harvest index to an extent of 6.2, 9.9 and 13.9%. Further evidence for the effects of sulphydryl compounds on translocation of sucrose from the source to sink in mustard was provided on the basis of the radio-isotopic studies (Srivastava et al. 2008). Thus, the sulphydryl compounds improve translocation of photosynthate and dry matter partitioning by improving phloem loading of sucrose (Sahu and Singh, 1995). The role of thiourea in the improvement of plant growth and grain yield of maize was further confirmed by studies conducted by Sahu et al. (1993) where seed soaking plus foliar spray of thiourea significantly improved leaf area index and number of green leaves per plant which ultimately led to increase in both biological and grain yields. Improvement in maize yield with thiourea treatments appeared to have resulted from increased photosynthetic efficiency and delayed leaf senescence possibly due to metabolic activities of cells supported by -SH group.

11.6 How Thiourea Imparts Crop Tolerance to Environmental Stresses

Thiourea is a thiol compound (Gyorgy et al. 1943; Jocelyn 1972), and there are ample evidences to support the role of thiol compounds in stress tolerance of plants. A variety of protein and nonprotein thiols and sulphur-containing molecules contribute to plant stress tolerance (Meyer and Hell 2005; Colville and Kranner 2010). The potential of thiol compounds in mitigation of water stress in pearl millet is evident from experiments involving presowing seed treatment with sulphydryl compounds such as dithiothreitol, thioglycollic acid, thiourea and cysteine. These compounds could improve the activities of key enzymes of superoxide scavenging system that involve superoxide dismutase, glutathione reductase and glutathione-S-transferase (Ramaswamy et al. 2007). In wheat, these compounds could minimize the effect of osmotic stress induced by polyethylene glycol on photosystem 1 and 2 activities (Nathawat et al. 2007). Beneficial effect of these compounds on primary photochemistry was evident in water-stressed pearl millet even when they were applied through foliar spray (D'Souza et al. 2009).

In many living organisms, thioredoxins (TRX) and glutaredoxins (GRX) are the main components of protein-based protection mechanisms that involve thiol-disulphide cycle. Thiourea being a redox regulatory molecule most probably influences thiol-disulphide cycle (Meyer et al. 2012). TRX and GRX are thiol oxidoreductases with a pair of cysteine residues that provide reducing power to a variety of stress-related enzymes such as thiol peroxidase, methionine sulfoxide reductase (MSR) and arsenate reductases. In addition, they may facilitate thiol-disulphide oxidoreductase activity to various protein targets. GRX are directly reduced by GSH to produce GSSG; however, the reduction of TRX requires reductases specific to compartments within the cell. In cytosol and in mitochondria, TRX is reduced by NADPH-dependent TRX reductases (Lemaire et al. 2007), but in plastid, ferredoxin/thioredoxin reductase (Schurmann and Buchanan, 2008) plays a key role.

On exposure to abiotic stresses, TRX increases either due to the high level of gene expression or the increase in activities at protein level (Zagorchev et al. 2013). In rice, Cu stress can lead to upregulation of *TRX* and *GRX* genes (Song et al. 2012). The existence of genetic variation was evident from difference in the fold increase between tolerant and sensitive lines. Similarly, GRX level was more in salt-resistant than in salt-sensitive barley genotypes (Fatehi et al. 2012). A significant genetic variation in the TRX gene expression was also observed in a genome-wide study in conditions of biotic and abiotic stress (Nuruzzaman et al. 2012). Stress tolerance through chaperone-like functions may be conferred by proteins related to TRX or GRX. For example, thioredoxin reductase type C (NTRC) in *Arabidopsis* can undergo heat shock-induced oligomerization and can exhibit chaperone functions to confer thermotolerance (Chae et al. 2013). NTRC is a typical example of a protective protein with numerous functions and roles, depending on the reduction of protein disulphide bonds or by extending chaperone functions to non-cysteine proteins (Chae et al. 2013). Thiourea has a role in the reduction of protein disulphide bonds as demonstrated by thiol-disulphide exchange occurring between thiourea and a disulphide bond in thyroid tissue (Maloof and Soodak 1961). Thiourea-based bifunctional organocatalysts are also capable of activating electrophilic reaction components through hydrogen bond donation that mimics a reductase activity (Procuranti and Connon 2007). Thus, it is highly probable that thiourea functions like thioredoxin reductase type C (NTRC) and regulates protein function and redox metabolism in plants under environmental stresses. In view of the above, if antioxidants are sprayed onto the crop plants, most of the damaging free radicals can be quenched, and the crop plants can be able to maintain metabolic homeostasis, which will then facilitate translocation and partitioning of assimilates for yield formation. Since sulphydryl compounds are strong antioxidants and also supply reactive sulphydryl group for the functioning of sucrose transport protein, they can more effectively improve assimilate partitioning and yield of crops under arid environment.

11.7 Effects of Thiourea on Plant Metabolism Under Environmental Stresses

11.7.1 Drought Stress

As a primary response to water stress, the plant hormone abscisic acid (ABA) accumulates in large quantities in plants (Wright and Hiron 1969). ABA accumulation is significantly inhibited by thiol compounds such as dithiothreitol (DTT) and cysteine (Cys) (Jia and Zhang 2000). Wheat originated from seed priming with salicylic acid combined with foliar application of thiourea exhibited stronger anti-drought effects, in terms of increased growth parameters, yield components, total soluble sugars, total carbohydrates, protein and improved nutritional contents of grains including P, K, Ca and Mg (Hassanein et al. 2012). Later, Hassanein et al. (2014–2015) further observed that treating wheat plants with TU, SA and their combination resulted in great increases in the activity of antioxidative system-associated enzymes such as superoxide dismutase (SOD) and catalase (CAT) accompanied by great reduction in peroxidase (POX) and ascorbate peroxidase (APX) activities. Maximum increase of SOD and CAT was observed by spraying with TU at 2.5 mM on SA-pretreated plants either under normal irrigation (44 and 32%, respectively) or drought conditions (57.8 and 47.8%, respectively). Thus, the improvement in wheat performance was assigned to enhancing antioxidant compounds (phenolics and flavonoids), membrane stability, and antioxidant enzymes (SOD and CAT) and by reducing hydrogen peroxide free radical product.

11.7.2 Salinity Stress

Seed soaking treatments with chemicals (salicylic acid, sodium thiosulphate and thiourea) could alleviate the salt stress (10 dS m^{-1}) and accelerate recovery of plant growth rate (Mahatma et al. 2009). This could be attributed to the decrease in lipid peroxidation and free amino acid content. Srivastava et al. (2010) reported that thiourea could alleviate stress even under a high degree of salinity through different signaling and effector mechanisms at an early stage. In mustard plant, TU supplementation to NaCl brought down levels of reactive oxygen species (ROS) to near control values as compared to that of NaCl stress. These positive effects could be related to the increase in the 1,1-diphenyl-2-picrylhydrazyl (DPPH) radical scavenging activity, reduced glutathione (GSH) and their ratio of reduced to oxidized glutathione (GSH/GSSG) and also in the activities of SOD and GR. These responses under salinity stress were accompanied by an increase in utilization of energy as indicated by ATP/ADP ratio and enhanced activity of ascorbate oxidase (AO), an important component of stress signaling (Srivastava et al. 2011). Thus, TU treatment was found to regulate redox and antioxidant mechanisms in mustard to alleviate oxidative stress induced by NaCl. While NaCl at 200 mM can inhibit the activity of amylase enzyme in germinating seeds (Sangeetha 2013), thiourea (20 mM) could minimize these inhibitory effects.

Salinity stress induced by 100 mM NaCl can substantially reduce accumulation of biomass, chlorophyll content and relative water content in the maize plants (Kaya et al. 2013). However, it increases the activities of CAT, SOD and polyphenol oxidase (PPO) and levels of hydrogen peroxide and electrolyte leakage, all associated with antioxidative system. However, it did not affect peroxidase (POD) activity. It was possible to reduce this adverse effect by foliar application of TU in case of salt-stressed maize compared to untreated maize plants. Thiourea (7 mM) was as effective as mannitol (30 mM) in alleviating salinity stress in maize plants as indicated by growth and physiological attributes. Treatment of seeds before sowing and foliar application of thiourea and nitric oxide in combination could alleviate salt-stressed maize seedlings (Kaya et al. 2015a). Foliar application of TU (500 ppm) resulted in considerable increases in the dry weight of salt-sensitive and salt-tolerant cultivars to an extent of 38% and 35%, respectively, relative to control (Kaya et al. 2015b). It partially improved the salt tolerance of maize plants by reducing Na and increasing N, K, Ca and P under salinity stress. In addition to this, the reduction in MP, MDA and H_2O_2 altered activities of antioxidant enzymes and increased photosynthetic pigments in response to TU and hence improved the growth of maize under saline stress (Kaya et al. 2015b).

11.7.3 Heat Stress

Heat stress during grain filling injures cellular membrane due to accelerated biochemical activity at cellular level which leads to accumulation of high levels of lipid peroxide and H_2O_2 contents (Asthir et al. 2013). However, heat-tolerant genotypes such as C 306 and C 273 exhibit less lipid peroxidation and membrane injury due to substantially high levels of antioxidants. Thiourea ameliorates the damages caused by high temperature by facilitating increase in antioxidant activity, total soluble proteins, amino acids and chlorophyll contents in all the genotypes. Impact of these biochemical events was reflected in substantial increase in plant height, peduncle length, peduncle weight and grain weight. The effect was so remarkable that heat-sensitive genotypes of wheat such as PBW 550 and PBW 343 exhibited higher grain weight despite greater injury to membranes relative to tolerant genotypes such as C 306 and C 273. Further, it was observed that application of thiourea twice as pre-sowing seed treatment and as foliar spray was more effective in improving the wheat performance as revealed by membrane stability, antioxidant potential and yield components recorded in the experiment (Asthir et al. 2013). High temperature (HT) significantly increases the activities of soluble acid, neutral and cell wall-bound invertases in addition to cellular level of reducing sugar and soluble protein, in both root and shoot of wheat (Asthir et al. 2015). However, the application of thiourea under HT stress further elevates invertase activities and levels of soluble protein because of the effects of -SH group that allows regulation of cellular carbon metabolism (Asthir et al. 2015).

Sunflower plants exposed to 35 or 45 °C for 12 h had reduced growth, chlorophyll content in leaves, relative leaf water content, oil content, leaf nutrient and nitrate reductase enzyme activity (Akladious 2014). Treatment with thiourea applied at 10 mM could improve plant growth by maintaining favourable levels of parameters mentioned above. It also induced non-enzymatic and enzymatic antioxidants. The studies at molecular and protein level have also confirmed the efficacy of thiourea in minimizing heat injury in sunflower plants (Akladious 2014). Similarly, foliar spray of 6 mM thiourea could help in the recovery of root to shoot ratio in terms of length and dry biomass in the heat-stressed maize (Khanna et al. 2015). As observed in other crops, thiourea specifically upregulated the activities of catalase and peroxidase in the roots and shoots and effectively reduced cellular level of H_2O_2.

11.8 Crop Response to Sulphydryl Bioregulator Thiourea

The role of thiourea in mitigating environmental stress has been ascribed to the bioregulatory role of its sulphydryl group configuration. There was remarkable increase in yield components and grain yield in response to foliar spray of thiourea (Ameta and Singh 2005). Thiourea at 0.2% spray at grain filling stage was most effective, resulting in higher number of grains/cob (24.7%) with increased grain weight by 43.3% over control, thus increasing the grain yield by 26.7%. Similarly, seed yield and dry matter production of mung bean was enhanced by presowing seed treatment with thiourea (500 ppm) followed by foliar application (1000 ppm) prior to flowering stage (Mathur et al. 2006). Thiourea leads to the increase in net photosynthesis and levels of many metabolites that promote plant growth and development. Thiourea could enhance the grain yield of moth bean by 89.2% (Sahu et al. 2006), while grain yield could be almost doubled in cluster bean. Thiourea application also consistently increased cowpea productivity to an extent of 26% under rainfed conditions (Anitha et al. 2004; Burman et al. 2004, 2007; Garg et al. 2006). However, maximum favourable effects were obtained with combined application of seed treatment and foliar spray as seen in other crops. All these studies provide sufficient evidences to support the view that the beneficial effects of thiourea could be due to its beneficial role in photosynthetic activity as well as nitrogen metabolism in addition to its important role in antioxidative mechanisms.

Under soil moisture deficit conditions, it was possible to get 68 and 373 kg of additional grain yield of rainfed pearl millet per ha with spray of 6% kaolin and dust mulching, respectively. On the other hand, 0.1% thiourea when sprayed at tillering and flowering resulted in an increase of grain yield by 26.6% (Sahu et al. 2006). In other experiments, where the highest productivity of pearl millet was obtained by minimising the rows of plants with dead furrow of 60 cm in between each "pair of rows" (Siddiqui et al. 2014), 0.1% thiourea spray on leaf canopy increased the grain yield by 10.7% over control regardless of crop stages. Grain filling in wheat is more sensitive to salinity than heat. Both of these stresses could be managed with foliar application of thiourea (10 mM) which improved net photosynthesis and grain yield (Anjum et al. 2008). In another study (Anjum et al. 2011), thiourea enhances the threshold of tolerance of wheat plants to salinity and to high temperature. Among the five varieties examined, thiourea could effectively enhance tolerance to high temperature and salinity in highly sensitive (S-24 and MH-97) varieties. These experiments provide sufficient evidences to conclude that it is possible to mitigate the adverse effects of the most important abiotic stresses such as salinity and high temperature with foliar spray of thiourea. Hence, this may be recommended for managing moderate salinity and high temperature stresses that otherwise limit wheat production. This can also be used for seed treatment to get the desired benefit (Abdelkader et al. 2012) as it facilitates antioxidant system in plants.

11.9   Way Forward

The ideal conditions for improving crop yield result from the optimization of all metabolic events at cellular level through macro- to micromanagement of the crop growth environment (Rossi et al. 2015). This needs optimization of rates and duration of all important processes, which are generally determined by genotype (G) but often largely influenced by the environment (GxE). Implicitly, the crop management (M) that influences microenvironment for crop growth must be rationally included in the yield equation: yield = GxExM. Plants have inbuilt mechanisms to counter the adverse effect of harsh environment. They are endowed with reducing-oxidizing (redox) systems in which catabolic and anabolic processes are driven by oxidative and reductive processes, respectively (Hong-bo et al. 2008). Since oxygen is the key element in such mechanism, its redox potential determines the scope for maintenance and regulation of metabolic functions. Under extreme environmental conditions, the pace with which plant metabolism scavenges harmful free radicals determines the extent of stress-induced damage (Noctor 2006; Bohnert and Jensen 1996; Foyer et al. 1994, 1997). ROS are key components contributing to cellular redox status through their participation in all processes controlled by redox reactions such as signal transduction, gene expression, protein synthesis and turnover and thiol-disulphide exchange reactions (Chaves et al. 2003; Shao et al. 2006, 2007a, b). ROS accumulation is closely linked to initialization of pre-emptive defence and adaptive responses that can lead either to stress acclimation or to cell death, depending on the degree of oxidative stress experienced. For example, responses of wheat to stresses are directed to acclimate and repair damage, which also occur in many organisms (Foyer and Nector 2005a, b; Miller et al. 2003). Redox signal transduction is a universal characteristic of aerobic life (Shao et al. 2005, 2008), and both oxidants and antioxidants fulfil signaling roles. Since kinase-dependent and kinase-independent pathways that are initiated by redox-sensitive receptors modulate thiol status (Hon-bo et al. 2008), thiol bioregulators are important agents for metabolic management and restoring cellular redox homeostasis in crop plants, both under optimum and stressful environments. Viewed in this context, thiourea, a thiol bioregulator, can be considered a novel crop management technology in the equation (Y = GxExM) mentioned above. It may be noted here that environmental stresses induced by climate change alter source-sink partitioning of assimilates as well as photosynthetic efficiency in crop plants (Paul and Foyer 2001; Nathawat et al. 2007). Thiourea, a sulphydryl compound, has been reported to improve assimilate partitioning and crop photosynthetic efficiency in maize (Sahu and Solanki 1991, Sahu et al. 1993). Recently, the role of thiourea as a ROS scavenger has been highlighted in regulating source-sink relationships and improving yield in Indian mustard (Pandey et al. 2013). In light of these findings, thiourea bioregulator holds considerable promise for achieving food security under the changing scenario of climate change in the world.

References

Abdelkader AF, Hassanein RA, Ali H (2012) Studies on effects of salicylic acid and thiourea on biochemical activities and yield production in wheat (*Triticum aestivum* var Gimaza 9) plants grown under drought stress. Afr J Biotechnol 11:12728–12739

Akladious SA (2014) Influence of thiourea application on some physiological and molecular criteria of sunflower (*Helianthus annuus* L.) plants under conditions of heat stress. Protoplasma 251:625–638

Ameta GS, Singh M (2005) Studies on use of thiourea spray for yield advantage in rainfed maize (*Zea mays* L.) Int J Trop Agric 23:307–310

Anitha S, Sreenivasan E, Purushothaman SM (2004) Effect of thiourea application on cowpea [*Vigna unguiculata* (L.) walp] productivity under rainfed conditions. J Trop Agric 42:53–54

Anjum F, Wahid A, Javed F, Arshad M (2008) Influence of foliar applied thiourea on flag leaf gas exchange and yield parameters of bread wheat (*Triticum aestivum*) cultivars under salinity and heat stresses. Int J Agric Biol 10:619–626

Anjum F, Wahid A, Farooq M, Javed F (2011) Potential of foliar applied thiourea in improving salt and high temperature. Int J Agric Biol 13:251–256

Ashraf M, Arhar HR, Harris PJC, Kwon TR (2008) Some perspective strategies for improving crop salt tolerance. Adv Agron 97:45–11

Asthir B, Thapar R, Farooq M, Bains NS (2013) Exogenous application of thiourea improves the performance of late sown wheat by inducing terminal heat resistance. Int J Agric Biol 15:1337–1342

Asthir B, Kaur R, Bains NS (2015) Variation of invertase in four wheat cultivars as influenced by thiourea and high temperature. Acta Physiol Plant 37:1712–1720

Barnell HR (1936) Seasonal changes in the carbohydrates of the wheat plant. New Phytol 35:229–266

Battisti DS, Naylor RL (2009) Historical warnings of food insecurity with unprecedented seasonal heat. Science 323:240–244

Bohnert HJ, Jensen RG (1996) Strategies for engineering water-stress tolerance in plants. Trends Biotechnol 14:89–97

Boyer JS (1982) Plant productivity and environment. Science 218:443–448

Brown ME, Funk CC (2008) Food security under climate change. Science 319:580–581

Burman U, Garg BK, Kathju S (2004) Interactive effects of thiourea and phosphorus on cluster-bean under water stress. Biol Plant 48:61–65

Burman U, Garg BK, Kathju S (2007) Interactive effects of phosphorus, nitrogen, and thiourea on clusterbean (*Cyamopsis tetragonoloba* L.) under rainfed conditions of the Indian arid zone. J Plant Nutr Soil Sci 170:803–810

Carr DJ, Wardlaw IF (1965) The supply of photosynthetic assimilates to the grain from the flag leaf and ear of wheat. Aust J Biol Sci 18:711–719

Cassman KG (1999) Ecological intensification of cereal production systems: yield potential, soil quality, and precision agriculture. Proc Natl Acad Sci U S A 96(11):5952–5959

Chae HB, Moon JC, Shin MR, Chi YH, Jung YJ, Lee SY, Nawkar GM, Jung HS, Hyun JK, Kim WY et al (2013) Thioredoxin reductase type C (NTRC) orchestrates enhanced thermo-tolerance to *Arabidopsis* by its redox-dependent holdase chaperone function. Mol Plant 6:323–336

Chaves MM, Maroco J, Pereira J (2003) Understanding plant responses to drought—from genes to the whole plant. Funct Plant Biol 30:239–264

Colville L, Kranner L (2010) Desiccation tolerant plants as model systems to study redox regulation of protein thiols. Plant Growth Regul 62:241–255

Conde A, Chaves MM, Geros H (2011) Membrane transport, sensing and signaling in plant adaptation to environmental stress. Plant Cell Physiol 52:1583–1602

D'Souza SF, Nathawat NS, Nair JS, Radha-Krishna P, Ramaswamy NK, Singh G, Sahu MP (2009) Enhancement of antioxidant enzyme activities and primary photochemical reactions in response to foliar application of thiols in water – stressed pearl millet. Acta Agronomica Hung 57:21–31

Evans LT, Wardlaw IF (1976) Aspects of the comparative physiology of grain yield in cereals. Adv Agron 28:301–359

Fatehi F, Hosseinzadeh A, Alizadeh H, Brimavandi T (2012) The proteome response of *Hordeum spontaneum* to salinity stress. Cereal Res Commun 40:1–10

Foyer C-H, Noctor G (2005a) Redox homeostasis and antioxidant signaling: a metabolic interface between stress perception and physiological responses. Plant Cell 17:1866–1875

Foyer C-H, Noctor G (2005b) Oxidant and antioxidant signaling in plants: a re-evaluation of the concept of oxidative stress in a physiological context. Plant Cell Environ 28:1056–1071

Foyer C-H, Lelandais M, Kunert KJ (1994) Photooxidative stress in plants. Physiol Plant 92:696–717

Foyer C-H, Lopez-Delgado H, Dat J-F et al (1997) Hydrogen peroxide-and glutathione- associated mechanisms of acclimatory stress tolerance and signaling. Physiol Plant 100:241–254

Garg BK, Burman U, Kathju S (2006) Influence of thiourea on photosynthesis, nitrogen metabolism and yield of clusterbean (*Cyamopsis tetragonoloba* (1.) Taub.) under rainfed conditions of Indian arid zone. Plant Growth Regul 48:237–245

Geiger DR, Giaquinta RT (1982) Translocation of photosynthate. In: Govindjee (ed) Photosynthesis vol. II. Development, carbon metabolism and plant productivity. Academic Press, New York, pp 345–386

Geiger DR, Servaites JC (1991) Carbon allocation and response to stress. In: Response of plants to multiple stresses. Academic Press, New York, pp 103–127

Giaquinta RT (1976) Evidence of phloem loading from the apoplast: chemical modification of membrane sulphydryl groups. Plant Physiol 53:872–875

Gifford RM, Evans LT (1981) Photosynthesis, carbon partitioning and yield. Annu Rev Plant Physiol 32:485–509

Gifford RM, Thorne JH, Hitz WD, Giaquinta RT (1984) Crop productivity and photoassimilate partitioning. Science 225:801–808

Godfray HCJ, Beddington JR, Crute IR, Haddad L, Lawrence D, Mair JF, Pretty J, Robinson S, Thomas SM, Toulmin C (2010) Food security: the challenge of feeding 9 billion people. Science 327:815–818

Gyorgy P, Stiller ET, Williamson MB (1943) Retardation of rancidity by sulfhydryl compounds. Science 98:518–520

Hassanein RA, Abdelkadr AF, Ali H, Amin AAE, Rashad EM (2012) Grain-priming and foliar pretreatment enhanced stress defense in wheat (*Triticum aestivum var. Gimaza* 9) plants cultivated in drought land. Aust J Crop Sci 6:121–129

Hassanein RA, Amin AAE, Rashad EM, Ali H (2014–2015) Effect of thiourea and salicylic acid on antioxidant defense of wheat plants under drought stress. Int J Chem Tech Res 7:346–354

Hong-bo S, Li-ye C, Mingan S, Jaleel CA, Hongmei M (2008) Higher plant antioxidants and redox signaling under environmental stresses. CR Biol 331:433–441

IPCC Summary for Policymakers Climate Change (2014) In: Field CB et al (eds) Impacts, adaptation, and vulnerability, part a: global and sectoral aspects. Cambridge University Press, New York

Jia W, Zhang J (2000) Water stress -induced abscisic acid accumulation in relation to reducing agents and sulfhydryl modifiers in maize plant. Plant Cell Environ 23:1389–1395

Jocelyn PC (1972) Biochemistry of -SH group: the occurrence, chemical properties and biological function of thiols and disulphides. Academic Press, London, pp 47, 122

Kaya C, Sonmez O, Aydemir S, Ashraf M, Dikiltas M (2013) Exogenous application of mannitol and thiourea regulates plant growth and oxidative stress responses in salt-stressed maize (*Zea mays* L.) J Plant Interact 8:234–241

Kaya C, Sonmez O, Ashraf M, Polat T, Tuna L, Aydemir S (2015a) Exogenous application of nitric oxide and thiourea regulates on growth and some key physiological processes in maize (*Zea mays* L.) plants under saline stress. Soil-Water J (special issue): 61–66

Kaya C, Ashraf M, Sonmez O (2015b) Promotive effect of exogenously applied thiourea on key physiological parameters and oxidative defense mechanism in salt-stressed *Zea mays* L. plants. Turk J Bot 39:1–10

Khanna P, Kaur K, Gupta AK (2015) Root biomass partitioning, differential antioxidant system and thiourea spray are responsible for heat tolerance in spring maize. Proc Natl Acad Sci India Sect B Biol Sci 87(2):351–359. doi:10.1007/s40011-015-0575-0

Lemaire SD, Michelet L, Zaffagnini M, Massot V, Issakidis-Bourguent E (2007) Thioredoxins in chloroplasts. Curr Genet 51:343–365

Lobell DB, Schlenker W, Costa-Roberts J (2011) Climate trends and global crop production since 1980. Science 333:616–620

Lupton FGH (1966) Translocation of photosynthetic assimilates in wheat. Ann Appl Biol 57:355–364

Mahatma MK, Bhatnagar R, Solanki RK, Mittal GK (2009) Effect of seed soaking treatments on salinity induced antioxidant enzymes activity, lipid peroxidation and free amino acid content in wheat (*Triticum aestivum* L.) leaves. Indian J Agric Biochem 22:108–112

Maloof F, Soodak M (1961) Cleavage of disulfide bonds in thyroid tissue by thiourea. J Biol Chem 236:1689–1692

Mathur N, Singh J, Bohra S, Bohra A, Vyas A (2006) Improved productivity of mungbean by application of thiourea under arid conditions. World J Agric Sci 2:185–187

Meyer AJ, Hell R (2005) Glutathione homeostasis and redox regulation by sulfhydryl groups. Photosynth Res 86:435–457

Meyer Y, Belin C, Delorme-Hinoux V, Reichheld JP, Riondet C (2012) Thioredoxin and glutaredoxin systems in plants: molecular mechanisms, crosstalks, and functional significance. Antioxid Redox Signal 17:1124–1160

Millar AH, Mittova V, Kiddle G (2003) Control of ascorbate synthesis by respiration and its implications for stress responses. Plant Physiol 133:443–447

Nathawat NS, Nair JS, Kumawat SM, Yadav NS, Singh G, Ramaswamy NK, Sahu MP, D'Souza SF (2007) Effect of seed soaking with thiols on antioxidant enzymes and photosystem activities in wheat subjected to water stress. Biol Plant 51:93–97

Noctor G (2006) Metabolic signaling in defence and stress: the central roles of soluble redox couples. Plant Cell Environ 29:409–425

Nuruzzaman M, Sharoni AM, Satoh K, Al-Shammari T, Shimizu T, Sasaya T, Omura T, Kikuchi S (2012) The thioredoxin gene family in rice: genome-wide identification and expression profiling under different biotic and abiotic treatments. Biochem Biophys Res Commun 423:417–423

Pandey M, Srivastava AK, D'Souza SF, Penna S (2013) Thiourea, a ROS scavenger, regulates source-to-sink relationship to enhance crop yield and oil content in *Brassica juncea* (L.) PLoS One 8:e73921

Paul MJ, Foyer C-H (2001) Sink regulation of photosynthesis. J Exp Bot 52:1383–1400

Porter JR, et al. (2014) In: Field CB et al. (eds) Climate change; impacts, adaptation, and vulnerability, part a: global and sectoral aspects. Cambridge University Press, New York, pp 485–533

Procuranti B, Connon SJ (2007) A reductase-mimicking thiourea organocatalyst incorporating a covalently bound NADH analogue: efficient 1,2-diketone reduction with *in situ* prosthetic group generation and recycling. Chem Commun 14(14):1421–1423

Ramaswamy NK, Nathawat NS, Nair JS, Sharma HR, Kumawat SM, Singh G, Sahu MP, D'Souza SF (2007) Effect of seed soaking with sulphydryl compounds on the photochemical efficiency and antioxidant defence system during the growth of pearl millet under water limiting environment. Photosynthetica 45:477–480

Rossi M, Bermudez L, Carrari F (2015) Crop yield: challenges from a metabolic perspective. Curr Opin Plant Biol 25:79–89

Sahu MP, Singh D (1995) Role of thiourea in improving productivity of wheat (*Triticum aestivum* L.) J Plant Growth Regul 14:169–173

Sahu MP, Solanki NS (1991) Role of sulphydryl compounds in improving dry matter partitioning and grain production of maize (*Zea mays* L.) J Agron Crop Sci 167:356–359

Sahu MP, Solanki NS, Dashora LN (1993) Effects of thiourea, thiamine and ascorbic acid on growth and yield of maize (*Zea mays* L.) J Agron Crop Sci 171:65–69

Sahu MP, Rathore PS, Kumawat SM, Singh G, Ramaswamy NK, D'Souza SF (2006) Role of sulphydryl bioregulators for improving productivity of arid zone crops. Rajasthan Agricultural University, Bikaner, Research Bulletin, RAU-BARC, Directorate of Research, p 20

Sangeetha R (2013) Effect of salinity induced stress and its alleviation on the activity of amylase in the germinating seeds of Zea mays. Int J Basic Life Sci 1:1–9

Schurmann P, Buchanan BB (2008) The Ferredoxin/Thioredoxin system of oxygenic photosynthesis. Antioxid Redox Signal 10:1235–1274

Shao HB, Liang ZS, Shao MA (2005) Adaptation of higher plants to stresses and stress signal transduction. Acta Ecol Sin 25:1871–1882

Shao HB, Chen XY, Chu LY et al (2006) Investigation on the relationship of proline with wheat anti-drought under soil water deficits. Biointerfaces 53:113–119

Shao HB, Chu LY, Wu G et al (2007a) Changes of some antioxidative physiological indices under soil water deficits among 10 wheat (Triticum aestivum L.) genotypes at tillering stage. Biointerfaces 54:143–149

Shao HB, Jang SY, Li FM et al (2007b) Some advances in plant stress physiology and their implications in the systems biology era. Biointerfaces 54:33–36

Shao HB, Chu LX, Cheruth AJ, Zhao CX (2008) Water-deficit-induced anatomical changes in higher plants. CR Biol 331:215–225

Sharma GK (1988) Effect of soil-applied sulphur and select chemicals on dry matter partitioning and grain production efficiency of pearl millet. M.Sc.(Ag.) thesis, Department of Agronomy, Rajasthan College of Agriculture, Udaipur, Rajasthan Agricultural University, Bikaner (India)

Siddiqui MZ, Yadav RA, Khan N (2014) Agro-techniques to overcome drought stress for yield maximization in pearlmillet (Pennisetum glaucum) under rainfed conditions. Plant Archiv 14:999–1000

Song Y, Cui J, Zhang H, Wang G, Zhao FJ, Shen Z (2012) Proteomic analysis of copper stress responses in the roots of two rice (Oryza sativa L.) varieties differing in Cu tolerance. Plant Soil 366(1/2):647–658. doi:10.1007/s 11104-012-1458-2

Srivastava AK, Nathawat NS, Ramaswamy NK, Sahu MP, Singh G, Nair JS, Paladi RK, D'Souza SF (2008) Evidence for thiol-induced enhanced in situ translocation of ^{14}C-sucrose from source to sink in Brassica juncea. Environ Exp Bot 64:250–255

Srivastava AK, Ramaswamy NK, Suprasanna P, D'Souza SF (2010) Genome-wide analysis of thiourea-modulated salinity-responsive transcripts in seeds of Brassica juncea: identification of signalling and effector components of stress tolerance. Ann Bot 106:663–674

Srivastava AK, Srivastava SP, Suprasanna P, D'Souza SF (2011) Thiourea orchestrates regulation of redox state and antioxidant responses to reduce the NaCl-induced oxidative damage in Indian mustard (Brassica juncea L.) Plant Physiol Biochem 49:676–686

Wang W, Vinocur B, Altman A (2007) Plant responses to drought, salinity and extreme temperature towards genetic engineering for stress tolerance. Planta 218:1–14

Wardlaw IF (1968) The control and pattern of movement of carbohydrates in plants. Bot Rev 34:79–105

Wardlaw IF (1980) Translocation and source – sink relationship. In: Carlson PS (ed) The biology of crop productivity. Academic Press, New York, pp 297–339

Wheeler T et al (2000) Temperature variability and the yield of annual crops. Agric Ecosyst Environ 82:159–167

Wright STC, Hiron RWP (1969) (+ −) abscisic acid, the growth inhibitor induced in detached wheat leaves by a period of wilting. Nature 224:719–720

Zagorchev L, Seal CE, Kranner L, Odjakova M (2013) A central role for thiols in plant tolerance to abiotic stress. Int J Mol Sci 14:7405–7432

Part III

Crop Based Mitigation Strategies

Improving Crop Adaptations to Climate Change: Contextualizing the Strategy

S. Naresh Kumar

Abstract

Climatic stresses have been affecting agricultural productivity and thereby present a major challenge for the food and nutritional security. The frequency and magnitude of these stresses are projected to increase and impact the crop yields at global level as well as in India. Genetic adaptation is identified as the most crucial factor for improving productivity in future climates. Contextualization of genetic improvement for changing climates is essential to improve the crop productivity as well as to conserve the natural resources. Serious reorientation of breeding efforts is required for a comprehensive genetic improvement programme that should address the challenges of changing climates and growing demand for food and nutritional quality. The approaches to be deployed for crop improvement should include characterization of projected climatic stresses, entire germplasm with projected climatic variability as background, utilization of entire genetic diversity and deploying multipronged approaches for genetic improvement. This chapter is aimed to contextualize the issues and approaches for breeding climate resilient varieties.

12.1 Introduction

Evolution has been the basis of species development. On the earth, about 1 million plant species exist, of which around 350,000 are accepted, while more than 240,000 are yet be resolved into 'accepted name' or 'synonym' (BGCI 2017). Out of these, only about 20,000 species are edible. Ever since settled agriculture was invented by

S.N. Kumar (✉)
Centre for Environmental Science and Climate Resilient Agriculture, ICAR-Indian Agricultural Research Institute, New Delhi 110012, India
e-mail: nareshkumar.soora@gmail.com

© Springer Nature Singapore Pte Ltd. 2017
P.S. Minhas et al. (eds.), *Abiotic Stress Management for Resilient Agriculture*,
DOI 10.1007/978-981-10-5744-1_12

human being over 10,000 years ago, numbers of species that are cultivated have been shrinking. Increase in population and demand for higher production of food grains led to expansion of area under agriculture on one hand while narrowing down the number of species to only a few on the other. Intensification of agriculture led to a further narrowing down of species that are being cultivated. Currently, only 20 species are providing over 90% of our food. Even among these, only a few varieties or hybrids are being cultivated causing extreme narrowing of gene pool in agroecosystems. Among the major species that are cultivated, probably rice is the only crop with larger diversity. Major cultivated crops have been subjected to improvement though human intervened conventional breeding or molecular breeding.

Natural selection has been the driving force for evolution of biological organisms on the earth. Organisms that could keep the pace of their evolution way ahead of the changing climatic conditions can dominate the ecosystems. Organisms that evolve at the pace of changing climates may sustain their existence till the time when the pace of climate change overtakes the pace of their evolution and eventually extinct. Other organisms which cannot keep the pace of evolution as that of the climate change will extinct. This basic principle is applicable to all organisms and cultivated species are no exception.

Climate, the mean state of weather over a long period (usually 30 years), of the earth has been dynamic and has been changing continuously. Out of 4.54 billion years of the earth's age, 3 million years has seen glacial and interglacial cycles. Last interglacial period occurred about 125,000 years ago. During this period the global mean surface temperatures were 1–2 °C warmer than present 15 °C. During the last glacial maximum, where the ice covered the earth's surface to the maximum extent about 21,000 years ago, the global mean surface temperatures were 4–7 °C cooler than present. The greenhouse gases (GHGs) are responsible for tapping of energy by the earth's atmosphere. Without GHGs, earth's surface mean annual temperature would have been -17 °C making earth unsuitable for living organisms. The presence of GHGs in the atmosphere and greenhouse effect causes warming of about 32 °C. This resulted in a global mean annual temperature of $+14.84$ °C in 2016 (NOAA 2017).

Human activities such as fossil fuel combustion and the GHG-emitting man-made technologies, on the one hand, and deforestation and land use change, on the other hand, have led to rapid accumulation of GHGs. All these GHGs have differential potential to warm the earth's surface, called global warming potential (GWP). The GWP is the measure of how much heat a greenhouse gas traps in the atmosphere relative to the amount of heat trapped by carbon dioxide. Carbon dioxide (CO_2), by definition, has a GWP of 1 regardless of the time period used, because it is the gas being used as the reference. The GWP of methane is 21 times and N_2O is 310 times of GWP of CO_2 in a 100-year period. The GWP of sulphur hexafluoride (SF6) (23,900), hydrofluorocarbons (HFCs, vary between 140 and 11,700 times depending the type of molecule) and perfluorocarbons (PFCs) (6500–9200) is extremely high for 100-year period. Collectively, their GWP is leading to increase in temperatures and climate change.

Climate change is projected to raise the global surface temperature in excess of 1.5 °C by 2100 relative to 1850–1900, and warming will continue beyond 2100 (IPCC AR-5 2013:2014). Increase of global mean surface temperatures for 2081–2100 relative to 1986–2005 is projected to be 0.3 °C to 1.7 °C in GHG mitigation scenario (representative concentration pathway, RCP2.6); between 1.1 °C and 2.6 °C (RCP4.5) and 1.4 °C to 3.1 °C (RCP6.0) in GHG stabilization scenarios and by 2.6 °C to 4.8 °C (RCP8.5) in GHG emission intensive scenario. Concurrently the atmospheric CO_2 concentrations are projected to increase to 421 ppm (RCP2.6), 538 ppm (RCP4.5), 670 ppm (RCP6.0) and 936 ppm (RCP 8.5) by the year 2100. Further, climate change is projected to increase frequency of extreme temperature events, extreme rainfall events and skewed monsoon leading to increased risk of drought-related water and food shortage. Further, the report suggests that the risk level can be moderately minimized with current adaptation and risk level with high/intensive adaptation can be minimized.

Climate change impacts on crops are projected to vary with the type of species, location and season of crop growth. Several studies have projected that global production of many crops may reduce to the tune of 40–60% due to rise in temperature and climate change by the end of the century (Rosenzweig et al. 2014). The magnitude and direction of climate change impacts have significant spatio-temporal variation for a given crop, with some regions gaining yield while other losing it based on the current climatic conditions (Naresh Kumar et al. 2011, 2013). However, adaptation to climate change will not only reduce the negative impacts but also maximize the positive effects. Adaptation of agriculture to climate change involves managing current and future climatic risks. This can be achieved through an integrated approach of (1) anticipatory research efforts, (2) management of natural resources, (3) use of technology and (4) proactive development and policy initiatives. Several low-cost technologies can reduce the negative impacts of climate change (Easterling et al. 2007). Among the natural resources, genetic resource is the major factor that determines the performance of agricultural productivity. Growing suitable crop variety in changing climates is identified as one of the essential and easy-to-adapt strategy for not only minimizing the negative impacts but also for harnessing the beneficial effects (Braun et al. 1996; Chapman et al. 2012).

Historically, crop improvement has been aimed to achieve high yield, resistance to disease and pest and tolerance to abiotic stresses. Screening germplasm for identification of donors having specific traits and their utilization in developing resistant and tolerant varieties has been the convention in crop improvement (Ortiz 2002; Xin-Guang et al. 2010b). Breeders have been successful in achieving their targets. However, changing climates have been throwing new challenges for crop improvement.

12.2   Why Genetic Adaptation in Changing Climates?

The centre of origin of many crops falls in the region between Tropic of Cancer and Capricorn. Climate change has been changing the climatic conditions of these regions at a much faster pace resulting in new or novel climates (Williams et al. 2007). Since climate is the primary factor for species distributions and ecosystem processes, the new climates may pose challenge to the existing species, while new species may emerge. Novel climates are projected to develop primarily in the tropics and subtropics, challenging large portion of existent biodiversity to evolve faster. Species that evolve faster can survive, while those who are slow will eventually become extinct. This calls for an immediate action of conservation of all biodiversity in these areas, in general and in biodiversity hotspots, in particular. It is quite probable that lack of such efforts had led to the collapse of civilizations due to the late Holocene droughts between 6000 and 1000 years ago. Droughts resulted in the collapse of empires and societies like the Akkadian Empire of Mesopotamia, c. 6200 years ago; the Classic Maya of Yucatan Peninsula, c. 1400 years ago (Ceccarelli et al. 2010); the Moche IV–V transformation of coastal Peru, c. 1700 years ago (de Menocal 2001); and the early bronze society in the southern part of the Fertile Crescent (Rosen 1990), to name a few. Not just of historical events, these examples have current relevance as well, particularly in the current world where the food habits of regions are gradually merging.

Crop species gene pools are collected, conserved, catalogued and characterized for the use in crop breeding. As mentioned earlier, most of the genetic enhancements are made to achieve higher yield, resistance to specific disease and pest and tolerance to some abiotic stresses. Efforts to increase quality also have led to nutritionally enhanced varieties. While all these were done with focussed screening of germplasm to identify lines with 'desirable' traits, they ignored or overlooked to analyse its performance with climatic variability as the backdrop. Therefore, there is a need to contextualize the breeding efforts in changing climates to improve the crop yields as they are projected to reduce in the changing climates.

Climate change is projected to affect the yield of several major crops across the world. For instance, the global yield of maize, wheat, rice and soybean is projected to be affected up to 20% in 2020 (2010–2039) scenario, 20–35% in 2050 (2040–2069) and 40–60% in 2090 (2070–2099) (Rosenzweig et al. 2014). A global analysis on wheat production indicated a decrease of about 6% yield with every 1 °C rise in temperature (Asseng et al. 2015). The impacts are variable over space and time, e.g. more effects would be visualized in tropical regions (IPCC 2014). The projected climate change events and major impacts on crop productivity in different continents (Table 12.1) indicate a need for concerted effort to adapt to climate change. The IPCC reported that each additional decade of climate change is expected to reduce mean yields by roughly 1%, while the anticipated increase in productivity per decade needed to keep pace with demand is roughly 14% (IPCC 2014).

In Indian region, areas encompassed by climatic stresses and magnitudes of crop loss have been increasing recently. These risks are projected to increase in future affecting food production if agriculture is not adapted to changing climates. The

Table 12.1 Major climate-related risks and projected impacts on agricultural systems in different continents

Region	Projected climate-related risks	Major projected impacts on crops agriculture
Africa	Warming, extreme temperatures, drying trend, sea level rise and change in mean and distribution of rainfall	Reduction in the length of growing season
		Yield reduction of 8% in Africa by 2050 averaged over crops, with wheat, maize, sorghum and millets more affected than cassava and sugarcane
		Fall in crop net revenues by up to 90% by 2100
		75 to 250 million people at risk of increased water stress by the 2020s and 350 to 600 million people by the 2050s
Australia	Warming, droughts, water stress	Agricultural production may decline by 2030 over much of southern and eastern Australia, and over parts of eastern New Zealand, due to increased drought and fire
		Change land use in southern Australia, with cropping becoming non-viable at the dry margins
		Production of Australian temperate fruits and nuts will drop on account of reduced winter chill
		Geographical spread of a major horticultural pest, the Queensland fruit fly (*Bactrocera tryoni*), may spread to other areas including the currently quarantined fruit fly-free zone
Islands	Drying trend, sea level rise, change in mean and distribution of rainfall and damaging cyclonic events, sea level rise, ocean acidification	Coastal agriculture to be affected due to salinization and sea level rise
		Loss to plantations such as coconuts due to cyclonic storms
		Subsistence and commercial agriculture on small islands will be adversely affected by climate change
		In mid- and high-latitude islands, higher temperatures and the retreat and loss of snow cover could enhance the spread of invasive species including alien microbes, fungi, plants and animals
South America	Warming, extreme temperatures, drying trend, sea level rise, change in mean and distribution of rainfall and damaging cyclonic events, snow cover	Generalized reductions in rice yields by the 2020s
		Reductions in land suitable for growing coffee in Brazil and reductions in coffee production in Mexico
		The incidence of the coffee disease and pest incidence in Brazil's coffee production area
		Risk of *Fusarium* head blight in wheat is very likely to increase in southern Brazil and in Uruguay

(continued)

Table 12.1 (continued)

Region	Projected climate-related risks	Major projected impacts on crops agriculture
North America	Warming, extreme temperatures, drying trend, sea level rise, change in mean and distribution of rainfall and damaging cyclonic events	Increased climate sensitivity is anticipated in the southeastern USA and in the USA corn belt making yield unpredictable
		Yields and/or quality of crops currently near climate thresholds (e.g. wine grapes in California) are likely to decrease
		Yields of cotton, soybeans and barley are likely to change
		Risk of extinctions of important species
		By the 2050s, 50% of agricultural lands in drier areas may be affected by desertification and salinization
Asia	Warming, extreme temperatures, drying trend, sea level rise, change in mean and distribution of rainfall and damaging cyclonic events	Monsson aberration-related crop loss
		Reduction water availability
		Crop yields could decrease by up to 30% in Central and South Asia
		More than 28 million hectares (ha) in arid and semiarid regions of South and East Asia will require substantial (at least 10%) increases in irrigation for a 1 °C increase in temperature
		Crop yields to reduce (wheat, rice, mustard, maize)
		Loss in agrobiodiversity
		Increased incidence of disease and pests
		Water scarcity related food insecurity
Europe	Warming, extreme temperatures, drying trend, sea level rise, extreme precipitation events, ocean acidification	Crop productivity is likely to decrease along the Mediterranean and in Southeast Europe
		Differences in water availability between regions are anticipated to increase
		Much of European flora is likely to become vulnerable, endangered or to extinct by year 2100

Synthesized from IPCC (2014) and FAO (2010)

spatio-temporal variation in direction and magnitude of climate change impacts vary with the nature of crops, and therefore the crop-wise impacts and adaptation gains are summarized below.

Rice Climate change is projected to reduce irrigated rice yields by ~4% in 2020 (2010–2039), ~7% in 2050 (2040–2069) and by ~10% in 2080 (2070–2099) climate scenarios. Whereas rainfed rice yields are likely to be reduced by ~6% in the 2020 scenario, yields may reduce only marginally (<2.5%) by 2050 and 2080. However, spatial variations exist for the magnitude of the impact, with some regions likely to be affected more than others. The study indicated that growing improved varieties with efficient agronomy can lead to an increase in all-India irrigated rice yields by about 17% over current values in the 2020 scenario, by 14% and by 8% in the 2050 and 2080 scenarios, respectively. Similarly, rainfed rice yield can be increased by ~20% in the 2020 and by ~35–38% in the 2050 and later scenarios (Naresh Kumar et al. 2013).

Wheat Wheat yield in India is projected to reduce by 6–23% by 2050 scenario based on management, if no adaptation is followed. Adaption by timely sowing of suitable variety and with input (fertilizer and irrigation) management may be a practical low-cost adaptation strategy to increase the yield (by >10%) in future climates (Naresh Kumar et al. 2014a). Central India is projected to lose yield despite this adaptation strategy warranting development of varieties highly tolerant to early and terminal heat stress.

Maize Climate change is projected to reduce irrigated maize yield by 18% in kharif season, but adaptation is projected to increase the yield up to 21% in 2020 scenario (Naresh Kumar et al. 2012).

Sorghum Climate change is projected to reduce rainfed sorghum yield by 2.5% in 2020 (2010–2039). However, it is projected that adaptation can increase the productivity by 8% in 2020 (Naresh Kumar et al. 2012).

Mustard In India, mustard yield is projected to reduce by ~2% in 2020 (2010–2039) if no adaptation is followed (Naresh Kumar et al. 2015). Adoption of a combination of improved agronomic management practices can improve the yield by ~17% with current varieties (Naresh Kumar et al. 2014b). However, with improved varieties, yield can be enhanced by ~25% in 2020 climate scenario.

Soybean Increase in soybean yield in the range of 8–13% under future climate scenarios (2030 and 2080) is projected (Naresh Kumar et al. 2012).

Groundnut The rainfed groundnut yield is projected to increase by 4–7%, except in the climate scenario of A1B 2080 under which yield is projected to reduce by −5% (Naresh Kumar et al. 2012).

Potato The potato yield is projected to reduce by ~2.5, ~6 and ~11% in 2020 (2010–2039), 2050 (2040–2069) and 2080 time periods, respectively, in the Indo-Gangetic Plains. Change in planting time is found to be the most important adaptation option for yield improvement by ~6% in 2020 (Naresh Kumar et al. 2015).

Cotton Cotton productivity in northern India is projected to marginally decline due to climate change, while in Central and southern India, productivity may increase implying that the overall productivity at the national level may not be affected (Hebbar et al. 2013).

Coconut Coconut productivity is projected to increase in western coastal region, Kerala, parts of Tamil Nadu, Karnataka and Maharashtra (with current level of water and management) while negative impacts are projected for Andhra Pradesh, Orissa, West Bengal, Gujarat and parts of Karnataka and Tamil Nadu due to climate change. On all-India basis, climate change is projected to increase coconut productivity by 1.9 to 6.8% in 2080 scenario. Adaptation can increase the productivity by ~33% in 2030, and by 25–32% in 2080 climate scenarios. Analysis further indicated that current productivity in India can be improved by 20% to almost double if all plantations in India are provided with location specific agronomic and genotype interventions (Naresh Kumar and Aggarwal 2014).

Horticultural Crops Climatic stresses such as extreme temperatures, hailstorms and heavy rainfall events damage horticultural crops. A 24 h flooding affects tomato crop and the flowering period is highly sensitive. In case of onion, bulb initiation stage is sensitive to flooding causing a 27 and 48% reduction in bulb size and yield, respectively (Rao et al. 2009). Productivity of temperate fruit crops such as apple is affected, and its cultivation is shifted to higher latitudes to 2500 mamsl from 1250 mamsl in Himachal Pradesh (Bhagat et al. 2009).

Climate change is projected to affect the quality in terms of reduced concentration of grain protein (under low fertilizer input conditions), and some minerals like zinc and iron due to elevated CO_2 (Porter et al. 2014). Elevated CO_2 caused reduction in the concentration of protein, secondary metabolites, while rise in temperature enhanced their concentration in pulse, several vegetable and fruit crops. Majority of studies indicate negative impacts; however, rise in temperature may decrease cold waves and frost events leading to reduced damage to frost-sensitive crops such as chickpea, mustard, potato and other vegetables.

12.3 Strategizing Crop Improvement for Adaptation

Crops have been adapting to external stresses; however, in the climate change scenario, the frequency of climatic stresses has been increasing posing serious threat to crop productivity. Several researchers have expressed concern that current breeding strategies will not be sufficient to meet the challenges of increased frequency of climatic risks. The basic strategy that has so far followed may need a thorough relook so that the climate resilient varieties are developed for meeting the ever-growing demand for food in changing climates (Rajeev et al. 2011; Stephen and Donald 2010a; Chikelu et al. 2012; Smith 2012). Only about 3% of the germplasm is being used for crop improvement. There is a need for involving wider germplasm pool for crop improvement. This implies that the germplasm characterization itself needs to be reoriented so that climatic stress responses are taken into consideration apart from other agronomic characteristics. Increased use of plant genetic resources is expected to play a major role in developing climate resilient agricultural systems (Lin 2011; Hodgkin and Bordoni 2012). Currently, the number of accessions of 612 genera and 34,446 agricultural species conserved ex situ worldwide has reached 7.4

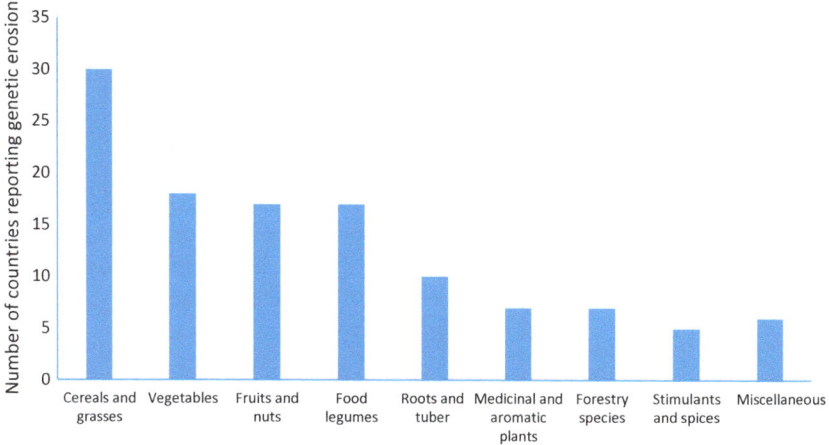

Fig. 12.1 Genetic erosion reported in agricultural species (Data used from The Second Report on The State of The World's Plant Genetic Resources for Food and Agriculture, FAO Report, 2010)

million. However, only less than 30 percent of the total numbers of accessions are estimated to be distinct. Though the number of accessions of minor crops and crop wild relatives (CWR) has increased, they are still generally underrepresented. Moreover, the germplasm erosion is reported from several countries from different types of agricultural crops (Fig. 12.1). Several breeding methods have been successfully employed for crop improvement. However, the use of crop wild relatives in crop improvement efforts is not optimized yet (Ortiz 2002). Genetic diversity available in crop wild relatives needs to be exploited for sustaining and improving the crop yield in dynamic biotic and abiotic stress events (Feuillet et al. 2008) in changing climates.

Under-exploitation of full genetic potential of edible species is a blessing in disguise as the demand for quantity and quality food will continue to increase in future. However, the rate of increase in biodiversity erosion is a major concern. As mentioned earlier, species which cannot adapt to the rate of climate change will become extinct immediately. Species that are able to adapt but at a rate slower than that of change in climate will extinct eventually. This implies that currently exploited species in agriculture and food production system must evolve faster. For this crop improvement, efforts need multipronged approach including conventional breeding, exploitation of full germplasm, molecular breeding, genetic engineering, distant hybridization and exploitation of crop wild relatives, among others. Among these, the use of currently available germplasm and crop wild relatives is of major concern because they are vulnerable to erosion causing serious loss of genetic diversity if extra care and precautions are not taken. Protecting areas of genetic diversity, national gene banks and ex situ and in situ conservation measures are being followed which help in conserving the genetic diversity.

A summary of the major climatic risks, crops in different regions and total germplasm available in different subregions indicate that major stresses and crops are almost common (Tables 12.1 and 12.2). Crops are being increasingly exposed to multiple stresses even in a crop season. Thus, exploiting the available entire germplasm is of utmost importance. Therefore, future crop improvement efforts need new initiatives and dimensions (Fig. 12.2) and some of them are briefly mentioned below:

• *Identification of current and anticipated climatic stresses and prioritizing the crops, traits and regions* for crop improvement. For instance, wheat is exposed to early and terminal heat stress in Central India and terminal heat stress in North India (Fig. 12.3). Simulation analysis indicated that timely sowing of

Table 12.2 Climatic stresses, major crops and number of total accessions in gene banks

Region	Climate-related risks	Subregion	No. of accessions	Major crops
Africa	Warming, extreme temperatures, drying trend, sea level rise and change in mean and distribution of rainfall	East Africa	145,644	Maize, sorghum, millets
		Central Africa	20,277	Sorghum, millets, maize, sugarcane, groundnuts
		West Africa	113,021	Sorghum, millets, maize, rice, cassava, groundnuts, pulses
		Southern Africa	70,650	Maize, millet, sorghum, cotton, pulses, sugarcane
Islands	Drying trend, sea level rise, change in mean and distribution of rainfall and damaging cyclonic events, sea level rise, ocean acidification	Indian Ocean Islands	4604	Rice, cassava, maize
America	Warming, extreme temperatures, drying trend, sea level rise, change in mean and distribution of rainfall and damaging cyclonic events, snow cover	South America	687,012	Maize, rice, sugarcane, soybean
		Central America and Mexico	303,021	Maize, rice, cassava, sorghum, sugarcane
		Caribbean	33,115	Sugarcane
America	Warming, extreme temperatures, drying trend, sea level rise, change in mean and distribution of rainfall and damaging cyclonic events	North America	708,107	Wheat, maize, soybean, cotton, sorghum, pulses

(continued)

Table 12.2 (continued)

Region	Climate-related risks	Subregion	No. of accessions	Major crops
Asia	Warming, extreme temperatures, drying trend, sea level rise, change in mean and distribution of rainfall and damaging cyclonic events	Pacific East Asia	1,036,946	Rice, wheat, groundnut, sugarcane, maize, soybean, potato
		Asia Pacific	252,455	Rice, maize, groundnut
		Pacific South Asia	714,562	Rice, wheat, soybean, maize, pulses, groundnut
		Pacific Southeast Asia	290,097	Rice, maze, oil palm, coconut
		Near East Central Asia	153,849	Wheat, barley, cotton
		Near East West Asia	165,930	Wheat, maize, pulses
		Near East South/East Mediterranean	141,015	Maize, wheat, pulse
Europe	Warming, extreme temperatures, drying trend, sea level rise, extreme precipitation events, ocean acidification	Europe	725,315	Wheat, maize, potato, barley, rapeseed

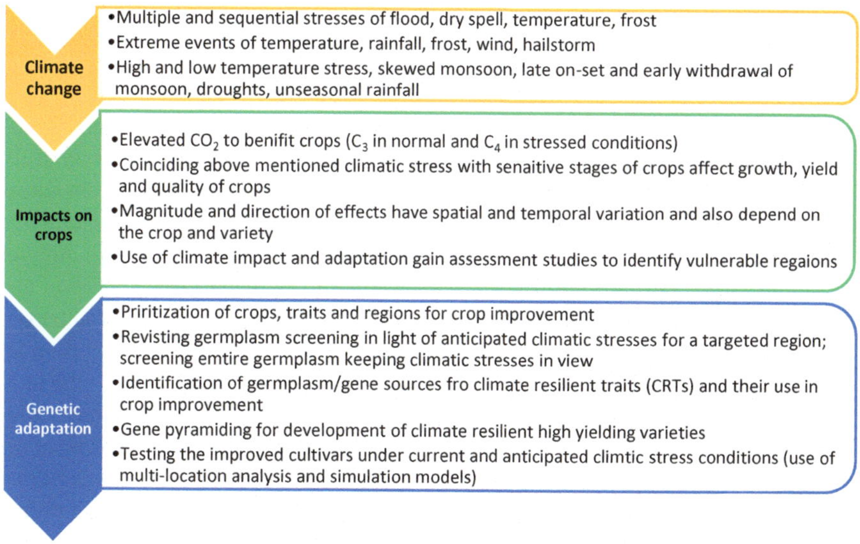

Climate change
- Multiple and sequential stresses of flood, dry spell, temperature, frost
- Extreme events of temperature, rainfall, frost, wind, hailstorm
- High and low temperature stress, skewed monsoon, late on-set and early withdrawal of monsoon, droughts, unseasonal rainfall

Impacts on crops
- Elevated CO_2 to benifit crops (C_3 in normal and C_4 in stressed conditions)
- Coinciding above mentioned climatic stress with senaitive stages of crops affect growth, yield and quality of crops
- Magnitude and direction of effects have spatial and temporal variation and also depend on the crop and variety
- Use of climate impact and adaptation gain assessment studies to identify vulnerable regaions

Genetic adaptation
- Priritization of crops, traits and regions for crop improvement
- Revisting germplasm screening in light of anticipated climatic stresses for a targeted region; screening emtire germplasm keeping climatic stresses in view
- Identification of germplasm/gene sources fro climate resilient traits (CRTs) and their use in crop improvement
- Gene pyramiding for development of climate resilient high yielding varieties
- Testing the improved cultivars under current and anticipated climtic stress conditions (use of multi-location analysis and simulation models)

Fig. 12.2 Basic steps in development of climate resilient cultivars

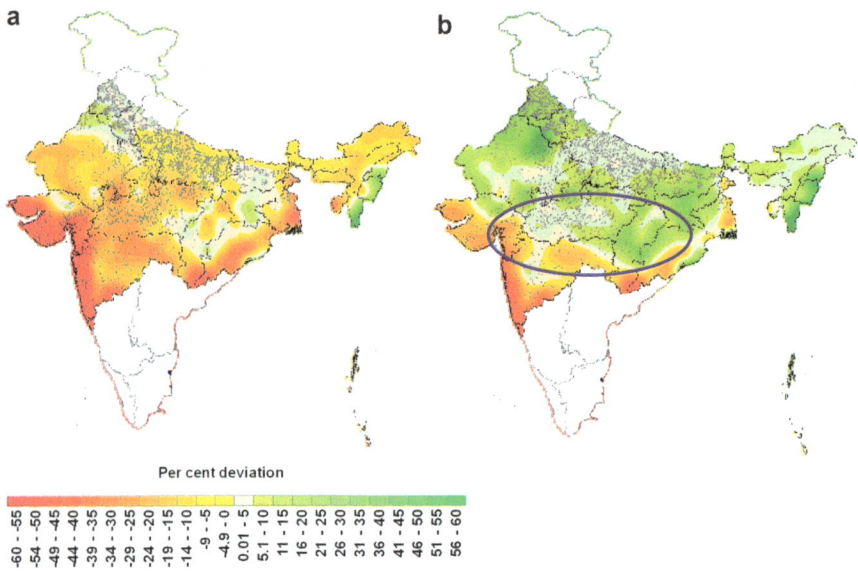

Per cent deviation

Fig. 12.3 Impact of climate change on wheat (**a**) without and (**b**) with adaptation in 2050 scenario (Naresh Kumar et al. 2014a)

wheat with suitable cultivars can improve the wheat yield despite climate change (Naresh Kumar et al. 2014a). However, yield in Central India is projected to be reduced if suitable cultivars are not developed. As per this analysis, development of short-duration and heat stress-tolerant varieties for Central India should be prioritized. Since development of a variety of an annual crop takes at least 8–15 years, it is essential to initiate concerted efforts immediately as by 2050 one may get about only 3–7 breeding cycles.

- *Characterizing entire germplasm of a species* in the backdrop of current and anticipated climatic stresses and conditions. There is a need to exploit all available genetic variability within a species with climatic stresses as backdrop. Targeted breeding, as is done in case of droughts, floods, high temperature and salinity stress, also needs to be extended to other types of climatic stresses (Table 12.3). Further, the traits that contribute to reduced GHG emission and enhanced CO_2 sequestration need to be identified. In addition to these, there is also a need to identify the climatic threshold for every sensitive phase of crop for developing climate resilient varieties. For instance, a rice analysis for India indicated that in areas with current seasonal (June–September) mean minimum temperatures of >23 °C, as in parts of Central, North and northeast regions of India, future temperatures constrain higher productivity of irrigated rice (Naresh Kumar et al. 2013). Moreover, high as well as low temperature stress coinciding pollination affects the pollen viability eventually reducing the rice yield. Similarly, in case of spring wheat varieties, yield would reduce in areas with mean seasonal

Table 12.3 Some climate-resilient traits (CRTs) for different types of abiotic stresses

Major target	Component traits	References
Drought-adaptive traits	Root architecture	Hammer et al. (2009)
	Accumulation and partitioning of water-soluble carbohydrates to storage organs	Steele et al. (2013)
	Abscisic acid concentration, stay green, canopy temperature, carbon isotope discrimination, stay green, ABA signalling for stomatal regulation	Rebetzke et al. (2008)
		Kholova et al. (2010)
		Lopes et al. (2014)
		Borrell et al. (2014)
		Ren et al. (2007)
Salinity tolerance	Na + accumulation via Na + exclusion	James et al. (2006)
Flooding tolerance	*Sub1* QTL	Xu et al. (2006)
Water logging tolerance	QTLs for submergence tolerance	Septiningsih et al. (2013)
	QTLs for adventitious root formation at the soil surface in maize	Mano et al. (2005)
Terminal drought	Early flowering	Hegde (2010)
Flowering time	Vernalization requirement (*VRN* genes), photoperiod sensitivity (*PPD* genes)	Bentley et al. (2011)
		Faure et al. (2007)
Developmental plasticity	Root elongation in water deficit condition and inhibition of shoot elongation	Spollen et al. (2008)
Cold tolerance	Winter hardiness	Pan et al. (1994)
	Vernalization response and frost tolerance	Galiba et al. (1995)
Heat stress tolerance	Pollen viability, starch accumulation during terminal heat stress in wheat, dehydration tolerance	Bita and Gerats (2013)
CO_2 sequestration	Fast-growing trees and perennial grasses	Garten et al. (2011) and Jansson et al. (2010)
Low greenhouse gas emission	Rice cultivars with low vascular transportation of methane	Gogoi et al. (2005); Nirmali et al. (2008); and Dubcovsky and Dvorak (2007)
	Nitrogen-use efficiency: NAM-B1 gene	

maximum and minimum temperatures more than 27 and 13 °C, respectively (Naresh Kumar et al. 2014a). Wheat is identified as the most important crop that needs to be focussed for crop improvement to beat climatic stress effects in South Asia (Lobell et al. 2008). In case of mustard, regions with mean seasonal temperature regimes above 25/10 °C are projected to lose yield due to temperature rise. As climatically suitable period for mustard cultivation may reduce in the future, short-duration (<130 days) cultivars with 63% pod-filling period will become more adaptable (Naresh Kumar et al. 2015). There has been a lot of literature available on this front with specific examples.

- In horticultural crops, the challenge is to retain and improve quality despite climatic stresses. For instance, breeding plantation crops need visionary approach as the plants live for up to 70 years, and one should take the anticipated future climatic stresses, technological improvements, land use change and socio-economic demand for quantity and quality into account. Thus, evaluation of germplasm taking into consideration the response to climatic extremes apart from other criteria becomes essential. Genotypic improvement strategies include population improvement through identified stress-tolerant plants, breeding for temperature stress tolerance of pollen and stigma, high retention of set fruits under climatic risks and improved source-sink balance and identification of multiple stress-tolerant cultivars. In addition, there is a need to understand the change in quality of produce with respect to climate change. Further, the biotic stresses that are anticipated to increase or emergence of new pests must also be taken into consideration while breeding climate resilient cultivars:

- *Identification of climate smart varieties* which can meet the challenges of future climates. For example, some of the major challenges include breeding:
  - Low-methane-emitting rice cultivars.
  - Rapid nitrogen uptake and its use efficient varieties for reducing the N_2O emission.
  - High water-use efficient varieties.
  - High carbon sequestration varieties for perennial crops, deep and high root volume varieties of annual crops.
  - Stress-tolerant varieties with high revival capacity.
  - Stress avoidance by phenological plasticity.
  - Multiple abiotic and biotic stress-tolerant/stress-resistant varieties.
  - Retaining quality of produce despite climate change.
  - In rice, low-methane-emitting cultivars, cultivars suitable for flood situation, water stagnation and salt tolerance become more important. In maize, cultivars with flood tolerance and endless gap between silking and tasselling gain importance. The farmers' varieties can be of immense source of tolerance gene pool for use. In plantation crops, identification of in situ tolerant trees (Naresh Kumar et al. 2002) and their use in population improvement programme becomes very important as they have been exposed to climatic stresses during their life cycle and still performed better in terms of physiological parameters and economic yield, indicating the presence of desirable genetic composition.
- *Utilization of crop wild relatives* in breeding programme for incorporating desirable genes for climatic stress tolerance. Though crop wild relatives may fail to adapt to new climatic conditions of their native habitats (Jarvis et al. 2008), they possess a gene pool which can be exploited for crop improvement. A number of reviews have taken stock of use of crop wild relatives in crop improvement for tolerance to abiotic stresses and quality (Radden et al. 2015).
- *Identification and utilization of genotypes from climate analogue analysis* for a quick intervention. The climate analogue analysis is a concept which is based on identification of areas where either the today's climate of a location corresponds to the future climate projected at another location or the projected future climate corresponds to the current climate of another site (Ramírez-Villegas et al. 2011).

Using this concept, testing the performance of genotypes in climate analogue sites can lead to identification of suitable genotypes for future climates. In addition, germplasm collection can also be rationalized suing climate analogues.

- *Geospatial analysis-based germplasm collection* to minimize the gaps in germplasm collection and also to minimize the duplication. Further, geospatial tagging and characterizing germplasm is possible in this approach. Geospatial software such as DIVA is extensively used for this purpose (Hijmans et al. 2001).
- *Distant hybridization* using interspecific and intergeneric breeding strategies (Liu et al. 2014).
- *Molecular breeding* helps in targeted crop improvement and is relatively faster than conventional breeding. In molecular or marker-assisted breeding, DNA markers are used as a substitute for phenotypic selection and to accelerate the release of improved cultivars.
- *Genetic engineering* has to be exploited for gene pyramiding to develop climate resilient varieties (Varshney et al. 2009; Scheben et al. 2016).
- *The use of omics science platforms* for crop improvement includes genomics, phenomic platform data, molecular data and bioinformatics tools. The growing number of available high-quality reference genomes and advances in population-level genotyping has contributed to improved understanding of genomic variation. These developments are leading towards plant pangenomics (Scheben et al. 2016).
- *The use of simulation models* in climate change research is indispensable for testing the performance of a cultivar in future environments. Several crop models such as DSSAT, InfoCrop, APSIM and Crosyst are being used for quantifying the impacts of climate change on crops (Assenge et al. 2013; Rosenzweig et al. 2014; Naresh Kumar et al. 2012, 2013, 2014, 2015). Further, combining the eco-physiological modelling and genetic mapping is becoming important approach in 'plant breeding through design' to predict the performance of genotype and recombinant inbred line population in terms of phenology and physiological traits (Yin et al. 2005).
- *Engineering agroecosystem genetic composition by varietal diversification* becomes essential in view of projected increase in climatic stresses and consequent biotic stresses. Varietal diversification can improve the horizontal resistance agroecosystems to climatic and consequential stresses.
- *Engineering agroecosystem genetic composition by crop diversification*: Out of over 20,000 species of edible plants in the world, fewer than 20 species provide 90% of our food. A quick analysis indicated that crop diversity ranged from 23 to 80 in major states of India (Fig. 12.4).

Diversification of food basket and food production systems helps in sustainable food and nutritional security systems in changing climates. There is a need to focus on socio-economic research to delineate the effects of globalization, markets, food habits, policy initiatives and crop diversification.

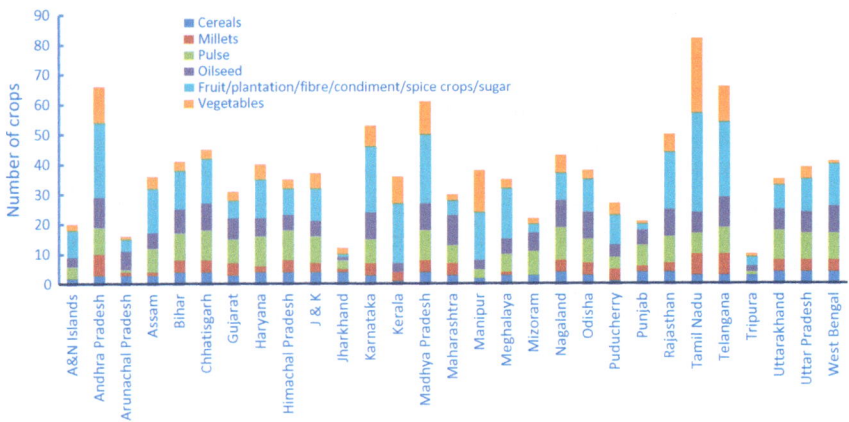

Fig. 12.4 Crop diversity in Indian states

12.4 Some Recent Examples of Breeding for Climate Change

- Common bean biodiversity has been used in plant breeding to develop both heat- and cold-tolerant varieties grown from the hot Durango region in Mexico to the cold high altitudes of Colombia and Peru.
- Corn genetic resources have been used in breeding varieties adapted to cultivation from sea level to over 3000 mamsl, as in Nepal.
- Protection of farmers' varieties as is done for varieties of millets and rice.
- Cultivar that can tolerate excessive heat during pollination for cowpea and corn and flooding early in the growing season for soybean and rice.
- Maize hybrids that show better synchronization of pollination and flowering under heat and water stress.
- Genome sequences are available for many crop species such as rice (Goff et al. 2002; Yu et al. 2002; IRGSP 2005), poplar (*Populus trichocarpa*) (Tuskan et al. 2006), sorghum (*Sorghum bicolor*) (Paterson et al. 2009), maize (Schnable et al. 2009), soybean (*Glycine max*) (Schmutz et al. 2010), cucumber (*Cucumis sativus*) (Huang et al. 2009), pigeon pea (*Cajanus cajan*) (http://www.icrisat.org/gt-bt/IIPG/home.html), wheat (http://www.genomeweb.com/sequencing/wheat-genome-sequenced-roches-454) and barley (*Hordeum vulgare*) (http://barleygenome.org/).
- Transgenic rice plants overexpressing *Arabidopsis* CBF3/DREB1A or ABF3 TF showed improved tolerance to drought and salinity without growth retardation (Oh et al. 2005).
- Using integrated biotechnology approaches, drought-tolerant maize cultivars were developed with about 20–50% higher yields under drought than the current cultivars. Several of them have already reached farmers' fields in Africa. The high NUE maize cultivars are also being developed (Varshney et al. 2011).

- To enhance adaptive phenotypic plasticity or yield stability of sorghum and pearl millet in variable climates, traits such as photoperiod-sensitive flowering, plastic tillering, flooding tolerance, seedling heat tolerance and phosphorus efficiency are identified for inducting into the cultivars for West Africa (Haussmann et al. 2012).
- Diagnostic markers for photoperiod sensitivity gene (*Ppd-D1*) and vernalization genes (*Vrn-A1*, *Vrn-B1* and *Vrn-D1*) were used for adaptation of wheat in Australia (Eagles et al. 2010).
- The *Sub1* rice tolerant to flood can survive total submersion for more than 2 weeks, with great benefits to farmers.
- Introduction of *Sub1* QTL resulted in rice varieties that can tolerate flooding for 12–14 days, and these varieties such as Swarna *Sub1* are cultivated in over one million hectares.

12.5   Conclusion

The strategy for adaptation to climate change has to be multidimensional with crops and cultivars as the central themes. There is a need for serious reorientation of breeding efforts to a comprehensive genetic improvement programme for sustaining the crop productivity in changing climates. Characterization of projected climatic stresses, characterization of entire germplasm with projected climatic variability as background, utilization of entire genetic diversity and deploying multipronged approaches for genetic improvement will ensure the enhanced crop production and quality to meet the demands of future climates and population.

References

Asseng S, Ewert F, Rosenzweig C, Jones JW, Hatfield JL, Ruane A, Boote KJ, Thorburn P, Rötter RP, Cammarano D, Brisson N, Basso B, Martre P, Aggarwa PK, Angulo C, Bertuzzi P, Biernath C, Challinor AJ, Doltra J, Gayler S, Goldberg R, Grant R, Heng L, Hooker J, Hunt LA, Ingwersen J, Izaurralde RC, Kersebaum KC, Müller C, Naresh-Kumar S, Nendel C, O'Leary G, Olesen JE, Osborne TM, Palosuo T, Priesack E, Ripoche D, Semenov MA, Shcherbak I, Steduto P, Stöckle C, Stratonovitch P, Streck T, Supit I, Tao F, Travasso M, Waha K, Wallach D, White JW, Williams JR, Wolf J (2013) Uncertainty in simulating wheat yields under climate change. Nat Clim Chang 3:827–832. doi:10.1038/nclimate1916
Asseng S, Ewert F, Martre P, Rötter RP, Lobell DB, Cammarano D, Kimball BA, Ottman MJ, Wall GW, White JW, Reynolds MP, Alderman PD, Prasad PVV, Aggarwal PK, Anothai J, Basso B, Biernath C, Challinor AJ, De-Sanctis G, Doltra J, Fereres E, Garcia-Vila M, Gayler S, Hoogenboom G, Hunt LA, Izaurralde RC, Jabloun M, Jones CD, Kersebaum KC, Koehler AK, Müller C, Naresh-Kumar N, Nendel C, O'Leary G, Olesen JE, Palosuo T, Priesack E, Eyshi Rezaei E, Ruane AC, Semenov MA, Shcherbak I, Stöckle C, Stratonovitch P, Streck T, Supit J, Tao F, Thorburn P, Waha K, Wang E, Wallach D, Wolf J, Zhao Z, Zhu Y (2015) Rising temperatures reduce global wheat production. Nat Clim Chang 5(2):143–147. http://dx.doi.org/10.1038/nclimate2470

Bentley AR, Turner AS, Gosman N, Leigh FJ, Maccaferri M, Dreisigacker S, Greenland A, Laurie DA (2011) Frequency of photoperiod-insensitive ppd-a1a alleles in tetraploid, hexaploid and synthetic hexaploid wheat germplasm. Plant Breed 130:10–15

BGCI (2017) Botanic gardens conservation international. https://www.bgci.org/policy/1521/. Accessed on 15 Feb 2017

Bhagat RM, Rana RS, Kalia V (2009) Weather changes related shift in apple belt in Himachal Pradesh; in global climate change and Indian agriculture-case studies from ICAR Network Project (ed). Aggarwal PK, ICAR Pub, New Delhi, pp 48–53

Bita CE, Gerats T (2013) Plant tolerance to high temperature in a changing environment: scientific fundamentals and production of heat stress-tolerant crops. Front Plant Sci 4:273. doi:10.3389/fpls.2013.00273

Borrell AK, Mullet JE, George-Jaeggli B, van Oosterom EJ, Hammer GL, Klein PE, Jordan DR (2014) Drought adaptation of stay-green sorghum is associated with canopy development, leaf anatomy, root growth, and water uptake. J Exp Bot 5(21):6251–6263

Braun HJ, Rajaram S, van Ginkel M (1996) CIMMYT's approach to breeding for wide adaptation. Euphytica 92:175–183

Ceccarelli S, Grando S, Maatougui M, Michael M, Slash M, Haghparast R, Rahmanian M, Taheri A, Al-Yassin A, Benbelkacem A, Labdi M, Mimounand H, Nachit M (2010) Climate change and agriculture paper: plant breeding and climate changes. J Agric Sci 148:627–637

Chapman SC, Chakraborty S, Dreccer MF, Howden SM (2012) Plant adaptation to climate change-opportunities and priorities in breeding. Crop Pasture Sci 63:251–268

Chikelu M, Eclio PG, Kakoli G (2012) Re-orienting crop improvement for the changing climate in 21st century. Agric Food Sec 1:1–17

DeMenocal PB (2001) Cultural responses to climate change during the late Holocene. Science 292:667–673

Dubcovsky J, Dvorak J (2007) Genome plasticity a key factor in the success of polyploid wheat under domestication. Science 316(5833):1862–1866

Eagles HA, Cane K, Kuchel H, Hollamby GJ, Vallance N, Eastwood RF, Gororo NN, Martin PJ (2010) Photoperiod and vernalization gene effects in southern Australian wheat. Crop and Pasture Sci 61:721–730

Easterling WE, Aggarwal PK, Batima P, Brander KM, Erda L, Howden SM, Kirilenko A, Morton J, Soussana J-F, Schmidhuber J, Tubiello FN (2007) Food, fibre and forest products. Climate Change 2007: impacts, adaptation and vulnerability. In: Parry ML, Canziani OF, Palutikof JP, van der Linden PJ, Hanson CE (eds) Contribution of working group II to the fourth assessment report of the intergovernmental panel on climate change. Cambridge University Press, Cambridge, pp 273–313

FAO (2010) The second report on the state of the world's plant genetic resources for food and agriculture. FAO, Rome

Faure S, Higgins J, Turner A, Laurie DA (2007) The flowering locus t-like gene family in barley (hordeum vulgare). Genetics 176:599–609

Feuillet C, Langridge P, Waugh R (2008) Cereal breeding takes a walk on the wild side. Trends Genet 24:24–32

Galiba G, Quarrie SA, Sutka J, Morgounov A, Snape JW (1995) Rflp mapping of the vernalization (vrn1) and frost-resistance (fr1) genes on chromosome 5a of wheat. Theorat Appl Sci 90:1174–1179

Garten CT, Wullschleger SD, Classen AT (2011) Review and model-based analysis of factors influencing soil carbon sequestration under hybrid poplar. Biomass Bioenergy 35:214–226

Goff SA et al (2002) A draft sequence of the rice genome (*Oryza sativa* L. ssp. japonica). Science 296:92–100

Hammer GL, Dong Z, McLean G, Doherty A, Messina C, Schusler J, Zinselmeier C, Paszkiewicz S, Cooper M (2009) Can changes in canopy and/or root system architecture explain historical maize yield trends in the us corn belt ? Crop Sci 49:299–312

Haussmann BIG, Fred-Rattunde H, Weltzien-Rattunde E, Traoré PSC, vom Brocke K, Parzies HK (2012) Breeding strategies for adaptation of pearl millet and sorghum to climate variability and change in west Africa. J Agron Crop Sci 198(5):327–339. doi: http://dx.doi.org/10.1111/j.1439-037X.2012.00526.x

Hebbar KB, Venugopalan MV, Prakash AH, Aggarwal PK (2013) Simulating the impacts of climate change on cotton production in India. Clim Chang 118(3-4):701–713

Hegde VS (2010) Genetics of flowering time in chickpea in a semi-arid environment. Plant Breed 129:683–687

Hijmans RJ, Cruz M, Rojas E, Guarino L (2001) DIVAGIS, version 1.4. A geographic information system for the management and analysis of genetic resources data. Manual. International Potato Center, Lima

Hodgkin T, Bordoni P (2012) Climate change and the conservation of plant genetic resources. J Crop Improv 26:329–345

Huang S et al (2009) The genome of the cucumber, *Cucumis sativus* L. Nat Genet 41:1275–1281

IPCC (2014) Climate change 2014: climate change impacts, adaptation and vulnerability summary for policymakers. Inter-Governmental Panel on Climate Change

IRGSP (2005) The international rice genome sequencing project. The map-based sequence of the rice genome. Nature 436:793–800

James RA, Davenport RJ, Munns R (2006) Physiological characterization of two genes for Na+ exclusion in durum wheat, nax1 and nax2. Plant Physiol 142:1537–1547

Jansson C, Wullschleger SD, Kalluri UC, Tuskan GA (2010) Phytosequestration: carbon biosequestration by plants and the prospects of genetic engineering. Bioscience 60:685–696

Jarvis A, Lane A, Hijmans RJ (2008) The effect of climate change on crop wild relatives. Agric Ecosyst Environ 126:13–23

Kholova J, Hash CT, Kakkera A, Kocova M, Vadez V (2010) Constitutive water-conserving mechanisms are correlated with the terminal drought tolerance of pearl millet (*Pennisetum glaucum* (l.) r. Br). J Exp Bot 61:369–377

Lin B (2011) Resilience in agriculture through crop diversification: adaptive management for environmental change. Bioscience 61:183–193

Liu D, Zhang H, Zhang L, Yuan Z, Hao M, Zheng Y (2014) Distant hybridization: a tool for interspecific manipulation of chromosomes. In: Pratap A, Kumar J (eds) Alien gene transfer in crop plants, Innovations, methods and risk assessment, vol 1, pp 25–42. doi:10.1007/978-1-4614-8585-8_2

Lobell DB, Marshall B, Claudia T, Mastrandrea MD, Falcon WP, Naylor RL (2008) Prioritizing climate change adaptation needs for food security in 2030. Science 319:607–610

Lopes MS, Rebetzke GJ, Reynolds M (2014) Integration of phenotyping and genetic platforms for a better understanding of wheat performance under drought. J Exp Bot 65(21):6167–6177

Mano Y, Muraki M, Fujimori M, Takamizo T, Kindiger B (2005) Identification of qtl controlling adventitious root formation during flooding conditions in teosinte (*Zea mays*) seedlings. Euphytica 142:33–42

Naresh-Kumar S, Aggarwal PK (2014) Climate change and coconut plantations in India: impacts and potential adaptation gains. Agric Syst 117:45–54. http://dx.doi.org/10.1016/j.agsy.2013.01.001

Naresh-Kumar S, Rajagopal V, Thomas TS, Vinu KC, Hanumanthappa M, Anil-Kumar, Sreenivasulu B, Nagvekar D (2002) Identification and characterization of *in situ* drought tolerant coconut palms in farmer's fields under different agro-climatic zones. In: Sreedharan K, Vinod-Kumar PK, Jayarama (eds) PLACROSYM XV Proceedings, pp 335–339

Naresh-Kumar S, Aggarwal PK, Swaroopa R, Jain S, Saxena R, Chauhan N (2011) Impact of climate change on crop productivity in Western Ghats, coastal and northeastern regions of India. Curr Sci 101(3):33–42

Naresh-Kumar S, Singh AK, Aggarwal PK, Rao VUM, Venkateswarlu B (2012) Climate change and Indian agriculture: salient achievements from ICAR Network Project. IARI Pub, New Delhi, 32p

Naresh-Kumar S, Aggarwal PK, Saxena R, Swarooparani DN, Jain S, Chauhan N (2013) An assessment of regional vulnerability of rice to climate change in India. Climate Change 118:683–699. doi:10.1007/s10584-013-0698-3

Naresh-Kumar S, Aggarwal PK, Swaroopa R, Saxena R, Chauhan N, Jain S (2014a) Vulnerability of wheat production to climate change in India. Clim Res 59:173–187. doi:10.3354/cr01212

Naresh-Kumar S, Aggarwal PK, Uttam-Kumar, Jain S, Swaroopa R, Chauhan N, Saxena R (2014b) Vulnerability of Indian mustard (*Brassica juncea* (L.) Czernj. Cosson) to climate variability and future adaptation strategies. Miti Adap Strat Global Change. doi:10.1007/s11027-014-9606-Z

Naresh-Kumar S, Govindakrishnan PM, Swaroopa R, Chauhan N, Jain S, Aggarwal PK (2015) Assessment of impact of climate change on potato and potential adaptation gains in the Indo-Gangetic Plains of India. Inter J Plant Prod 9(1):151–170

Nirmali G, Baruah KK, Gupta PK (2008) Selection of rice genotypes for lower methane emission. Agron Sustain Dev 28:181–186. doi:10.1051/agro:2008005

NOAA (2017) https://www.ncdc.noaa.gov/sotc/global/201704

Oh SJ, Song IKS, Kim YS, Jang HJ, Kim SY, Kim M, Y-Ik K, Nahm BH, Kim J-K (2005) Arabidopsis CBF3/DREB1A and ABF3 in transgenic rice increased tolerance to abiotic stress without stunting growth. Plant Physiol 138:341–351

Ortiz R (2002) Germplasm enhancement to sustain genetic gains in crop improvement. In: Engels JMM, Ramanatha-Rao V, Brown AHD, Jakson M (eds) Managing plant genetic diversity. IPGRI-Wallingford, Rome, Italy. CAB International, Wallingford, pp 275–290

Pan A, Hayes PM, Chen F, Chen THH, Blake T, Wright S, Karsai I, Bedo Z (1994) Genetic-analysis of the components of winterhardiness in barley (*Hordeum vulgare* L). Theor Appl Genet 89:900–910

Paterson AH, Bowers JE, Bruggmann R, Dubchak I, Grimwood J, Gundlach H, Haberer G, Hellsten U, Mitros T, Poliakov A, Schmutz J, Spannagl M, Tang H, Wang H, Wicker T, Bharti AK, Chapman J, Feltus AF, Gowik U, Grigoriev IV, Lyons E, Maher CA, Martis M, Narechania A, Otillar RP, Penning BW, Salamov AA, Wang Y, Zhang L, Carpita NC, Freeling M, Gingle AR, Hash CT, Keller B, Klein P, Kresovich S, McCann MC, Ming R, Peterson DG, Mehboob-ur-Rahman WD, Westhoff P, Mayer KFX, Messing J, Rokhsar DS (2009) The *Sorghum bicolor* genome and the diversification of grasses. Nature 457:551–556

Porter JR, Xie L, Challinor AJ, Cochrane K, Howden SM, Iqbal MM, Travasso MI (2014) Food security and food production systems. In: Climate change 2014: impacts, adaptation, and vulnerability, Chapter 7. Cambridge University Press, Cambridge, pp 485–533

Rajeev KV, Kailash CB, Pramod KA, Swapan KD, Peter QC (2011) Agricultural biotechnology for crop improvement in a variable climate: hope or hype? Trends Plant Sci 15(7):363–372

Ramírez-Villegas J, Lau C, Köhler A-K, Signer J, Jarvis A, Arnell N, Osborne T, Hooker J (2011) Climate analogues: finding tomorrow's agriculture today. Working paper no. 12. CGIAR Research Program on Climate Change, Agriculture and Food Security (CCAFS), Cali, Colombia. Available online at: www.ccafs.cgiar.org

Rao NKS, Laxman RH, Bhatt RM (2009) Impact of elevated carbon dioxide on growth and yield of onion and tomato. In: Aggarwal PK (ed) Global climate change and Indian agriculture-case studies from ICAR Network Project. ICAR, New Delhi, pp 35–37

Rebetzke GJ, van Herwaarden AF, Jenkins C, Weiss M, Lewis D, Ruuska S, Tabe L, Fettell NA, Richards RA (2008) Quantitative trait loci for water-soluble carbohydrates and associations with agronomic traits in wheat. Aust J Agric Res 59:891–905

Redden R, Yadav SS, Maxted N, Dulloo E, Gurino L, Smith P (2015) Crop wild relatives and climate change. Wiley Blackwell Publishing, Hoboken

Ren H, Wei K, Jia W, Davies WJ, Zhang J (2007) Modulation of root signals in relation to stomatal sensitivity to root-sourced abscisic acid in drought-affected plants. J Integr Plant Biol 49:1410–1420

Rosen AM (1990) Environmental change at the end of the early Bronze Age Palestine. In: De Miroschedji P (ed) L'urbanisation de la Palestine à l'âge du Bronze ancient. BAR International, Oxford, pp 247–255

Rosenzweig C, Elliottb J, Deryngd D, Ruane AC, Müller C, Arneth A, Boote KJ, Folberth C, Glotter M, Khabarov N, Neumann K, Piontek F, Pugh TAM, Schmid E, Stehfest E, Yang H, Jones JW (2014) Assessing agricultural risks of climate change in the 21st century in a global gridded crop model inter-comparison. PNAS 111(9):3268–3273

Scheben A, Yuan Y, Edwards D (2016) Advances in genomics for adapting crops to climate change. Curr Plant Biol 6(2016):2–10

Schmutz J et al (2010) Genome sequence of the paleopolyploid soybean. Nature 463:178–183

Schnable PS et al (2009) The B73 maize genome: complexity, diversity, and dynamics. Science 326:1112–1115

Septiningsih EM, Ignacio JC, Sendon PM, Sanchez DL, Ismail AM, Mackill DJ (2013) Qtl mapping and confirmation for tolerance of anaerobic conditions during germination derived from the rice landrace ma-zhan red. Theor Appl Genet 126(5):1357–1366

Smith P (2012) Delivering food security without increasing pressure on land. Glob Food Sec 2:18–23

Spollen WG, Henderson D, Schachtman DP, Davis GE, Springer GK, Sharp RE, Nguyen HT, Tao W, Valliyodan B, Chen K, Hejlek LG, Kim J-J, Lenoble ME, Zhu J, Bohnert HJ (2008) Spatial distribution of transcript changes in the maize primary root elongation zone at low water potential. BMC Plant Biol 8:32–32

Steele KA, Price AH, Witcombe JR, Shrestha R, Singh BN, Gibbons JM, Virk DS (2013) Qtls associated with root traits increase yield in upland rice when transferred through marker-assisted selection. Theor Appl Genet 126:101–108

Stephen PL, Donald RO (2010a) More than taking the heat: crops and global change. Curr Opin Plant Breed 13:241–248

Tuskan GA, DiFazio S, Jansson S, Bohlmann J, Grigoriev I, Hellsten U, Putnam N, Ralph S, Rombauts S, Salamov A, Schein J, Sterck L, Aerts A, Bhalerao RR, Bhalerao RP, Blaudez D, Boerjan W, Brun A, Brunner A, Busov V, Campbell M, Carlson J, Chalot M, Chapman J, Chen GL, Cooper D, Coutinho PM, Couturier J, Covert S, Cronk Q, Cunningham R, Davis J, Degroeve S, De'jardin A, de Pamphilis C, Detter J, Dirks B, Dubchak I, Duplessis S, Ehlting J, Ellis B, Gendler K, Goodstein D, Gribskov M, Grimwood J, Groover A, Gunter L, Hamberger B, Heinze B, Helariutta Y, Henrissat B, Holligan D, Holt R, Huang W, Islam-Faridi N, Jones S, Jones-Rhoades M, Jorgensen R, Joshi C, Kangasjärvi J, Karlsson J, Kelleher C, Kirkpatrick R, Kirst M, Kohler A, Kalluri U, Larimer F, Leebens-Mack J, Leplé J-C, Locascio P, Lou Y, Lucas S, Martin F, Montanini B, Napoli C, Nelson DR, Nelson C, Nieminen K, Nilsson O, Pereda V, Peter G, Philippe R, Pilate G, Poliakov A, Razumovskaya J, Richardson P, Rinaldi C, Ritland K, Rouzé P, Ryaboy D, Schmutz J, Schrader J, Segerman B, Shin H, Siddiqui A, Sterky F, Terry A, Tsai C-J, Uberbacher E, Unneberg P, Vahala J, Wall K, Wessler S, Yang G, Yin T, Douglas C, Marra M, Sandberg G, Vande Peer G, Rokhsar D (2006) The genome of black cottonwood, *Populus trichocarpa* (Torr. & Gray). Science 313:1596–1604

Varshney RK, Nayak SN, May GD, Jackson SA (2009) Next-generation sequencing technologies and their implications for crop genetics and breeding. Trends Biotechnol 27:522–530

Varshney RK, Bansal KC, Aggarwal PK, Datta SK, Craufurd PQ (2011) Agricultural biotechnology for crop improvement in a variable climate: hope or hype? Trends Plant Sci 16(7):363–371

Williams JW, Stephen TJ, Kutzbach JE (2007) Projected distributions of novel and disappearing climates by 2100 AD. PNAS 104(14):5738–5742. doi:10.1073/pnas.0606292104

Xin-Guang, Stephen PL, Donald RO (2010b) Improving photosynthetic efficiency of for greater yield. Annu Rev Plant Biol 61:235–261

Xu KN, Xu X, Fukao T, Canlas P, Maghirang-Rodriguez R, Heuer S, Ismail AM, Bailey-Serres J, Ronald PC, Mackill DJ (2006) Sub1a is an ethylene-response-factor-like gene that confers submergence tolerance to rice. Nature (London) 442

Yin X, Struik PC, Eeuwijk FA, Stam P, Tang J (2005) QTL analysis and QTL-based prediction of flowering phenology in recombinant inbred lines of barley. J Exp Bot 56:967–976

Yu J, Hu S, Wang J, Wong GK, Li S, Liu B, Deng Y, Dai L, Zhou Y, Zhang X, Cao M, Liu J, Sun J, Tang J, Chen Y, Huang X, Lin W, Ye C, Tong W, Cong L, Geng J, Han Y, Li L, Li W, Hu G, Huang X, Li W, Li J, Liu Z, Li L, Liu J, Qi Q, Liu J, Li L, Li T, Wang X, Lu H, Wu T, Zhu M, Ni P, Han H, Dong W, Ren X, Feng X, Cui P, Li X, Wang H, Xu X, Zhai W, Xu Z, Zhang J, He S, Zhang J, Xu J, Zhang K, Zheng X, Dong J, Zeng W, Tao L, Ye J, Tan J, Ren X, Chen X, He J, Liu D, Tian W, Tian C, Xia H, Bao Q, Li G, Gao H, Cao T, Wang J, Zhao W, Li P, Chen W, Wang X, Zhang Y, Hu J, Wang J, Liu S, Yang J, Zhang G, Xiong Y, Li Z, Mao L, Zhou C, Zhu Z, Chen R, Hao B, Zheng W, Chen S, Guo W, Li G, Liu S, Tao M, Wang J, Zhu L, Yuan L, Yang H (2002) A draft sequence of the rice genome (*Oryza sativa* L.ssp. indica). Science 296:79–92

Biotechnological Applications for Improvement of Drought Tolerance

13

Monika Dalal and T.R. Sharma

Abstract

Biotechnology has contributed significantly toward understanding the fundamentals of plant biology. It is steadily making headway in to the agricultural sector for improving quality and sustaining yield under different environmental stress conditions. However, progress in developing transgenic plants with enhanced drought tolerance is relatively slow. Drought tolerance is a quantitative trait. Plants respond to water deficit by different physiological, molecular, and biochemical mechanisms. Several genes and gene networks have been identified and have been shown to confer drought tolerance in different plant species. However, most of these studies on drought tolerance are limited to controlled conditions of lab or greenhouses. There are a few studies where evaluation of transgenic plants has been carried out in field or near-field conditions, and yield advantage has been demonstrated under drought stress. Since performance of crop plants under field condition is essential, this article describes studies on evaluation of transgenic crop plants for drought tolerance under field conditions.

13.1 Introduction

Since 1980s when the first transgenic plant was developed, plant biotechnology has come a long way. Decoding of complete *Arabidopsis* genome and development of a floral dip method of genetic transformation in *Arabidopsis* were the stepping stones in the field of plant molecular biology and biotechnology. The advent of genomics

M. Dalal • T.R. Sharma (✉)
ICAR-National Research Centre on Plant Biotechnology,
Pusa Campus, New Delhi 110012, India
e-mail: monikadalal@hotmail.com; trsharma1965@gmail.com

© Springer Nature Singapore Pte Ltd. 2017
P.S. Minhas et al. (eds.), *Abiotic Stress Management for Resilient Agriculture*,
DOI 10.1007/978-981-10-5744-1_13

Table 13.1 List of commercially approved genetically modified events in different crops

Trait	No. of events	Plant species
Abiotic stress tolerance	8	Maize, soybean, sugarcane
Altered growth/ yield	3	Maize, soybean, eucalyptus
Disease resistance	26	Bean, papaya, plum, potato, squash, sweet pepper, tomato
Herbicide tolerance	237	Alfalfa, argentine canola (*B. napus*), carnation, chicory, cotton, creeping bent grass, flax, maize, polish canola (*B. rapa*), potato, rice, soybean, sugar beet, tobacco, wheat
Insect resistance	198	Cotton, eggplant, maize, poplar, potato, rice, soybean, tomato
Modified product quality	80	Alfalfa, apple, argentine canola (*B. napus*), carnation, maize, melon, petunia, potato, rice, rose, soybean, tobacco, tomato
Pollination control system	25	Argentine canola (*B. napus*), chicory, maize

Source: http://www.isaaa.org/gmapprovaldatabase/commercialtraitlist/default.asp as on 21.2.16

and proteomics also accelerated the gene discovery, understanding of basic plant biological processes, and the responses to different environmental stresses. Though plants differ from each other on various aspects including evolutionary period, the core components of gene functions and mechanisms have been found to be relatively conserved among plants (Nakashima et al. 2009). Therefore, extrapolating the fundamental knowledge from model plants like *Arabidopsis* to crop plants became relatively easier. In addition, the ability to introduce useful genes from heterologous systems to the crop plants not only diversifies the gene pool but was proved immensely useful for crop improvement. As of now, there are total 577 events of different crops under commercially approved genetically modified (GM) plants category (Table 13.1).

Progress in developing transgenic crop plants with enhanced abiotic stress tolerance has been invariably slow. The eight events under abiotic stress tolerance are confined to only three crops: maize, soybean, and sugarcane. Though several genes and gene networks involved in abiotic stress response have been identified, and many of these have been shown to confer tolerance in different plant species (Valliyodan and Nguyen 2006; Yamaguchi-Shinozaki and Shinozaki 2006; Reguera et al. 2012), only few events reached commercial cultivation stage. There are several reasons for this slow pace. Initially, the recalcitrant nature of major food crops, that is, rice, wheat, and maize, for both tissue culture and transformation remained a limiting step. Further the plant growth and survival were the major criteria for analyzing drought tolerance in transgenic plants (Umezawa et al. 2006). Drought tolerance was mostly evaluated during vegetative stage, and even if analyzed at reproductive stage, it was mainly under controlled conditions. The yield of plants in controlled environment or greenhouses cannot be equated with crop yield in field. The natural environmental conditions are more dynamic where plants often suffer more than one stress at a time. Despite all the odds, there is a dire need to dissect the strategy of enhancing the drought tolerance of crop plants. This would require amalgamation of knowledge on plant molecular biology, biotechnology, plant physiology and biochemistry, and finally agronomy and crop management at the field level to succeed in this challenging endeavor.

13.2 Drought and Mechanism of Tolerance

Drought can be defined in terms of meteorological, hydrological, agricultural, and socioeconomic aspects. In general, it refers to precipitation deficit over an extended period of time which can be for a season or more, leading to shortage of water thereby causing adverse impacts on plants and animals. Drought is a normal and recurrent phenomenon worldwide, but the frequency, intensity, and duration of drought may increase in the future due to global climate change (Dai 2013).

Agricultural drought refers to soil moisture deficit that affects optimum growth and yield of the plants. The water uptake from soil and its transport in the plant are driven by the soil-plant-atmosphere continuum (SPAC). Plants experience water deficit or drought stress when the demand for water through transpiration exceeds the supply. The demand for water varies with growth stage of plants. The soil moisture required for germination is different from that required during grain-filling stage. In general, reproductive stage of plants is more susceptible to drought. The effect of water deficit also varies depending on the plant species and soil properties. Moreover, crops may suffer one or more spells of soil moisture deficit during a growing season due to shortfall in precipitation/irrigation or increased evaporation. Plants have evolved several morphological, physiological, biochemical, and molecular processes through which they sense, respond, and adapt to drought stress. These processes can be categorized into two mechanisms, viz., dehydration avoidance and dehydration tolerance. Dehydration avoidance is the ability of plants to sustain high plant water status or cellular hydration under drought conditions by enhanced extraction of soil moisture through deep root system and to reduce water loss by closing the stomata and through protective plant cuticle, leaf rolling, senescence of older leaves, etc. Dehydration tolerance is defined as the capacity to sustain plant function even at relatively low tissue water potential. Dehydration tolerance in plants depends on modification of metabolism, production of organic compatible solutes (proline, sugars, polyols, betaine, etc.), and expression of genes involved in membrane integrity, cellular homeostasis (ionic, osmotic, and metabolic homeostasis), stress damage control, repair, etc.

A plant may have a combination of these adaptive traits for drought tolerance. Moreover there are species and genotype-specific differences in drought adaptation. In spite of the complex nature of the trait, there are a few studies where significant improvement in drought tolerance and grain yield has been reported. The major criterion for success of agricultural crop whether it is bred through genetics or genetic engineering is performance under field conditions. The following section describes those studies where transgenic crop plants have been subjected to field evaluation and analyzed for yield or yield components.

13.3   Strategies for Engineering Drought Tolerance

Under drought stress, two groups of genes are expressed. The first group codes for functional proteins that protect plant cells against water deficit. This includes chaperones, late embryogenesis abundant (LEA) proteins, sugar and proline, transporters, detoxification enzymes, etc. The second group codes for regulatory proteins involved in regulation of signal transduction and transcriptional regulation of stress response (Shinozaki and Yamaguchi-Shinozaki 2007). Majority of attempts to develop drought-tolerant plants relied on manipulation of single gene coding for either functional or regulatory protein which primarily imparted the cellular tolerance (Table 13.2). However, realizing the contribution of different component traits in drought tolerance, efforts were also made to modulate physiological processes and morphological adaptive traits in plants (Fig. 13.1).

Table 13.2 Genes conferring drought tolerance in field conditions

Genes	Crop	Attributes	References
Mn-SOD	Alfalfa	Two- to fivefold forage yield and higher survival	McKersie et al. (1996)
HVA1 (*LEA3*)	Wheat	Improved biomass and yield attributes	Sivamani et al. (2000) and Bahieldin et al. (2005)
OsLEA3–1	Rice	Higher grain yield and spikelet fertility	Xiao et al. (2007)
CspA; *CspB*	Maize	Increase in the yield (10.2–30.8%)	Castiglioni et al. (2008)
(OCPI1	Rice	Higher grain yield	Huang et al. (2007)
BetA	Sugarcane	–	Waltz (2014); http://www.isaaa.org/gmapprovaldatabase/
BetA	Maize	Higher cell membrane stability and photosynthetic rate	Quan et al. (2004)
TsVP	Maize	Improved grain weight per ear and 1000-grain weight	Li et al. (2008)
ZmNF-YB2	Maize	Yield advantage (~50%), improved stomatal conductance, reduced wilting, and maintenance of photosynthesis	Nelson et al. (2007)
SNAC1	Rice	Higher spikelet fertility and enhanced seed setting	Hu et al. (2006)
OsNAC10		Increasing grain yield	Jeong et al. (2010)
AP37	Rice	Yield advantage, lesser decrease in spikelet and spikelet filling rate	Oh et al. (2009)
LOS5; *ZAT10*	Rice	Higher yield and higher spikelet fertility	Xiao et al. (2009)

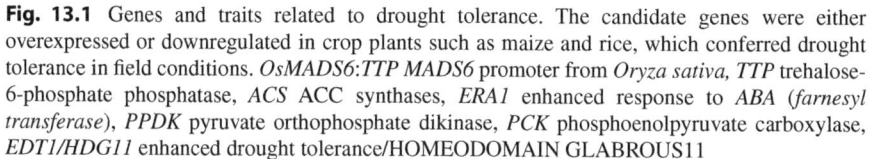

| Candidate genes | Traits for drought tolerance |

OsMads6::TTP — Sugar transport to seeds

RNAi - ACS — Higher seed set/ reduced kernel abortion

ERA1 — stomatal conductance and reduced transpiration

C4 PPDK and C4 PCK — Photosynthetic capacity (Pathway)

HARDY — Biomass and WUE

EDT1/HDG11 — Reduced stomatal density and higher WUE

EDT1/HDG11 — Rooting depth

Fig. 13.1 Genes and traits related to drought tolerance. The candidate genes were either overexpressed or downregulated in crop plants such as maize and rice, which conferred drought tolerance in field conditions. *OsMADS6:TTP MADS6* promoter from *Oryza sativa, TTP* trehalose-6-phosphate phosphatase, *ACS* ACC synthases, *ERA1* enhanced response to *ABA (farnesyl transferase)*, *PPDK* pyruvate orthophosphate dikinase, *PCK* phosphoenolpyruvate carboxylase, *EDT1/HDG11* enhanced drought tolerance/HOMEODOMAIN GLABROUS11

13.4   Manipulating Functional Genes

One of the pioneering studies reporting field trial for evaluating drought stress tolerance in plants through transgenic approach was reported by McKersie et al. (1996). The strategy involved overexpression of antioxidant enzyme *Mn-superoxide dismutase (Mn-SOD)* in subcellular compartments such as mitochondria and chloroplast of alfalfa (*Medicago sativa*) plants. Antioxidant enzymes protect cellular components from ROS (oxidative stress). Overexpression of *Mn-superoxide dismutase (Mn-SOD)* in alfalfa *(Medicago sativa)* plants reduced cellular damage from water-deficit stress and increased regrowth from crowns. In a 3-year field trial, alfalfa transgenic lines showed a significantly improved (two- to fivefold) herbage yield and higher survival as compared to that of wild type (McKersie et al. 1996).

A protein-coding gene family that was successful in imparting drought tolerance in many plants is late embryogenesis abundant (LEA) protein. LEA proteins accumulate during seed desiccation and in vegetative tissues when plants experience water deficit (Dalal et al. 2008). An ABA-responsive group 3 LEA protein gene from barley (*HVA*1) driven by maize ubiquitin (*ubi*1) promoter was introduced into spring wheat (*Triticum aestivum* L.). T4 progenies of transgenic lines were tested in nine field experiments over six cropping seasons. Among these, three lines performed better than wild-type (WT) control in terms of plant height, biomass, and yield attributes (Sivamani et al. 2000; Bahieldin et al. 2005). Though results were variable from year to year, yet it showed the potential of *HVA*1 to confer drought stress tolerance in crop plants. Another drought and salt stress-responsive group 3 LEA protein, *OsLEA3–1*, driven by three different promoters, viz., *CaMV35S*, *Actin1*, and *HVA1*, was transformed into rice. The transgenic plants expressing *OsLEA3–1* by a drought-inducible *HVA1-like* promoter and constitutive promoter *CaMV35S* showed improved drought tolerance than those with *Actin 1* driven *OsLEA3–1* gene. The homozygous transgenic lines (with *OsLEA3-35S* and *OsLEA3-HVA* constructs) were tested in the field for drought tolerance. The transgenic lines showed higher grain yield (15.2–17.8 g) than the wild type (12.1 g) under drought stress (Xiao et al. 2007). The spikelet fertility of these transgenic plants was higher than the WT plants, while yield components such as number of spikelet per plant and grain weight did not show significant difference in transgenic and WT under both normal and stress conditions (Xiao et al. 2007). Thus, the improvement in yield of transgenic plants under drought stress conditions was mainly due to relatively higher spikelet fertility.

In addition to LEA proteins, another class of chaperones is RNA chaperone which stabilizes and promotes efficient folding of the RNA under stressful conditions. In maize, heterologous expression of two different RNA chaperones, *CspA* from *E. coli* and *CspB* from *B. subtilis* (cold shock protein A and B), improved stress tolerance at both vegetative and reproductive stages (Castiglioni et al. 2008). The field trials of 20 independent events (10 *CspA* and 10 *CspB* events) for grain yield under water-deficit stress and nonstress conditions revealed an average yield increase of 4.6 and 7.5% with *CspA* and *CspB* transgenic plants, respectively, than controls under water stress. The best performing *CspA* and *CspB* transgenic events recorded a significant increase in the yield ranging from 10.2 to 30.8% under water stress. One of the *CspB*-expressing events (*CspB-Zm* event 1) was consistent for drought tolerance over several field trials. The performance of this *CspB* event in three different hybrid backgrounds was evaluated under water-deficit stress, and irrespective of hybrid background, transgenic event outperformed at least by 0.5 Mg ha^{-1} as compared to wild-type control under both the vegetative and reproductive stages water-deficit stress (Castiglioni et al. 2008). The *CspB* event (MON-8746Ø-4) developed by Monsanto has been registered with trade name Genuity® DroughtGard™ and approved in 14 countries for food, feed, or cultivation (http://www.isaaa.org/gmapprovaldatabase/event/). Huang et al. (2007) developed rice transgenics overexpressing a proteinase inhibitor gene *chymotrypsin inhibitor-like 1* (*OCPII*) from rice. The transgenic rice lines were evaluated at reproductive stage and showed significantly higher grain yield (5.4–6.1 g per plant) than the negative transgenic control and WT (1.9–2.1 g per plant) even under severe drought stress conditions with soil water content of about 16% (Huang et al. 2007).

One mechanism of cellular adaptation in plants is osmoregulation. The cells synthesize and accumulate compatible solutes such as amines (polyamines and glycine betaine), amino acids (proline), sugars (trehalose, fructan), sugar alcohols (mannitol), etc. Accumulation of these solutes enables maintenance of cellular water potential and cell turgor and also helps in stabilizing membranes and/or scavenges reactive oxygen species (ROS) (Reguera et al. 2012). Overexpression of genes leading to synthesis of osmolytes has been shown to confer drought tolerance in many species (Chen and Murata 2002, Reguera et al. 2012). One such successful event to reach commercial approval is reported in sugarcane. The scientists from PT Perkebunan Nusantara XI and University of Jember (Indonesia) have developed transgenic sugarcane expressing *BetA* gene from *Rhizobium meliloti* (Waltz 2014). *BetA* codes for choline dehydrogenase which converts choline into betaine aldehyde, which is then converted to osmoprotectant glycine betaine by the enzyme betaine aldehyde dehydrogenase. There are currently three events of drought-tolerant GM sugarcane from Indonesia that have been approved by the National Genetically Modified Product Biosafety Commission. Two events (NXI-4 T and NXI-6 T) harbor the *BetA* gene from *Rhizobium meliloti*, while one event (NXI-1 T) harbors *BetA* from *E. coli* (http://www.isaaa.org/gmapprovaldatabase/event/). In maize, overexpression of *betA* gene from *Escherichia coli* resulted in drought tolerance at germination, seedling, and at reproductive stages. In protected field experiment, the grain yield of some of the transgenic plants was 10–23% higher than that of wild-type plants after 3 weeks pre-flowering drought treatment. The enhanced glycine betaine accumulation in transgenic maize was associated with higher cell membrane stability and photosynthetic rate in transgenic plants as compared to wild-type plants (Quan et al. 2004).

Some plants maintain their turgor at low solute potentials by actively transporting the organic acids or solutes in to the vacuole against their electrochemical gradients. The electrochemical gradients across the vacuolar membrane are established by two proton pumps, namely, a vacuolar H^+-adenosine triphosphatase (VH$^+$-ATPase) and a vacuolar H^+-pyrophosphatase (V-H$^+$-PPase) (Martinoia et al. 2007). Li et al. (2008) demonstrated that by constitutive overexpression of a potassium-dependent *TsVP* (vacuolar H^+-pyrophosphatase) gene from a halophyte *T. halophile*, the drought tolerance of maize plants can be significantly increased. Transgenic maize plants recorded better yield attributes such as grain weight per ear, number of grains per row, and 1000-grain weight as compared to wild-type (WT) plants after 6 weeks of drought stress in the field.

13.5 Manipulating Regulatory Genes

With the basic understanding of the gene networks and mechanisms governing abiotic stress tolerance, it was realized that drought tolerance being a multigenic trait would need coherent functioning of several genes and networks to attain tolerance. This led to another strategy that is to manipulate a master switch. These master switches are transcription factors that regulate expression of several downstream genes. There were regulatory genes that were found to respond to multiple stresses, e.g., drought and low-temperature stress at the transcriptional level (reviewed in Shinozaki and Yamaguchi-Shinozaki 2007). Therefore, the rationale was to target multiple stress tolerance by manipulating upstream regulatory protein. The pioneering work started by overexpressing dehydration-responsive element-binding protein 1 (*DREB1*) in *Arabidopsis* (Liu et al. 1998). However, constitutive expression had pleiotropic effects on plants; hence, expression was regulated by stress-responsive promoter, e.g., *rd29A* (Kasuga et al. 1999). Since then several members of *DREB* family have been identified and shown to confer tolerance to different abiotic stresses in several plant species (Lata and Prasad 2011).

Several other transcription regulatory gene families such as *AP2/ERF*, *bZIP*, *NAC*, *MYB*, *zinc finger*, and *NFY* have been engineered for drought tolerance in many plant species. Nelson et al. (2007) demonstrated that constitutive expression of *ZmNF-YB2* subunit in maize improved stress-related responses such as stomatal conductance, leaf temperature, reduced wilting, and maintenance of photosynthesis. During 2-year field trials, the best performing transgenic line showed yield advantage of about 50% over that of control (avg 74 bushels acre^{-1}) under severe drought stress conditions. In rice, overexpression of stress-responsive *NAC1* (*SNAC1*) gene enhanced the seed setting by 22–34% as compared to that of control under severe stress (soil moisture content 15%) in the field (Hu et al. 2006). The increase seed set was related to higher spikelet fertility (7.4–22.3% and 23.0–34.6%) in transgenic plants as compared to control under both moderate and severe drought stresses, respectively. The enhanced drought tolerance of the transgenic plants was partly due to the increased stomatal closure and/or ABA sensitivity which prevented water loss from the plants (Hu et al. 2006). Another member of the NAC family, *OsNAC10*, was overexpressed in rice by constitutive (*GOS2*) as well as root-specific promoter (*RCc3*). The plants harboring *RCc3:OsNAC10* showed increasing grain yield of up to 42% under drought and maximum 14% increase under normal conditions over control (Jeong et al. 2010).

In rice, there are at least 42 APETELA2 (AP2) domains containing genes that are induced by one or more abiotic stress conditions (Oh et al. 2009). Among these, overexpression of *AP37* driven by cytochrome c gene (*OsCc1*) promoter in rice imparted drought and salinity stress at vegetative stage. At reproductive stage, the transgenic plants showed a yield advantage of 16%–57% over controls under severe drought conditions. Since there was lesser reduction in the number of spikelet and filling rate in the transgenic plants under drought stress, it was suggested that *AP37* enhanced drought tolerance by protecting the development of panicle and spikelet in transgenic rice (Oh et al. 2009).

In an ambitious project, efficacy of seven well-characterized genes for drought and salt tolerance from model plant was compared in the same rice background (rice cultivar Zhonghua 11) (Xiao et al. 2009). Independent transgenic rice plants with seven genes, viz., *CBF3*, *SOS2*, *NCED2*, *NPK1*, *LOS5*, *ZAT10*, and *NHX1*, cloned under the control of constitutive promoter (*Actin1*) and stress-inducible promoter (*OsHVA22*) were developed and evaluated for reproductive stage drought tolerance in field (Xiao et al. 2009). The drought tolerance was evaluated based on yield per plant and spikelet fertility. Among these, transgenic lines of *LOS5* and *ZAT10* with both the promoters showed significantly higher yield as well as higher spikelet fertility and thus showed more efficacy for drought tolerance in field than the rest of the five genes and promoter combinations examined (Xiao et al. 2009).

13.6 Modulation of Physiological and Morphological Traits

Abscisic acid (ABA) is known as a stress hormone. Under drought stress, there is a severalfold increase in the ABA accumulation in the leaves. ABA regulates stomatal closure, transpiration, and expression of several drought response genes. Once stress is relieved, the ABA concentration in leaves reduces to normal levels. Thus, ABA synthesis or accumulation can be one of the targets for enhancing drought tolerance. Wang et al. (2005) employed an antisense technique to downregulate *ERA1* (β-subunit of farnesyl transferase), a negative regulator of ABA synthesis, under the control of stress-inducible *rd29A* promoter in canola (*B. napus*). Transgenic canola lines showed enhanced ABA sensitivity, significant reduction in stomatal conductance, and water transpiration under drought stress conditions. Two of the events, YPT1 and YPT2, showed up to 15 and 16% higher grain yield as compared to nontransgenic control during 3 consecutive years of field trial. The transgenic plants were also more resistant to drought-induced seed abortion during flowering. These transgenic lines were designated as yield protection technology (YPT) lines (Wang et al. 2005). The robustness of this strategy was further proved by shoot-specific downregulation of α-subunit of farnesyl transferase in canola by Wang et al. (2009). According to the developer company (Performance Plants, Kingston, Ontario), in 5 years of field trials, YPT®-protected canola delivered up to 26% more seed yield with moderate drought stress without any yield penalty under normal conditions. YPT® has been successful in corn and petunia, and it is currently being deployed in soybean, rice, sorghum, cotton, and turf grass (http://www.performanceplants.com).

Photosynthesis is one of the main physiological processes which get negatively affected under water deficit as net carbon uptake gets affected by stomatal closure. This affects the photo-assimilate partitioning in the plants and moreover affects the grain filling and development. With the aim of improving the photosynthesis rate and grain yield under drought stress in rice, Gu et al. (2013) developed transgenic rice harboring maize C4 *pyruvate orthophosphate dikinase enzyme* (*PPDK*) gene independently or in combination with maize C_4-specific *phosphoenolpyruvate carboxylase* (*PCK*) gene. The transgenic plants showed 20–40% higher photosynthetic rate under well water condition, 45–60 and 80–120% higher photosynthetic rate under moderate and severe drought conditions, respectively. Similarly grain yield was improved by 16.1, 20.2, and 20.0% under the well-watered, moderate, and severe drought conditions, respectively. Furthermore the transgenic plants had higher leaf water content, stomatal conductance, transpiration efficiency, root oxidation activity, and a stronger active oxygen-scavenging system than the WT under all the three treatments (Gu et al. 2013). Thus, increase in photosynthesis and hence photo-assimilate had a positive effect on other plant processes and also reduced the generation of ROS.

Increasing the water use efficiency (WUE) of the plants can be one of the strategies to save water under drought stress. In physiological terms, WUE is referred to as transpiration efficiency, that is, ratio of photosynthesis (A) to transpiration (T). There are few candidate genes that have been demonstrated to improve the water use efficiency and hence the drought tolerance of the plants. Among these, *ERECTA*, a leucine-rich repeat receptor-like kinase (LRR-RLK) was initially shown to enhance WUE in *Arabidopsis* (Masle et al. 2005). Recently extensive field trials conducted at multiple locations demonstrated that overexpression of *AtERECTA* (ER) in rice and tomato confers thermotolerance (Shen et al. 2015). The transgenic tomato plants showed decreased stomatal density, decreased stomatal conductance, and increased transpiration efficiency, while the transgenic rice plants had significantly higher seed setting (55–70%) than the control transgenic line (~35%) under heat stress (Shen et al. 2015). However, authors suggested that thermotolerance was achieved irrespective of water loss. Another promising candidate gene is *HRD* gene, an AP2/ERF-like transcription factor that was first characterized in *Arabidopsis* and rice for conferring drought tolerance by increasing biomass and WUE (Karaba et al. 2007). *HARDY* gene from *Arabidopsis* was overexpressed in *Trifolium alexandrinum* L. (Egyptian clover) which is an important forage crops in semiarid regions (Abogadallah et al. 2011). In a combined field trial of drought and salt stress, the transgenic plants performed better than the wild-type plants in terms of biomass.

HVA1 gene was also shown to enhance drought tolerance by increasing the WUE in wheat (Bahieldin et al. 2005). Enhanced WUE was one of the criteria for selecting transgenic wheat plants expressing *DREB1A* under the control of *rd29A* promoter in greenhouse conditions (Saint Pierre et al. 2012). However, under field conditions, these transgenic lines did not impart any yield advantage over wild type under water deficit (Saint Pierre et al. 2012). Transgenic rice harboring enhanced drought tolerance/HOMEODOMAIN GLABROUS11 (*EDT1/HDG11*), a homeodomain-leucine zipper transcription factor from *Arabidopsis*, showed several stress-adaptive responses such as higher levels of abscisic acid, proline, soluble sugar, and reactive oxygen species-scavenging enzyme activities and extensive root system, reduced stomatal density, and higher WUE (Yu et al. 2013). In field, the transgenic lines yielded significantly higher (about 16%) than the wild-type control under both normal and drought conditions. The increased grain yield was majorly attributed to larger panicle size and higher seed-setting rate in transgenic rice plants (Yu et al. 2013).

Kernel abortion is one of the major problem in maize under drought stress conditions. Since higher ethylene levels in kernels lead to kernel abortion, an RNAi strategy was employed to target the *ACC synthases* (*ACS*) expression and reduce its synthesis (Habben et al. 2014). With this method, about 57 and 49% reduction in ethylene could be achieved in transgenic under low- and high-stress conditions as compared to the wild type. In field, seven transgenic events registered higher grain yield compared with the WT hybrid average of 106.2 bu. acre^{-1}. An increase of 9.3 bu. acre^{-1} over WT was with the event (DP-E29) after a flowering time drought stress. The transgenic events showed decrease in the anthesis-silking interval and increase in kernel number/ear as compared to wild type. Over 2-year field trials, two promising events (DP-E29 and DP-E21) were identified based on consistent positive yield response compared with WT under drought stress (Habben et al. 2014).

The problem of reduced supply of carbohydrates (sugar) to the developing kernel also leads to reduced seed set. Nuccio et al. (2015) overexpressed a rice *trehalose-6-phosphate phosphatase* (*TPP*) in developing maize ears using a floral promoter *OsMads6*. TPP is a trehalose biosynthesis pathway enzyme which produces trehalose from trehalose-6-phosphate (T6P). Trehalose-6-phosphate acts as a sugar signal that regulates growth and development. Overexpression of *TPP* increased the sucrose concentration in spikes by 20%. In multisite and multi-season field trials, the *OsMads6-Tpp1* transgenic plants showed yield increase up to 49% under non-drought or mild drought conditions and from 31 to 123% under more severe drought conditions, as compared to that of nontransgenic controls (Nuccio et al. 2015).

Among morphological traits, epicuticular wax present on the cuticle, the outermost surfaces of plants, is one of the drought adaptation traits that reduces the cuticular conductance of the plants and also protects against different environmental stress. *Glossy 1* (GL1) gene is one of the genes involved in wax synthesis. Rice transgenic plants overexpressing *OsGL1–2*, a homolog of *GL1* from rice, were evaluated at reproductive stage for drought tolerance in refined field (Islam et al. 2009). The transgenic plants showed slower chlorophyll leaching from the overexpression plant leaves than that from the WT leaves, thereby suggesting a decrease in cuticular permeability in the transgenic plants.

13.7 Conclusions

The success of plant biotechnology is evident by 100-fold increase in cultivation of transgenic crops since their introduction in 1996. Such tremendous increase and acceptance of transgenic crops are possible due to yield advantages offered by these crops. The commercially grown transgenic plants have shown economic, environmental, and social benefits by increasing the yield, by decreasing the pesticide or herbicide use, and by increasing the income of the farmers. Drought tolerance in crops still remains a major concern. The number of transgenic crops with drought tolerance trait is not so encouraging. However, there are promising reports of engineering drought tolerance and yield gains under moderate and severe drought stress conditions in field. There is definitely no single magical gene or physiological trait that may impart complete drought tolerance, and not all transgenic lines engineered with same gene are equally tolerant. Thus, there is a need to understand not only the molecular mechanisms but also the physiological basis of drought tolerance, and together these can be deployed for designing crops that are climate resilient. There are excellent examples of single gene modification in plants outperforming the wild type consistently over multilocation and multi-season field trials. However, pyramiding the genes conferring cellular tolerance and modulating the physiological traits may prove more versatile. Furthermore depending on the site of gene insertion and copy number and several other factors, the fate of gene and its expression varies accross the transgenic events. Therefore, it is essential to generate several hundreds of transgenic plants, to screen them under greenhouse conditions, and finally at field level to select the best performer. The success of new techniques such as genome editing has been demonstrated in some crops (Jiang et al. 2013; Shan et al. 2013; Svitashev et al. 2015) and can be exploited for modifying several genes at a time. Furthermore, it would be better to combine best with the best, that is, to improve the trait in the elite or high-yielding background. Finally, one cannot expect yield without water. It is important to keep the trade-off between water and yield in such a way that yields are sustained under drought stress.

Acknowledgment Authors are thankful to the Indian Council of Agricultural Research for financial assistance to various research programmes at NRCPB.

References

Abogadallah GM, Nada RM, Malinowski R, Quick P (2011) Overexpression of HARDY, an AP2/ERF gene from *Arabidopsis*, improves drought and salt tolerance by reducing transpiration and sodium uptake in transgenic *Trifolium alexandrinum* L. Planta 233(6):1265–1276

Bahieldin A, Mahfouz HT, Eissa HF, Saleh OM, Ramadan AM, Ahmed IA, Dyer WE, El-Itriby HA, Madkour MA (2005) Field evaluation of transgenic wheat plants stably expressing the HVA1 gene for drought tolerance. Physiol Plant 123(4):421–427

Castiglioni P, Warner D, Bensen RJ, Anstrom DC, Harrison J, Stoecker M, Abad M, Kumar G, Salvador S, D'Ordine R, Navarro S, Back S, Fernandes M, Targolli J, Dasgupta S, Bonin C, Luethy MH, Heard JE (2008) Bacterial RNA chaperones confer abiotic stress tolerance in plants and improved grain yield in maize under water-limited conditions. Plant Physiol 147(2):446–455

Chen TH, Murata N (2002) Enhancement of tolerance of abiotic stress by metabolic engineering of betaines and other compatible solutes. Curr Opin Plant Biol 5(3):250–257

Dai A (2013) Increasing drought under global warming in observations and models. Nat Clim Change 3(1):52–58

Dalal M, Tayal D, Chinnusamy V, Bansal KC (2008) Abiotic stress and ABA-inducible group 4 LEA from *Brassica napus* plays a key role in salt and drought tolerance. J Biotechnol 139(2):137–145

Gu J-F, Qiu M, Yang J-C (2013) Enhanced tolerance to drought in transgenic rice plants overexpressing C4 photosynthesis enzymes. Crop J 1(2):105–114

Habben JE, Bao X, Bate NJ, DeBruin JL, Dolan D, Hasegawa D, Helentjaris TG, Lafitte RH, Lovan N, Mo H, Reimann K, Schussler JR (2014) Transgenic alteration of ethylene biosynthesis increases grain yield in maize under field drought-stress conditions. Plant Biotechnol J 12(6):685–693

Hu H, Dai M, Yao J, Xiao B, Li X, Zhang Q, Xiong L (2006) Overexpressing a NAM, ATAF, and CUC (NAC) transcription factor enhances drought resistance and salt tolerance in rice. Proc Natl Acad Sci U S A 103(35):12987–12992

Huang Y, Xiao B, Xiong L (2007) Characterization of a stress responsive proteinase inhibitor gene with positive effect in improving drought resistance in rice. Planta 226(1):73–85

Islam MA, Du H, Ning J, Ye H, Xiong L (2009) Characterization of *Glossy1*-homologous genes in rice involved in leaf wax accumulation and drought resistance. Plant Mol Biol 70(4):443–456

Jeong JS, Kim YS, Baek KH, Jung H, Ha SH, Do Choi Y, Kim M, Reuzeau C, Kim JK (2010) Root-specific expression of *OsNAC10* improves drought tolerance and grain yield in rice under field drought conditions. Plant Physiol 153(1):185–197

Jiang W, Zhou H, Bi H, Fromm M, Yang B, Weeks DP (2013) Demonstration of CRISPR/Cas9/sgRNA-mediated targeted gene modification in *Arabidopsis*, tobacco, sorghum and rice. Nucleic Acids Res 41:e188

Karaba A, Dixit S, Greco R, Aharoni A, Trijatmiko KR, Marsch-Martinez N, Krishnan A, Nataraja KN, Udayakumar M, Pereira A (2007) Improvement of water use efficiency in rice by expression of HARDY, an *Arabidopsis* drought and salt tolerance gene. Proc Natl Acad Sci U S A 104(39):15270–15275

Kasuga M, Liu Q, Miura S, Yamaguchi-Shinozaki K, Shinozaki K (1999) Improving plant drought, salt, and freezing tolerance by gene transfer of a single stress-inducible transcription factor. Nat Biotechnol 17(3):287–291

Lata C, Prasad M (2011) Role of DREBs in regulation of abiotic stress responses in plants. J Exp Bot 62(14):4731–4748

Li B, Wei A, Song C, Li N, Zhang J (2008) Heterologous expression of the TsVP gene improves the drought resistance of maize. Plant Biotechnol J 6(2):146–159

Liu Q, Kasuga M, Sakuma Y, Abe H, Miura S, Yamaguchi-Shinozaki K, Shinozaki K (1998) Two transcription factors, DREB1 and DREB2, with an EREBP/AP2 DNA binding domain separate two cellular signal transduction pathways in drought- and low-temperature-responsive gene expression, respectively, in *Arabidopsis*. Plant Cell 10(8):1391–1406

Martinoia E, Maeshima M, Neuhaus HE (2007) Vacuolar transporters and their essential role in plant metabolism. J Exp Bot 58(1):83–102

Masle J, Gilmore SR, Farquhar GD (2005) The ERECTA gene regulates plant transpiration efficiency in *Arabidopsis*. Nature 436:866–870

McKersie BD, Bowley SR, Harjanto E, Leprince O (1996) Water-deficit tolerance and field performance of transgenic alfalfa overexpressing superoxide dismutase. Plant Physiol 111(4):1177–1181

Nakashima K, Ito Y, Yamaguchi-Shinozaki K (2009) Transcriptional regulatory networks in response to abiotic stresses in *Arabidopsis* and grasses. Plant Physiol 149:88–95

Nelson DE, Repetti PP, Adams TR, Creelman RA, Wu J, Warner DC, Anstrom DC, Bensen RJ, Castiglioni PP, Donnarummo MG, Hinchey BS, Kumimoto RW, Maszle DR, Canales RD, Krolikowski KA, Dotson SB, Gutterson N, Ratcliffe OJ, Heard JE (2007) Plant nuclear factor Y (NF-Y) B subunits confer drought tolerance and lead to improved corn yields on water-limited acres. Proc Natl Acad Sci U S A 104(42):16450–16455

Nuccio ML, Wu J, Mowers R, Zhou H-P, Meghji M, Primavesi LF, Paul MJ, Chen X, Gao Y, Haque E, Basu SS, Lagrimini LM (2015) Expression of trehalose-6-phosphate phosphatase in maize ears improves yield in well-watered and drought conditions. Nat Biotechnol 33(8):862–869

Oh SJ, Kim YS, Kwon CW, Park HK, Jeong JS, Kim JK (2009) Overexpression of the transcription factor AP37 in rice improves grain yield under drought conditions. Plant Physiol 150(3):1368–1379

Quan R, Shang M, Zhang H, Zhao Y, Zhang J (2004) Engineering of enhanced glycine betaine synthesis improves drought tolerance in maize. Plant Biotechnol J 2(6):477–486

Reguera M, Peleg Z, Blumwald E (2012) Targeting metabolic pathways for genetic engineering abiotic stress-tolerance in crops. Biochim Biophys Acta 1819(2):186–194

Saint Pierre C, Crossa JL, Bonnett D, Yamaguchi-Shinozaki K, Reynolds MP (2012) Phenotyping transgenic wheat for drought resistance. J Exp Bot 63(5):1799–1808

Shan Q, Wang Y, Li J, Zhang Y, Chen K, Liang Z, Zhang K, Liu J, Xi JJ, Qiu J-L, Gao C (2013) Targeted genome modification of crop plants using a CRISPR-Cas system. Nat Biotechnol 31(8):686–688

Shen H, Zhong X, Zhao F, Wang Y, Yan B, Li Q, Chen G, Mao B, Wang J, Li Y, Xiao G, He Y, Xiao H, Li J, He Z (2015) Overexpression of receptor-like kinase ERECTA improves thermotolerance in rice and tomato. Nat Biotechnol 33(9):996–1003

Shinozaki K, Yamaguchi-Shinozaki K (2007) Gene networks involved in drought stress response and tolerance. J Exp Bot 58(2):221–227

Sivamani E, Bahieldin A, Wraith JM, Al-Niemi T, Dyer WE, Ho TD, Qu R (2000) Improved biomass productivity and water use efficiency under water deficit conditions in transgenic wheat constitutively expressing the barley HVA1 gene. Plant Sci 12 155(1):1–9

Svitashev S, Young JK, Schwartz C, Gao H, Falco SC, Cigan AM (2015) Targeted mutagenesis, precise gene editing, and site-specific gene insertion in maize using Cas9 and guide RNA. Plant Physiol 169(2):931–945

Umezawa T, Fujita M, Fujita Y, Yamaguchi-Shinozaki K, Shinozaki K (2006) Engineering drought tolerance in plants: discovering and tailoring genes to unlock the future. Curr Opin Biotechnol 17(2):113–122

Valliyodan B, Nguyen HT (2006) Understanding regulatory networks and engineering for enhanced drought tolerance in plants. Curr Opin Plant Biol 9(2):189–195

Waltz E (2014) Beating the heat. Nat Biotechnol 32:610–613

Wang Y, Ying J, Kuzma M, Chalifoux M, Sample A, McArthur C, Uchacz T, Sarvas C, Wan J, Dennis DT, McCourt P, Huang Y (2005) Molecular tailoring of farnesylation for plant drought tolerance and yield protection. Plant J 43(3):413–424

Wang Y, Beaith M, Chalifoux M, Ying J, Uchacz T, Sarvas C, Griffiths R, Kuzma M, Wan J, Huang Y (2009) Shoot-specific down-regulation of protein farnesyl transferase (alpha-subunit) for yield protection against drought in canola. Mol Plant 2(1):191–200

Xiao B, Huang Y, Tang N, Xiong L (2007) Over-expression of a LEA gene in rice improves drought resistance under the field conditions. Theor Appl Genet 115(1):35–46

Xiao B-Z, Chen X, Xiang C-B, Tang N, Zhang Q-F, Xiong L-Z (2009) Evaluation of seven function-known candidate genes for their effects on improving drought resistance of transgenic rice under field conditions. Mol Plant 2(1):73–83

Yamaguchi-Shinozaki K, Shinozaki K (2006) Transcriptional regulatory networks in cellular responses and tolerance to dehydration and cold stresses. Annu Rev Plant Biol 57:781–803

Yu L, Chen X, Wang Z, Wang S, Wang Y, Zhu Q, Li S, Xiang C (2013) *Arabidopsis* enhanced drought tolerance1/*HOMEODOMAIN GLABROUS11* confers drought tolerance in transgenic rice without yield penalty. Plant Physiol 162(3):1378–1391

Managing Abiotic Stresses in Wheat

14

V. Tiwari, H.M. Mamrutha, S. Sareen, S. Sheoran, R. Tiwari,
P. Sharma, C. Singh, G. Singh, and Jagadish Rane

Abstract

Wheat, a major staple crop of the world as well as of India, provides food and nutritional security to millions of the global populace. While the rate of genetic gain in productivity during the recent years has not been as impressive as in the past, the cultivars under development are being tailored to meet the demand for higher production together with the challenges imposed by several abiotic stresses such as high temperature, restricted access to irrigation water, drought, salinity/alkalinity, waterlogging, mineral deficiency, crop lodging and preharvest sprouting. Since the conventional approaches being practiced for wheat improvement will not be sufficient to achieve the productivity targets, it is essential to integrate the modern approaches leveraged by advances in phenomics, molecular biology, functional genomics, etc. Furthermore, stress mitigation options particularly through agronomic interventions are also essential to stabilize the productivity in wheat. Recent efforts being attempted in this direction have been highlighted in this article.

V. Tiwari (✉) • H.M. Mamrutha • S. Sareen • S. Sheoran • R. Tiwari • P. Sharma
C. Singh • G. Singh
ICAR-Indian Institute of Wheat & Barley Research, Karnal 132001, Haryana, India
e-mail: pici.dwr@gmail.com; mamruthamadhu@gmail.com; sareen9@hotmail.com;
sheoransonia@yahoo.co.in; tiwari64@yahoo.co.in; neprads@gmail.com;
n_charansingh@hotmail.com; gysingh@gmail.com

J. Rane
National Institute of Abiotic Stress Management, Indian Council for Agricultural Research,
Baramati, Maharashtra, India
e-mail: jagrane@hotmail.com

© Springer Nature Singapore Pte Ltd. 2017
P.S. Minhas et al. (eds.), *Abiotic Stress Management for Resilient Agriculture*,
DOI 10.1007/978-981-10-5744-1_14

14.1   Introduction

Wheat (*Triticum aestivum* L.) is the second most important cereal crop after rice in providing food and nutritional security to the masses in India. Wheat production in India has progressively scaled new heights over the years with phenomenal increase in area, production and productivity. The Indian wheat programme has been very vibrant during the last four decades, but there is no scope for complacency and the programme needs to be more responsive to the new emerging challenges posed by climatic changes. Vulnerability to abiotic stresses is evident from a substantial dip in wheat production to 86.53 million tonnes, due to erratic rains, during 2014–2015 over the high production of 95.85 M tonnes obtained during the previous year 2013–2014. The country produced 93.50 M tonnes of wheat during 2015–2016 crop season.

Wheat is grown in India under six diverse agroclimatic zones, wherein the Indo-Gangetic plains (IGP) comprising the north-western plains zone (NWPZ) and the north-eastern plains zone (NEPZ) are the major wheat tract covering over 20 million hectares, followed by the central zone (CZ) and the peninsular zone (PZ). The northern hills zone (NHZ) and southern hills zone (SHZ) are the two other wheat-producing zones falling in the hills agroecosystem. This classification of zones for wheat cultivation has been based on climatic conditions, soil types and duration of the wheat crop. During the growing season, expected changes in climatic factors, viz. precipitation/winter rains, minimum and maximum temperature, wind velocity and its direction, sunshine hours, etc., need to be considered while developing strategies for enhancement of production and productivity in wheat. It is obvious that any significant change in the climate will have an impact on agriculture and food production. However, realizing the vast variation in the climatic conditions in different wheat-growing regions of India, the ongoing programmes are tuned to address even the micro-niches besides the region-specific needs.

14.2   Abiotic Stresses Affecting Wheat Production

Although more than 95% of the wheat area sown in India has access to irrigation, major parts face water deficits due to restrictions in the quantity of water availability particularly at critical plant growth stages. Wheat grown in the rest of the area is dependent on rains and hence highly vulnerable to water stress which reduces the productivity and production (Fig. 14.1).

The productivity in wheat is most affected by water deficits. The effect of drought on wheat becomes more conspicuous when the south-west monsoon fails to precipitate sufficient soil moisture essential for early establishment of the crop. Drought stress may occur throughout the growing season, early or late in the season. Crop yield is reduced mostly when water stress occurs at heading time, but its effect on yield is highest when it occurs after anthesis. It is evident from restricted irrigation experiments that the grain yield obtained in many wheat cultivars under rainfed condition can be doubled by providing a single post-sown irrigation. Therefore, improvement in drought tolerance ability of wheat genotypes is crucial in view of

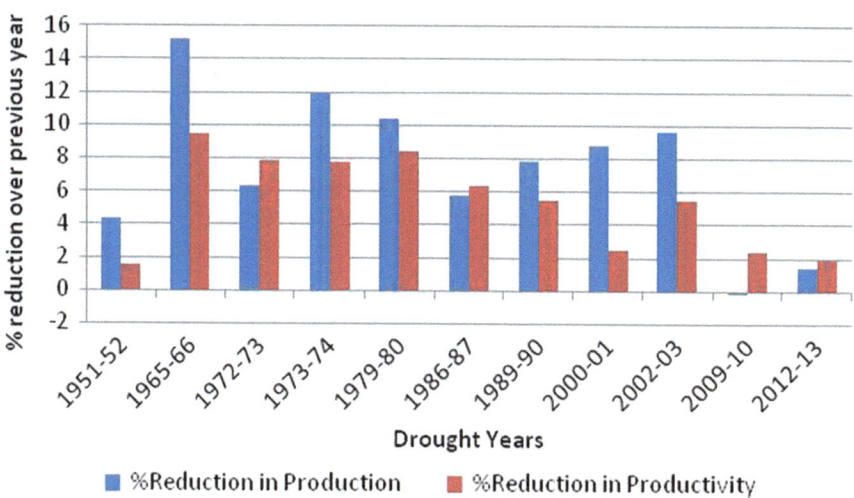

Fig. 14.1 Percent reduction in wheat yield during drought years in India (Based on DOAC data)

the restricted access to irrigation water due to the diversion of more water to nonagricultural uses. Early drought in wheat at seedling establishment stage reduces the number of plants per unit area and tillers per plant; mid-season drought at CRI to ear emergence stage reduces the total biomass, number of productive tillers and grain number per spike; and terminal drought at ear emergence to physiological maturity reduces the current photosynthesis, spike fertility and grain weight. The losses in the present varieties of wheat due to moisture stress in central India can range from 11.6 to 43.6% if no irrigation is provided to the crop (Tiwari et al. 2015).

While the most optimum temperature for growth and development of the wheat plant is considered to be around 21° to 24 °C, the optimal temperature range may vary depending on the prevailing agroclimatic situation at the place where the crop is grown. The whole wheat-cropped area in India experiences heat. While the central and peninsular parts experience heat stress all through the crop season, significant parts of north-western and north-eastern plains experience terminal heat. The trend by farmers in north-western and central India towards early sowing of wheat to take advantage of residual moisture is demanding development of wheat genotypes for both early and terminal heat tolerance (Misra and Varghese 2012).

The temperature often rises above 30 °C during the subsequent stages of its growth, thereby severely affecting the formation and filling of grains (Al-Khatib and Paulsen 1984; Randall and Moss 1990; Stone et al. 1995; Wardlaw and Moncur 1995; Rane et al. 2007). High temperature imposed before anthesis can also decrease yield (Wardlaw et al. 1989; Hunt et al. 1991). Under controlled experiments, grain yield of wheat per spike was reduced by 3–4% per 1 °C increase in temperature above 15 °C. (Wardlaw et al. 1989). The effect of short periods of exposure to high temperatures (>30 °C) is thought to be equivalent to 2–3 °C warming in the seasonal mean temperature (Wheeler et al. 1996). Also yield reduction (up to 23%) has been reported from as little as 4 days of exposure to very high temperatures (Randall and

Moss 1990; Stone and Nicolas 1994). The effect of high temperature is likely to assume much larger proportion considering the current trends and future predictions about global warming continue. According to the fifth assessment report of the Inter-Governmental Panel on Climate Change (2014), the globally averaged combined land and ocean surface temperature data show a warming of 0.85 °C (0.65–1.06) over the period 1880 to 2012, and temperature is projected to rise over the twenty-first century. Besides high temperature, wheat plants also suffer damage due to cold or frost injury when there is a sudden dip in the temperature in northern India and mountainous and sub-mountainous regions of the northern hills zone. However, the loss due to cold or frost injury in these areas is not that severe as high heat conditions.

About 6.73 million ha land in India is salt affected, out of which 3.77 and 2.96 million ha are covered by sodic and saline soils, respectively. The Indo-Gangetic plains include 2.5 million hectares of sodic soils and 2.2 million hectares affected by seepage water from irrigation canals (CSSRI 1997). Salinity- and alkalinity-affected soils lead to reduction in crop yield. The soil where water stands on the surface for a prolonged period of time or the available water fraction in the soil surface layer is at least 20% higher than the field water capacity falls in the category of waterlogged soil. Waterlogging adversely affects production in about 4.5 million hectares in irrigated soils of the Indo-Gangetic plains of northern India (CSSRI 1997). The combined effect of salt and waterlogging stresses significantly reduces wheat yield by causing reduction in grain weight, length of spike and spikelet number and also shows more adverse effect than salt stress alone in the case of compacted soils. Due to the increased frequency of extreme climate events like heavy rains and storms, particularly where water table is high and soils are sodic, waterlogging has become an important constraint to crop production globally.

Preharvest sprouting (PHS) is also of major concern for wheat cultivation in eastern and far- eastern parts of India due to early cessation of rains around maturity time during the month of March. The PHS in wheat is characterized by premature germination of kernels in a mature spike prior to harvest due to early breakage of seed dormancy under moist weather conditions that persist after physiological maturity. PHS causes loss in yield due to decrease in thousand grain weight and also ultimately affects the end product quality. Bread baked from sprouted wheat grain express smaller volume and a compact interior. This decrease in the quality is mainly due to early α-amylase activity which can be characterized by Hagberg falling number. The erratic rainfall patterns in such areas, where temperature and moisture during grain development adversely affect the expression of dormancy, lead to an increase in problems like PHS and call for the development of wheat genotypes with a balanced degree of seed dormancy.

Crop lodging caused by storms and hail is another important factor which severely affects the wheat production. The development of genotypes with good stem structural strength for lodging tolerance is now gaining more attention. As productivity enhancement is targeted in high-fertility environments, genotypes with strong stem strength are needed to sustain under lodging situations. In addition, mineral deficiency is also likely to emerge as a major abiotic constraint in zones under intensive cultivation (Arvind et al. 2015).

14.3 Adaptive Traits Imparting Tolerance

Under natural condition, the plants adapt to abiotic stress condition mainly by three important mechanisms, viz. stress avoidance, stress escape and stress tolerance. These mechanisms are in turn governed by many associated traits. In avoidance mechanism, the plant avoids the stress through such traits like reduction in leaf area, increased pubescence, leaf rolling and leaf reflectance with epicuticular wax accumulation. In escape mechanism, plants sense the future occurrence of stress and adjust the phenology to complete their life cycle early so that the effect of stress is not evident in the genotypes. While, in tolerance mechanism, the plant experiences stress but is able to withstand the stress condition through a number of adaptive features like deep root system, accumulating osmolytes, maintaining membrane integrity and relative water content, etc.

To improve grain yield under stressed environment, it is essential to improve the adaptation to abiotic stresses. Hence, traits contributing to such adaptation are crucial. Observing the plant morphological and anatomical traits such as coleoptile length, leaf phyllotaxy, orientation and angle, pubescence, wax, stem length, peduncle length and root traits vis-à-vis the abiotic stresses helps in underlining the stress tolerance adaptations. The past arguments that empirical selection for grain yield can indirectly place selection pressure for adaptive traits are getting diluted as is evident from recent slowdown in genetic gain in the yield potential of wheat. This necessitates trait based selection for improvement in adaptation to abiotic stresses. It is known that every genotype has different adaptive mechanisms with different associated traits. This can be achieved with the knowledge gained about adaptive traits, feasibility of using them as traits for selection and their inheritance pattern. In spite of new developments in phenotyping techniques, still it is not always true that a single trait will account for all the variations present in the population for tolerance to abiotic stress. Hence, there is a need to develop a protocol which is able to capture numerous traits associated with abiotic stresses.

Water stress mostly leads to stunted plant growth, and leaf wilting is the first visible sign to observe stressed plants. Drought stress reduces the number of days to heading, peduncle length and plant fresh and dry biomass production. The stomatal closure under drought condition to prevent transpirational water loss also limits photosynthesis through limiting CO_2 uptake by leaves (Cornic 2000). Water stress also causes reduction in the relative water content. Although a number of morphological traits and physiological parameters have been put forward for identifying drought-tolerant genotypes of wheat, none of these parameters could become practically feasible for selecting individual plants from the breeding generations. Hence, selection of individual segregants is based on survival under natural drought conditions which in effect may or may not result in improvement in grain yield. Nevertheless, on limited scale, the segregating material can be exposed to artificial moisture stress under controlled conditions to select drought-tolerant and suscepti-ble plants based on seedling survival. Recently, the carbon isotope discrimination (CID) technique has been successfully utilized in identification of drought-tolerant genotypes (Rebetzke et al. 2002).

The magnitude of damage in wheat due to high temperature depends on the background ambient temperature, stage of plant development and the genotype. Extremes of temperature prevailing at sensitive developmental stages are especially detrimental. Temperature above 30 °C around anthesis time can affect pollen forma-tion and reduce yield. For wheat, there are two critical stages that are sensitive to temperature during reproductive growth, viz. the first at micro- or mega-sporogenesis (approximately 12 to 3 days prior to anthesis), leading to loss of fertility, and the second at pollination/fertilization stage leading to decreased levels of pollen shed, pollen reception on stigma, pollen tube growth and fertilization as well as early abortion. It is established that many physiological traits/parameters, viz. grain-filling duration (GFD) and canopy temperature (CT), have a strong correlation with terminal heat tolerance (THT). It is assumed that reduced GFD will avoid the dam-age due to terminal heat stress, while a low CT will help the plant to withstand ter-minal heat stress (Sharma et al. 2015). In view of this, both GFD and CT have been used while selecting for higher yield under high temperature at the time of maturity and thus can be effectively utilized for QTL analysis as parameters for terminal heat tolerance.

The abiotic stresses like heat, drought and salt concentration can be quantified by recording some physiological, biochemical and developmental traits in wheat plants as given below.

Early Vigour A greater biomass/plant canopy during early growth stages provides more efficient water utilization by shading the soil and thereby minimizes evapo-transpiration from the soil. Thus, retained residual moisture will be utilized by the plants under severe stress condition to survive. Indirectly, early vigour is attributed to a deeper root system. Under abiotic stress condition, early vigour and delayed leaf senescence showed positive correlation with grain yield. Genetic variability exists for early vigour in wheat (Rane et al. 2002), and this trait may be used as a suitable selection criterion under a range of moisture conditions. It can be measured quickly and easily through visual score or using NDVI and can also be used for quick evaluation of large populations.

Osmotic Adjustment (OA) Osmotic adjustment maintains cell water contents by increasing the osmotic force that can be exerted by cells on their surroundings and thus increasing water uptake. OA involves the net accumulation of organic or inor-ganic solutes/osmolytes, total soluble sugars, total free amino acids, proline, gly-cinebetaine, sodium, chloride and potassium in cells in response to a fall in the water potential. As a consequence of this net accumulation, the cell osmotic poten-tial is lowered, and turgor pressure tends to be maintained along with more water in leaf cells with greater OA resulting in higher turgor as compared with leaves having less OA. Osmoregulation and turgor maintenance allow continuous growth of roots and uptake of moisture from the soil (Sharp and Davies 1979). Free amino acids and proline accumulation contributed significantly for osmotic adjustment in wheat flag leaves under salinity stress contributing to higher grain yield (Bandeh-hagh et al. 2008).

Relative Water Content (RWC) Assessment of water loss from excised leaves has been indicated as a good criterion for characterizing drought resistance. RWC is related with grain yield under irrigation and dryland conditions. Drought-tolerant genotypes have higher RWC, and this trait can be used for screening and identification of genotypes (Sonia et al. 2015).

Membrane Leakage A major impact of plant environmental stress is cellular membrane modification which results in its abnormal function or total dysfunction. The exact structural and functional modification caused by stress is not fully resolved. However, the cellular membrane dysfunction due to stress is well expressed through increased permeability and leakage of ions, which can be readily measured by the efflux of electrolytes. Hence, the estimation of membrane dysfunction under stress by measuring cellular electrolyte leakage from affected leaf tissue into an aqueous medium is finding growing use as a measure of cell membrane stability (CMS) and as a screen for stress resistance. The CMS is positively associated and explains above 70% of drought tolerance as compared to other physiological traits. The membrane leakage showed negative relation with heat tolerance for 2 consecutive years in a set of spring wheat genotypes under late sown condition (Sharma et al. 2015).

Chlorophyll Content Index Chlorophyll loss is associated with environmental stress, and the variation in total chlorophyll/carotenoids ratio may be a good indicator of stress in plants. High chlorophyll content is a desirable characteristic as it indicates a low degree of photoinhibition of the photosynthetic apparatus. Water stress condition causes reduction in chlorophyll content up to 20%. Hence, the flag leaf chlorophyll content and its stability over time can indicate resilience of the photosynthetic activity.

Canopy Temperature The canopy temperature reflects transpirational cooling efficiency of the plants. The relationship between canopy temperature, air temperature and transpiration is complex, and it involves atmospheric conditions (vapour pressure deficit, air temperature and wind velocity), soil (mainly available soil moisture) and plants (canopy size, canopy architecture and leaf adjustments to water deficit). These variables are considered when canopy temperature is used to develop the crop water stress index (CWSI) which is used for scheduling irrigation in crops (Lopes and Reynolds 2010). It has been shown that bread wheat genotypes with cool canopy temperatures under drought and heat stress have been associated with increased plant access to water as a result of deeper roots.

Chlorophyll Fluorescence The Fv/Fm measurements during drought stress reveal the effects of co-occurring stresses (heat stress, photo inhibition, etc.) or to the early phases of leaf senescence. Measurements of the slow and the fast chlorophyll fluorescence kinetics have been observed to be sensitive to drought stress. The decrease of effective PS-II quantum yield (FPSII) and ETR in drought-stressed leaves as compared to well-hydrated leaves is mainly due to lack of CO_2 inside the leaf caused by stomatal closure. The chlorophyll fluorescence has also exhibited positive correlation with grain yield under terminal heat stress condition (Pandey et al. 2014). Chlorophyll fluorescence tool could reveal that spikes of durum wheat are more tolerant to desiccation in relation to bread wheat (Rane, unpublished).

Pollen Viability The efficiency of pollen transfer and viability of the pollen grains determine the reproductive success in plant species. It has been reported that the viability of pollen grains under field conditions is highly variable indicating that differences in micro-environments may have a profound effect on pollen viability. Drought and heat stresses limit crop pollination by reducing pollen grain production, increasing pollen grain sterility and decreasing pollen grain germination and pollen tube growth (Al-Ghzawi et al. 2009). Drought stress also reduces the megagametophyte fertility and decreases the setting of seed and its development.

Stem Reserve Mobilization Stem reserves are sugars such as fructans, sucrose, glucose and fructose which get accumulated in the stem as reserves. Water soluble carbohydrates (WSC) accumulate up to anthesis time and are partitioned to the stem from where they are later available as a reservoir for remobilization to the developing grains. These reserves are an important source of carbon for grain filling as demand frequently exceeds current assimilation, potentially contributing 10–20% of the grain yield under favourable conditions. Stem reserve mobilization has been shown to be adaptive for drought, heat and/or disease tolerance when current assimilation during grain filling causes a greater demand for stem reserves during grain filling. When wheat plants were shaded during grain filling, up to 0.93 g of grain was produced per gram of assimilates exported from the stem (Kiniry 1993). Genetic variability for stem reserve mobilization exists in wheat, and it can be explored for improvement of productivity under high temperature and water deficit environments (Nagarajan et al. 1998; Rane et al. 2003; Nagarajan and Rane 2002).

14.4 Phenotyping Strategies

Over the past few decades, rapid developments have taken place for phenotyping the tolerance to abiotic stresses. The focus has been mainly on the traits that contribute to plant survival under stress conditions, e.g. root architecture, transpiration efficiency/carbon isotope discrimination, stomatal conductance, canopy temperature, osmotic adjustment, stay green habit, etc. Phenotyping for evaluating the impact of both drought and heat stress conditions in wheat is usually done by comparison with the irrigated condition. For studying heat stress in open fields, late sowing of the crop complemented with recommended number of irrigations is used. The late sowing under irrigated condition enables plants to experience high temperature stress without suffering any drought stress. However, both drought and heat stresses are difficult to be simulated precisely at the field scale. Multilocation testing in the target environments is another option towards phenotyping stresses since they help to obtain response to the stresses under varied natural conditions (Rane et al. 2007). There is lack of screening techniques which are sufficiently rapid, simple, repeatable, not environment dependent with a good predictive value of yield under stress and also applicable in the field. The severity, duration and timing of stress, as well as responses of plants after stress removal and interaction between stress and other factors are extremely and equally important.

Abiotic stresses at a given growth phase are likely to affect the proper development of plant organs leading to a reduction in the yield. The reproductive phase is usually considered to be the most stress-sensitive stage of wheat. Drought causes impaired germination and poor crop stand establishment. Early vigour and rapid ground covering have been proposed as important traits for screening for early drought tolerance. During drought stress, morphological traits including leaf (shape, expansion, area, size, senescence, pubescence, waxiness and cuticle tolerance) and root (dry weight, density and length) are affected. Early maturity, reduced plant height and leaf area are related to the intensity of drought stress. Late onset and/or a slower rate of leaf senescence have an advantage in overcoming water stress. Water stress at pre-anthesis stage reduces the time taken for anthesis, while at post-anthesis stage, it shortens the duration of grain filling. Post-anthesis drought stress is detrimental to grain yield regardless of the severity of stress. Following heading, drought has little impact on the rate of kernel development, but it reduces the weight of grains by shortening the duration of grain development (Jain et al. 2013).

The size of the root system in wheat can be a selection target for drought and heat tolerance. Deeper roots enable plants to explore higher soil volumes and thus remain hydrated under drought and permit cooling of the canopy under heat stress. The evaluation of grain yield performance in areas of frequent stress remains the most widely applied criterion for characterizing adaptation of genotypes under stressful conditions. However, in stress environments, yield *per se* is not always the most suitable selection trait. However, an approach based on evaluation of some physiological traits involved in stress tolerance would be a better option. Thus, increasing grain weight and grain size might be a way worthy to be followed to improve grain yield, especially in case of early water stress which affects mainly spikelet and floret initiation, thereby limiting the grain number per unit area, whereas grain weight is affected by terminal drought. Grain-filling duration under drought and heat stress may be reduced, but it might be partly compensated by a higher grain-filling rate if abundant carbohydrates are available from the leaf photosynthesis or from stem or leaf reserves. The grain-filling rate may be slightly increased, and the duration strongly decreases with heat stress (Garg et al. 2013).

The important target traits which have been used as indicators of stress tolerance include reduced plant height which is associated with high harvest index; reduced number of days to anthesis and maturity which enables the crop to evade terminal

Fig. 14.2 Rainout shelter deployed at IIWBR, Karnal (**a**). Precise phenotyping for drought stress in ROS. (**b**) Screening of genotypes for root traits in large >1.5 m PVC pipes (**c**)

stress; root architectural traits such as even distribution and root length density which enable effective water uptake; seedling traits associated with vigorous seedling establishment such as coleoptile length and early ground cover which reduces evaporative losses; and traits associated with reduced evaporative losses and photoassimilate production such as leaf rolling, flag leaf persistence, stomatal conductance and canopy temperature. However, these traits should have a positive correlation with yield under stress condition. The ultimate criteria for selection of a genotype should be the capability to integrate its adaptive mechanisms to optimize yield without falling on a single trait. A range of stress tolerance indices including yield, morphological, and physiological traits has been suggested for use in screening wheat genotypes under stress conditions.

14.4.1 Precision Field Phenotyping for Drought Tolerance

The precision phenotyping for drought stress is done under rainout shelter (ROS) to identify the real drought-tolerant wheat genotype. Lack of uniformity of plant stand while conducting a field experiment can substantially contribute to errors in the prediction of association between plant phenotypic traits and the genotypes. Among the several factors that can contribute to experimental errors, inconsistent seed depth and plant spacing often occur due to lack of precision when seeds are sown by hand or seed drills. An improved planting method was devised for sowing of field experiments as well as in ROS. The method involved a tool designed for dibbling seeds and a protocol to place seeds uniformly in the soil (Fig. 14.2) (Sharma et al. 2016). We studied advantage of the new methods over conventional methods of sowing, viz. seed drill and by hand. Compared with conventional methods, the new method improved the consistency in plant spacing and depth substantially as indicated by reduction in standard deviation at least by three times (unpublished). The germplasm accessions and mapping populations for drought tolerance are being evaluated based on root characters, delayed senescence, canopy temperature, chlorophyll content, relative water content, 1000-grain weight and grain yield in comparison to irrigated conditions. The *ex situ* root phenotyping precisely can be done using

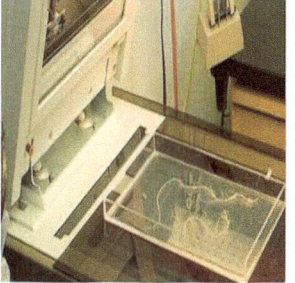

Fig. 14.3 Direct root coring to phenotype deep roots

>1.5 m long PVC pipes both under stress and control conditions. The in situ root phenotyping is done by using root corer, extracting the roots, washing the roots and later scanning using root scanner WinRhizo (Fig. 14.3).

14.4.2 Precision Phenotyping for Heat Tolerance

Various attempts have been made to understand traits associated with high tempera-ture tolerance in wheat through experiments conducted both in the field and under controlled environmental facilities. Many of these attempts were not conclusive due to lack of sufficient precision in simulating the ambient temperature dynamics and micro-environments prevailing in the field or due to lack of repeatability of results in the field, often featured by inconsistent exposure of genotypes to desired level of temperature during evaluation. These bottlenecks severely affect the prediction of the relationship between plant phenotype and genes. Hence, we attempted to develop a method for phenotyping wheat genotypes for high temperature tolerance by integrating a novel design of temperature-controlled phenotyping facility (TCPF) as a novel inexpensive tool to ensure uniform crop stand. The novel TCPF has been designed that allows screening of several wheat genotypes in larger plot size (as in the fields) at a desired temperature at any stage of crop growth while allowing plants to grow in the natural environment during the rest of the period (Fig. 14.4).

The size of the structure is approximately 100 ft. by 35 ft. Motorized control units of this unique system allows roofs and walls to slide down to open the struc-ture during initial growth stages allowing the plants to respond according to the prevailing open environment conditions. The structure is closed with the help of sliding roof and windows to seal the whole unit so that the genotypes can be screened at any desired temperature and at any desired crop growth stage. For increasing temperature, a boiler-based heating system is utilized in which warm water runs through a network of pipelines hanging from the roof with several inlets and outlets to avoid formation of temperature gradient from one end to another end in the struc-ture. The temperature regulation in the structure is precise and is linked to ambient temperature so that desired difference between the temperature inside and outside the structure is maintained in relation to the diurnal cycle during the heat stress

Fig. 14.4 Temperature-controlled phenotyping facility

treatment. Cooling is done using split air conditioners integrated and automatically governed by the control panel. Desired humidity level is maintained through mist system that releases fine water droplets, and the desired soil moisture is obtained through drip irrigation system. The structure allows the termination of required temperature stress treatment to expose the crop again to natural environment conditions.

Studies clearly revealed the advantage of the integrated method over conventional methods in differentiating high temperature responses of a large number of genotypes of wheat. The reduction in error and the lowest coefficient of variation (CV) for the plant traits measured in the new method relative to other methods indicated possibility of enhancing precision in phenotyping responses of plants under field condition. The repeatable assessment of plant response to stress in terms of growth, physiology and productivity supported the view that the novel tool developed for ensuring uniform crop establishment and the TCPF together can enhance the precision of phenotyping crop genotypes. There was high consistency in grouping heat tolerant and susceptible wheat genotypes in TCPF as against late sown field screening.

14.4.3 Characterization of Wheat for Waterlogging Tolerance

The problem of waterlogging becomes more acute if the fields are not properly levelled and normal irrigation schedules are followed by unseasonal heavy rainfall. Waterlogging occurring at any growth stage in case of wheat usually causes degradation of chlorophyll in leaves and also protein content in wheat grains. It also decreases the concentrations of nitrogen, phosphorous and potassium in the shoot of wheat plants. Waterlogging also causes reductions in biomass accumulation in both shoot and root and hence affects final grain yield.

14.4.4 Traits for Characterization of Preharvest Sprouting Tolerance

PHS-resistant/PHS-tolerant wheat cultivars and land races have been identified globally, and both red- and white-seeded spring wheat cultivars are known to carry resistance to PHS. There are a number of methods to measure the PHS. The falling number (FN) is most commonly used approach to quantify PHS which indirectly measures the activity of the enzyme α-amylase that breaks down starch in germinating grains. Two other important traits for the characterization of PHS are GI (germination index) and SI (susceptibility index). GI values are deduced from seed germination tests in petri dishes and constitute a direct measure of seed dormancy. The SI values obtained via artificial wetting of intact wheat spikes help in detecting dormancy and properties of the inflorescence that affect PHS (DePauw et al. 2012).

14.5 Genetic Understanding of Traits

The use of selection indices is more efficient than direct selection for grain yield alone, and the relative efficiencies could be better when two or more traits are merged than using each of the single traits independently. Correlation studies also increase the possibility of indirect selection for different traits. High estimates of broad sense heritability for tillers per plant, plant height, spike length, grains per spike, grain yield and 1000-grain weight, root length and shoot length indicate the occurrence of additive gene effects. Among the variability parameters, estimates of PCV were generally higher than that of GCV for most of the traits, indicating thereby the role of environment in total variability. Under stress conditions, tillers m^{-2} and 1000-grains weight have expressed positive correlation with grain yield. Path analysis indicated that 1000-grain weight had the highest direct effect on grain yield followed by tillers m^{-2} under both stress and non-stress conditions. Based on positive direct effect along with positive genotypic correlation with grain yield, 1000-grain weight may be considered to be a suitable selection trait under irrigated and rainfed conditions. Genotypic correlation study showed the importance of tillers and 1000-grain weight under drought stress conditions (Sareen et al. 2014). Grain yield is correlated to biomass and number of fertile tillers per unit area. Significant positive correlations with grain yield were also observed for early as well as late ground cover, dry matter yield and number of fertile tillers per unit area.

The additive-dominance model is inadequate to account for the inheritance of most of the traits and the environmental conditions. Additive, additive x additive and additive x dominance gene effects were higher than the dominance and dominance x dominance gene effects, proving the important role of additive gene effects for most studied traits. Polygenes with mainly additive effects were involved in the control of stem diameter which was also positively correlated under both drought and drought plus heat stress with stem weight and stem density. Stem diameter was significantly associated with 1000-kernel weight and grain yield per spike in three environments. Such strong association of stem diameter with single-grain mass and grain yield per spike under stress indicated the important role this character plays in sustaining grain filling through provision of greater capacity for assimilation in the stem before mobilizing it to grains (Sallam et al. 2014). Early vigour is genetically fixed and positively related to large kernel size. Moreover, physiological traits are prone to variation within a trial and between environments, therefore having only intermediate heritability.

In a study involving diverse wheat germplasm which were screened under alkaline and waterlogged soils in ten environments over 2 years at three locations in India (IIWBR, CSSRI & NDUAT), the performance of genotypes (plant height, tillers, 1000-grains weight and grain yield) over years, conditions and locations was adversely influenced by alkalinity and waterlogging (unpublished). Genotype x environment interaction was evident from the negligible genetic correlation between the grain yields observed at CSSRI and the same at NDUAT under waterlogged conditions for both the years ($r^2 = 0.00\text{--}0.01$). However, the pattern of reduction in grain yields of different genotypes due to waterlogging was consistent over the years at same location as evident from the genetic correlation values ($r^2 = 0.80\text{--}0.90$ for CSSRI and 0.50 for NDUAT). At NDUAT, where waterlogging resulted in the most severe reduction in grain yield, there was no significant difference in the pattern of genetic variability observed under waterlogged and non-waterlogged conditions indicating that enhanced efforts are needed for genetic improvement of tolerance to waterlogging (Singh et al. 2014).

14.6 Molecular Approaches

It is being realized that the success of molecular markers depends on the phenotyping methods used to characterize the plant responses. This can only be achieved by repeated experiments and increasing replications within the experiments, particularly when the genotypes are evaluated under field condition. This may often be considered an impractical task when a large number of genotypes have to be characterized with conventional methods. There is little or no evidence that a single phenotyping approach, whether conducted in pots, glasshouse or laboratory, is effective for germplasm improvement related to abiotic stresses. Trethowan et al. (2005) and Rebetzke et al. (2014) have strongly advocated the use of field-based screening in environments rigorously controlled for the timing and amount of water availability. Therefore, the importance of accurate phenotyping for drought tolerance has been realized.

The traits associated with improved performance of wheat under drought have complex genetic control with each trait controlled by many genes, each gene having small effect. QTLs in large number have already been reported for several traits associated with drought tolerance including coleoptile length, CID or Δ, water soluble carbohydrates, root system, grain yield and related traits recorded under water stress (Sheoran et al. 2015a, b). Most QTLs for drought tolerance in wheat have been identified through yield and yield component measurements under water-limited conditions. Genetic analysis has led to the identification of two major QTLs for grain yield under water-limited condition, including a QTL each on chromosomes 4AL and 7AL. The QTL on 4AL was co-localized with QTLs for spike density, grains per m^{-2}, grain-filling rate and biomass production (Kirgwi et al. 2007), while QTL for 7A was co-localized with QTLs for grain weight per ear, flag leaf chlorophyll content and flag leaf width (Quarrie et al. 2005). The genome-wide association mapping approach has been applied recently for QTL detection in wheat. Although the development of gene-based molecular markers and genome sequencing in wheat should accelerate positional cloning, the genomic regions associated with individual QTL are still very large and are usually unsuitable for screening in a breeding programme.

Development of functional markers for useful alleles utilizing DREB genes is crucial for crop improvementstrategies. ASM developed for distinguishing drought tolerance in bread wheat (Sareen et al. 2015) was validated in 21 wheat accessions already field phenotyped. Bi-allelic variation with this primer was observed in genotypes IC36761A, IC57586, IC30276A and IC28665. On the basis of drought susceptibility index analysis, linear regression between AS-PCR and the phenotypic traits like GY, TGW and GW was found to be 1.8%, 41.5% and 20.4%, respectively. The highest correlation was observed between SNP and TGW.

Tolerant lines identified would thus lead to higher production and productivity under abiotic stress situations that adversely affect the wheat crop. These lines may also be utilized in hybridization programme for developing next-generation mapping (MAGIC and NAM) populations for identification of fine stress tolerance QTLs. Thus, the strategy involving trait-specific germplasm, precision phenotyping and the selection criteria based upon indices will be rewarding for increasing grain yield in wheat under harsh environments (Pandey et al. 2015).

PHS tolerance is a complex trait, and its genetics needs to be dissected using modern methods of QTL analysis. One facet of this complex trait deals with the balance of plant growth regulators, abscisic acid (ABA), gibberellic acid (GA) and their control to α-amylase enzyme activity. The level of *viviparous-1* (*Vp-1*) gene expression in immature embryos positively regulates ABA sensitivity and promotes seed dormancy (Kumar et al. 2015). Molecular studies for PHS indicated that chromosomes 3A, 3B, 3D and 4A have been considered potential region for PHS tolerance/dormancy. PHS resistance could be improved in spring wheat by pyramiding PHS resistance QTLs from different sources. Furthermore, identification and incorporation of genetic factors underlying resistant genotypes are keys to improve resistance/tolerance to preharvest sprouting in future cultivars (McCaig and DePauw 1992; Flinthan 2000; Mares et al. 2005; DePauw et al. 2012).

Modern plant science is dominated by genomics, and it has tremendous capacity to provide deep insights into the genetic makeup of plants. However, this needs complementation by precise characterization of response of plants to stresses which is now encompassed in the emerging field of phenomics. In addition, other components of *omics* science such as proteomics and metabolomics with bioinformatics as tool for integration of all these sciences are in place for genetic improvement of abiotic stress tolerance. These are equally applicable in important crops like wheat.

14.6.1 Transcriptomics

With the advent of microarray, DNA chip technologies, subtraction libraries, cDNA-AFLP, serial analysis of gene expression (SAGE) and RNA sequencing (RNA-seq), genome-wide transcript profiling has been widely used to identify drought-responsive genes in wheat. Next-generation transcriptome sequencing is also used for analysis of gene expression, the structure of genomic loci and sequence variation present at expressed gene loci. The expression profile of transcription factors involved in abiotic stress has been studied in wheat. The differential contributions of homeologous genes to abiotic stress response in hexaploid wheat on a genome-wide scale have been observed wherein large proportion (68.4%) of wheat homeologous genes exhibited partitioned gene expression in a temporal and stress-specific manner when subjected to heat stress, drought stress and their combination (Liu et al. 2015). Activity of antioxidant enzymes, viz. superoxide dismutase (SOD), ascorbate peroxidase (APX) and catalase (CAT), and the expression of their genes were studied in wheat genotypes under controlled severe drought (Sheoran et al. 2015a, b). The results indicated a unique pattern of activity and gene expression of antioxidant enzymes suggesting existence of genetic variation in drought responses of wheat at molecular and biochemical level.

14.6.2 Proteomics

The key proteins/enzymes and metabolic pathways identified from drought-tolerant wheat lines could be potentially targeted for designing drought-tolerant varieties of wheat. A series of proteomic experiments have been carried out in wheat to elucidate differential stem proteome patterns in two divergent wheat landraces (N49 and N14) under terminal drought stress. The tolerant landrace (N49) was more efficient at remobilizing stem reserves than the sensitive landrace (N14). The maximum number of differentially expressed proteins was noted at 20 days after anthesis in N49 when active remobilization of dry matter was recorded, thus suggesting potential participation of these proteins in efficient stem reserve remobilization (Bazargani et al. 2011).

14.6.3 Metabolomics

Plants react to abiotic stresses by altering the composition and concentration of metabolites so that they can acclimatize to adverse environmental changes.

Metabolic profiling has accelerated discovery of stress signal transduction molecules and compounds that are integral part of plant response to abiotic stresses. To accelerate the trait-based analysis of complex biochemical process, it is necessary to assess metabolic profiling along with NGS techniques, transcriptomics and proteomics underlying cellular biochemical events across diverse conditions. The metabolite profiling of given mapping population can be combined with the genetic linkage makeup maps to obtain greater insights into the genetic map of complex traits, thereby rendering metabolomics particularly relevant to crop breeding. Experiments have identified multiple metabolite QTLs in wheat under drought stress and pinpointed some genomic segments that control both agronomic traits and specific metabolites.

14.6.4 Identification and Validation of Stress-Induced Micro-RNA

Micro-RNAs (miRNAs) are a class of short endogenous non-coding small RNA molecules of about 18–22 nucleotides in length. Computational predictions have raised the number of miRNAs in wheat significantly using an EST-based approach. Hence, a combinatorial approach which is amalgamation of bioinformatics software and PERL script was used to identify new miRNA to add to the growing database of wheat miRNA. Identification of miRNAs was initiated by mining the EST (expressed sequence tags) database available at the National Center for Biotechnology Information. Pandey et al. (2013) investigated that as many as 4677 mature miRNA sequences belonging to 50 miRNA families from different plant species were used to predict miRNA in wheat. These authors further identified five abiotic stress-responsive new miRNAs. Also four previously identified miRNAs, i.e. *Ta-miR1122*, *miR1117*, *Ta-miR1134* and *Ta-miR1133*, were predicted in newly identified EST sequence, and 14 potential target genes were subsequently predicted, most of which seem to encode ubiquitin-carrier protein, serine/threonine protein kinase, 40S ribosomal protein, F-box/kelch-repeat protein and BTB/POZ domain-containing protein, transcription factors which are involved in growth, development, metabolism and stress response. Among the predicted miRNAs, expression of *miR855* in wheat for salt tolerance has been validated (Pandey et al. 2013). The result has increased the number of miRNAs in wheat, which should be useful for further investigation into the biological functions and evolution of miRNAs in wheat and other plant species. In order to understand the differential regulatory mechanism in wheat genotype C-306, expression profile of selected abiotic stress-responsive miRNAs involved in adaption to drought was examined. The drought-stressed C-306 genotype resulted in differential expression of six miRNAs. The accumulation of *miR393*, *miR1029* and *miR172* was significantly higher; however, drought had no major effect on the expression profiling of *miR529* as compared to mock-treated plants. These findings indicate that a diverse set of miRNAs could play an important role in mitigating drought stress responses in wheat. Using NGS, 30 novel miRNAs have been mined and work on further validation of selective drought-specific miRNAs, and their targets are being explored in wheat genotypes exhibiting contrasting responses to drought (unpublished).

14.6.5 Transformation of Abiotic Stress Genes for Imparting Stress Tolerance

Wheat is a less explored cereal crop for the development of transgenics. For the development of transgenic wheat, there is a necessity for having a robust regeneration and transformation protocol. Most of the earlier reported transformation systems are in Chinese spring and Bobwhite genotypes, and they are highly genotype dependent. Hence, at ICAR-IIWBR a robust wheat transformation system was developed in recently released Indian wheat genotypes with a transformation efficiency of 14% (unpublished). The first wheat transformation was reported by Cheng et al. (1997) using marker gene *uidA* through *A. tumefaciens*. The HDR77 wheat variety was transformed with *AtCBF3* gene using particle bombardment method and T_1 plants showed tolerance under moisture stress conditions (Kasirajan et al. 2013). The overexpression of *TaNF-YB4* gene enhanced the grain yield in wheat (Yadav et al. 2015). In the last few years, a number of attempts have been made to generate salt-tolerant genotypes. The *Vigna aconitifolia*_1-pyrroline-5-carboxylate synthetase '*P5CS*' gene-encoding enzyme required for the biosynthesis of proline was delivered into wheat, and the resultant transgenics showed significant salinity tolerance (Sawahel and Hassan 2002). The overexpression of transcription factor *TaNAC69* driven under barley drought-inducible *HvDhn4s* promoter enhanced shoot and root biomass of transgenic lines under combined mild salt stress and drought conditions (Xue et al. 2011).

14.7 Mitigation of Abiotic Stresses

14.7.1 Role of Phytohormones in Stress Amelioration

Hormones play an important role in plant adaptations to adverse environmental conditions. Crosstalk in hormone signalling reflects the ability to integrate different inputs and respond appropriately. There are six main groups of hormones, namely auxin, cytokinin (CK), gibberellic acid (GA), abscisic acid (ABA), ethylene and brassinosteroids.

Among all plant hormones, ABA is the most critical and hence termed as 'stress hormone'. Stress-induced senescence and abscission are the key processes mediated by ABA. Under water deficit conditions, ABA-modified root architecture contributes to the development of deeper root system along with enhancing hydraulic conductivity in plants, maintenance of cell turgor which finally contributes to desiccation tolerance. Other hormones, such as auxin, ethylene and cytokinins (CKs) , may alter the effect and biosynthesis of ABA. Under water and temperature stress, ethylene regulates root growth and development by limiting organ expansion. The higher ABA concentration ingrains might result from autosynthesis within the grain and partly by the translocation from leaves and roots during drying. ABA increases the endogenous content of proline under drought conditions. CKs play a supportive role during water deficit conditions by stimulating osmotic adjustment. Auxin is positively associated, and ABA is negatively correlated with the activity of *expansin* protein under oxidative stress condition. The ability of 28-homo-brassinolide to confer resistance to soil moisture stress in wheat is also established (Sairam 1994).

Hormonal homeostasis, stability, content, biosynthesis and compartmentalization are altered under heat stress. Brassinosteroids are considered as hormones with pleiotropic effects as they influence varied developmental processes like growth, germination of seeds, rhizogenesis, flowering and senescence. Brassinosteroids also confer resistance to plants against various abiotic stresses. They increase the tolerance to high temperature in wheat leaves (Seeta-Ram et al. 2002). The tolerance in plants to high temperature due to application of brassinosteroids is associated with the induction of *de novo* polypeptide (heat shock protein) synthesis. In a dwarf wheat variety, high temperature-induced decrease in cytokinin content was found to be responsible for reduced kernel filling and its dry weight (Wilkinson and Davies 2010). The external application of salicylic acid (SA) under field condition alleviates terminal heat stress in wheat (Mamrutha et al. 2015; Ratnakumar et al. 2016).

The ABA concentration increases in different plant parts in response to stress under salinity. ABA acts as the main signalling molecule under salinity stress. Presowing treatment of wheat seeds with growth regulators (IAA, GA) inhibits the effect of salinity. It has been reported that plant growth-stimulating compounds like gibberellic acid, zeatin and ethephon also help to alleviate the impact of salinity (Afzal et al. 2005). It was hypothesized that CKs could increase salt tolerance in wheat plants by interacting with other plant hormones, especially auxins and ABA. Exogenous application of kinetin overcame the effects of salinity stress on the growth of wheat seedlings. Gibberellic acid (GA_4) has been reported to be helpful in enhancing wheat growth under saline conditions (Parasher and Varma 1988). Exogenously applied phytohormones act as bioactivators of carbohydrates and alleviate the salt stress by acting as an osmoregulator. Salicylic acid (SA)-induced resistance of seedlings against salinity was associated with increase in chlorophyll content because SA is linked with the enhanced activity of photosynthetic pigments like Chl*a*, *b* and carotenoids in salt-stressed plant leaves. SA also provides a pool of compatible osmolytes under saline conditions. Exogenous applications of SA enhanced the accumulation of sugars and hence contribute towards yield enhancement in wheat. Proline also plays a supportive role by osmotic modifications along with SA-mediated defence-related role under salinity stress in wheat. SA action against salinity stress includes the development of antistress programmes and acceleration of normalization of growth processes even after the period of stress. SA can protect enzymes like nitrate reductase critical for nitrogen assimilation in wheat (Rane et al. 1995; Ratnakumar et al. 2016).

The plant hormones are of interest to produce shorter (2–15 cm), thicker and stronger stems which reduce lodging mainly by altering plant's biosynthesis GA (chlormequat chloride) or ethylene (ethephon). Changes in internode lignin content accompanied by those in cytokinin, IPA and t-zeatin suggest the role of CKs in the regulation of lignin deposition in wheat under waterlogging condition. The accumulated lignin contributes to mechanical strength/resistance to lodging, tolerance to biotic and abiotic stresses and feedstock quality of wheat straw (Nguyen et al. 2016).

14.7.2 Agronomic Interventions

While genetic improvement is often prioritized for addressing abiotic stresses, the agronomic techniques are also helpful in the management and mitigation of abiotic stresses. Though many of the factors causing stress remain beyond control, these can be alleviated to some extent for obtaining high productivity. Wheat yields in warm environments can be raised significantly by modifying agronomic practices involving conservation agriculture which involves significant reductions in tillage, surface retention of adequate crop residues and adopting diversified and economically viable crop rotations. The main longer-term productivity benefit of conservation agriculture practices comes from its potential to reverse the widespread and chronic soil degradation that threatens yields in the intensive wheat-based cropping systems. Resource-conserving practices like zero tillage (ZT) allow early sowing of wheat after rice harvest so that the crop is able to fill the grains before the hot weather ensues. As average temperatures in the region rise, early sowing will become even more important for wheat (Ortiz-Monasterio et al. 1994).

The agronomic interventions and management options for mitigating various abiotic stresses are briefly discussed hereunder.

Heat Stress Using sprinkler irrigation to cool down the canopy in the afternoon whenever the temperature goes beyond 30 °C improves productivity. For addressing the early heat stress at tillering stage, need-based light irrigation can be applied. In addition, adopting conservation agriculture practices helps in mitigating the temperature stress by moderating temperature variations, conserving soil moisture and improving soil organic matter status. Early sowing or timely sowing of the crop helps in escaping terminal heat stress and also leads to saving water required for pre-sowing irrigation in wheat crop by utilizing the residual moisture available in soil after harvesting the previous crop (Rane et al. 2007).

Drought Stress Scheduling irrigation according to growth stages of crop, use of efficient sowing methods like bed planting, irrigation based on soil moisture status and providing extra irrigation results in higher grain yield by mitigating the effects of moisture stress. Different methods of irrigation like furrow irrigation in furrow-irrigated raised bed (FIRB) planting system, sprinkler and drip irrigation help in efficient management of scarce water resources. These methods can also be utilized to use saline water for irrigation without its deleterious effects on the crop growth and development. Conservation agriculture practices also help significantly by reducing soil surface evaporation, moderating temperature and enhancing water availability for plant uptake, thereby mitigating the moisture stress effects.

Salt Stress Management practices for the safe use of saline water for irrigation comprise the use of land preparation methods for uniform distribution of water and to increase infiltration, leaching and removal of salts; special planting methods to minimize salt accumulation in the vicinity of the seed; irrigation to maintain a high level of soil moisture and periodic leaching of salts in the soil; and special treatments such as tillage and additions of chemical amendments, organic matter and growing green manure crops to maintain soil permeability and tilth. The use of micro-irrigation systems, such as drip and sprinklers, help in better control on salt and water distribution and thus enhance the use efficiency of saline water especially for high value crops. Saline soils are reclaimed by scraping of salts by using mechanical ways; flushing is used in crusting and low permeability soils to wash away the surface salts using good quality water to remove salts, and leaching works well on saline soils having good structure and internal drainage. Leaving crop residue at the soil surface, as in conservation agriculture, helps to reduce soil evaporation losses to decrease the accumulation of surface salts. The alkali/sodic soils are generally reclaimed by leaching of the excess sodium, deep ploughing to incorporate the calcareous subsoil into the topsoil and adding acidifying minerals like pyrite and gypsum.

Waterlogging Adopting conservation agriculture can be beneficial as addition of organic manure improves soil physical factors, reduces soil surface crusting, enhances plant rooting and alleviates the effects of pan formation on yield. Increase in organic content of soil helps in reclaiming problem of waterlogging due to increase in infiltration rate. Bed and furrow system or sprinkler irrigation on soils prone to waterlogging has been shown to reduce the problem significantly. Furrows also help to drain fields or keep a large portion of the root system out of waterlogged soils. It has been observed that adopting conservation agriculture and furrow-irrigated raised-bed planting system also helps to reduce crop lodging.

Nutrient Deficiency/Toxicity The nutrient management options like integrated nutrient management (INM), site-specific nutrient management, enhancing precision in nitrogen application based on NDVI and fertilizer application on the basis of soil test values help in mitigating nutrient deficiencies and toxicity in the soil as well as in plants (Yaduvanshi and Sharma 2008).

14.8 The Way Forward

Tolerance to abiotic stresses is essential to improve the productivity of wheat under harsh environments and reduce food insecurity threatened by climate changes. There are several reports indicating the existence of genetic variation in traits associated with tolerance to abiotic stresses. In this regard it is necessary to characterize the available germplasm and mapping populations for the known traits responsible for providing tolerance to different stresses. Assessment of the level of stress tolerance in popular cultivars not only in terms of yield and its components but also for relevant stress tolerance traits can help in fixing the targets for improvement. However, there is an urgent need for developing a database of screened germplasm and cultivars. The information on reduction in values for the traits in popular cultivars and genetic stocks when grown under optimal and suboptimal conditions in different agroecosystems should also be compiled. The rapid advances being made in genomics have to be utilized for identification of new genes and markers for genetic improvement of wheat for tolerance to drought, high temperature, waterlogging, etc. The already identified genes and markers need to be validated for their suitability in a particular genetic background. Focus is also required towards developing progenies with combination of traits rather than individual traits for different stresses and also towards finding markers so that they can be easily introgressed or tracked in breeding populations. This is essential as the future genetic gain in productivity improvement is likely to emerge from a combination of traits which can be stacked in a desired agronomic background by molecular approaches.

Since agroecologies where wheat is grown are diverse, genotypes adapted to specific locations need to be identified together with agronomic options for mitigation of stress for boosting wheat production in the future. Further, there have been promising results from experiments to identify hormones and bioregulators to alleviate abiotic stresses. The protocols for their use need to be optimized so that they can be integrated in agronomic packages for management of abiotic stresses in wheat. While this can be accomplished by experts in resource management, the desired level of resistance/tolerance to stresses in wheat genotypes can be incorporated through a close linkage between plant breeders and experts undertaking genotyping/phenotyping.

References

Afzal I, Basra S, Iqbal A (2005) The effect of seed soaking with plant growth regulators on seedling vigor of wheat under salinity stress. J Stress Physiol Biochem 1:6–14

Al-Ghzawi AA, Zaitoun S, Gosheh HZ, Alqudah AM (2009) The impacts of drought stress on bee attractively and flower pollination of *Trigonella moabitica* (fabaceae). Arch Agron Soil Sci 55(6):683–692

Al-Khatib K, Paulsen GM (1984) Mode of high temperature injury to wheat during grain development. Physiol Plant 61:363–368

Arvind K, Shukla R, Malik S, Tiwari PK, Prakash C, Behera SK, Yadav H, Narwal RP (2015) Status of micronutrient deficiencies in soils of Haryana. Impact on crop productivity and human health. Indian J Fert 11(5):16–27

Bazargani MM, Sarhadi E, Bushehri AS, Matros A, Mock H, Naghavi M, Hajihoseini V, Mardi M, Hajirezaei M, Moradi F, Ehdaie B, Salekdeh GH (2011) A proteomics view on the role of drought-induced senescence and oxidative stress defense in enhanced stem reserves remobilization in wheat. J Proteome 74:1959–1973

Bandeh-Hagh A, Toorchi M, Mohammadi A, Chaparzadeh N, Salekdeh GH, Kazemnia H (2008) Growth and osmotic adjustment of canola genotypes in response to salinity. J Food Agric Environ 6(2):201–208

Cheng M, Pang JE, Zhou S, Hironaka H, Duncan CM, Conner DR, Wan T (1997) Genetic transformation of wheat mediated by *Agrobacterium tumefaciens*. Plant Physiol 115:971–980

Cornic G (2000) Drought stress inhibits photosynthesis by decreasing stomatal aperture—not by affecting ATP synthesis. Trends Plant Sci 5:187–188

CSSRI (1997) Vision 2020 – CSSRI perspective plan. CSSRI, Karnal

DePauw RM, Knox RE, Singh AK, Fox S, Humphreys DG, Hucl P (2012) Developing standardized methods for breeding pre-harvest sprouting resistant wheat, challenges and successes in Canadian wheat. Euphytica 188:7–14

Flintham JE (2000) Different genetic components control coat imposed and embryo-imposed dormancy in wheat. Seed Sci Res 10:43–50

Garg D, Sareen S, Dala S, Tiwari R, Singh R (2013) Grain filling duration and temperature pattern influence the performance of wheat genotypes under late planting. Cereal Res Comm 41(3):500–507

Hunt LA, Vander Poorten G, Pararajasingham S (1991) Postanthesis temperature effects on duration and rate of grain filling in some winter and spring wheats. Canad J Plant Sci 71:609–617

Jain N, Ramya P, Krishna H, Ammasidha B, Prashant Kumar KC, Rai N, Todkar L, Vijay P, Pandey M, Kumar A, Bisht K, Ramya KT, Jadon V, Datta S, Singh PK. Singh GP. Vinod Prabhu K (2013) Genomic approaches for improvement of drought and heat in wheat. In: Recent trends on production strategies of wheat in India, pp 31–37

Kasirajan L, Boomiraj K, Bansal KC (2013) Optimization of genetic transformation protocol mediated by biolistic method in some elite genotypes of wheat (*Triticum aestivum* L.) African J Biotechnol 12(6):531–538

Kiniry JR (1993) Non-structural carbohydrate utilization by wheat shaded during grain growth. Agron J 85:844–848

Kirigwi FM, Van Ginkel M, Brown-Guedira G, Gill BS, Paulsen GM, Fritz AK (2007) Markers associated with a QTL for grain yield in wheat under drought. Mol Breed 20:401–413

Kumar S, Knox RE, Clarke FR, Pozniak CJ, DePauw RM, Cuthbert RD, Fox S (2015) Maximizing the identification of QTL for pre-harvest sprouting resistance using seed dormancy measure in a white-grained hexaploid wheat production. Euphytica 205:287–309

Liu Z, Xin M, Qin J, Peng H, Ni Z, Yao Y, Sun Q (2015) Temporal transcriptome profiling reveals expression partitioning of homeologous genes contributing to heat and drought acclimation in wheat (*Triticum aestivum* L.) BMC Plant Biol 15:152

Lopes MS, Reynolds MP (2010) Partitioning of assimilates to deeper roots is associated with cooler canopies and increased yield under drought in wheat. Funct Plant Biol 37:147–156

Mamrutha HM, Kumar R, Yadav VK, Venkatesh K, Tiwari V (2015) External application of salicylic acid as an option for mitigating terminal heat stress in wheat. Wheat Barley Newslett 9(1&2):12

Mares DJ, Mrva K, Cheong J, Williams K, Watson B, Storlie E, Sutherland M, Zou Y (2005) A QTL located on chromosome 4A associated with dormancy in white and red grained wheats of diverse origin. Theor Appl Genet 111:1357–1364

McCaig TN, DePauw RM (1992) Breeding for pre-harvest sprouting tolerance in white-seed-coat spring wheat. Crop Sci 32:19–23

Misra SC, Varghese P (2012) Breeding for heat tolerance in wheat. In: Singh SS, Hanchinal RR, Singh G, Sharma RK, Tyagi BS, Saharan MS, Sharma I (eds) Wheat: productivity enhancement under changing climate. Narosa Publishing House, New Delhi, p 398

Nagarajan S, Rane J, Maheshwari M, Gambhir PN (1998) Effect of post–anthesis water stress on accumulation of dry matter, carbon and nitrogen and their partitioning in wheat varieties differing in drought tolerance. J Agron Crop Sci 183:129–136

Nagarajan S, Rane J (2002) Relationship of simulated water stress using senescing agent with yield performance of wheat genotypes under drought stress. Indian J Plant Physiol 7(4):333–337

Nguyen TN, Son SH, Jordan MC, Levin DB, Ayele BT (2016) Lignin biosynthesis in wheat (*Triticum aestivum* L.): its response to water logging and association with hormonal levels. BMC Plant Biol 16:28

Ortiz-Monasterio R, Sayre JI, Pena KD, Fischer RA (1994) Improving the nitrogen use efficiency of irrigated spring wheat in the Yaqui Valley of Mexico. 15th World Cong. Soil Sci 5b:348–349

Pandey B, Gupta OP, Pandey DM, Sharma I, Sharma P (2013) Identification of new micro RNA and their targets in wheat using computational approach. Plant Signal Behav 8:e23932-1-9

Pandey GC, Mamrutha HM, Tiwari R, Sareen S, Bhatia S, Tiwari V, Sharma I (2015) Physiological traits associated with heat tolerance in bread wheat. Physiol Mol Biol Plants 21:93–99

Parasher A, Varma SK (1988) Effect of pre-sowing seed soaking in gibberellic acid on growth of wheat (*Triticum aestivum* L.) under different saline conditions. Indian J Biol Sci 26:473–475

Quarrie SA, Steed A, Calestani C, Semikhodskii A, Lebreton C (2005) High-density genetic map of hexaploid wheat (*Triticum aestivum* L.) from the cross Chinese Spring x SQ1 and its use to compare QTLs for grain yield across a range of environments. Theor Appl Genetics 110:865–880

Randall PJ, Moss HJ (1990) Some effects of temperature regime during grain filling on wheat quality. Aust J Agric Res 41:603–617

Rane J, Lakkineni KC, Kumar P, Abrol YP (1995) Salicylic acid protects nitrate reductase activity of wheat (*Triticum aestivum* L.) leaves. Plant Physiol Biochem 22(2):119–121

Rane J, Rao NVPRG, Nagarajan S (2002) Association between early vigour and root traits in wheat (*Triticum aestivum*) under moisture stress. Indian J Agric Sci 72:474–476

Rane J, Chauhan H, Shoran J (2003) Post anthesis stem reserve mobilization in wheat genotypes tolerant and susceptible to high temperature. Indian J Plant Physiol (special issue): 383–385

Rane J, Pannu RK, Sohu VS, Saini RS, Mishra B, Shoran J, Crossa J, Vargas M, Joshi AK (2007) Performance of yield and stability of advanced wheat genotypes under heat stress environments of the Indo-Gangetic Plains. Crop Sci 47:1561–1573

Ratnakumar P, Mir K, Minhas PS, Farooq MA, Sultana R, Per TS, Deokate PP, Khan NA, Singh Y, Rane J (2016) Can plant bio-regulators minimize crop productivity losses caused by drought, salinity and heat stress? An integrated review. J Appl Bot Food Qual 89:113–125

Rebetzke GJ, Condon AG, Richards RA, Farquahr GD (2002) Selection for reduced carbon isotope discrimination increases aerial biomass and grain yield of rain fed bread wheat. Crop Sci 42:739–745

Rebetzke GJ, Fischer RA, van Herwaarden AF, Bonnett DG, Chenu K, Rattey AR, Fettell NF (2014) Plot size matters: interference from intergenotypic competition in plant phenotyping studies. Funct Plant Biol 41:107–118

Sairam SK (1994) Effects of homo-brassinolide application on plant metabolism and grain yield under irrigated and moisture-stress conditions of two wheat varieties. Plant Growth Reg 14(2):173–181

Sallam A, El-Sayed H, Hashad M, Omara M (2014) Inheritance of stem diameter and its relationship to heat and drought tolerance in wheat (*Triticum aestivum* L.) J Plant Breed Crop Sci 6(1):11–23

Sareen S, Tyagi BS, Sarial AK, Tiwari V, Sharma I (2014) Trait analysis, diversity and genotype by environment interaction in some wheat landraces evaluated under drought and heat stress conditions. Chilean J Agric Res 74(2):135–142

Sareen S, Kundu S, Malik R, Dhillon OP, Singh SS (2015) Exploring indigenous wheat (*Triticum aestivum*) germplasm accessions for terminal heat tolerance. Indian J Agric Sci 85(2):194–198

Sawahel WA, Hassan AH (2002) Generation of transgenic wheat plants producing high levels of the osmoprotectant proline. Biotechnol Lett 24:721–725

Seeta-Ram SR, Vidya BV, Sujatha E, Anuradha S (2002) Brassinosteroids – a new class of phytohormones. Curr Sci 82(10):1239–1245

Sharma D, Mamrutha HM, Gupta VK, Tiwari R, Singh R (2015) Association of SSCP variants of HSP genes with physiological and yield traits under heat stress in wheat. Res Crops 16(1):139–146

Sharma D, Singh R, Rane J, Gupta VK, Mamrutha HM, Tiwari R (2016) Mapping quantitative trait loci associated with grain filling duration and grain number under terminal heat stress in bread wheat (*Triticum aestivum* L.) Plant Breed 135(5):538–545

Sharp RE, Davies WJ (1979) Solute regulation and growth by roots and shoots of water stressed maize plants. Planta 147:43–49

Sheoran S, Thakur V, Narwal S, Turen R, Mamrutha HM, Singh V, Tiwari V, Sharma I (2015a) Differential activity and expression profile of antioxidant enzymes and physiological changes in wheat (*Triticum aestivum* L.) under drought. Appl J Biochem Biotech 177(6):1282–1298

Sheoran S, Malik R, Narwal S, Tyagi BS, Mittal M, Khaurb AS, Tiwari V, Sharma I (2015b) Genetic and molecular dissection of drought tolerance. J Wheat Barley Res 7(2):1–13

Singh G, Kulshreshtha N, Singh BN, Setter TL, Singh MK, Saharan MS, Tyagi BS, Ajay V, Indu S (2014) Germplasm characterization, association and clustering for salinity and water logging tolerance in bread wheat (*Triticum aestivum* L.) Indian J Agric Sci 84(9):1102–1110

Stone PJ, Savin R, Wardlaw IF, Nicolas ME (1995) The influence of recovery temperature on the effects of a brief heat shock on wheat: I. Grain growth. Aust J Plant Physiol 22:945–954

Stone PJ, Nicolas ME (1994) Wheat cultivars vary widely in their responses of grain yield and quality to short periods of postanthesis heat stress. Aust J Plant Physiol 21:887–900

Trethowan RM, Reynolds MW, Sayre K, Ortiz-Monasterio I (2005) Adapting wheat cultivars to resource conserving farming practices and human nutritional needs. Ann Appl Biol 146:405–413

Tiwari R, Sheoran S, Rane J (2015) Wheat improvement for drought and heat tolerance. In: Shukla RS, Mishra PC, Chatrath R, Gupta RK, Tomar SS, Sharma I (eds), Recent trends on production strategies of wheat in India, pp 39–58

Wardlaw IF, Dawson IA, Munibi P, Fewster R (1989) The tolerance of wheat to high temperatures during reproductive growth: I. Survey procedures and general response patterns. Aust J Agric Res 40:1–13

Wardlaw IF, Moncur L (1995) The response of wheat to high temperature following anthesis I. The rate and duration of kernel filling. Aust J Plant Physiol 22:391–397

Wheeler TR, Batts G, Ellis RH, Haley P, Morison JH (1996) Growth and yield of winter wheat (*Triticum aestivum*) crops in response to CO2 and temperature. J Agric Sci (Camb) 127:37–48

Wilkinson S, Davies WJ (2010) Drought, ozone, ABA and ethylene: new insights from cell to plant to community. Plant Cell Environ 33:510–525

Xue GP, Way HM, Richardson T, Drenth J, Joyce PA, McIntyre CL (2011) Overexpression of *TaNAC69* leads to enhanced transcript levels of stress up-regulated genes and dehydration tolerance in bread wheat. Mol Plant 4(4):697–712

Yadav D, Shavrukov Y, Bazanova N, Chirkova L, Borisjuk N, Kovalchuk N, Ismagul A, Parent B, Langridge P, Hrmova M, Lopato S (2015) Constitutive overexpression of the TaNF-YB4 gene in transgenic wheat significantly improves grain yield. J Exp Bot 66(21):6635–6650

Yaduvanshi NPS, Sharma DR (2008) Tillage and residual organic manures/chemical amendment effects on soil organic matter and yield of wheat under sodic water irrigation. Soil Tillage Res 98(1):11–16

Breeding Rice Varieties for Abiotic Stress Tolerance: Challenges and Opportunities

15

Vishnu V. Nachimuthu, Robin Sabariappan,
Raveendran Muthurajan, and Arvind Kumar

Abstract

Climate change-induced abiotic stresses are considered as notable threat to world food security affecting crop, livestock, and fisheries production which are all fundamental for sustainable development of human life. Impact of climate variability affecting water availability, nutrient levels, soil moisture, temperature, and tropical ozone in crop yield is measured in various studies. Rice, the critical crop in maintaining food security, has high vulnerability to increased frequency and intensity of extreme weather events which affects crop growth at macro- and microenvironment. Meanwhile, rising temperatures and consequent rise in sea level can make farming riskier by increasing salinity in the cultivatable lands. Decrease in productivity of rice is mainly related to extreme environmental conditions such as water deficit, high temperature, submergence, salinity, cold, and accumulation of heavy metals apart from higher incidence of pathogens and pests. Crop germplasm, wild relatives, and other species serve as main genetic sources for tolerance. These can be useful in crop breeding as they have had adaptation and acclimation responses developed through natural selection process. Hence, identification of genetic loci, mechanism, and signaling path-

V.V. Nachimuthu • A. Kumar (✉)
South Asia Breeding Hub, International Rice Research Institute, ICRISAT Campus,
Patencheru 502342, Hyderabad, Telengana, India
e-mail: v.varthini@irri.org; a.kumar@irri.org

R. Sabariappan
Center for Plant Breeding and Genetics, Tamil Nadu Agricultural University,
Coimbatore 641003, Tamil Nadu, India
e-mail: robin.tnau@gmail.com

R. Muthurajan
Centre for Plant Molecular Biology & Biotechnology, Tamil Nadu Agricultural University,
Coimbatore 641003, Tamil Nadu, India
e-mail: raveendrantnau@gmail.com

© Springer Nature Singapore Pte Ltd. 2017
P.S. Minhas et al. (eds.), *Abiotic Stress Management for Resilient Agriculture*,
DOI 10.1007/978-981-10-5744-1_15

ways provides a paradigm to improve yield under diverse ecosystem. By using marker-assisted selection, beneficial alleles from wild relatives can be introgressed to develop climate-ready rice varieties.

15.1 Introduction

Global food production needs to be increased by at least 50% to meet out the food and fuel requirement of projected 9.1 billion people during 2050 (Yamori et al. 2013). This warrants an increase of one billion tonnes of annual cereals and around 200 million tonnes of meat. Predictions based on prevailing growth trends in major food crops, viz., rice, wheat, maize, soybean, etc., suggest that prevailing productivity growth will not be sufficient to meet the projected food demands (Ray et al. 2013). Achieving this target needs overcoming challenges, viz., yield plateau, declining land, water and labor resources, and predicted adverse effects of global climate change. Impacts of these key determinants are being witnessed all over the world, but vulnerability of agriculture in developing countries is of major concern in view of the large populations being dependent on agriculture, excessive pressure on natural resources, and poor coping mechanisms. Climate change has already started affecting yield of wheat and paddy in parts of India due to increased temperature, water stress, and reduction in number of rainy days. Its impacts are likely to aggravate yield losses in major agricultural crops thereby expected to affect food security. The projected agricultural productivity loss due to changing climate by 2100 is about 30–40% (Cline 2007; Aggarwal 2008). Predicted increase of 1–2 °C in global mean temperature along with reduced water availability may reduce cereal yields in tropical regions (Rosenzweig et al. 2013). The reduction in annual productivity may be to the tune of 22, 3, 9, and 2 million tonnes of maize, rice, wheat, and soybean, respectively (Ray et al. 2015).

Rice is the second most important cereal food crop next to wheat, and its production has to be increased by at least 30% by 2030 and 50–70% by 2050. Abiotic factors such as drought at the beginning of the cropping season, flash flooding at the reproductive stage, coastal salinity, susceptibility of fine-grained cultivars to diseases, etc. are main factors that affect the productivity of the most of the rice-growing areas. Annual rice production is susceptible to ~32% fluctuation due to climatic factors, viz., higher precipitation variation in South Asia and higher temperature variation in Southeast and East Asia rice-growing tracts (Ray et al. 2015). Enumerating the shifting pattern of abiotic and biotic stresses caused by climate change, the productivity of major calorie providers such as rice, wheat maize, and soybean has to be expanded to 87% more by 2050 to bottle up emerging global food shortage (Kromdijk and Long 2016).

Considering the magnitude of problems due to climate change, developing climate-smart rice varieties either through conventional breeding or molecular breeding or genetic engineering approaches provides for ways to tackle current predicament in rice production. Climate-smart varietal breeding aims at improving yield and its yield stability across multiple stressors induced by climate extremes. For rice production, flooding, water stress, high temperature, salinity, cold, and altered pest and disease outbreak pose major challenges for its production.

The green revolution during the 1960s was focused on higher productivity with better fertilizer responsiveness through introduction of modern semidwarf and high-yielding varieties in rice and wheat. As a consequence, two- to threefold increase in productivity has been achieved through modern plant breeding tools during the last three to four decades. However, yields are stagnating around 8 Mg ha^{-1} for the past 10–15 years even under well-managed conditions. As more than 40% of the rice-growing areas are either rainfed lowland or upland cultivation, performance of the high-yielding varieties is affected frequently due to vagaries of monsoon and other biotic/abiotic stresses. Current scenario of climate change has shifted the focus of rice breeders from just yield to combining yield with multiple stress tolerance for surviving under extreme climate events. Therefore, multidisciplinary programs have been launched for developing climate-smart rice varieties adapted to changing climatic conditions. Different innovative breeding techniques are being integrated to exploit genetic variation to develop climate-smart rice varieties exhibiting enhanced tolerance against multiple stresses coupled with high yield. An overview of the global efforts in developing rice varieties adapted to climate-induced stresses and thereby preventing yield decline by extreme weather events is discussed below.

15.2 Breeding for Enhanced Drought Tolerance

Plant drought stress is defined as "Water deficit at any plant growth stage – though with more impact during reproductive and grain filling stages" (Reynolds et al. 2016) – that results in 10% yield loss compared to an adequately-watered control. Major symptoms of drought include accelerated plant development resulting in reduced biomass, decreased seed set due to reduced panicle size or failure of pollination or reduced grain size, and reduction in harvest index (HI) due to early grain filling and premature senescence. Plant's adaptation to soil moisture deficit is controlled by several mechanisms, viz., escape, avoidance, or tolerance strategies (O'Toole and Chang 1979). Drought escape mechanism is defined as the ability of the plant to complete its life cycle before enduring water starvation. Rapid phenological development and developmental plasticity pave way for plant's escape from drought. Avoidance strategy makes the plant to withstand drought by maintaining higher internal water status either through increased uptake of water from soil or through reduced water loss through the canopy. Drought tolerance is the ability of the plant to survive even at low tissue water content. Translocation of assimilates capable of maintaining osmotic status is considered as an adaptive trait associated with dehydration tolerance (Arrandeau 1989). Breeding for drought tolerance still remains a challenge for breeders due to its complex genetic nature with higher environmental plasticity and involvement of multiple metabolic pathways. Rice germplasm harbors large amount of functional genetic diversity for drought tolerance-related traits/mechanism, and careful exploitation of genetic variation will allow us to integrate diverse traits for improving resilience of rice against climate change. Most of the conventional approaches in drought breeding emphasize on secondary traits such as root architecture and mass, physiological parameters such as water use efficiency, relative water content, osmotic adjustment, etc. Selection for secondary traits such as root architecture (Courtois et al. 2009; Uga et al. 2013), leaf rolling (Price et al. 1997), stomatal conductance (Price et al. 1997; Khowaja and Price 2008), relative water content, osmotic adjustment (Kamoshita et al. 2002; Nguyen et al. 2004), cell membrane stability (Tripathy et al. 2000), epicuticular wax (Srinivasan et al. 2008), stem reserve mobilization, and canopy temperature (Prince et al. 2015) has been used to develop drought-tolerant lines. Integrating all these parameters with better yield under water deficit condition was a great challenge for plant breeders. Because of low heritability and difficulty in precision phenotyping of these traits, attempts were made to use "yield under stress" as a criterion for selection in breeding programs despite the moderate heritability of yield traits (Venuprasad et al. 2007, 2008; Kumar et al. 2008, 2009). This enabled the scientists to select drought-tolerant rice genotypes directly based on yield under drought stress (Kumar et al. 2014; Lafitte et al. 2004; Lanceras et al. 2004; Bernier et al. 2007; Venuprasad et al. 2007; Kumar et al. 2008).

Completion of rice genome sequencing project and advancements in genotyping procedures have enabled us to dissect out the genetic basis of drought tolerance through linkage mapping and association mapping. QTL mapping for drought tolerance was carried out for different primary and secondary traits using various populations to dissect genetic basis of this multi-loci controlling trait (Kamoshita et al. 2008). QTLs and markers identified have been utilized via marker-assisted selection (MAS), marker-assisted backcross breeding (MABB), and marker-assisted recurrent selection (MARS) to develop drought-tolerant rice varieties.

Marker-assisted selection for better root traits (QTLs on chromosomes 2, 9, and 11) in a population derived between Kalinga III and Azucena resulted in the identification of a superior drought-tolerant rice genotype, e.g., *Birsa Vikas Dhan* 111, that was released in Jharkhand, India (Steele et al. 2006, 2007). Keeping grain yield under stress as a criterion for genetic mapping has resulted in the discovery of several mega-effect QTLs in tolerant genotypes, viz., Apo, Way Rarem, Vandana, Nagina 22, etc. Bernier et al. (2007) have reported the first large-effect QTL, namely, $qDTY_{12.1}$, associated with grain yield (explaining 33% variance under severe upland stress) under water deficit condition in a population derived between Vandana x Way Rarem. Further molecular genetics studies using the QTL-NILs revealed that this mega-effect QTL is offering enhanced drought tolerance through the production of more number of lateral roots and efficient transpiration control (Henry et al. 2014), and unexpectedly the susceptible parent Way Rarem contributed positively (Dixit et al. 2012).

The same locus was also reported in the genetic background of a cross between IR74371-46-1-1 and Sabitri (Mishra et al. 2013). Later, a major-effect QTL $qDTY_{1.1}$ on chromosome 1 was identified in a cross involving a drought-tolerant Nagina 22 with high-yielding cultivars, i.e., Swarna, IR64, and MTU1010. This is one of the most consistent QTLs for grain yield under drought which coincide with *sd1* (semidwarf) region and QTL for plant height and flowering under water deficit condition (Vikram et al. 2011). Several other major-effect QTLs, namely, $qDTY_{2.1}$, $qDTY_{3.1}$, $qDTY_{3.2}$, $qDTY_{6.1}$, and $qDTY_{10.1}$, were reported to be associated with grain yield under water deficit conditions through the studies conducted at the International Rice Research Institute, Philippines (Kumar et al. 2013), by utilizing mapping population(s) developed between Swarna and Apo (Venuprasad et al. 2009). $qDTY_{3.2}$ was identified in the region of *HD9* that affects flowering in Vandana/Way Rarem population whereas its effect on grain yield under drought stress established from N22/Swarna population (Vikram et al. 2011). Most of the QTLs associated with grain yield under stress were found to be linked with QTLs controlling plant height and days to flowering. Co-localization of these traits is not preferred to breed varieties for various ecosystems; therefore, a suitable MAB strategy to break the linkage with these traits needs to be designed for transferring these QTLs. Pyramiding of these desirable QTLs will lead to enhanced level of tolerance. Swamy and Kumar (2013) demonstrated pyramiding of different combinations of grain yield under stress QTLs in the genetic background of IR64 and reported that BC lines harboring combination $qDTY_{2.2}$ and $qDTY_{4.1}$ yielded higher than lines with $qDTY_{2.2}$, $qDTY_{4.1}$, $qDTY_{9.1}$, and $qDTY_{10.1}$. Vikram et al. (2016) have studied the linkage and pleiotropic effects of the QTLs, i.e., $qDTY_{1.1}$, $qDTY_{3.2}$, and $qDTY_{12.1}$, and concluded that all these three QTLs linked with QTLs for plant height and flowering. $qDTY_{3.2}$ has shown positive interaction with other QTLs by reducing flowering duration, thereby helping in avoiding drought and leading to increase in grain yield. Using these QTLs, several commercial high-yielding varieties, viz., IR64, Swarna, Vandana, Sabitri, Samba Mahsuri, TDK1, and Anjali, were improved for their drought tolerance ability (Kumar et al. 2014).

To overcome the limitations in biparental linkage mapping, linkage disequilibrium mapping/association mapping strategy has been developed which takes advantage of historical recombination that provides allelic richness and better resolution to detect natural variation. Linkage disequilibrium, nonrandom associations among different genetic loci, provides a way to detect robust QTLs by using dense markers evenly distributed throughout the genome. Association analysis is done either by considering candidate gene responsible for the trait (Candidate gene association studies) or whole genome scan for detecting QTLs. In rice, association analysis is mainly done with landraces, elite cultivars, and other germplasm lines which provide abundant variations for different traits. Association mapping needs population structure and LD information for curtailing false positives. In rice, five major groups, i.e., *indica*, aus, aromatic, temperate *japonica*, and tropical *japonica*, were identified in many studies (Garris et al. 2005; Xu et al. 2012). Phung et al. (2014) has reported that LD decay was faster in *indica* subpopulation than *japonica* subpopulation. Hence *indica* subpopulation need high marker density for genomic analysis thereby provides better resolution for association analysis. Association mapping was carried out by various researchers for drought tolerance in rice with the focus on secondary traits such as root architecture. Phung et al. (2016) further identified two significant associations for root thickness on chromosome 2 and for crown root number on chromosome 11 in a panel of 200 accessions. Association analysis was performed using mixed model with population structure and kinship values for whole panel as well as for *japonica* and *indica* accessions subpanels separately. Courtois et al. (2013) has performed association analysis in *japonica* panel of 167 accessions with the identification of 51 significant associations with different root traits such as maximum root length, deep root biomass, root dry mass, shoot biomass, and root to shoot ratio. Deep rooting, an important drought avoidance trait, has been mapped to a major QTL at chromosome 2 by genome-wide association mapping as well as family-based linkage mapping (Lou et al. 2015). Genome-wide association mapping was performed by Zaniab Al-Shugeairy and Robinson (2015) for the trait drought recovery with 328 rice cultivars which reveals a locus at chromosome 2 has significant association with this trait. Therefore, various genomic regions identified for drought-related traits through association analysis provide an effective way to narrow down the wide genomic regions influencing this quantitative trait. Next-generation breeding tools, viz., nested association mapping (NAM) and multi-parent advanced generation intercrosses (MAGIC), are the renewed strategies for simultaneous tagging of genomic regions controlling drought tolerance-related traits and identification of superior recombinants.

15.3   Breeding for Heat Stress Tolerance

High temperature stress (HTS) is one of the important yield-limiting factors for most of the agriculturally important crops, and frequent occurrence of short-term heat shocks (extremely high temperatures) had been a common phenomenon during recent years (Mackay 2008). Predictions on global climate models suggest that a mean increase of 1.0–3.7 °C is expected to occur in majority of agricultural areas by the end of the twenty-first century (IPCC 2013). High temperature affects grain yield especially when it coincides with reproductive stage by affecting fertility of gametes. Reynolds et al. (2016) have reported two critical stages, viz., micro- or megasporogenesis and fertilization, which are highly sensitive to rise in temperature which can lead to loss of male and female gamete fertility, reduced pollen deposition and growth, poor fertilization, as well as early embryo abortion. The adverse effect of high temperature on crop yield can be minimized by breeding heat-tolerant genotypes utilizing functional alleles.

Even though rice is a heat-loving plant, its performance is affected when temperature is beyond threshold levels. Analysis of the temperature and rice yield during 1992–2003 at the International Rice Research Institute (IRRI) showed that rice grain yield declined by 10% for each 1 °C increase in the temperature during growing season (Peng et al. 2004). Recent studies from the Yangtze River in China showed that an estimated 3 million ha of rice were damaged and about 5.18 million t of paddy rice lost in 2003 due to a heat wave with HDT >38 °C lasting more than 20 days (Li et al. 2004; Xia and Qi 2004; Yang et al. 2004). Likewise, IR64, one of the widely grown rice varieties, which performed well over the last three decades under local hot tropical conditions in Pakistan, suffered a 30% yield decline during a heat wave in 2007 (Dr. Mari, IRRI, Dokri).

Multi-crop Experts' Committee Meeting conducted by the US Agency for International Development (USAID) and the Bill & Melinda Gates Foundation during 2013 has described heat stress as "Supra-optimal temperatures occurring at any plant growth stage that can result in ≥10% yield loss. This is typically characterized by accelerated plant development resulting in reduced photosynthetic area, plant biomass, and seed set. Where heat stress occurs during grain-filling, reduced grain weight (and therefore HI) will result from inhibition of starch synthesis, increased starch breakdown, and/or premature and rapid increase in senescence" (Reynolds et al. 2016).

The process of evolution has developed various adaptive mechanisms, viz., avoidance and tolerance in plants to survive under high temperature stress. Mechanism of high temperature stress avoidance includes flowering time alteration to favor flowering during the cooler period of the day, altered leaf orientation, excessive rooting, and reducing temperature by transpiration (Sailaja et al. 2015). Tolerance mechanisms against heat stress involve rapid alterations in molecular processes, i.e., production of heat shock proteins (HSP), late embryogenesis abundant (LEA) proteins, and metabolites, membrane modifications, accumulation of free radical scavengers, and altered cytoskeleton (Bita and Gerats 2013). Existence of natural genetic variation for heat-adaptive traits in crop germplasm makes the breeding a viable option for achieving yield stability in heat stress environment. Reynolds et al. (2016) has identified priority traits for high temperature stress tolerance (Table 15.1).

Table 15.1 Traits for high temperature tolerance

Trait class	Specific trait
Photosynthesis/biomass/metabolism	Canopy temperature
	Normalized difference vegetative index (NDVI)
	Final biomass
	Night respiration
	Chlorophyll fluorescence
	Starch synthesis
	Membrane thermostability
	Spectral indices for pigments (chlorophyll and carotenoid)
Fertility/source-sink partitioning	Anthesis/flowering time
	Spikelet fertility/pollen viability/pollen shedding level
	Grain filling
	Harvest index
	Phenology
	Tillering
	Assimilate remobilization
	Plant growth regulators
	Coleoptile length

Source: Reynolds et al. (2016).

Global heat stress breeding programs have successfully identified donors for different strategies to increase resilience to heat stress damage (Ishimaru et al. 2010; Jagadish et al. 2010; Ye et al. 2012). Firstly, cultivation of early maturing versions of elite genotypes suitable for summer cultivation will enable the rice plant escape from hotter months. Secondly, genetic manipulation of early-morning flowering (EMF) will shift the flowering patterns to cooler hours of the morning. Thirdly, true heat tolerance will be induced through maintained reproductive success in spite of higher temperatures coinciding with flowering. Breeding by conventional techniques for heat stress has been a challenging task due to limited information on genetic determinants, biological mechanisms, and effective screening methods. In general, genotypes are screened under multiple hot environments and selection based on its yield stability across various environments. Advancements in development of controlled climate growth chambers, phytotron facilities, etc. have enabled the scientists to devise rapid protocols for screening large number of genotypes against high temperature stress (Jagadish et al. 2010).

Intensive research efforts across the globe have resulted in the identification of few heat-tolerant rice lines, viz., Nagina 22, Dular, Kasalath, Giza178, and Todorokiwase (Tenorio et al. 2013). In general *indica* genotypes were found to exhibit better heat tolerance than *japonica* genotypes (Zhao et al. 2016). By using these tolerant genotypes, few major QTLs have been identified for various high temperature adaptive traits at various growth stages of rice. Through several studies using independent biparental mapping populations, around 52 main-effect QTLs and 25 epistatic QTLs explaining phenotypic variance of 2.27–50.11% have been identified (Cao et al. 2002, 2003; Zhu et al. 2005, 2006; Zhao et al. 2006; Chen et al. 2008; Kui et al. 2008; Zhang et al. 2008; Jagadish et al. 2010; Pan et al. 2011; Xiao et al. 2011). Cao et al. (2003) has identified few QTLs for seed setting under heat stress in a doubled haploid population derived between IR64 x Azucena and the most significant QTLs explaining largest phenotypic variance for seed setting was identified in IR64. Zhang et al. (2009) have identified two SSR markers, RM3735 on chromosome 4 and RM3586 on chromosome 3, to be linked to heat stress tolerance-related traits. The most significant heat-tolerant QTL explaining up to 50% of phenotype variance was found to be contributed by a heat-tolerant *indica* landrace Kasalath (Zhu et al. 2006). Wei et al. (2013) has identified a dominant locus *OsHTAS* on chromosome 9 which influences heat tolerance at seedling stage in rice. Using Giza178 as heat-tolerant donor, six QTLs were identified for spikelet fertility under HT stress in two different mapping populations derived from IR64 cross and Milyang 23 cross. One of the QTLs *qHTSF4.1* explaining phenotypic variance of 17.6% was found to increase spikelet fertility in various heat-tolerant lines (Ye et al. 2015). Zhao et al. (2016) used chromosome segment substitution lines derived from Sasanishiki (heat susceptible) and Habataki (heat tolerant) for identifying eleven QTLs for spikelet fertility, daily flowering time and pollen shedding level (Table 15.2).

A mutant NH219 of Nagina 22 was found to exhibit enhanced level of tolerance against HT stress, and single-marker analysis among F_2 segregates of IR64 × NH219 revealed significant association of RM1089 with number of tillers and yield per

Table 15.2 QTLs identified for heat tolerance in rice, particular quantities trait loci (QTL) on chromosome (Chr) number

QTL	Chr.	Marker	Position (Mb)	Population	Donor	References
qSFht2	2	RM1234-RM3850	11.3–35.4	SasanishikiXHabataki CSSL	Habataki	Zhao et al. (2016)
qtl2.2		C601	30.3	Bala × Azucena RIL	Azucena	Jagadish et al. (2010)
qHt2		RM183-RM106	25.1	T219 × T226 RIL	T226	Chen et al. (2008)
qhts-2		RM406-525	28.3–35.2	Zhongyouzao 8 × Toyonishiki RIL	Toyonishiki	Zhang et al. (2008)
qSFht4.2	4	RM3916-RM2431	28.6–34.9	SasanishikiXHabataki CSSL	Habataki	Zhao et al. (2016)
qhr4–3		RG214-RG143	31.7–33.7	IR64 × Azucena DH	IR64	Cao et al. (2003)
qPSLht5	5	RM1248-RM4915	0–4.3	SasanishikiXHabataki CSSL	Habataki	Zhao et al. (2016)
qhts-5		RM405-274	3.1–26.9	Zhongyouzao 8 × Toyonishiki RIL	ZYZ8	Zhang et al. (2008)
qDFT3	3	RM3766-RM3513	6.9–25.1	SasanishikiXHabataki CSSL	Sasanishiki	Zhao et al. (2016)
qhr3–1		RZ892-RG100	5.2–12.9	IR64 × Azucena DH	Azucena	Cao et al. 2003)
qhts-3		RM157b-282	12.41	Zhongyouzao 8 × Toyonishiki RIL	Toyonishiki	Zhang et al. (2008)
qDFT8	8	RM5891-RM4997	26.6–28.2	SasanishikiXHabataki CSSL	Sasanishiki	Zhao et al. (2016)
qtl8.3		RG598	27.6	Bala × Azucena RIL	Bala	Jagadish et al. (2010)
qDFT10.1	10	RM6737-RM6673	18.7–23.0	SasanishikiXHabataki CSSL	Sasanishiki	
qtl10.1		C16	20.8	Bala × Azucena RIL	Bala	Jagadish et al. (2010)
SSPF10, SSPc-10,	11	RM6132-RM6100	18.8–18.9	996 × 4628 RIL	996	Xiao et al. (2011)
qDFT11		RM1355-RM2191	17.2–24.7	SasanishikiXHabataki CSSL	Sasanishiki	Zhao et al. (2016)
qtl11.1		G1465	24.2	Bala × Azucena RIL	Azucena	Jagadish et al. (2010)
qPSLht1	1	RM1196-RM6581	21.9–31.5	SasanishikiXHabataki CSSL	Sasanishiki	Zhao et al. (2016)
qhr1		RG381-RZ19	30.9	IR64 × Azucena DH	Azucena	Cao et al. (2003)

plant, RM423 with leaf senescence, RM584 with leaf width, and RM229 with yield per plant (Poli et al. 2013). In an African rice cultivar, a major QTL TT1 involved in the degradation of ubiquitinated proteins during heat stress influencing thermotolerance was identified (Li et al. 2015). Many putative QTLs identified in various studies were overlapped or adjacent with each other suggesting the existence of heat-tolerant alleles widely across the chromosomes. Fine mapping of these QTLs will provide extensive information on underlying candidate genes and metabolism involved in high temperature tolerance stress. Fine mapping of these QTLs will provide extensive information on underlying candidate genes and metabolism involved in high temperature stress. These QTLs can be introgressed into commercial varieties for developing heat-tolerant lines.

15.4   Breeding for Submergence Tolerance

Submergence is becoming a second major abiotic stress affecting rice production under rainfed lowlands of Asia. During recent years, extent of damage caused by submergence stress has increased due to extreme weather events such as unexpected heavy rains that inundate rice areas. Consequences of submergence range from low light intensity, limited oxygen availability (hypoxia), soil nutrient effusion, physical injury, and increasing pest and disease access. Rice is extremely susceptible to anaerobic conditions during all its growth stages starting from germination (anaerobic germination), early vegetative growth, and post-flowering (Angaji et al. 2010). Flooding during germination and early seedling growth affects seedling establishment in direct seeded rice, both in rainfed and irrigated areas, because of their high sensitivity to hypoxia at this stage (Ismail et al. 2009). Hence, developing submergence-tolerant rice varieties is inevitable to sustain rice production in South and Southeast Asia. Limited genetic variation has been reported for submergence tolerance, and landraces or traditional genotypes, viz., FR13A, Goda Heenathi, Thavalu, Kurkaruppan, etc., were found to exhibit high level of tolerance to submergence (Miro and Ismail 2013). As these genotypes are poor yielders and inferior in their grain quality, developing high-yielding varieties that are submergence tolerant and readily accepted by farmers will help in sustaining increased rice productivity in target regions.

Understanding of the physiological and biochemical bases of submergence tolerance has progressed well in recent years, making it possible to design efficient phenotyping protocols, and has paved way for further genetic and molecular studies, to discover genes underlying component traits associated with tolerance. High-yielding rice cultivars with enhanced tolerance to repeated flooding, better regeneration capacity, and faster growth after flooding to produce sufficient biomass in a shorter period and to minimize delay in flowering when submerged are needed for flood-prone areas. Thorough biochemical studies involving susceptible and tolerant rice genotypes revealed that maintenance of high levels of stored carbohydrates in the seedlings prior to submergence coupled with minimum shoot elongation and retention of chlorophyll were contributing for enhanced tolerance.

Several QTLs controlling submergence tolerance have been reported by many authors including a single major quantitative trait locus (QTL) on chromosome 9 (Xu and Mackill 1996; Nandi et al. 1997; Toojinda et al. 2003). Most of these reports were based on the exploitation of the landrace FR13A, one of the most submergence-tolerant cultivar originated in eastern India. A major QTL named Sub1 with a LOD score of 36 and an R^2 value of 69% (Xu and Mackill 1996) was found to confer tolerance against complete submergence for up to 2 weeks. Further fine-mapping of Sub1 using back cross inbred lines has resulted in the identification of a genomic region of approx. 0.06 cM (Xu et al. 2000). Sequencing the Sub1 region in an FR13A-derived line revealed the presence of three genes encoding putative ethylene responsive factors (ERF), Sub1A, Sub1B, and Sub1C. Out of these three putative candidates, Sub1A was subsequently identified as the major determinant of submergence tolerance (Xu et al. 2006). Cloning and characterization of a major QTL *Sub1* and subsequent identification of candidate genes underlying the effect of this QTL not only generated knowledge on the molecular mechanisms controlling submergence tolerance but also helped to design tightly linked gene-based markers for molecular breeding programs aimed at developing submergence-tolerant rice cultivars (Siangliw et al. 2003; Toojinda et al. 2005; Neeraja et al. 2007).

Considerable genetic variation exists in rice germplasm for anaerobic germination under flooding, and genotypes, viz., Khao Hlan On, Ma-Zhan Red, Khaiyan, Kalonji, Kharsu, and Nanhi, exhibited better germination under anaerobic conditions leading to better crop stand (Miro and Ismail 2013). Several genomic regions influencing anaerobic germination under flooding were identified by exploiting tolerant landraces. Two major QTLs, viz., qAG9-2 on chromosome 9 (Angaji et al. 2010) and qAG7-1 on chromosome 7 (Septiningsih et al. 2013), were reported using Khao Hlan On and Ma-Zhan Red, respectively. Recently, *qAG-9-2* was fine-mapped, and a candidate gene *OsTPP7* encoding a trehalose-6-phosphate phosphatase was identified to control anaerobic germination. Functional characterization of *OsTPP7* suggested its involvement in enhancing starch mobilization to drive embryo germination and coleoptile elongation (Kretzschmar et al. 2015). Both Sub1 and qAG9-2 are being extensively deployed in rice breeding programs to enhance submergence tolerance at both germination and vegetative stages. In 2003, IRRI initiated a program to introduce the *Sub1* QTL into six mega-varieties, like Swarna, Samba Mahsuri, BR11, IR64, CR1009, and TDK1, using marker-assisted breeding which has been extended to other popular high-yielding varieties that are grown in flood-prone regions of Asia and Africa (Septiningsih et al. 2009). The productivity and quality of these Sub1 varieties are indistinguishable from that of the parental cultivars under non-submerged conditions based on field trials and farmers' experience. *Sub1* introgressed lines showed a yield advantage of 1–3.5 Mg ha^{-1} based on the duration and flood conditions.

15.5   Breeding for Salinity Tolerance

Rice production is often constrained due to problems associated with soil and irrigation water. Among the soil factors, salinity remains at the top by affecting rice productivity significantly. It has been estimated that about 100 million ha of agricultural area suitable for rice cultivation in South and Southeast Asia is not under rice cultivation due to soil-associated problems (Senadhira et al. 1994). Sustained increase in rice production to meet the emerging demand in rice consumption warrants bringing in these areas under rice cultivation. Salinity is due to accumulation of soluble salts in the soil affecting crop growth and productivity. Salinity affects plant growth by reduced water uptake, cellular toxicity, and imbalance among nutrients inside cells (Munns et al. 2006). Rice productivity in both coastal area or salinity prone inlands is affected significantly thereby hindering harvest of potential yield of popular varieties.

Rice is sensitive to salinity throughout its life cycle starting from germination to reproductive phase. Limited genetic variation was found to exist in rice germplasm, and several studies have been conducted to identify intricate networks of salt tolerance mechanisms. Research across several salinity-acclimatized crops has revealed existence of salt tolerance mechanisms, viz., salt exclusion, control of net Na+ uptake across the plasma membrane (intracellular influx), and tonoplast (vacuolar and pre-vacuolar compartmentalization) of both root and shoot cells to minimize cytosolic and organellar ion toxicity, K^+/Na^+ ion homeostasis, vacuolar osmotic adjustment through biosynthesis of osmo-protectants, and compatible solutes for turgor maintenance and hormonal modifications (Munns and Tester 2008).

Survey of genetic variability in rice germplasm for salinity tolerance-related traits has led to the identification of tolerant cultivars such as Pokkali and Nona Bokra. Still, breeding for salinity tolerance has been slow in progress due to complex nature of genetic and physiological mechanisms underlying salinity tolerance. Key objectives in breeding for salt tolerance in rice include earliness, exclusion of sodium at root and shoot level, enhancing tissue tolerance, improved potassium uptake, upregulation of antioxidants, stomatal responsiveness, enhanced source partitioning, and high yield, most of which are quantitative traits. With a view to determine genetic basis of salinity tolerance in rice, an RIL population was developed from a cross between the salt-susceptible variety "IR29" and salt-tolerant "Pokkali." An important QTL *Saltol* associated with the sodium and potassium ratio in shoots and salinity tolerance at the seedling stage was identified in chromosome 1 (Gregorio 1997; Bonilla et al. 2002). Similarly, several QTLs controlling salt tolerance traits were identified after mapping analysis of an $F_{2:3}$ populations as developed from a cross between the susceptible *japonica* cultivar Koshihikari and salt-tolerant *indica* cultivar Nona Bokra. It includes major QTLs for Shoot K^+ concentration (*qSKC-1*) in chromosome 1 as it is associated with shoot K^+ ion concentration and Shoot Na+ concentration (*qSNC-7*) in chromosome 7 (Lin et al. 2004). Subsequent cloning reveals OsHKT1;5 (previously *SKC1*) gene that encodes sodium transporter that helps to control K^+ homeostasis under salinity by unloading Na+ from xylem. Existence of allelic variation in this region is said to confer salinity tolerance in rice seedlings by maintaining Na^+ and K^+ in shoots and leaf blades. Some *OsHKT1;5* alleles effectively maintain shoot K^+/Na^+ homeostasis owing to specific amino acid variations that enhance K^+ over Na^+ transport into root xylem sap and that alter either protein transmembrane stability or phosphorylation that affects function (Ren et al. 2005). Besides *Saltol* locus, many studies have identified various QTLs for different tolerant traits (Mangrauthia et al. 2014). Numerous genes have also been identified through functional genomics studies of salt stress responses which have shown improved tolerance either by over- or underexpression. Availability of NGS techniques allowed us to clone major genes for developing SNP and InDel markers which can be utilized to transfer these QTLs through marker-assisted backcross approach for developing improved rice varieties for salt stress environments.

15.6   Breeding for Cold Tolerance

Climate variability with increased occurrence of extreme temperatures is destructive to crop growth and productivity. One of the extreme temperature stresses is cold stress caused by suboptimal temperature to cause significant yield reduction in rice. Generally, rice growth is affected at all stages of development when temperature reaches below 15 °C (Howarth and Ougham 1993; Fujino et al. 2004). Cold stress comes under two categories, namely, chilling and freezing based on temperature of 0–15 °C and below 0 °C, respectively. Cold temperature impairs various physiological and metabolic processes that affect crop development based on developmental stage and intensity of low temperature stress. It affects seedling stage by inhibiting seed germination, reducing seedling vigor, and inducing leaf discoloration that weakens photosynthesis by decreasing total chlorophyll content. Besides, it reduces plant height and causes leaf necrosis, chlorosis, and mottled chlorosis. During panicle development and booting stage, spikelet sterility along with irregular grain maturity and poor grain quality caused by cold stress has the highest impact on rice productivity.

Few rice cultivars mostly belonging to *japonica* subspecies have possessed tolerance to low temperature stress (Mackill and Lei 1997). *Javanica* cultivars, an ecotype of *japonica* such as Silewah, Lambayeque 1, and Padi Labou Alumbis, have contributed cold tolerance genes for *japonica* breeding lines (Saito et al. 2001). Several genetic analyses have revealed the complex nature of cold tolerance because of its interaction nature with environment. Genetic analysis by Nishimura and Hamamura (1993) revealed the dominant digenic control tolerance of low temperature at the reproductive stage. Later, Nagasawa et al. (1994) have identified cold tolerance as a complex quantitative trait controlled by four or more genes. In order to breed cold-tolerant cultivars by understanding its genetic mechanism, many QTL analyses were carried out in various biparental mapping populations as well as germplasm panel in different developmental stages, i.e. germination, seedling, and reproductive stages (Table 15.3). The *japonica* cultivars have better cold tolerance than *indica* cultivars at both germination and booting stage. QTL analysis carried out in different populations suggests that QTL hotspot for cold tolerance existed in

Table 15.3 QTLs identified for cold tolerance in rice

Genetic loci	Traits	Chromosome no.	References
Ctb1, Ctb2	Spikelet fertility/undeveloped spikelet	4	Saito et al. (2001)
qCT-7	Spikelet sterility/culm length	7	Takeuchi et al. (2001)
qCTB2a, qCTB3	Spikelet fertility/undeveloped spikelet	2, 3	Andaya and Mackill (2003b)
qCTS12a	Seedling growth	12	Andaya and Mackill (2003a)
Ctb1	Spikelet fertility	4	Saito et al. (2004)
Dth, cl, fer, pe, dc	Days to heading/culm length/ spikelet fertility/ panicle neck exertion/discoloration	1, 3, 5, 6, 7, 8, 9, 11	Oh et al. (2004)
qCTB-1-1, qCTB-10-2	Booting stage	1, 10	Dai et al. (2004)
qSV-3-1/2, -5, -8-1/2	Seedling growth	3, 5, 8	Zhang et al. (2005)
qLVG2, qLVG7-2, qCIVG7-2	Vigor of germination	2, 7	Han et al. (2006)
qCTS12	Seedling growth	12	Andaya and Tai (2006)
qCTS4	Seedling growth	4	Andaya and Tai (2007)
qCTS-2	Seedling growth	2	Lou et al. (2007)
qLTG3-1	Vigor of germination	3	Fujino et al. (2008)
qCTB-1-1, -4-1/2, -5-1/2, -10-1/2, -11-1	Spikelet fertility	1, 4, 5, 10, 11	Xu et al. (2008)
qSCT-1, qSCT-5, qSCT-6	Seedling growth	1, 5, 6	Jiang et al. (2008)
qCTP11, qCTP12	Vigor of germination	11, 12	Baruah et al. (2009)
qPSST-3, -7, -9	Spikelet fertility/reproductive stage	3, 7, 8, 9, 11	Suh et al. (2010)
Ctb1	Spikelet fertility/undeveloped spikelet	4	Saito et al. (2010)
qCtss11	Seedling growth	11	Koseki et al. (2010)
qCTB-5-1/2/3, -7	Vigor of germination	5, 7	Lin et al. (2010)
qCTS4a, qCTS4b	Seedling growth	4	Suh et al. (2012)
qLTB3	Seed fertility	3	Shirasawa et al. (2012)

(continued)

Table 15.3 (continued)

Genetic loci	Traits	Chromosome no.	References
24 loci	Booting stage (association analysis)	3, 4, 6, 8, 9, 10, 11	Cui et al. (2013)
qCTG6, qCTG7-1, qCTG7-2, qCTG-8, qCTG-11, qCTS5(1), qCTS6(1), qCTS11(1)-1, qCTS11(1)-2, qCTS11(2)-1, qCTS11(2)-2, qCTS2(2), qCTS7(2), qCTS8(2)	Germination/seedling stage	1, 2, 6, 7, 8, 11	Ranawake et al. (2014)
51 QTLs, qCTSSR6-3 (both *indica* and *japonica*)	Germination/booting stage (assoc. analysis)	All chromosomes	Pan et al. (2015)

Modified from Zhang et al. (2014)

chromosome 11 which harbors several QTLs affecting all developmental stages, and it has synteny with chromosome 5A of wheat and 5H of barley that has the cluster of *FR2* (Frost Resistance 2) and *CBF* genes (also known as *DREB*) (Ranawake et al. 2014). Nonetheless, complete understanding of genetics and molecular basis of cold tolerance is lacking due to unreliable phenotyping methods and complex environmental interaction of this stress. Therefore, it is crucial to explore on phenomics to increase the reliability of QTL studies for breeding cold tolerance in rice. After careful validations, identified QTLs may be useful to hasten the development of improved cold-tolerant genotypes through marker-assisted breeding approach.

15.7 Transgenic Breeding

In rice, transgenic breeding, one of the genetic engineering tools, has proven its effect as a viable option for developing stress-tolerant lines. It is mainly carried out either by quantitative expression alteration or site specificity in expression of desirable genes. Transgenic breeding are mainly aimed at developing plants with genes encoding transcription factors such as DREB, NAC proteins, MYB, and MYC and genes involving the production and accumulation of osmolytes, late embryogenesis abundant protein genes, heat shock proteins, reactive oxygen scavengers, aquaporins, and transporter proteins (Wang et al. 2016). For abiotic stress tolerance, DREB (dehydration-responsive element-binding protein) transcription factor is widely utilized for improving stress tolerance for drought, heat, salinity, and cold (Reddy et al. 2016). With the availability of genetic information on signal transduction and pathways of various complex traits, transgenic breeding has tremendous potential for creating climate-smart rice varieties by transferring multiple genes for different stress tolerance.

15.8   Multi-stress Tolerance Breeding

Anthropogenic climate change can potentially enhance different biotic and abiotic stresses in agriculture. But global warming, the main threat of this century, causes climate change effects with extreme weather events of heavy downpour, low temperature, and high temperature associated with water deficit. The rapid weather phenomenon can expose crops at any growth stage to multiple kinds of stresses in a crop season. This necessitates development of new crop cultivars possessing multiple stress tolerance, more yield, and yield stability across different environments. Genomic advances favor deciphering genetic information for different complex traits, which can pave way for pyramiding multiple QTLs/genes for developing multi-stress-tolerant lines. In rice, attempts have been made by various groups to pyramid QTLs/genes for developing climate-smart rice cultivars. Shamsudin et al. (2016) has introgressed three QTLs for drought tolerance, i.e. $qDTY_{2.2}$, $qDTY_{3.1}$, and $qDTY_{12.1}$, in the background of an elite Malaysian rice cultivar MR219. Basmati rice cultivars were constantly improved by multiple genes/QTLs for bacterial leaf blight, brown plant hopper, and blast and gall midge diseases. The International Rice Research Institute is involved in developing climate-smart rice varieties by pyramiding multiple genes/QTLs conferring biotic and abiotic stress tolerance in different varietal backgrounds suitable for different rice-growing parts of the world (Arvind Kumar, Personal communication). Renu-Singh et al. (2016) reported about a project involving successful pyramiding of various QTLs for drought, namely, qDTY1.1, $qDTY_{2.1}$, $qDTY_{2.2}$, $qDTY_{3.1}$, $qDTY_{3.2}$, $qDTY_{9.1}$, and $qDTY_{12.1}$, into submergence-tolerant lines, i.e. Swarna-Sub1, Samba Mahsuri-Sub1, and IR64-Sub1. In India, there is a multi-institutional network project funded by the Department of Biotechnology, Government of India, for introgressing QTLs for drought, submergence, cold tolerance, blast, bacterial blight, and gall midge into various high-yielding mega-varieties for developing climate-resilient rice varieties using marker-assisted breeding approach (Agarwal et al. 2016).

15.9   Conclusion

Climate change impact on food security is one of the major challenges in present-day life. Various crop modeling studies coupling weather parameters and crop response conducted in rice, wheat, maize, soybean, oilseeds, cassava, sugarcane, banana, common bean, and millets show its negative impact on crop yield. This information emphasizes the need to include climate change factors in breeding objectives to increase food production. Rice, being the staple food for more than half of world population, has to maintain its production and productivity in order to promote food security. But rising levels of biotic and abiotic stresses due to varying weather pattern possess great challenge to maintain rice production. Several multi-stress-tolerant rice cultivars are being developed by various research groups utilizing various omics tools for understanding its genetic and physiological mechanism in order to cope up with the constraints possessed by climate change. Dissecting genetic basis of different traits through QTL analysis and gene identification can accelerate the breeding process of climate-ready rice. Marker-assisted selection serves as an effective tool to introgress positive alleles in desirable genetic background. Thus, to reduce food security risks from global climate variation, multi-stress-tolerant cultivar breeding with various genomics tools would spur the efforts for stress resilience in rice.

References

Agarwal P, Parida SK, Raghuvanshi S, Kapoor S, Khurana P, Khurana JP, Tyagi AK (2016) Rice improvement through genome-based functional analysis and molecular breeding in India. Rice 9(1):1–17

Aggarwal P (2008) Global climate change and Indian agriculture: impacts, adaptation and mitigation. Ind J Agric Sci 78(11):911–919

Andaya V, Mackill D (2003a) Mapping of QTLs associated with cold tolerance during the vegetative stage in rice. J Exp Bot 54(392):2579–2585

Andaya V, Mackill D (2003b) QTLs conferring cold tolerance at the booting stage of rice using recombinant inbred lines from a *japonica* × *indica* cross. Theor Appl Genet 106(6):1084–1090

Andaya V, Tai T (2006) Fine mapping of the qCTS12 locus, a major QTL for seedling cold tolerance in rice. Theor Appl Genet 113(3):467–475

Andaya VC, Tai TH (2007) Fine mapping of the qCTS4 locus associated with seedling cold tolerance in rice (*Oryza sativa* L.) Mol Breed 20(4):349–358

Angaji SA, Septiningsih EM, Mackill D, Ismail AM (2010) QTLs associated with tolerance of flooding during germination in rice (*Oryza sativa* L.) Euphytica 172(2):159–168

Arrandeau MA (1989) Breeding strategies for drought resistance. In: Baker EWG (ed) Drought resistance in cereals. CAB International, London, pp 107–116

Baruah AR, Ishigo-Oka N, Adachi M, Oguma Y, Tokizono Y, Onishi K, Sano Y (2009) Cold tolerance at the early growth stage in wild and cultivated rice. Euphytica 165(3):459–470

Bernier J, Kumar A, Ramaiah V, Spaner D, Atlin G (2007) A large-effect QTL for grain yield under reproductive-stage drought stress in upland rice. Crop Sci 47(2):507–516

Bita C, Gerats T (2013) Plant tolerance to high temperature in a changing environment: scientific fundamentals and production of heat stress-tolerant crops. Front Plant Sci 4:273–284

Bonilla P, Mackell D, Deal K, Gregorio G (2002) RFLP and SSLP mapping of salinity tolerance genes in chromosome 1 of rice (*Oryza sativa* L.) using recombinant inbred lines. Philippine Agricultural Scientist (Philippines) Report, p. 176

Cao LY, Zhu J, Zhao ST, He LB, Yan QC (2002) Mapping QTL for heat tolerance in a DH population from indicajaponica cross of rice (*Oryza sativa* L.) J Agric. Biotech 10:210–214

Cao L, Zhao J, Zhan X, Li D, He L, Cheng S (2003) Mapping QTLs for heat tolerance and correlation between heat tolerance and photosynthetic rate in rice. Chin J Rice Sci 17:223–227

Chen Q, Yu S, Li C, Mou T-M (2008) Identification of QTLs for heat tolerance at flowering stage in rice. Sci Agric Sin 41(2):315–321

Cline WR (2007) Global warming and agriculture: impact estimates by country. Peterson Institute, Washington

Courtois B, Ahmadi N, Khowaja F, Price AH, Rami J-F, Frouin J, Hamelin C, Ruiz M (2009) Rice root genetic architecture: meta-analysis from a drought QTL database. Rice 2(2–3):115–128

Courtois B, Audebert A, Dardou A, Roques S, Ghneim-Herrera T, Droc G, Frouin J, Rouan L, Gozé E, Kilian A (2013) Genome-wide association mapping of root traits in a japonica rice panel. PLoS One 8(11):e78037

Cui D, Xu C-Y, Tang C-F, Yang C-G, Yu T-Q, Xin-xiang A, Cao G-L, Xu F-R, Zhang J-G, Han L-Z (2013) Genetic structure and association mapping of cold tolerance in improved japonica rice germplasm at the booting stage. Euphytica 193(3):369–382

Dai L, Lin X, Ye C, Ise K, Saito K, Kato A, Xu F, Yu T, Zhang D (2004) Identification of quantitative trait loci controlling cold tolerance at the reproductive stage in Yunnan landrace of rice, Kunmingxiaobaigu. Breed Sci 54(3):253–258

Dixit S, Swamy BM, Vikram P, Ahmed H, Cruz MS, Amante M, Atri D, Leung H, Kumar A (2012) Fine mapping of QTLs for rice grain yield under drought reveals sub-QTLs conferring a response to variable drought severities. Theor Appl Genet 125(1):155–169

Fujino K, Sekiguchi H, Sato T, Kiuchi H, Nonoue Y, Takeuchi Y, Ando T, Lin S, Yano M (2004) Mapping of quantitative trait loci controlling low-temperature germinability in rice (*Oryza sativa* L.) Theor Appl Genet 108(5):794–799

Fujino K, Sekiguchi H, Matsuda Y, Sugimoto K, Ono K, Yano M (2008) Molecular identification of a major quantitative trait locus, qLTG3-1, controlling low-temperature germinability in rice. Proc Natl Acad Sci 105(34):12623–12628

Garris AJ, Tai TH, Coburn J, Kresovich S, McCouch S (2005) Genetic structure and diversity in Oryza sativa L. Genetics 169(3):1631–1638

Gregorio GB (1997) Tagging salinity tolerance genes in rice using amplified fragment length polymorphism (AFLP). PhD Thesis. pp. 1–185

Han LZ, Zhang YY, Qiao YL, Cao GL, Zhang SY, Kim JH, Koh HJ (2006) Genetic and QTL analysis for lowtemperature vigor of germination in rice. Acta Genet Sin 33(11):998–1006. doi:10.1016/S0379-4172(06)60135-2

Henry A, Dixit S, Mandal NP, Anantha M, Torres R, Kumar A (2014) Grain yield and physiological traits of rice lines with the drought yield QTL qDTY12. 1 showed different responses to drought and soil characteristics in upland environments. Funct Plant Biol 41(11):1066–1077

Howarth CJ, Ougham HJ (1993) Gene expression under temperature stress. New Phytol 125(1):1–26

IPCC (2013) The physical science basis. In: Tignor M, Allen SK, Boschung J, Nauels A, Xia Y, Bex V, Midgley PM (eds) Contribution of working group I to the fifth assessment report of the intergovernmental panel on climate change. Cambridge University Press, Cambridge, pp 1–35

Ishimaru T, Hirabayashi H, Ida M, Takai T, San-Oh YA, Yoshinaga S, Ando I, Ogawa T, Kondo M (2010) A genetic resource for early-morning flowering trait of wild rice *Oryza officinalis* to mitigate high temperature-induced spikelet sterility at anthesis. Ann Bot 106(3):515–520

Ismail AM, Ella ES, Vergara GV, Mackill DJ (2009) Mechanisms associated with tolerance to flooding during germination and early seedling growth in rice (Oryza sativa). Ann Bot 103:197–209

Jagadish S, Cairns J, Lafitte R, Wheeler TR, Price A, Craufurd PQ (2010) Genetic analysis of heat tolerance at anthesis in rice. Crop Sci 50(5):1633–1641

Jiang L, Xun M, Wang J, Wan J (2008) QTL analysis of cold tolerance at seedling stage in rice (*Oryza sativa* L.) using recombination inbred lines. J Cereal Sci 48(1):173–179

Kamoshita A, Wade L, Ali M, Pathan M, Zhang J, Sarkarung S, Nguyen H (2002) Mapping QTLs for root morphology of a rice population adapted to rainfed lowland conditions. Theor Appl Genet 104(5):880–893

Kamoshita A, Babu RC, Boopathi NM, Fukai S (2008) Phenotypic and genotypic analysis of drought-resistance traits for development of rice cultivars adapted to rainfed environments. Field crop Res 109(1):1–23

Khowaja FS, Price AH (2008) QTL mapping rolling, stomatal conductance and dimension traits of excised leaves in the Bala× Azucena recombinant inbred population of rice. Field Crops Res 106(3):248–257

Koseki M, Kitazawa N, Yonebayashi S, Maehara Y, Wang Z-X, Minobe Y (2010) Identification and fine mapping of a major quantitative trait locus originating from wild rice, controlling cold tolerance at the seedling stage. Mol Genet Genomics 284(1):45–54

Kretzschmar T, Pelayo MAF, Trijatmiko KR, Gabunada LFM, Alam R, Jimenez R, Mendioro MS, Slamet-Loedin IH, Sreenivasulu N, Bailey-Serres J (2015) A trehalose-6-phosphate phosphatase enhances anaerobic germination tolerance in rice. Nat Plants 1:15124

Kromdijk J, Long SP (2016) One crop breeding cycle from starvation? How engineering crop photosynthesis for rising CO2 and temperature could be one important route to alleviation. In: Proc R Soc B, vol 1826. The Royal Society, DOI: 10.1098/rspb.2015.2578

Kui L, Tan L, Tu J, Lu Y, Sun C (2008) Identification of QTLs associated with heat tolerance of Yuanjiang common wild rice (*Oryza rufipogon* Griff.) at flowering stage. J Agric Biotechnol 16:461–464

Kumar A, Bernier J, Verulkar S, Lafitte H, Atlin G (2008) Breeding for drought tolerance: direct selection for yield, response to selection and use of drought-tolerant donors in upland and lowland-adapted populations. Field Crop Res 107(3):221–231

Kumar A, Verulkar S, Dixit S, Chauhan B, Bernier J, Venuprasad R, Zhao D, Shrivastava M (2009) Yield and yield-attributing traits of rice (*Oryza sativa* L.) under lowland drought and suitability of early vigor as a selection criterion. Field Crops Res 114(1):99–107

Kumar A, Dixit S, Henry A (2013) Marker-assisted introgression of major QTLs for grain yield under drought in Rice. In: Varshney RK, Tuberosa R (eds) Translational genomics for crop breeding: abiotic stress, yield and quality, vol 2. Wiley, Chichester, p 47

Kumar A, Dixit S, Ram T, Yadaw R, Mishra K, Mandal N (2014) Breeding high-yielding drought-tolerant rice: genetic variations and conventional and molecular approaches. J Expt Bot 65(21):6265–6278

Lafitte H, Price A, Courtois B (2004) Yield response to water deficit in an upland rice mapping population: associations among traits and genetic markers. Theor Appl Genet 109(6):1237–1246

Lanceras JC, Pantuwan G, Jongdee B, Toojinda T (2004) Quantitative trait loci associated with drought tolerance at reproductive stage in rice. Plant Physiol 135(1):384–399

Li CY, Peng CH, Zhao QB, Xie P, Chen W (2004) Characteristic analysis of the abnormal high temperature in 2003 midsummer in Wuhan City. J Central China Normal Univ 38(03):379–381

Li X-M, Chao D-Y, Wu Y, Huang X, Chen K, Cui L-G, Su L, Ye W-W, Chen H, Chen H-C (2015) Natural alleles of a proteasome [alpha] 2 subunit gene contribute to thermotolerance and adaptation of African rice. Nat Genet 47(7):827–833

Lin HX, Zhu MZ, Yano M et al (2004) QTLs for Na and K uptake of the shoots and roots controlling rice salt tolerance. TAG Theor Appl Genet 108:253–260. doi:10.1007/s00122-003-1421-y

Lou Q, Chen L, Sun Z, Xing Y, Li J, Xu X, Mei H, Luo L (2007) A major QTL associated with cold tolerance at seedling stage in rice (*Oryza sativa* L.) Euphytica 158(1–2):87–94

Lou Q, Chen L, Mei H, Wei H, Feng F, Wang P, Xia H, Li T, Luo L (2015) Quantitative trait locus mapping of deep rooting by linkage and association analysis in rice. J Exp Bot 66(15):4749–4757

Mackay A (2008) Climate change 2007: impacts, adaptation and vulnerability. Contribution of working group II to the fourth assessment report of the intergovernmental panel on climate change. J Environ Qual 37:2407. doi:10.2134/jeq2008.0015br

Mackill DJ, Lei X (1997) Genetic variation for traits related to temperate adaptation of rice cultivars. Crop Sci 37(4):1340–1346

Mangrauthia SK, Revathi P, Agarwal S, Singh AK, Bhadana V (2014) Breeding and transgenic approaches for development of abiotic stress tolerance in rice. In: Improvement of Crops in the Era of Climatic Changes. Springer, pp 153–190

Miro B, Ismail AM (2013) Tolerance of anaerobic conditions caused by flooding during germination and early growth in rice (*Oryza sativa* L.) Front Plant Sci 4:269. doi:10.3389/fpls.2013.00269

Mishra KK, Vikram P, Yadaw RB, Swamy BM, Dixit S, Cruz MTS, Maturan P, Marker S, Kumar A (2013) qDTY 12.1: a locus with a consistent effect on grain yield under drought in rice. BMC Genet 14(1):1–12

Munns R, Tester M (2008) Mechanisms of salinity tolerance. Annu Rev Plant Biol 59:651–681

Munns R, James RA, Läuchli A (2006) Approaches to increasing the salt tolerance of wheat and other cereals. J Exp Bot 57(5):1025–1043

Nagasawa N, Kawamoto T, Matsunaga K, Sasaki T, Nagato Y, Hinata K (1994) Cool temperature sensitive mutants at the booting stage of Rice. Jpn J Breed 44(1):53–57

Nandi S, Subudhi P, Senadhira D, Manigbas N, Sen-Mandi S, Huang N (1997) Mapping QTLs for submergence tolerance in rice by AFLP analysis and selective genotyping. Mol Gen Genet MGG 255(1):1–8

Neeraja C, Maghirang-Rodriguez R, Pamplona A, Heuer S, Collard B, Septiningsih E, Vergara G, Sanchez D, Xu K, Ismail A (2007) A marker-assisted backcross approach for developing submergence-tolerant rice cultivars. Theor Appl Genet 115(6):767–776

Nguyen T, Klueva N, Chamareck V, Aarti A, Magpantay G, Millena A, Pathan M, Nguyen H (2004) Saturation mapping of QTL regions and identification of putative candidate genes for drought tolerance in rice. Mol Genet Genomics 272(1):35–46

Nishimura M, Hamamura K (1993) Diallel analysis of cool tolerance at the booting stage in rice varieties from Hokkaido. Japanese J Breed 43:557–566

Oh C-S, Choi Y-H, Lee S-J, Yoon D-B, Moon H-P, Ahn S-N (2004) Mapping of quantitative trait loci for cold tolerance in weedy rice. Breed Sci 54(4):373–380

O'Toole JC, Chang TT (1979) Drought resistance in cereals—rice: a case study. In: Mussell, Staples (eds) Stress physiology in crop plants. Wiley, New York, pp 373–405

Pan Y, Luo L-H, Deng H-B, Zhang G-L, Tang W-B, Chen L-Y, Xiao Y-H (2011) Quantitative trait loci associated with pollen fertility under high temperature stress at flowering stage in rice. Chinese J Rice Sci 25:99–102

Pan Y, Zhang H, Zhang D, Li J, Xiong H, Yu J, Li J, Rashid MAR, Li G, Ma X (2015) Genetic analysis of cold tolerance at the germination and booting stages in rice by association mapping. PLoS One 10(3):e0120590

Peng S, Huang J, Sheehy JE, Laza RC, Visperas RM, Zhong X, Centeno GS, Khush GS, Cassman KG (2004) Rice yields decline with higher night temperature from global warming. Proc Natl Acad Sci USA 101(27):9971–9975

Phung NTP, Mai CD, Mournet P, Frouin J, Droc G, Ta NK, Jouannic S, Lê LT, Do VN, Gantet P (2014) Characterization of a panel of Vietnamese rice varieties using DArT and SNP markers for association mapping purposes. BMC Plant Biol 14(1):371–379

Phung NTP, Mai CD, Hoang GT, Truong HTM, Lavarenne J, Gonin M, Le Nguyen K, Ha TT, Do VN, Gantet P (2016) Genome-wide association mapping for root traits in a panel of rice accessions from Vietnam. BMC Plant Biol 16(1):16–64

Poli Y, Basava RK, Panigrahy M, Vinukonda VP, Dokula NR, Voleti SR, Desiraju S, Neelamraju S (2013) Characterization of a Nagina22 rice mutant for heat tolerance and mapping of yield traits. Rice 6(1):36–42

Price A, Young E, Tomos A (1997) Quantitative trait loci associated with stomatal conductance, leaf rolling and heading date mapped in upland rice (Oryza sativa). New Phytol 137(1):83–91

Prince SJ, Beena R, Gomez SM, Senthivel S, Babu RC (2015) Mapping consistent rice (Oryza sativa L.) yield QTLs under drought stress in target rainfed environments. Rice 8(1):53–59

Ranawake AL, Manangkil OE, Yoshida S, Ishii T, Mori N, Nakamura C (2014) Mapping QTLs for cold tolerance at germination and the early seedling stage in rice (Oryza sativa L.) Biotechnol Biotechnol Equip 28(6):989–998

Ray DK, Mueller ND, West PC, Foley JA (2013) Yield trends are insufficient to double global crop production by 2050. PLoS One 8(6):e66428

Ray DK, Gerber JS, Macdonald GK, West PC (2015) Climate variation explains a third of global crop yield variability. Nat Commun 6:5989. doi: 10.1038/ncomms6989

Reddy SSS, Singh B, Peter A, Rao TV (2016) Production of transgenic local rice cultivars (Oryza sativa L.) for improved drought tolerance using Agrobacterium mediated transformation. Saudi J Biol Sci 23(4):16–26

Ren Z-H, Gao J-P, Li L-G, Cai X-L, Huang W, Chao D-Y, Zhu M-Z, Wang Z-Y, Luan S, Lin H-X (2005) A rice quantitative trait locus for salt tolerance encodes a sodium transporter. Nat Genet 37(10):1141–1146

Renu S, Singh Y, Xalaxo S, Verulkar S, Yadav N, Singh S, Singh N, Prasad K, Kondayya K, Rao PR (2016) From QTL to variety-harnessing the benefits of QTLs for drought, flood and salt tolerance in mega rice varieties of India through a multi-institutional network. Plant Sci 242:278–287

Reynolds MP, Quilligan E, Aggarwal PK, Bansal KC, Cavalieri AJ, Chapman SC, Chapotin SM, Datta SK, Duveiller E, Gill KS (2016) An integrated approach to maintaining cereal productivity under climate change. Glob Food Sec 8:9–18

Rosenzweig C, Elliott J, Deryng D (2013) Assessing agricultural risks of climate change in the 21st century in a global gridded crop model intercomparison. PNAS 111(9):1–6

Sailaja B, Subrahmanyam D, Neelamraju S, Vishnukiran T, Rao YV, Vijayalakshmi P, Voleti SR, Bhadana VP, Mangrauthia SK (2015) Integrated physiological, biochemical, and molecular analysis identifies important traits and mechanisms associated with differential response of rice genotypes to elevated temperature. Front Plant Sci 6:1044

Saito K, Miura K, Nagano K, Hayano-Saito Y, Araki H, Kato A (2001) Identification of two closely linked quantitative trait loci for cold tolerance on chromosome 4 of rice and their association with anther length. Theor Appl Genet 103(6–7):862–868

Saito K, Hayano-Saito Y, Maruyama-Funatsuki W, Sato Y, Kato A (2004) Physical mapping and putative candidate gene identification of a quantitative trait locus Ctb1 for cold tolerance at the booting stage of rice. Theor Appl Genet 109(3):515–522

Saito K, Hayano-Saito Y, Kuroki M, Sato Y (2010) Map-based cloning of the rice cold tolerance gene Ctb1. Plant Sci 179(1):97–102. doi:10.1016/j.plantsci.2010.04.004

Senadhira D, Neue H, Akbar M (1994) Development of improved donors for salinity tolerance in rice through somaclonal variation. SABRAO J 26(1–2):19–25

Septiningsih EM, Pamplona AM, Sanchez DL, Neeraja CN, Vergara GV, Heuer S, Ismail AM, Mackill DJ (2009) Development of submergence-tolerant rice cultivars: the Sub1 locus and beyond. Ann Bot 103(2):151–160

Septiningsih EM, Ignacio JCI, Sendon PM, Sanchez DL, Ismail AM, Mackill DJ (2013) QTL mapping and confirmation for tolerance of anaerobic conditions during germination derived from the rice landrace Ma-Zhan red. Theor Appl Genet 126(5):1357–1366

Shamsudin NAA, Swamy BM, Ratnam W, Cruz MTS, Raman A, Kumar A (2016) Marker assisted pyramiding of drought yield QTLs into a popular Malaysian rice cultivar, MR219. BMC Genet 17(1). doi:10.1186/s12863-016-0334-0

Shirasawa S, Endo T, Nakagomi K, Yamaguchi M, Nishio T (2012) Delimitation of a QTL region controlling cold tolerance at booting stage of a cultivar,'Lijiangxintuanheigu', in rice (Oryza sativa L.) Theor Appl Genet 124(5):937–946

Siangliw M, Toojinda T, Tragoonrung S, Vanavichit A (2003) Thai jasmine rice carrying QTLch9 (SubQTL) is submergence tolerant. Ann Bot 91(2):255–261

Srinivasan S, Gomez SM, Kumar SS, Ganesh S, Biji K, Senthil A, Babu RC (2008) QTLs linked to leaf epicuticular wax, physio-morphological and plant production traits under drought stress in rice (Oryza sativa L.) Plant Growth Regul 56(3):245–256

Steele K, Price A, Shashidhar H, Witcombe J (2006) Marker-assisted selection to introgress rice QTLs controlling root traits into an Indian upland rice variety. Theor Appl Genet 112(2):208–221

Steele K, Virk D, Kumar R, Prasad S, Witcombe J (2007) Field evaluation of upland rice lines selected for QTLs controlling root traits. Field Crop Res 101(2):180–186

Suh J, Jeung J, Lee J, Choi Y, Yea J, Virk P, Mackill D, Jena K (2010) Identification and analysis of QTLs controlling cold tolerance at the reproductive stage and validation of effective QTLs in cold-tolerant genotypes of rice (Oryza sativa L.) Theor Appl Genet 120(5):985–995

Suh J, Lee C, Lee J, Kim J, Kim S, Cho Y, Park S, Shin J, Kim Y, Jena K (2012) Identification of quantitative trait loci for seedling cold tolerance using RILs derived from a cross between japonica and tropical japonica rice cultivars. Euphytica 184(1):101–108

Swamy BM, Kumar A (2013) Genomics-based precision breeding approaches to improve drought tolerance in rice. Biotechnol Adv 31:1308–1318. doi:10.1016/j.biotechadv.2013.05.004

Takeuchi Y, Hayasaka H, Chiba B, Tanaka I, Shimano T, Yamagishi M, Nagano K, Sasaki T, Yano M (2001) Mapping quantitative trait loci controlling cool-temperature tolerance at booting stage in temperate Japonica rice. Breed Sci 51(3):191–197

Tenorio F, Ye C, Redoña E, Sierra S, Laza M, Argayoso M (2013) Screening rice genetic resources for heat tolerance. SABRAO J Breed Genet 45(3):371–381

Toojinda T, Siangliw M, Tragoonrung S, Vanavichit A (2003) Molecular genetics of submergence tolerance in rice: QTL analysis of key traits. Ann Bot 91(2):243–253

Toojinda T, Tragoonrung S, Vanavichit A, Siangliw JL, Pa-In N, Jantaboon J, Siangliw M, Fukai S (2005) Molecular breeding for rainfed lowland rice in the Mekong region. Plant Prod Sci 8(3):330–333

Tripathy J, Zhang J, Robin S, Nguyen TT, Nguyen H (2000) QTLs for cell-membrane stability mapped in rice (*Oryza sativa* L.) under drought stress. TAG Theor Appl Genet 100(8):1197–1202

Uga Y, Sugimoto K, Ogawa S et al (2013) Control of root system architecture by DEEPER ROOTING 1 increases rice yield under drought conditions. Nat Genet 45:1097–1102. doi:10.1038/ng.2725

Venuprasad R, Lafitte H, Atlin G (2007) Response to direct selection for grain yield under drought stress in rice. Crop Sci 47(1):285–293

Venuprasad R, Cruz MS, Amante M, Magbanua R, Kumar A, Atlin G (2008) Response to two cycles of divergent selection for grain yield under drought stress in four rice breeding populations. Field Crops Res 107(3):232–244

Venuprasad R, Dalid C, Del Valle M, Zhao D, Espiritu M, Cruz MS, Amante M, Kumar A, Atlin G (2009) Identification and characterization of large-effect quantitative trait loci for grain yield under lowland drought stress in rice using bulk-segregant analysis. Theor Appl Genet 120(1):177–190

Vikram P, Swamy BM, Dixit S, Ahmed HU, Cruz MTS, Singh AK, Kumar A (2011) qDTY 1.1, a major QTL for rice grain yield under reproductive-stage drought stress with a consistent effect in multiple elite genetic backgrounds. BMC Genet 12(1):12–89

Vikram P, Swamy BM, Dixit S, Trinidad J, Cruz MTS, Maturan PC, Amante M, Kumar A (2016) Linkages and interactions analysis of major effect drought grain yield QTLs in rice. PLoS One 11(3):e0151532

Wang H, Wang H, Shao H, Tang X (2016) Recent advances in utilizing transcription factors to improve plant abiotic stress tolerance by transgenic technology. Front Plant Sci 7:7–67

Wei H, Liu JP, Wang Y, Huang NR, Zhang XB, Wang LC, Zhang JW, Tu JM, Zhong XH (2013) A dominant major locus in chromosome 9 of rice (Oryza sativa L.) confers tolerance to 48°C high temperature at seedling stage. J Hered 104:287–294

Xia M, Qi H (2004) Effects of high temperature on the seed setting percent of hybrid rice bred with four male sterile lines. Hubei Agric Sci 2:21–22

Xiao Y, Pan Y, Luo L, Zhang G, Deng H, Dai L, Liu X, Tang W, Chen L, Wang G-L (2011) Quantitative trait loci associated with seed set under high temperature stress at the flowering stage in rice (*Oryza sativa* L.) Euphytica 178(3):331–338

Xu K, Mackill DJ (1996) A major locus for submergence tolerance mapped on rice chromosome 9. Mol Breed 2(3):219–224

Xu K, Xu X, Ronald P, Mackill D (2000) A high-resolution linkage map of the vicinity of the rice submergence tolerance locus Sub1. Mol Gen Genet MGG 263(4):681–689

Xu K, Xu X, Fukao T, Canlas P, Maghirang-Rodriguez R, Heuer S, Ismail AM, Bailey-Serres J, Ronald PC, Mackill DJ (2006) Sub1A is an ethylene-response-factor-like gene that confers submergence tolerance to rice. Nature 442(7103):705–708

Xu L-M, Zhou L, Zeng Y-W, Wang F-M, Zhang H-L, Shen S-Q, Li Z-C (2008) Identification and mapping of quantitative trait loci for cold tolerance at the booting stage in a japonica rice near-isogenic line. Plant Sci 174(3):340–347

Xu X, Liu X, Ge S, Jensen JD, Hu F, Li X, Dong Y, Gutenkunst RN, Fang L, Huang L (2012) Resequencing 50 accessions of cultivated and wild rice yields markers for identifying agronomically important genes. Nature Biotech 30(1):105–111

Yang H, Huang Z, Jiang Z, Wang X (2004) High temperature damage and its protective technologies of early and middle season rice in Anhui province. J Anhui Agric Sci 32(1):3–4

Yamori W, Hikosaka K, Way DA (2013) Temperature response of photosynthesis in C3, C4, and CAM plants: temperature acclimation and temperature adaptation. Photosynth Res 119:101–117. doi: 10.1007/s11120-013-9874-6

Ye C, Argayoso MA, Redona ED, Sierra SN, Laza MA, Dilla CJ, Mo Y, Thomson MJ, Chin J, Delaviña CB (2012) Mapping QTL for heat tolerance at flowering stage in rice using SNP markers. Plant Breed 131(1):33–41

Ye C, Tenorio FA, Argayoso MA, Laza MA, Koh H-J, Redoña ED, Jagadish KS, Gregorio GB (2015) Identifying and confirming quantitative trait loci associated with heat tolerance at flowering stage in different rice populations. BMC Genet 16(1):1

Zaniab Al-Shugeairy AHP, Robinson D (2015) Genome wide association mapping for drought recovery trait in Rice (*Oryza sativa* L.) Inter J Appl Agric Sci 1(1):11–18. doi:10.11648/j.ijaas.20150101.12

Zhang Z-H, Qu X-S, Wan S, Chen L-H, Zhu Y-G (2005) Comparison of QTL controlling seedling vigour under different temperature conditions using recombinant inbred lines in rice (*Oryza sativa*). Ann Bot 95(3):423–429

Zhang T, Yang L, Jiang K, Huang M, Sun Q, Chen W, Zheng J (2008) QTL mapping for heat tolerance of the tassel period of rice. Mol Plant Breed 6:867–873

Zhang G, Chen L, Xiao G, Xiao Y, Chen X, Zhang S (2009) Bulked segregant analysis to detect QTL related to heat tolerance in rice using SSR markers. Agric Sci China 8:482–487. doi:10.1016/S1671-2927(08)60235-7

Zhang Q, Chen Q, Wang S et al (2014) Rice and cold stress: methods for its evaluation and summary of cold tolerance-related quantitative trait loci. Rice 7(1). doi:10.1186/s12284-014-0024-3

Zhao Z, Zhang L, Xiao Y, Zhang W, Zhai H, Wan J (2006) Identification of QTLs for heat tolerance at the booting stage in rice. Acta Agron Sin 32:640–644

Zhao L, Lei J, Huang Y, Zhu S, Chen H, Huang R, Peng Z, Tu Q, Shen X, Yan S (2016) Mapping quantitative trait loci for heat tolerance at anthesis in rice using chromosomal segment substitution lines. Breed Sci 66(3):358–366

Zhu C, Jiang L, Zhang W, Wang C, Zhai H, Wan J (2005) Identifying QTLs for thermo–tolerance of amylose content and gel consistency in rice. Zhongguo Shuidao Kexue 20(3):248–252

Zhu CL, Jiang L, Zhang WW, Wang CM, Zhai HQ, Wan JM (2006) Identifying QTLs for thermo-tolerance of amylose content and gel consistency in rice. Chinese J Rice Sci 20:248–252

Abiotic Stress Tolerance in Barley

16

A.S. Kharub, Jogendra Singh, Chuni Lal, and Vishnu Kumar

Abstract

Barley (*Hordeum vulgare* L.) is one of the primitive and oldest domesticated cereals and preferred by gladiators due to its nutritional properties. Primarily this crop is used for feed and food purposes, whereas its enzymatic and husk properties make it unique for malting and brewing purposes. Barley is known to be a climate resilient crop and can thrive well under adverse conditions of cultivation. In some regions of the world, where winter crops like wheat are difficult to be grown or their cultivation is uneconomical due to the harsh climatic conditions and/or problematic soils, barley is the only option to sustain the populace of these regions. In India, this is generally grown under rainfed conditions where this crop uses the residual moisture efficiently. Despite the fact that barley is endowed with resilience to harsh climates, several abiotic stress factors inhibit the performance of the crop to its fullest genetic potential. Some of these stresses are drought, salinity and alkalinity, water logging, lodging, etc. Morphological and physiological traits linked to drought resistance in barley are early growth vigour, root development, tillering, grain weight, stay green habit, leaf water potential, stomata size, membrane stability, leaf rolling, waxiness, leaf temperature, carbon isotope discrimination and the accumulation of metabolites such as proline and betaine. Breeders have resorted to the development of semidwarf barley varieties through incorporation of dwarfing genes and to enhance stem strength to minimize the risk of lodging. Field screening for tolerance to salinity or alkalinity has not been very reliable; therefore, field screening supplemented with laboratory screening has been advocated in barley to breed varieties tolerant to salinity or alkalinity.

A.S. Kharub (✉) • J. Singh • C. Lal • V. Kumar
ICAR-Indian Institute on Wheat and Barley Research, Karnal 132001, Haryana, India
e-mail: askharub@gmail.com; jogendrasail@yahoo.co.in; chunilal_nrcg@rediffmail.com; vishnupbg@gmail.com

16.1 Introduction

Barley (*Hordeum vulgare* L.) is one of the first domesticated cereals, which contributes nearly 11–12% of the coarse cereal production worldwide (Kumar et al. 2014, 2016a). It ranks fourth in total acreage under cereal production in the world with 49.0 million hectares cultivated area with a production of approximately 140.0 million metric tons (FAOSTAT 2016; Kumar et al. 2016b). It is a poor man's crop because it is grown in marginal lands and requires low inputs such as fertilizers, irrigations, insecticides, herbicides, etc. In addition, crop can be cultivated in most severe conditions like elevation, aridity, salinity, poor soil fertility and poor agronomic management. Thus the wider adaptability to uncongenial and harsh environments has led to its widespread cultivation across 100 countries of the world (Bothmer et al. 1995). In India, it is generally grown as rain-fed crop, which thrives well on residual moisture. The crop is grown in the states of Rajasthan, Uttar Pradesh, Madhya Pradesh, Punjab, Haryana, Bihar, Himachal Pradesh, Uttarakhand and Jammu and Kashmir in the hills (Kumar et al. 2013).

Abiotic stresses, i.e., drought, heat, salinity, sodicity, water logging and cold sensitivity, adversely affect plant growth, plant architecture, yield, biomass, etc. However, drought and salinity are the most devastating stresses affecting large acreage in India. It is estimated that more than six million ha area suffered due to salt concentrations, and similarly drought hampers crop production in all the zones, which cover a sizeable area. Inclusion of responding traits is required to minimize the realized and potential yield gaps in abiotic stress-prone environments. To increase the yield under harsh environments, the new molecular tools as QTL analysis, association analysis, genotyping by sequencing, transcriptome analysis and proteomics can play a pivotal role for the improvement of barley. Dry agroecologies of arid and semiarid regions are more exposed to salt accumulation ultimately leading to salinity and alkalinity problems. The inbuilt resistance mechanism is one of the best approach, which is eco-friendly, sustainable, economic and with no environmental hazards. However, the crop management practices are integral and essential part and cannot be ignored especially for abiotic stress management. Considerable improvements in drought and salt tolerance in different crops including barley have been achieved. This could only happen due to the concerted efforts in understanding physiological mechanism and underlying genetics and changes in different biochemical pathways.

The optimum growth and development of barley is influenced by a number of abiotic stress factors that can prevent the plant from expressing to its fullest genetic potential. Globally, abiotic stresses are the most serious threats to the challenging task of feeding a rapidly growing human population. The plant responses caused by these stresses include alterations in transpiration, photosynthetic ability, respiration and hormonal response through the development of genetic defence mechanisms. The duration of the stress, the crop growth stage at the onset of the stress and inbuilt ability of the plant to sustain the negative effects of the stress determine the level of damage the stress can cause. Genetic variability plays a primary role in determining adaptation to abiotic stresses and, resultantly, the expansion of barley genotypes/

varieties to adverse climatic conditions. Occurrence, intensity, timing and duration of stresses vary from location to location and in the same location from year to year. Keeping above in view, some of the aspects related to abiotic stresses like moisture (water logging/drought), temperature, lodging and mineral (saline/alkaline) are discussed below.

16.2   Drought Stress

Drought stress is one of the major abiotic stresses, which occurs worldwide, with devastating effects on crop growth and yields (Ludlow and Muchow 1990). This stress affects about 40–60% agricultural lands of the world (Shahryari and Mollasadeghi 2011). It is considered that severe drought occurs on an average once in every 5 years in most of the tropical countries. Scarcity of moisture leading to water stress is of a common occurrence in rain-fed areas due to low to very low rainfall and poor irrigation (Wang et al. 2005). To meet the requirements of ever-increasing world population, it is necessary to develop drought-tolerant crop varieties. However, the progress in this regard is largely hampered by the physiological and genetic complexity of traits associated with drought tolerance. The mechanisms of drought resistance have been classified as escape, avoidance and tolerance by stress physiologists (Levitt 1972). Drought tolerance is the ability of the plant to survive with decreased water potential of the plant tissue (Jenks and Hasegawa 2014). The reduced water potential may be accompanied with changed tissue water content, depending on the severity of the water deficit and the species of the plant involved. Drought tolerance allows the plant to function or to survive under harsh water deficit conditions.

Large-scale attempts have been made to understand and improve drought tolerance in existing cultivars and to develop the novel drought-tolerant cultivars. The use of multidisciplinary approach to identify and introgress alleles for drought tolerance from wild and landraces may be helpful for successful development of drought-tolerant barley varieties. The use of wild barley as a source of novel genes for crop improvement essentially remains untapped (Eglinton et al. 2016). Landraces and wild species of barley have been recognized by the International Center for Agricultural Research in the Dry Areas (ICARDA) breeding programme as rich source of genes for adaptation to environments where drought stress is common. Barley is well known for its low consumption of water per unit weight of dry matter produced than other cereals and can be grown with limited irrigation. Compared to other cereals, barley is well adapted to drought as it possesses high water use efficiency which is essential for environments where drought is common (Stanca et al. 1992). A plant with a robust and deep root system can have sufficient moisture due to water uptake from lower levels. Stress factors interact with each other and often affect plants in several ways, and the collective effect of action of stress factors may vary with the sum of effects of individual stress factors (Mittler 2002, 2006).

Based on the plant growth phase affected by water deficit, three patterns of drought stress may occur: (i) before flowering, (ii) grain development and

(iii) throughout the life cycle (Reynolds et al. 2005). To overcome this stress due to water deficits, the plants have generally evolved three basic strategies, i.e., escape, avoidance and tolerance (Levitt 1980; Chaves et al. 2003; Larcher 2003). Drought escape is the mechanism to cope up with the drought stress and the completion of plant life cycle before the initiation of drought. During drought avoidance, plant cell maintains high water status even during water moisture conditions. The mechanisms of high pollen viability, stomata closure, inhibition of shoot growth, high root growth, membrane stability and the accumulation of osmotic components are some of the phenomenon during drought tolerance (Blum 2005). The increased production of free radicals, molecular chaperones (protein protectors) and proteases are some of the actions for drought tolerance (Bartels and Sunkar 2005).

Drought events are often associated with high temperatures which impose an additional level of stress to plants. The predicted climatic changes will also modify the annual temperature profile (Tubiello et al. 2000), which in turn will adversely affect the availability of soil moisture and irrigation water. This might imply that in the long term, a consequent variation in the sowing date, growth habit and/or heading time can be expected. Constraints in modifying some of the environmental factors suggest that fitting crops to the changed conditions of environment would be more sustainable strategy than modifying the environment to fit the crops. Barley is one hope to mitigate the adverse effects of increased temperature, uncertainty of monsoon and uneven distribution of precipitation brought about by the climatic changes as barley and its wild relatives possess natural potential which is not common in most of the cereals.

It has always been difficult to mimic drought stress conditions and screen germplasm and breeding populations for selecting drought-tolerant varieties. However, a number of field studies have been conducted in barley to select barley genotypes tolerant to moisture-deficit conditions. During screening of barley genotypes for drought tolerance for agro-physiological traits under optimum and drought stress conditions, Vaeiz et al. (2010) noticed that days to heading and days to maturity influenced the yield the most during water stress conditions. Negative correlation was observed between yield, days to heading and days to maturity under drought stress. However, 1000-grain weight, grains per spike, relative water potential and stay green were related with yield. Yield was also associated with osmotic adjustment, while it was least related to plant height. Zare et al. (2011) noticed that grain yield was the most affected trait under drought stress condition, and it was probably due to reduction in biological yield, seed number/spike and spike number/plant. Moffat et al. (1990) also reported that drought stress at the reproductive growth stage reduced grain yield in wheat significantly. On the other hand, the least percent reduction in magnitude of traits under drought stress condition was observed for harvest index, suggesting that reduction in shoot dry weight was lesser than in grain yield. 1000-grain weight was adversely affected under drought stress, and this decrease in the 1000-grain weight was probably due to reduction in the assimilates and also the translocation of assimilates from the source to sink. Under drought conditions, three important plant attributes, namely, number of spikes per metre square, number of grains per spike and 1000-grain weight, determine the final grain yield of barley. The grain filling duration and growth cycle also contribute to the grain yield in barley (Garcia-del-Moral et al. 1991). Barley crop reveals large genotypic variability and effects of genotype by environment interactions (GEI) of several traits like carbon isotope discrimination; osmotic adjustments; TE, transpiration efficiency; WUE, water use efficiency; and affecting the drought resistance. New molecular approaches like marker-assisted selection and advanced backcross QTL analysis have provided options before the breeders for successful introgression of wild genes to the cultivated ones (Tanksley and Nelson 1996).

Drought resistance is a complex trait governed by several genes and depends upon the genetic structure of a population, critical growth stage at the time of occurrence of drought, the prevailing environmental conditions and other edaphic factors affecting the final crop. Agronomic yield under drought stress conditions is determined by both constitutive and inductive quantitative trait loci, which affects yield independently and only under water deficit conditions (Collins et al. 2008). Recently, a complete barley genome annotation has been published (The International Barley Genome Sequencing Consortium 2012). Barley has a very big and distinct gene pool including several landraces adapted to adverse climatic conditions. It can be attributed to the fact that this crop was domesticated in the area called Fertile Crescent about 10,000 years ago (Araus et al. 2007). The knowledge of the target environment, plant ideotype and the suitable breeding methodology based upon nature of gene action, mode of pollination and transfer of the QTLs underlying drought resistance with marker-assisted selection will be certainly helpful to combat drought stress.

Many morphological and physiological characteristics are associated with drought resistance as reported in different studies. The major focus has been on the traits, viz., drought susceptibility index (DSI), canopy temperature depression (CTD), water use efficiency (WUE), xylem diameter, root-shoot lengths and ratio, chlorophyll content, stay green habit, leaf and culm waxiness, leaf breadth, leaf senescence, 1000-grain wt., membrane stability, osmotic adjustment, etc. For the phenotyping, the novel high-throughput phenotyping platforms (HTPPs) by utilizing robotics, remote sensing, drones and other cutting-edge science technologies are emerging fast for the precise characterization in large mapping populations. Araus and Cairns (2014) reported an overview of modern and precise phenotyping methodologies for the assessment of drought-associated traits. With a limited study on barley, several dozen of QTLs for different traits associated with drought tolerance have been mapped (Diab et al. 2004; Teulat et al. 2001, 2002; Tondelli et al. 2006).

The traits associated with plant mechanism to cope up with water deficit conditions have been identified and discussed in different studies. Among them, traits such as stomatal closure, waxy leaf and plant surface, small plant size, early maturity and reduced leaf area help in reducing total seasonal evapotranspiration and also reducing yield potential (Fischer and Wood 1979). Awns contribute significantly in spike photosynthesis, and the genotypes with longer awns are preferred for drought-prone areas (Martin et al. 1976). In barley, the traits, i.e., early heading, early growth vigour, long peduncle, leaf and culm waxiness and short grain filling duration, were found to be prominent for drought stress (Acevedo and Ceccarelli 1989;

Ceccarelli et al. 2004). Cattivelli et al. (1994) reported additional morphological and physiological traits linked to drought resistance in barley such as tillering, root development, plant vigour, leaf water potential, stomata size, membrane stability, desiccation tolerance, leaf rolling, waxiness, leaf temperature, carbon isotope discrimination and the accumulation of metabolites such as proline and betaine.

Strategy to deal with drought stress in barley would be the introduction of new alleles into elite breeding germplasm lines through the exploitation of landraces grown in harsh environments. The wild progenitor species and primitive landraces of barley offer rich sources of genetic variation for improvement of this crop (Nevo 1992). The approach for identification of donors for genes for drought resistance should involve genetic fingerprinting of wild barley sources *(Hordeum spontaneum* C. Koch), which are largely untapped and may be helpful from different regions (Forster et al. 1997). These gene pools can be exploited using conventional breeding procedures aided with biotechnological tools for greater precision in selecting desired genotypes. Genetic fingerprinting of *H. spontaneum* has revealed genetic markers associated with site of origin, ecogeographic factors and also experimentally imposed stresses (Forster et al. 2000). Such landraces are supposed to carry genes enabling them to cope with the difficult environments. Identification of genes involved for drought tolerance by TILLING and ecoTILLING in barley landraces and its progenitor and using most suitable germplasm for pre-breeding barley for resistance to drought stress will be beneficial to breed barley for water-limited conditions.

Barley is a good source of feed and fodder. It is a key animal feed and fodder in dry areas of India (Kharub et al. 2013; Kumar et al. 2013). In India, barley network centres under AICW&BIP have got a major research component on barley improvement for grain/feed purposes. A large number of varieties have been developed/released either by CVRC for different zones or by respective SVRCs for their respective states addressing different production conditions and agroclimatic situations under AICW&BIP. Subsequently, a large number of improved varieties with resistance to biotic and abiotic stresses were released for feed and food purposes for different agro-ecologies. The rain-fed varieties which yield better under low-moisture conditions have been developed for different areas (Table 16.1).

16.3   Lodging Resistance

Lodging is also a critical abiotic stress that occurs in barley when plant stem is unable to support its own weight, and it is highly associated with yield loss (Pinthus 1974). According to Ennos (1991), there are two forms of lodging in cereals: stem fragility and poor root development. Early lodging can increase moisture in the plant canopy, which adds risk of fungal diseases. Early lodging in barley causes immense decline in grain yield potential than lodging at later growth stages (Brigs 1990). However, level of yield loss depends on cultivar, growth stage and lodging severity (Jedel and Helm 1991). Matusinsky et al. (2015) observed that increased tiller biomass weight and increased plant height during early growth stages were

Table 16.1 Barley varieties released for feed and food purposes for rain-fed areas

Cultural practices	Variety	Year	Area of adaptation
Rain fed (plains)	PL 419	1995	Punjab
	K 560	1997	NEPZ
	Getanjali (K1149)	1997	Uttar Pradesh
	K 603	2000	NEPZ
	RD 2624	2003	NWPZ
	JB58	2005	Madhya Pradesh
	RD 2660	2006	NWPZ
Rain fed (hills)	BHS352	2003	NHZ
	BHS380	2010	NHZ
	VLB118	2014	NHZ
	BHS400	2014	NHZ

significant. Lodging risk can be reduced by introducing dwarfing genes to produce semidwarf varieties (Kucyzynska et al. 2013; Ren et al. 2014) and by increasing stem strength (Ma 2009). Genotypes accompanied with semidwarf plant height provide good response to the high inputs such as fertilizers and irrigation. Presently, utilization of dwarfing gene in breeding process is vital for the development of modern varieties. In barley, more than 30 semidwarf types have been identified so far. However, only few of them have been successfully used in breeding programmes. Kucyzynska et al. (2013) noticed that cultivars with dwarfing gene had improved lodging resistance and a higher harvest index. Zhang et al. (2003) screened 500 barley germplasm lines for lodging resistance adopting the scale 1–3 for lodging. Among 500 germplasm lines evaluated, 73 (14.6%) were very resistant to lodging, 74 (14.8%) showed medium resistance, and remaining 353 (70.6%) were severely lodged. It is reported that most of the varieties resistant or medium resistant to lodging were bred with relatively shorter plant height.

16.4 Salinity Stress

Among abiotic stresses, soil salinity is one of the important stresses which affects large areas of the world's cultivated land and causes significant reductions in food grain production. Due to salt stress, approximately 20% of agricultural land and 50% of crop yield in the world have been affected (Flowers and Yeo 1995). Salt-affected soils commonly contain a mixture of cations of sodium, calcium, magnesium and potassium and anions of chloride, sulphate, bicarbonate, carbonate and sometimes borate and nitrate (Dan 2010) and could lead to the impairment of plant growth and development. Among cereals, barley is one of the most salt-tolerant crops (Maas and Hoffman 1977) due to traits such as fast growth and phenological development which help in early flowering and maturity.

Development of salt-tolerant varieties of barley is one of the cheapest sources to contain the harmful effects of salts and to sustain yields in salt-affected soils. The landraces of barley through conventional breeding may be used to incorporate

Table 16.2 Barley varieties for cultivation under saline-alkaline soils

Variety	Year of release	Recommended for conditions	Area
DL 88	1997	Irrigated (timely/late sowing)	NWPZ
RD 2552	1999	Irrigated (timely sowing)	NWPZ and NEPZ
NDB 1173	2005	Irrigated (timely sowing)	NWPZ and NEPZ
NDB1445	2014	Saline/alkaline	Uttar Pradesh
RD2794	2016	Saline/alkaline	NWPZ, NEPZ

genes for salinity tolerance. The other approach is to exploit the wild species such as foxtail barley (*Hordeum jubatum*) for salinity tolerance (Israelsen et al. 2014). However, it is a very challenging task to screen the materials for such conditions due to very high variability in the field condition affected by salinity and alkalinity. As a result the field screening for salinity tolerance sometimes cannot be fully reliable, because of non-repetitive performance due to soil heterogeneity. Supplemented with in vitro screening for salinity-alkalinity tolerance, several varieties have been developed and released which perform well under unfavourable conditions (Verma et al. 2012).

Barley and wheat are largely affected during early seedling growth stages than the later growth stages (Dan 2010). Tavakkoli et al. (2011) studied effects of Na^+ and Cl^- ions under salinity stress and reported that the reductions in growth and photosynthesis were large under NaCl and mainly were due to the combined effects of Na^+ and Cl^- stress. Forster et al. (1994) reported the varietal differences in the extent of accumulation of Na^+ and Cl^- in leaves, but relationship between Na^+ and Cl^- accumulation and salt tolerance was lacking in barley than the wheat and rice (Colmer et al. 2005). Soil salinity not only delays but also reduces flowering and yield of crops (Hayward and Wadleigh 1949). Effect of soil salinity on grain filling and grain development in barley was observed by Gill (1979), and barley cultivars showed wide differences in yield attributes under normal and saline conditions. Under salt stress condition, grain yield was affected due to reduced efficiency to fill the grains and also due to disturbed starch-sugar balance.

Efforts towards improvement of salinity tolerance through plant biotechnology have also been made. Since salinity tolerance is a complex trait, QTL mapping is commonly used to identify the potential genetic loci that could be associated to salinity tolerance. In barley, many QTLs involved in salinity tolerance have been detected (Mano and Takeda 1997; Xue et al. 2009). Nonetheless, some of the barley varieties which are grown in saline/alkaline regions (Table 16.2) can also serve as source of tolerance for developing salt-tolerant varieties.

16.5   Water Logging

Very limited reports are available in case of stress caused by water logging in barley. Water logging is one of the most hazardous abiotic stress which results in losses in grain yield about 20–25% in the barley crop depending upon the extent of plant damage and it may exceed up to 50% due to water logging (Setter et al. 1999).

16.6 Conclusion

Barley is mainly grown as a rain-dependent *rabi* crop which utilizes the residual moisture of precipitation received during the regular rainy season. It can grow well under harsh conditions of cultivation, and growing of such climate-resilient crops is very important in the present context than it was ever due to the fast-changing climate. Though much of the barley cultivation area has been replaced by wheat due to augmentation of irrigation facilities during and after the green revolution, this crop continues to be an important source of green fodder during the winter season when not many alternatives are available. Barley is a crop that can produce more from lesser water, or the productivity of barley per unit of water used is higher compared to other cereals. Ingression of salinity, particularly in the coastal and plain areas, is an ever-increasing problem which renders the soils unfit for cultivation of many crops, but barley can be grown and produced easily. The remarkable intrinsic ability of barley to sustain salinity stress at a level much higher than that of other crops is another avenue to expand area under barley. The future research and development efforts in barley need to be steered towards the development of enhanced resilience to drought and salinity. Availability of such varieties can create more opportunities of livelihood and income for farmers in marginal areas featured by drought and salinity.

References

Acevedo E, Ceccarelli S (1989) Role of physiologist-breeder in a breeding program for drought resistance conditions. In: Baker FWG (ed) Drought resistance in cereals. CAB International, Wallingford, pp 117–139

Araus JL, Cairns JE (2014) Field high-throughput phenotyping: the new crop breeding frontier. Trends Plant Sci 19:52–61

Araus JL, Ferrio JP, Buxó R, Voltas J (2007) The historical perspective of dryland agriculture: lessons learned from 10 000 years of wheat cultivation. J Exp Bot 58:131–145

Bartels D, Sunkar R (2005) Drought and salt tolerance in plants. Crit Rev Plant Sci 24(1):23–58

Blum A (2005) Drought resistance, water-use efficiency, and yield potential – are they compatible, dissonant, or mutually exclusive? Aust J Agric Res 56:1159–1168

Bothmer R, Jacobsen N, Baden C, Jørgensen RB, Linde-Laursen I (1995) An Ecogeographical study of the genus Hordeum, Systematic and Ecogeographical Studies on Crop Genepools, 2nd edn. IBPGR, Rome

Brigs KG (1990) Studies of recovery from artificially induced lodging in several six row barley cultivars. Can J Plant Sci 70(1):173–181

Cattivelli L, Delogu G, Terzi V, Stanca AM (1994) Progress in barley breeding. In: Slafer GA (ed) Genetic improvement of field crops. Marcel Dekker Inc, New York, pp 95–181

Ceccarelli S, Grando S, Baum M, Udupa S (2004) Breeding for drought resistance in a changing climate. In: Roberts CA (ed) Challenges and strategies for dryland agriculture. Crop Science Society of America Inc and American Society of Agronomy Inc, Madison, pp 167–190

Chaves MM, Maroco JP, Pereira JS (2003) Understanding plant responses to drought – from genes to the whole plant. Funct Plant Biol 30:239–264

Collins NC, Tardieu F, Tuberosa R (2008) Quantitative trait loci and crop performance under abiotic stress: where do we stand? Plant Physiol 147:469–486

Colmer TD, Munnus R, Flowers TJ (2005) Improving salt tolerance of wheat and barley: future prospects. Aust J Exp Agric 45:1425–1443

Dan Ogle (2010) Plants for saline to sodic soil conditions. USDA. Natural Resources Conservation Service. Boise, Idaho. Technical Note No.9A, p10

Diab AA, Teulat B, This D, Ozturk NZ, Benscher D, Sorrells ME (2004) Identification of drought-inducible genes and differentially expressed sequence tags in barley. Theor Appl Genet 109:1417–1425

Eglinton JK, Evans DE, Brown AHD, Langridge P, McDonald G, Jefferies SP, Barr AR (2016) The use of wild barley (*Hordeum vulgare* ssp. *spontaneum*) in breeding for quality and adaptation. The regional institute- Australian Barley Technical Symposium, pp1–9

Ennos AR (1991) The mechanics of anchorage in wheat *Triticum aestivum* L. II Anchorage of mature wheat against lodging. J Exp Bot 42(245):1607–1613

FAOSTAT (2016) FAOSTAT. Food and agriculture organization (FAO) of the United Nations, Rome. http://faostat3.fao.org. (Accessed Jan 2016)

Fischer RA, Wood JR (1979) Drought resistance in spring wheat cultivars. III. Yield associations with morpho-physiological traits. Aust J Agric Res 30:1001–1020

Flowers TJ, Yeo AR (1995) Breeding for salinity resistance in crop plants: where next? Aust J Plant Physiol 22:875–884

Forster BP, Pakniyat H, Macaulay M, Matheson W, Phillips MS, Thomas WTB, Powell W (1994) Variation in the leaf sodium content of the *Hordeum vulgare* cultivar Maythorpe and its derived mutant cv. Golden Promise Heredity 73:249–253

Forster BP, Russel JR, Ellis RP, Handley LL, Robinson D, Hackett CA, Nevos E, Waugh R, Gordeon DC, Keith R, Powell W (1997) Locating genotypes and genes for abiotic stress tolerance in barley: a strategy using maps, markers and the wild species. New Phytol 137:141–147

Forster BP, Ellis RP, Thomas WTB, Newton AC, Tuberosa R, This D, Ei-Enein RA, Bahri MH, Ben-Salem M (2000) The development and application of molecular markers for abiotic stress tolerance in barley. J Exp Bot 51:19–27

Garcia-del-Moral LF, Ramos JM, Garcia-del-Moral MB, Jimenez-Tejada MP (1991) Ontogenetic approach to grain production in spring barley based on path-coefficient analysis. Crop Sci 31:1179–1185

Gill KS (1979) Effect of soil salinity on grain filling and grain development in barley. Biol Plant 21(4):241–244

Hayward HE, Wadleigh CH (1949) Plant growth in saline and alkaline soil. Adv Agron 1:1–38

Israelsen KR, Ransom CV, Waldon BL (2014) Salinity tolerance of foxtail barley. Weed Sci 59(4):500–505

Jedel PE, Helm JH (1991) Lodging effects on a semi dwarf and two standard barley cultivars. Agron J 83(1):158–161

Jenks MA, Hasegawa PM (2014) Drought tolerance mechanism and their molecular basis. In: Plant Abiotic stress. Willey Blackwell, p15–46

Kharub AS, Verma RPS, Kumar D, Kumar V, Selvakumar R, Sharma I (2013) Dual purpose barley (*Hordeum vulgare* L.) in India: performance and potential. J Wheat Res 5(1):55–58

Kucyzynska A, Surma M, Adamski T, Mikolajczak K, Krystkowiak K, Ogrodowicz P (2013) Effects of the semi-dwarfing sdw1/denso gene in barley. J Appl Genet 54(4):381–390

Kumar V, Kumar R, Verma RPS, Verma A, Sharma I (2013) Recent trends in breeder seed production of barley (*H. vulgare* L.) in India. Indian J Agric Sci 83(5):576–578

Kumar V, Khippal A, Singh J, Selvakumar R, Malik R, Kumar D, Kharub AS, Verma RPS, Sharma I (2014) Barley research in India: retrospect & prospects. J Wheat Res 6(1):1–20

Kumar V, Kharub AS, Verma RPS, Verma A (2016a) AMMI, GGE biplots and regression analysis to comprehend the G X E interaction in multi-environment barley trials. Indian J Genet 76(2):202–204

Kumar V, Kharub AS, Verma RPS, Verma A (2016b) Applicability of joint regression and biplots models for stability analysis in multi-environment barley trials. Indian J Agric Sci 86(11):1443–1448

Larcher W (2003) Physiological plant ecology, 4th edn. Springer Verlag, Berlin–Heidelberg

Levitt J (1972) Responses of plants to environmental stress. Academic Press, New York

Levitt J (1980) Responses of plants to environmental stress. Volume II: water, radiation, salt, and other stresses, 2nd edn. Academic Press, New York

Ludlow MM, Muchow RC (1990) A critical evaluation of traits for improving crop yields in water limited environments. Adv Agron 43:107–153

Ma QH (2009) The expression of caffeic acid 3-O-methyltransferase in two wheat genotypes differing in lodging resistance. J Exp Bot 60(9):2763–2771

Maas EV, Hoffman GJ (1977) Crop salt tolerance – current assessment. J Irrig Drain Div ASCE 103:1309–1313

Mano Y, Takeda K (1997) Mapping quantitative trait loci for salt tolerance at germination and seedling stage in barley (*Hordeum vulgare* L.) Euphytica 94:263–272

Martin JH, Warren HL, David L (1976) Principles of field crop production, 3rd edn. Macmillan Publishing Co. Inc., New York, pp 503–510

Matusinsky P, Svobodova I, Misa P (2015) Spring barley stand structure as an indicator of lodging risk. Zemdirbyste Agriculture 102(3):273–280

Mittler R (2002) Oxidative stress, antioxidants and stress tolerance. Trends Plant Sci 7:405–410

Mittler R (2006) Abiotic stress, the field environment and stress combination. Trends Plant Sci 11:15–19

Moffat J, Sears MRG, Paulsen GM (1990) Wheat high tolerance during reproductive growth. I. Evaluation by chlorophyll influence. Crop Sci 30:881–885

Nevo E (1992) Origin, evolution, population genetics and resources for breeding of wild barley, *Hordeum spontaneum* in the fertile crescent. In: Shewry PR (ed) Barley genetics, biochemistry, molecular biology and biotechnology. CAB International, Wallingford, pp 19–43

Pinthus MJ (1974) Lodging in wheat, barley and oats: the phenomenon, its causes and preventative measures. Adv Agron 25:209–263

Ren XF, Sun DF, Dong WB, Sun GL, Li CD (2014) Molecular detection of QTL controlling plant height components in a doubled haploid barley population. Genet Mol Res 13(2):3089–3099

Reynolds MP, Mujeeb-Kazi A, Sawkins M (2005) Prospects for utilizing plant-adaptive mechanisms to improve wheat and other crops in drought- and salinity-prone environments. Ann Appl Biol 146:239–259

Setter TL, Burgess P, Water I, Kuo J (1999) Genetic diversity of barley and wheat for waterlogging tolerance in Western Australia, Proceeding of the 9th Australian barley Technical Symposium, Melbourne

Shahryari R, Mollasadeghi V (2011) Introduction of two principal components for screening of wheat genotypes under end seasonal drought. Adv Environ Biol 5(3):519–522

Stanca AM, Terzi V, Cattivelli L (1992) Biochemical and molecular studies of stress tolerance in barley. Chapter 13. In: Shewry PR (ed) Barley: genetics, biochemistry, molecular biology and biotechnology. CAB International, Wallingford, pp 277–288

Tanksley SD, Nelson JC (1996) Advanced backcross QTL analysis: a method for the simultaneous discovery and transfer of valuable QTLs from unadapted germplasm into elite lines. Theor Appl Genet 92:191–203

Tavakkoli E, Fatehi F, Coventry S, Rengasamy P, McDonald GK (2011) Additive effects of Na+ and Cl− ions on barley growth under salinity stress. J Exp Bot 62(6):2189–2203

Teulat B, Borries C, This D (2001) New QTLs identified for plant water status, water-soluble carbohydrate and osmotic adjustment in a barley population grown in a growth-chamber under two water regimes. Theor Appl Genet 103:161–170

Teulat B, Merah O, Sirault X, Borries C, Waugh R, This D (2002) QTLs for grain carbon isotope discrimination in field-grown barley. Theor Appl Genet 106:118–126

The International Barley Genome Sequencing Consortium (2012) A physical, genetic and functional sequence assembly of the barley genome. Nature 491:711–717

Tondelli A, Francia E, Barabaschi D, Aprile A, Skinner JS, Stockinger EJ, Stanca AM, Pecchioni N (2006) Mapping regulatory genes as candidates for cold and drought stress tolerance in barley. Theor Appl Genet 112:445–454

Tubiello FN, Donatelli M, Rosenzweig C, Stockle CO (2000) Effects of climate change and elevated CO_2 on cropping systems: model predictions at two Italian locations. Eur J Agron 12:179–189

Vaeiz B, Bavei V, Shiran B (2010) Screening of barley genotypes for drought tolerance by agrophysiological traits in field condition. Afr J Agric Res 5(9):881–892

Verma RPS, Kumar V, Sarkar B, Kharub AS, Kumar D, Selvakumar R, Malik R, Sharma I (2012) Barley Cultivars Released in India: Names, Parentages, Origins and Adaptations. Directorate of Wheat Research, Karnal 132 001 (Haryana). Research Bulletin No. 29, pp26

Wang FZ, Wang QB, Kwon SY, Kwak SS, Su WA (2005) Enhanced drought tolerance of transgenic rice plants expressing a pea manganese superoxide dismutase. J Plant Physiol 162:465–472

Xue DW, Huang YZ, Zhang XQ, Wei K, Westcott S (2009) Identification of QTLs associated with salinity tolerance at late growth stage in barley. Euphytica 169:187–196

Zare M, Azizi MH, Bazrafshan F (2011) Effect of drought stress on some agronomic traits in ten barley cultivars. Tech J Eng Appl Sci 1(3):57–62

Zhang J, Zhou, M, Mendham NJ (2003) Screening barley germplasm for stress tolerance and disease resistance. In: proceedings 11th Australian barley Technical Symposium, 2003, Adelaide, South Australia [http://ecite.utas.edu.au/29253]

Sugarcane Crop: Its Tolerance Towards Abiotic Stresses

A.K. Shrivastava, A.D. Pathak, Varucha Misra,
Sangeeta Srivastava, M. Swapna, and S.P. Shukla

Abstract

Sugarcane (*Saccharum* species hybrids) is a long-duration, high-water-requiring cash crop cultivated under different agroecological conditions. Besides experiencing vagaries of weather all the year round, the climate change, the order of the day, further aggravates effects of these abiotic stresses affecting sugarcane growth, development, sugar synthesis, its accumulation and recovery, ratooning ability and availability of the seed cane for succeeding planting. The relatively more resilience of sugarcane to abiotic stresses appears to be due to some natural endowments like a good deal of compensatory ability, C_4 photosynthesis, higher temperature optima for most of the physiological activities (but for sugar accumulation), higher water-use efficiency, use of genetic complements from *Saccharum spontaneum* imparting tolerance to various abiotic stresses (in breeding varieties) and carbon-managing ability. Besides, some of the physiological interventions like inducing drought hardiness, training roots to penetrate deeper, reducing the heat load by trash mulching, increasing the age of the crop at the advent of drought/floods, organic matter amendment in the soil, nutrient management, managing rhizospheric salinity/alkalinity, etc., also contribute to its stress tolerance. Besides a large number of genes, molecular markers and *miR-NAs* associated with these stress responses contribute to resilience of sugarcane to abiotic stresses. Such efforts have led to development of a transgenic utilizing *betA* gene, imparting drought tolerance, for commercial cultivation.

A.K. Shrivastava (✉) • A.D. Pathak • V. Misra • S. Srivastava • M. Swapna • S.P. Shukla
ICAR-Indian Institute of Sugarcane Research, Lucknow 226 002, Uttar Pradesh, India
e-mail: shrivastavaashokindu@gmail.com; pathkashwini@rediffmail.com;
misra.varucha@gmail.com; sangeeta_iisr@yahoo.co.in; sugarswapna@gmail.com;
somendraprasads@gmail.com

© Springer Nature Singapore Pte Ltd. 2017
P.S. Minhas et al. (eds.), *Abiotic Stress Management for Resilient Agriculture*,
DOI 10.1007/978-981-10-5744-1_17

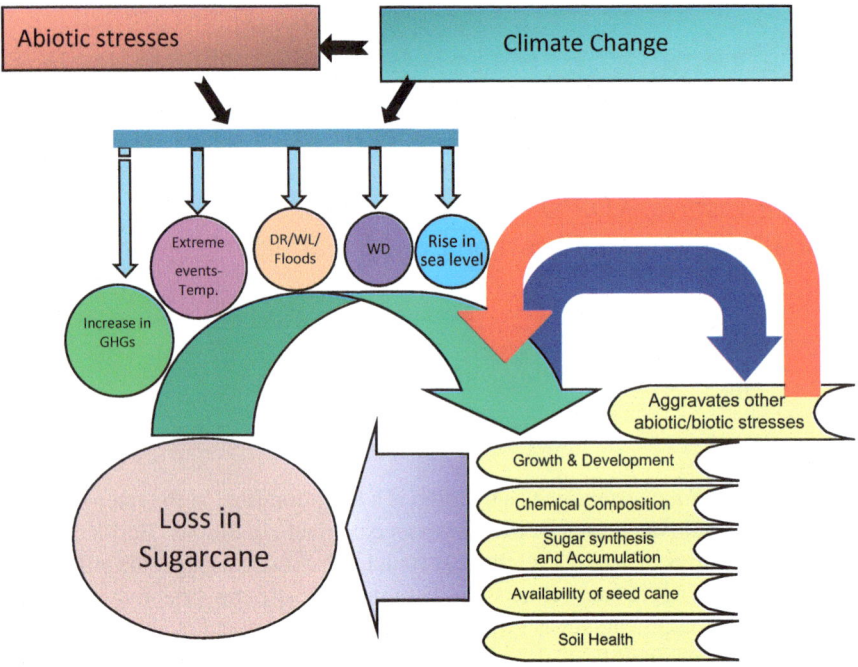

Fig. 17.1 Influence of abiotic stresses and climate change induced abiotic stresses on sugarcane

17.1   Introduction

Sugarcane, being a long-duration (12 months or more) crop, experiences all of the variations of weather prevailing all the year round in different seasons. The critical stages (tillering, grand growth phase, ripening and initiation of ratoon) in its life cycle lie far apart temporally, and it is practically impossible to provide optimal conditions for these to harness its optimal production potential. Further, due to over-exploitation of natural resources as also human activities, our ecosphere has become more prone to abiotic stresses like drought (as also shortage of irrigation water), waterlogging/flooding, salinity, temperature extremes, soil-related problems, emerging nutrient deficiencies, pollution, etc. In India's coastal areas (especially in the East Coast Zone), cyclone and winds uproot and damage sugarcane and influence productivity. In addition, use of high-yielding varieties, high-intensity irrigation and fertilizer consumption, particularly lopsided use of N and pesticides has also provoked some of these tribulations (Shrivastava and Srivastava 2006). Abiotic stresses, as such or the ones induced by climate changes, affect sugarcane's growth and development, its chemical composition, synthesis of sugar as well as its accumulation in stalks and ultimately affect sugarcane/sugar productivity *per se*. Seed cane availability for successive planting of sugarcane is also affected. Certain other abiotic and biotic stresses are also provoked by the primary impinging abiotic stresses and increase the intensity of the effects (Fig. 17.1; Table 17.1). As per

Table 17.1 Abiotic/biotic stresses likely aggravated by the climate change

Climate change induced stress	Aggravated stress	
	Abiotic stress	Biotic stress
Drought	Salinity	*Diseases*: wilt, smut, leaf scald
		Insect pests: termites, shoot borer, *Pyrilla*, mealy bugs, white flies, scale insect, thrips, etc.
Waterlogging	Salinity, alkalinity, acidity, Fe toxicity, nutrient imbalance and deficiency of N and K	*Diseases*: red rot, wilt syndrome and pineapple disease
		Insect pests: Whitefly (in ratoon), cut worm, scale insect and Gurdaspur borer
Salinity	Salt blight, toxicity of boron	Shoot borer (*C. infuscatellus*)
Low temperature	Reduced water conductance, formation of *frost heaves*, localized partial salt stress and banded chlorosis	Stem borer (in peninsular zone)
High temperature	Drought	*Diseases*: Incidence increases
		Insects pests: Stem borer, root borer

Source: Shrivastava et al. (2016)

projection by the IPCC, global mean annual surface air temperatures may increase 1.8–4.0 °C, by the turn of this century, and the future may witness hot extremes, heat waves and heavy precipitation (Aggarwal 2010). UK Met Office has predicted 2016 to be the warmest year ever recorded. Thus the climate change-induced advent of abiotic stresses may accentuate further. Abiotic stresses experienced by the sugarcane crop in India in various zones are given in Table 17.2.

17.2 Sugarcane Tolerance to Abiotic Stresses

The tolerance of sugarcane to abiotic stresses is achieved by some of its natural endowments, physiological and biotechnological interventions, *etc.*

17.2.1 Natural Endowments

To mitigate losses due to climate change-induced abiotic stresses, sugarcane has been bestowed with seven natural endowments like a unique compensatory ability, relatively higher temperature optima for most of the physiological processes (except for sugar storage), its ability not to saturate with light in hot and dry conditions, ability to manage carbon (of CO_2 in the atmosphere), presence of osmoprotectants, use of *S. spontaneum* as parent and allopolyploidy (Fig. 17.2).

The compensatory ability of sugarcane is due to its unique "Compensatory Physiologic Continuum" attributed by its tillering (performance of individual tillers, their age and late shoots), two types of roots (sett roots and shoots roots as well

Table 17.2 Abiotic stress experienced by the sugarcane crop in various sugarcane zones in India

Zone	Area covered	Abiotic stress experienced by the crop
North-west	Punjab, Haryana, western, Central Uttar Pradesh, Rajasthan and Uttarakhand	High and low temperatures, frost, salinity and alkalinity
North-central	Eastern Uttar Pradesh, Bihar and West Bengal	Relatively less pre-monsoon rainfall and high rainfall in later part of the monsoon, floods
North-east	Assam and Nagaland	Floods, waterlogging, less development of water resources and irrigation, acid soils
East coastal	Orissa, coastal Andhra and Tamil Nadu	Cyclonic winds before summer and winter monsoons uproot and damage cane, ingression of sea water, temperatures favourable for growth but not the ripening
Peninsular	Maharashtra, Karnataka, Gujarat, Madhya Pradesh, interior Andhra Pradesh, peninsular area of Tamil Nadu	Variability in soil, seepage and poor drainage, now severe drought, salinity and acid soils in some parts (Marathwada and Vidarbha in Maharashtra and parts of Karnataka)

Source: Shrivastava et al. (2016)

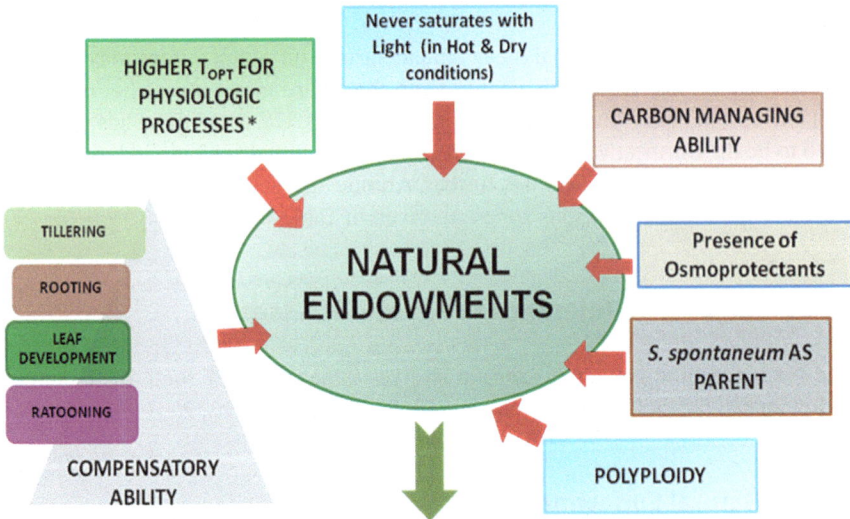

Partial resilience to Climate change induced abiotic and biotic stresses

(*T$_{opt}$ for sugar accumulation is 17°C)

Fig. 17.2 Natural endowments to sugarcane contributing partial resilience to climate change induced abiotic and biotic stresses

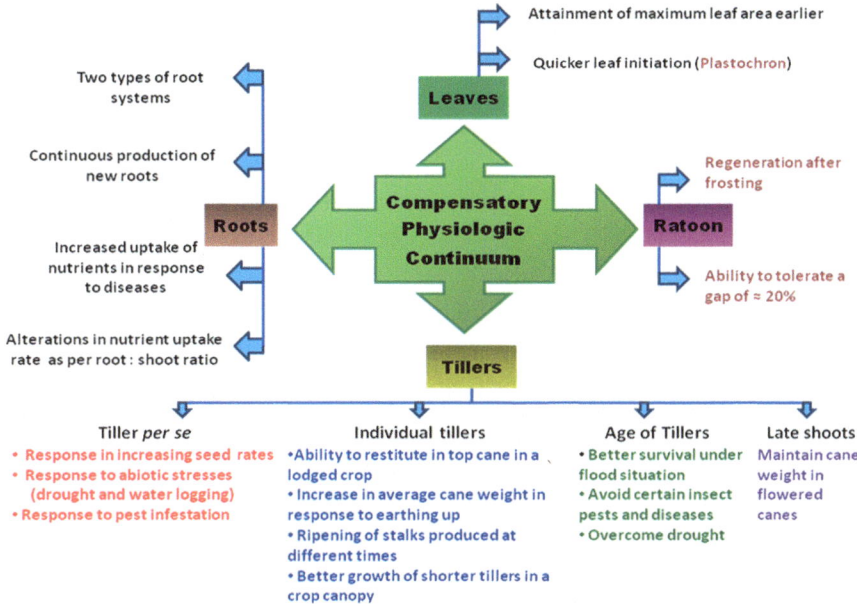

Fig. 17.3 Compensatory ability in sugarcane and expression of its components (Source: Shrivastava et al. 2009)

as their emergence from root primordia, at different times), development of leaves and the vast ratooning potential. Such an ability may enable sugarcane for bare survival, sustenance or mopping-up of certain losses/deficiencies resulting from damage by insect pests or even mechanical means, delay in planting or advent of certain stress conditions by raising the magnitude of some physiological character-istics/morphologic component(s) to counterbalance the loss or the deficiency incurred (Fig. 17.3). Higher optimal temperatures for most of the growth processes (except sugar storage) are another natural endowment to this crop (Table 17.3) which reduces sensitivity to increasing temperatures as compared to a C_3 plant hav-ing relatively lower optimal temperatures for most of its growth processes.

Sugarcane, like other C_4 plants, is bestowed with higher light saturation, and it never saturates with light in hot and dry conditions (http://hyperphysics.phy-astr.gsu.edu/hbase/biology/phoc.html, 13.05.2016). As per Mona Loa Laboratory, Hawaii, the current CO_2 concentration in the atmosphere is 403.8 ppm, while its safe limit is 350 ppm. Sugarcane has been bestowed with a unique ability to manage CO_2 concentration by its low CO_2 compensation point (0–10 ppm), high rates of photosynthesis (39–47 mg $CO_2 dm^{-2} h^{-1}$) and high carbon sequestration ability. Carbon sequestration, a long-term storage of carbon dioxide (or other forms of car-bon) in plants and soil, is another natural mode to counteract CO_2 emissions and resultant global warming to mitigate their effects (https://en.wikipedia.org/wiki/Carbon_sequestration, 21.08.2015). Like many other grasses, sugarcane also pos-sesses such a unique sequestration process for making plant tone or phytoliths. This

Table 17.3 Optimal temperatures for germination, growth, sugar synthesis, its accumulation and flowering of sugarcane

Process	Optimum temperature/range (°C)
Germination	
Germination of true seed	30–35[1]
Germination of setts	22–36[1]
Sprouting	26–33[2]
Growth processes	
Root growth	35 (in soil)[3]
Shoot growth	36 (in soil)[3]
Tillering	33.3–34.4[3]
Stalk elongation	30[4]
Photosynthesis	
Carbon assimilation	30[5]
Sugar synthesis, transport and storage	
Sugar synthesis	30[6]
Sugar transport	30–33[7]
Sugar storage; ripening	17[8]; 12–14[9]
Initiation and development of floral primordia	
Early flowering varieties	T_{Min} 27.2; T_{Max} 30.0[10]
Late flowering varieties	T_{Min} 21.1 °C; T_{Max} 23.9[10]

Source: [1]Anon. (1987); [2]Blume (1985); [3]Mathur and Haider (1940); [4]Brandes and Artschwager (1958); [5]Singh and Lal (1935); [6]Hartt (1940); [7]Julian et al. (1989); [8]Rojeff (2002);[9] Fageria et al. (2010); [10]Singh (1985)

process extracts around 300 Mt of CO_2 yr^{-1} from the atmosphere and stores it in the soil for thousands of years (Parr and Sullivan 2007; Parr et al. 2009). This is also called phyto-occluded carbon (PhytOC). Parr and Sullivan (2005) have estimated the PhytOC yield of a sugarcane crop to be 18.1 g C m^{-2} yr^{-1}. This amounts to 181 kg C sequestered ha^{-1} $yr.^{-1}$. In a study carried out at the ICAR-Indian Institute of Sugarcane Research, Lucknow, cultivation of sugarcane in a multi-ratooning production system accrued 3.3–3.4 Mg ha^{-1} yr^{-1} carbon with the amendment of organic matter in the soil; however, the rate of carbon sequestration was only 1.6 Mg ha^{-1} yr^{-1} when the recommended nutrients were supplied through fertilizers (Suman et al. 2009). Monoculture of sugarcane over a period of 50 years in Mauritius did not significantly influence soil organic carbon (SOC) as compared to that of the virgin land in 0–50 cm profile. The long-term sugarcane cultivation resulted in a depletion of original SOC by 34–70%. However, this loss was fully compensated by C input from sugarcane residues itself so that there was no net change in SOC (Gunshiam et al. 2014). Hence sugarcane cultivation, on long-term basis, maintains the soil fertility as well as productivity.

Sugarcane contains some osmo-protectants like trehalose, glycine betaine, proline and myo-inositol whose concentration increases under drought (Silva et al. 2011). These not only protect the metabolic system but also overcome, to some extent, the morphological alteration during drought. Earlier *Saccharum officinarum* varieties were affected by abiotic and biotic stresses which hampered their productivity, and survival of sugar industry was at stake. In India, a breakthrough was achieved in sugarcane breeding by crossing *S. officinarum* (*Vellai*) and *S. spontaneum* which generated the first hybrid Co 205 which produced 1.5-fold cane yield than the prevailing indigenous varieties. Specifically *S. spontaneum* possessed tolerance to drought, waterlogging, cold and comparatively low-nutrient requirements (Anon 1987).

Allopolyploidy in sugarcane with genomic contributions from *Saccharum* species is endowed with many favourable adaptability characters, stress resistance/tolerance factors and yield-contributing economic traits (Premachandran et al. 2011). Generally polyploids are more tolerant to drought, cold (Levin 1983), mutagenesis and irradiation and resistant to insect pests and diseases (Lewis 1980). Some of the cases are there where polyploidy has not shown these traits (Stebbins 1985). Polyploidy confers gene redundancy which diversifies the gene function by altering redundant copies of essential genes (Comai 2005); this is perhaps responsible for such behaviour.

All these natural endowments to the sugarcane impart partial resilience against the prevailing abiotic stresses as well as the climate change-induced multiple abiotic stresses and sustain the productivity to some extent under such conditions.

17.2.2 Variability in Germplasm and Breeding Efforts

Saccharum species and related genera also exhibit multiple stress tolerance which could be advantageous for breeding climate-resilient sugarcane varieties (Table 17.4).

Table 17.4 Tolerance to abiotic stresses associated with *Saccharum* species and related genera

Species/ Genera	Character- tolerance to	Reference
S. spontaneum	Drought, waterlogging, cold, low nutrient requirement	Krishnamurthi (1989), Anon (1987), Brandes (1939) and Earle (1928)
S. barberi	Salinity, cold	Krishnamurthi (1989), Sreenivasan et al. (2001) and Ramana et al. (1985)
S. robustum	Waterlogging, salinity	Krishnamurthi (1989)
S. sinense	Salinity	Sreenivasan et al. (2001) and Ramana et al. (1985)
Erianthus spp.	Drought, cold, salinity and robust growth under low input conditions, low nutrient requirement	Sreenivasan et al. (2001), Roach and Daniels (1987) and Krishnamurthi (1989)
Narenga spp.	Drought	Krishnamurthi (1989)
Miscanthus spp. *Miscanthus nepalensis*	Cold (performance at high altitudes)	Krishnamurthi (1989) and Anon (1987)

Source: Shrivastava et al. (2016)

Among various *Saccharum* species maintained at the World Germplasm Collection, at SBI Research Station Cannanore, the following clones/indigenous varieties have been reported to be tolerant to salinity (Ramana et al. 1985).

S. barberi: Katha (Coimbatore), Kewali-14-G, Khatuia-124, Kuswar, Lalri, Nargori and Pathri

S. sinense: Khakai, Panshahi, Reha and Uba seedling

S. robustum: IJ-76-422, IJ-76-470, 28-Ng 251, 57-Ng-201, 57 Ng -231, Ng-77-34, Ng-77-55, Ng-77-136, Ng-77-160, Ng-77-167, Ng-77-170, Ng-77-221 and Ng-77-237

Among the indigenous canes growing in India, *Hemja, Khari, Khagari* and *Ikri* were tolerant to drought and waterlogging. Among these, *Hemja* was well adapted to early drought and late waterlogging, and *Khagari* grew well even under 6 f of water stagnating over 3 months. *Katha* was widely adaptable and tolerant to drought, flooding and, to a lesser extent, frost. *Erianthus* clones endowed with abundant tillering, high biomass production, robust growth with low inputs and efficient root system which imparted tolerance to drought, salinity and also resistance to some of the insect pests. Further, in the progeny, sucrose contents did not decline when it was used as a pollen parent (Sreenivasan et al. 2001), and thus it could be an excellent donor imparting multiple stress tolerance.

Some of these could be beneficially utilized in breeding programmes for imparting requisite tolerance trait to the progeny to develop a desired variety. Some examples of the use of *S. spontaneum* clones are development of Co 205 in 1918 in India which sustained sugar industry in earlier days, use of *Mandalaya* – a clone from Burma that is associated with the success of Australian "Early CCS Canes Programme" – and use of US56--15-8, another *S. spontaneum* that is in the parentage of a Louisiana variety LCP 85-384, which is not only cold tolerant and required less N but also gives high cane and sugar yield and exhibits an early ripening behaviour (Jackson 2005). Tew (1987) and Shrivastava and Srivastava (2006, 2012) have listed several sugarcane varieties exhibiting multiple abiotic stress tolerance (mainly drought, rainfed conditions, waterlogging, salinity and low temperature). Their response to tolerance to various abiotic stresses is depicted in Fig. 17.4. The use of some of these genotypes for cultivation as such or the use as parents, in breeding programmes, could impart multiple stress tolerance for sustaining sugarcane productivity.

In India, under the auspices of the All India Co-ordinated Research Project on Sugarcane, some improved sugarcane varieties have been developed for tolerance to abiotic stresses for various sugarcane zones (Table 17.5). Although in sugarcane many physiological and biochemical characteristics have been identified to evaluate sugarcane varieties for tolerance to a particular abiotic stress, attributes like *trehalose* and *glycine-betaine* contents are related to tolerance of more than one abiotic stress (Shrivastava and Srivastava 2012). Further under the emanating climate change scenario, there is a need to evaluate germplasm comprising *Saccharum* and related genera as well as sugarcane varieties and also to elucidate physiological and biochemical characteristics associated with them. Breeding programmes must be

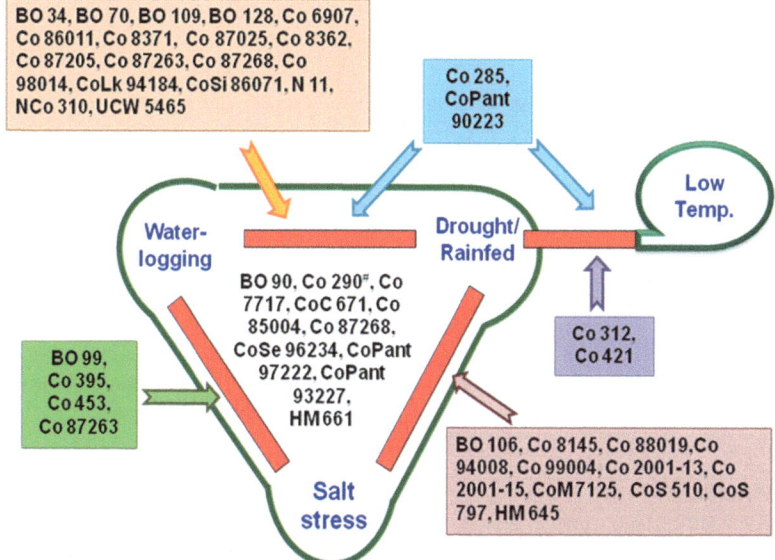

#Performs well under most adverse conditions

Fig. 17.4 Sugarcane varieties exhibiting tolerance to more than one and multiple abiotic stresses (Modified from Shrivastava and Srivastava 2006)

tailored to breed multiple stress-tolerant varieties, as in nature more than one stress often occur together with or the prevalence of one stress may accentuate or aggravate some other stresses.

17.3   Physiological Interventions

Some of the physiological interventions are helpful in imparting some tolerance to abiotic stresses and sustain productivity under such situations to some extent. Inducing drought hardiness is by soaking setts in saturated lime solution (80 kg lime 1000 l^{-1}) for 2 h a day before planting (Anon 2008–2009) and by withholding water for 30 days at 90 days after planting (Rajkumar and Kambar 1999). Similarly the application of calcium silicate improved tolerance of sugarcane to freeze damage (Ulloa and Anderson 1991). Training of the roots to proliferate vertically deep (rather than superficially and laterally) as practiced in trench planting in tropical India (Naidu et al. 1983) and delaying the first irrigation after planting in subtropical India (Dr. R.S. Verma, personal communication), to some extent, impart some degree of tolerance to drought in the ensuing summer season. Reducing the heat load by trash mulching reduces soil water evaporation and sustains productivity under drought conditions in tropical India (Durai 1997; Jayabal and Chockalingam 1990; Kathiresan and Balasubramanian 1991; Manoharan et al. 1990; Parameswaran

Table 17.5 Abiotic stress tolerant sugarcane varieties identified by the AICRP (Sugarcane) and released by the Central Varietal Release Committee (CVRC)

Variety	Tolerance to abiotic stress
Peninsular zone (Madhya Pradesh, Gujarat, Karnataka, Kerala, interior Andhra Pradesh and plateau region of Tamil Nadu	
Co 94008 (Shyama) – early	DR, salinity, WA
Co 8371 (Bhima)	DR, WL
Co 87025 (Kalyani)	DR, WL
Co 87044 (Uttara)	DR
CoM 88121 (Krishna)	DR, MQL
Co 91010 (Dhanush)	DR, RF
Co 99004 (Damodar)	DR, salinity
Co 2001–13 (Sulabh)	DR, salinity stress
Co 2001–15 (Mangal)	DR, salinity and lodging.
North central zone (eastern Uttar Pradesh, Bihar and West Bengal)	
Co 87263 (Sarayu) – early	WL, RF, LIC
Co 87268 (Moti) – early	DR, WL, high soil pH
CoSe 96234 (Rashmi) – early	Stress conditions, in general
CoSe 96436 (Jalpari)	WL
BO 128 (Pramod)	Saline-sodic soils
CoLk 94184 (Birendra) – early	DR, WL
Co 0232 (Kamal) – early	WL
Co 0233 (Kosi)	WL and lodging
BO 146	DR, WL
North-west zone (Punjab, Haryana, Rajasthan, western and central Uttar Pradesh)	
CoH 92201 (Haryana-92) – early	LPC
CoS 95255 (Rachna) – early	LPC
CoPant 90223 (pant 90223)	DR, WL, LT
CoPant 93227(pant 93227)	LIC, suboptimal environments
CoH 119	DR
Co 98014 (Karan-1) – early	DR, WL
CoPant 97222	DR, WL, salinity
CoJ 20193	Late planting
Co 0118 (Karan-2) – early	DR, WL
Co 0238 (Karan-4) – early	DR, WL, LT, better ratooning
Co 0239 (Karan-6) – early	DR, WL
CoPK 05191	DR
Co 86249 (Bhavani)	WA
CoC 01061 – early	DR

DR drought, *WL* waterlogging, *RF* rainfed conditions, *WA* wider adaptability, *LIC* low-input conditions, *MQL* maintains juice quality for long, *LPC* late-planted conditions [Courtesy: Dr. O.K. Sinha, Project Co-ordinator, AICRP (Sugarcane)]

and Ramakrishnan 1987; Ramakrishan et al. 1988). In Australia, under dry conditions, use of green cane trash blanket improved cane yields by better moisture retention and other associated features like prevention of soil erosion and weed management (Page et al. 1986). The use of moisture absorbers in the soil like FYM, pressmud, coir waste (Manoharan et al. 1990), Jalshakti, Stocksorb and Alcosorb (Bendigeri and Pawar 1997; Durai 1997) maintained better moisture regimes and sustained sugarcane productivity under drought.

Foliar spray of plant growth regulators (PGRs, like Ethrel (100 ppm), before the onset of drought, sustained sugarcane productivity by imparting some drought tolerance (Li and Solomon 2003). Application of 50 or 100 ppm gibberellic acid was helpful in reducing post-frost losses (Avtar 1993). Increasing the physiological age of the crop (by early planting or planting pregerminated setts, rayungans, *etc.*) at the advent of drought in peninsular India (Prasada-Rao 1989) or waterlogging in Kerala (Parthasarathy 1972; Prasada-Rao 1989) and Odisha (Nayak et al. 1997) and subtropical India (Srivastava et al. 1988) so that the plants are sufficiently taller/stout to bear the blunt of drought/flood and sustained sugarcane productivity, to some extent. In saline conditions, pregerminated settlings are good planting material (Srivastava et al. 1988).

Organic matter amendment in the soil (as FYM, pressmud, coir waste, *etc.*) sustained sugarcane productivity under drought by making moisture available (Manoharan et al. 1990) and waterlogged soils by improving soil aeration, *etc.* (Singh et al. 1991). Amending bulk manures and green manuring by *Sesbania aculeata* enhanced the productivity of saline soils (Kanzaria and Patel 1985; Perur and Mithyantha 1985; Patil and Ghonsikar 1985; Singh et al. 1985). In alkaline soils, planting cane in deep (20–25 cm) furrows and packing trash in the furrows immediately after the germination and tiller initiation improved sugarcane yields (Panje et al. 1966).

Nutrients also impart tolerance to stress conditions. A fertilized crop of sugarcane was relatively less prone to drought (Perez et al. 1996), and plants grown in a soil with K are relatively less sensitive to variations in irrigation (Stiles and Cocking 1969). K has been important in enzyme activation, anion neutralization, membrane transport and maintaining osmotic potential (Clarkson and Hanson 1980). In sugarcane, use of K (Rutherford 1989) and P (Sato et al. 2010) both perhaps alleviated the negative effects of water deficit on sugarcane through accumulation of an osmoprotectant. In tropical India, spray of K alone or in combination with urea at periodic intervals sustained sugarcane productivity under drought conditions (Anon 1987; Annadurai et al. 2002; Jayabal and Chockalingam 1990; Naidu et al. 1983). Increase in P content in the stalk was related to tolerance to flooding (Pandey 1964). Application of additional P may increase plant growth under saline conditions (Awad et al. 1990), but its high concentrations are injurious (Nieman and Clark 1976).

Planting on ridges (45–60 cm high from soil surface and 90–120 cm from the bottom of the furrow) to prevent water from entering root zone is helpful in flooded conditions (Smith 1978) when water level in the field is around 0.6 m. Improving drainage by using slotted PVC pipes laid with gravel envelopes (Patel et al. 2000) or by network of surface drains (Lal et al. 2000), low frequency of irrigation (Singh et al. 1985), deep tillage and subsurface drainage (Smith 1978) sustained productivity to some extent under waterlogged conditions.

For alkali soil conditions, additions of gypsum (1.7 Mg ha^{-1} to remove 1 meq Na$^+$ 100 g^{-1} soil) are helpful (Yadav 1987), and its use in combination with rice husk and flushing with water was even more beneficial (Raman et al. 1985). Addition of molasses up to 4.8 Mg ha^{-1} can also be used to partially reclaim alkaline soils after which sugarcane may be planted (Bhan 1963). With adjustment of the ridge direction from east to west, in subtropical India, in saline conditions, planting sugarcane on slope of the ridges is beneficial (Srivastava et al. 1988). Crop rotations are also helpful under saline conditions (Zende and Hapase 1986; Patil and Ghonsikar 1985).

17.4   Molecular Interventions: Genes, Markers, *miRNAs* and Transgenics

The genetic complexities of sugarcane and regulatory mechanisms for various abiotic stresses complicate the analyses of various traits. The advances in molecular biology and other areas like functional genomics have enabled the study of response of sugarcane to various abiotic stresses like water deficit and waterlogging tolerance, heat tolerance, cold tolerance, salinity tolerance, etc. and the manipulation of associated pathways, to some extent.

Increasing incidents of these abiotic stresses such as drought, waterlogging, cold and salinity due to changing climatic scenario induce changes in growth and development of sugarcane and affect sugarcane and sugar productivity. Recently Abberton et al. (2015) have reiterated the role of genomics for enhancing crop productivity, reducing cost-intensive agricultural inputs, augmenting crop biodiversity, stabilizing yield as well as inculcating the resilience to climate change. For agriculturally important traits, approaches like identifying gene sequence(s) and epigenetic variations may increase the diversity of desirable alleles for use and help breeding for new climate-resilient ideotype(s). The advancement in genomic and transcriptomic technologies and freely accessible information from plant genetic databases has opened up the gates for candidate gene approach to identify potential genes for stress tolerance. Several stress-related candidate genes including transcription factors, osmo-protectants, antioxidants, signalling genes, stress-inducible proteins and ion transporters are available for tolerance to various abiotic and biotic stresses in sugarcane. Among these, DREB (dehydration-responsive transcription factor), ERD4 (early response to dehydration protein 4), GolS (galactional synthase), HSP (heat shock proteins), LEA (late embryogenesis abundant) protein, *NHX* (sodium proton antiporter), P5CS (pyrroline-5-carboxylase synthetase), RAB (responsive to abscisic acid), SodERF3 (sugarcane ethylene-responsive factor), SUT1 (sucrose transporter1), heat stress-induced DHNs, some genes encoding for radicals like O$^-$/OH$^-$ and reduction of H$_2$O$_2$ by peroxidase/catalase under heat stress, peroxidase/catalase, stress-related clusters showing differential expression (>twofold) during biotic and abiotic stress conditions, cold-inducible ESTs (expressed sequence tags), NADP-MDH (nicotinamide adenine dinucleotide phosphate-malate dehydrogenase), PPDK (pyruvate orthophosphate dikinase) and NADP-ME proteins and dehydrin-like proteins which protect membranes against chilling stress, osmolytes like proline and glycine betaine and other stress-inducible proteins, dehydrin, osmotin, annexin, choline oxidase, trehalose and PDH (proline dehydrogenase), have been identified in sugarcane in response to drought/water deficit, temperature, PEG (polyethylene glycol) and salinity-induced stress conditions (Nogueira et al. 2003; Molinari et al. 2007; Wahid and Close 2007; Chagas et al. 2008; Guimarães et al. 2008; Patade et al. 2008; Trujillo et al. 2009; Gupta et al. 2010; Jangpromma et al. 2010; Iskandar et al. 2011; Nair 2011; Prabu et al. 2011; Srivastava et al. 2012; Shrivastava et al. 2016).

Recently, Park et al. (2015) identified nearly 600 differentially expressed genes for activity of the transmembrane transporter in sugarcane exposed to chilling (low temperature) stress, with ~2.5-fold increase in expression of SspNIP2 (*Saccharum* homologue of a NOD26-like major intrinsic protein gene). Sugarcane transgenics overexpressing PDH45, a DEAD-box helicase gene isolated from pea, exhibited an upregulation of DREB2-induced downstream stress-related genes and improved tolerance to drought and salinity (Augustine et al. 2015). In response to foliar application of salicylic acid, expression of genes responsible for synthesis/expression of trehalose 5-phosphate and sucrose-phosphate were involved in response to drought in sugarcane (Almeida et al. 2003). Sugarcane drought-responsive gene 1(Scdr1) is upregulated in response to drought, but it is not associated with drought tolerance of some of the varieties of sugarcane. However, transformation of tobacco, using this gene, conferred tolerance to multiple abiotic stresses like drought, salinity and oxidative stresses (Begcy et al. 2012). Utilization of these specific stress-induced genes and signalling cascades for inculcating stress resistance/tolerance in elite sugarcane varieties by their overexpressing or acting upstream in response to a certain stress or multiple stresses may lead to development of climate-resilient sugarcane varieties which may sustain or even improve sugarcane/sugar productivity in the climate change scenario. Not only this but some stress-upregulated genes in sugarcane like Scdr1 may confer tolerance to multiple abiotic stresses in some other plants like tobacco.

Molecular markers have been widely used for studying the genetic diversity among sugarcane genotypes in response to certain abiotic stresses imposed (Ming et al. 2006). Molecular marker systems like RAPD, RFLP, SSRs both genomic and EST derived, STS, ISSR and TRAP have been used to study the genetic diversity among sugarcane clones for tolerance to drought, waterlogging salinity, cold, etc. (Fahmy-Eman et al. 2008; Sharma 2009; Hemaprabha and Simon 2012; Markad et al. 2014). Drought-specific candidate genes for superoxide dismutase and indole-3-glycerol phosphate synthase were identified for use in marker-assisted selection for drought tolerance in sugarcane (Hemaprabha and Simon 2010). Kido et al. (2012) identified 75,404 unigenes in sugarcane, of which 213 were drought-responsive and upregulated under drought stress. A SCAR (sequence characterized amplified region) marker was found highly effective in identifying drought-tolerant genotypes of sugarcane (Srivastava et al. 2012). Sequence-specific STS primers of DREB 2 were used by Khan et al. (2013) to study diversity with respect to drought tolerance. Cloning and characterization of two homologues of Δ' pyroline-5-carboxylate synthase

(P5CS) resulted in identification of SoP5CS1 as a potential marker, to screen sugarcane varieties for drought tolerance (Iskandar et al. 2014). Transcription factors such as CBF/DREB are important in stress as they regulate gene expression in response to drought, salinity and cold by binding to drought-responsive cis-acting elements. Upregulation of ten candidate genes, viz., SOD (superoxide dismutase), DHAR (dehydroascorbate reductase), cAPX (cytosolic ascorbate peroxidase), GST 1 (glutathione-S-transferases) and GSHS1 (glutathione synthase 1), Prokin (protein kinase), LEA 3, IGS (indole-3-glycerol-phosphate synthase), PIN 1 (polar auxin transport gene 1) and DREB (dehydration-responsive element binding) proteins in droughtresistant clones indicated their role in drought resistance in sugarcane (Hemaprabha and Simon 2012). Six putative candidate genes ABF 2, CIPK 14, LEA 3, MYB 2, RD 28, RGS1 and SNRK 2.5 belonging to ABA-dependent pathway which were present in the tolerant species but absent in susceptible clones of *S. officinarum* indicated their role in imparting drought tolerance to sugarcane (Priji and Hemaprabha 2015). A QTL mapping study using a panel of 80 genotypes exhibited 21 markertrait associations involving SSR markers and 11 marker-trait associations involving TRAP markers with drought-related physiological traits (Sharma 2009).

Up- and downregulation of gene expressions have been studied in sugarcane in response to abiotic stresses. High-throughput transcriptome profiling combined with in silico methods has been used to study the variable expression of genes in response to abiotic stresses in sugarcane. Microarray as a tool for expression studies has been put to use in sugarcane by several researchers to understand the differential gene expression during water deficit (Rocha et al. 2007; Rodrigues et al. 2009, 2011; Iskandar et al. 2011), salinity stress and cold stress (Nogueira et al. 2003). The SUCEST (sugarcane EST sequencing project from Brazil) have identified 2,38,000 ESTs from vegetative and reproductive parts of sugarcane which have been utilized for studying differential expression of genes under abiotic stress conditions (Kurama et al. 2002; Nogueira et al. 2003; Rocha et al. 2007; Rodrigues et al. 2009, 2011; Zingaretti et al. 2012). Osmo-protectants like betaine aldehyde dH (BADH), P5CS, P5CR, myo-inositol-1-phosphate synthase, trehalose-6-phosphate synthase, trehalose phosphate protein *etc.*, and transcription factor orthologues like Myb, WRKY, NAC, DREB proteins, etc., have been found to be differentially expressed under water-deficit conditions (Silva et al. 2011; Menossi et al. 2008). Expression studies of drought-response genes during sucrose accumulation and water deficit (Iskandar et al. 2011) indicated a change in expression of some genes associated with sucrose accumulation in stress-related conditions. Under water stress, LEA and dehydrin transcripts were expressed 100- and 1000-fold, respectively. Expression of a number of genes involved in water stress response was correlated with sucrose content, but a different regulatory mechanism is presumed to operate under water deficit.

Cold stress seemed to activate novel cold response pathways with the identification of 20 novel genes (Nogueira et al. 2003). This was apart from the 25 genes whose expression was repressed on exposure of sugarcane plants to 4 °C and another 34 genes that were upregulated. Upregulation of a putative endonuclease gene, involved in DNA repair, was also observed in response to cold (Vicentini and Menossi 2007). Guerzoni et al. (2014) confirmed the role of P5CS and PDH (proline dehydrogenases) genes functioning in proline biosynthesis and catabolism in providing salinity stress tolerance in sugarcane. Expression profiling of orthologues of four abiotic-stress-responsive genes encoding ERD4, GolS, LEA3 and P5CS under cold, salinity and moisture-deficit stress conditions in sugarcane revealed strong expression of GolS and P5CS under salt stress, ERD4 during drought stress and LEA3 during cold stress (McQualter and Dookun-Saumtally 2007). Short-term salt- or PEG-induced osmotic stress resulted in expression of sugarcane shaggy-like protein kinase (SuSK) involved in signalling of hormone and affecting response of plants under stress conditions (Patade et al. 2012).

As in many other crops, in sugarcane also, *miRNAs* have been speculated to have a major role in regulating the plant response to drought and other abiotic stresses (Zanca et al. 2010). miRNA-mediated regulation of drought response has been studied in sugarcane by many workers (Ferreira et al. 2012; Gentile et al. 2013, 2015; Thiebaut et al. 2012; Lin et al. 2014). Eighteen miRNA families have been identified in cultivars differing in their drought tolerance (Ferreira et al. 2012); seven of these were differentially expressed. Four of the drought-tolerant and drought-sensitive genotypes were used by Thiebaut et al. (2012), for studying the miRNAs linked to drought stress. 67 miRNAs were identified specifically in the water-deficit assay, with 20 being shared by the tolerant and susceptible assays. Lin et al. (2014) identified 11 differentially expressed miRNAs in a drought-resistant cultivar ROC 22. Gentile et al. (2013) analysed micro-transcriptome regulating drought response of two cultivars of sugarcane that differed in their response to drought. Thirteen miRNAs were differentially expressed in drought-stressed plants, and seven miRNAs were differentially expressed in both the cultivars. Patade and Suprasanna (2010) studied expression of mature miRNA, miR159, in sugarcane leaves exposed to salt- and PEG-induced stress. Short-term PEG stress led to significant upregulation in the leaves. Members of the MYB transcription factor family were predicted to be the potential targets. All these studies confirm that miRNAs indeed have a role in regulating the drought response in sugarcane and helped in identifying, to some extent, the target genes for these regulatory sequences.

The application of these potential candidate genes in developing abiotic stress-tolerant sugarcane crops using a transgenic approach through the transfer of genes from various sources is rapidly increasing in sugarcane. Major realms of such researches are resistance to insect pests or herbicides, enhanced sucrose content or sucrose accumulation with antisense soluble acid invertase gene. Attempts are being made to develop sugarcane transgenics having abiotic stress tolerance. Q117 over-expressing *Arabidopsis* CBF4 gene (C-repeat-binding factor 4 protein regulating gene for drought adaptability) led to increased expression of ERD4 and P5CS but had no significant effect on expression levels of GolS and LEA3, suggesting thereby the presence of active abiotic-stress-inducible pathways in sugarcane (McQualter and Dookun-Saumtally 2007). Genetic transformation mediated by *Agrobacterium tumefaciens* with two plasmids LBA4404 pB1 121 construct GLY1 conferred stress tolerance in sugarcane (Shaik et al. 2007). Further tolerance to drought and salinity in sugarcane was introduced using *Arabidopsis* vascular pyrophosphatase (AVP1) gene by *A. tumefaciens*-mediated transformation (Kumar et al. 2014). In 2011, *Empresa Brasileira* de *Pesquisa Agropecuaria*, EMBRAPA, Brazil, has developed sugarcane transgenics using DREB2A gene for drought tolerance which were evaluated as well as field tested (www.isaaa.org/kc/cropbiotechupdate/, 12.05.2016). A DEAD-box helicase gene isolated from pea (PDH45) provided tolerance to soil moisture stress and salinity in transgenic sugarcane, and its overexpression resulted in upregulation of DREB2-induced downstream stress-related genes which enhanced abiotic stress tolerance (Augustine et al. 2015). Ramiro et al. (2016) showed that transgenic expression of Bax inhibitor-1 (AtBI-1), a highly conserved cell death suppressor, from *Arabidopsis thaliana* conferred water stress tolerance when exposed to long-term (>20 days) water stress. Besides, a glycosylation inhibitor protein – tunicamycin – also increased tolerance of the sugarcane to induction of endoplasmic reticulum stress.

In Indonesia, PT Perkebunan Nusantara (a state-owned sugar milling conglomerate), University of Jember (East Java) and Ajinomoto Co., Inc., Japan, have developed a transgenic sugarcane using the betA gene from the *Rhizobium meliloti* which produces glycine-betaine, an osmo-protectant imparting drought tolerance. This GM sugarcane variety, under drought conditions, produced 20–30% more sugar over the conventional varieties, and by being approved for commercial cultivation in Indonesia by the national genetically Modified Product Biosafety Commission of Indonesia (Marshall 2014; Waltz 2014), it has become the world's first commercialized GM sugarcane (Anon 2013).

Thus a large number of genes associated with abiotic stress tolerance, molecular markers and miRNAs associated with response to some of abiotic stresses affecting sugarcane have been identified. Some transgenics have also been developed for tolerance to abiotic stresses, and GM sugarcane utilizing betA gene has been commercialized in Indonesia. With the availability of efficient genetic transformation systems in sugarcane, the days are not far behind when it would be possible to impart multiple stress tolerance and improve economically important traits through appropriate gene transfer.

17.5 Conclusions

Sugarcane (*Saccharum* species hybrids), an important cash crop in our country, is a long-duration (10–12 months), high-water-requiring (20 million Lha^{-1} $yr.^{-1}$) crop growing under different agroecological conditions in subtropical and tropical belts experiencing most of the abiotic stresses (including drought, waterlogging, salinity, low as well as high temperatures and associated western disturbance, cyclones and ingression of sea water in coastal regions), etc. Besides experiencing vagaries of weather all the year round, the present scenario of climate change further aggravates abiotic stresses affecting sugarcane and sugar productivity, *per se*, through influencing the soil supporting its growth and development, ratooning ability, sugar synthesis and its accumulation, harvest and recovery and also the availability of seed cane for planting to raise the next crop.

Sugarcane has been uniquely bestowed with certain natural endowments like C_4 photosynthesis associated with relatively higher temperature optima for most of the physiological processes (except for sugar accumulation) and higher water-use efficiency, a good deal of compensatory ability, use of genetic complements from *Saccharum spontaneum* (to impart tolerance to various abiotic and biotic stresses) in breeding programmes and carbon managing ability by high rates of photosynthesis, low CO_2 compensation point and carbon sequestration (due to *Phytolith* formation) which make it relatively more resilient to climate change-induced abiotic stresses.

- There is good deal of tolerance to abiotic stresses among various clones in *Saccharum* species and related genera; and some of these have been exploited to improve adaptations to abiotic stresses, but more needs to be done.
- Some of the physiological interventions like inducing drought hardiness, training the roots to go vertically deeper, reducing the heat load by trash mulching, increasing the age of the crop at the advent of drought/floods, organic matter amendment in the soil, nutrient management, managing rhizospheric salinity/alkalinity, *etc.* have paid dividends.
- A large number of genes, molecular markers and miRNAs associated with tolerance to abiotic stresses in sugarcane have been identified; and such efforts have culminated into a GM sugarcane variety carrying *betA* gene, released for cultivation in Indonesia.

Acknowledgement Authors are grateful to the ICAR-Indian Institute of Sugarcane Research (IISR), Lucknow and NICRA Project for necessary support.

References

Abberton M, Batley J, Bentley A, Bryant J, Cai H, Cockram J, de Oliveira AC, Cseke LJ, Dempewolf H, de Pace C, Edwards D, Gepts P, Greenland A, Hall A, Henry R, Hori K, Howe GT, Hughes S, Humphreys M, Lightfoot D, Marshall A, Mayes S, Nguyen HT, Ogbonnaya FC, Ortiz R, Paterson AH, Tuberosa R, Valliyodan B, Varshney RK, Yano M (2015) Global agricultural intensification during climate change: a role for genomics. Plant Biotechnol J 14(4):1095–1098

Aggarwal PK (2010) Managing impacts of climate change. The Hindu Survey of Indian Agriculture, pp 64–65

Almeida JCV, Sanomya R, Leite CF, Cassinelli NF (2003) Eficiência agronômica de sulfomethuron-methil como maturador na cultura da cana-de-açúcar. Stab 21(3):36–37

Annadurai G, Shin-Juang R, Jong-Lee D (2002) Use of cellulose-based wastes for adsorption of dyes from aqueous solutions. J Hazard Mater 92:263–274

Anonymous (2008–2009) Annual Report, DARE, Ministry of Agriculture, Government of India & Indian Council of Agricultural Research, Krishi Bhavan New Delhi, pp 47

Anonymous (2013) Indonesia approves first GM sugarcane. Crop Biotech Update (May 22, issue), International Service for the Acquisition of Agri-Biotech Applications (http:// www.isaaa.org / kc/cropbiotechupdate/article/default.asp?ID=10989)

Anonymous (Coimbatore, India) (1987) Research achievements: 1912–1987. Sugarcane Breeding Institute, Coimbatore, p 121

Augustine SM, Narayan JA, Syamaladevi DP, Appunu C, Chakravarthi M, Ravichandran V, Tuteja N, Subramonian N (2015) Introduction of pea DNA helicase 45 into sugarcane (*Saccharum* spp. hybrid) enhances cell membrane thermostability and upregulation of stress-responsive genes leads to abiotic stress tolerance. Mol Biotechnol 57:475–488

Avtar S (1993) Economics of intercropping in autumn planted sugarcane. J Sugarcane Tech 8(1):71–72

Awad AS, Edwards DG, Campbell LC (1990) Phosphorus enhancement of salt tolerance of tomato. Crop Sci 30(1):123–128

Begcy K, Eduardo DMAG, Carolina GL, Sonia MZ, Glaucia MS, Marcelo M (2012) A novel stress-induced sugarcane Gene confers tolerance to drought, salt and oxidative stress in transgenic tobacco plants. PLoS One 7(9):e44697. doi:10.1371/journal.pone.0044697

Bendigeri AV, Pawar MW (1997) Use of superabsorbent (Stocksorb) under moisture stress in sugarcane cultivation. Bharatiya Sugar 22(5):23–40

Bhan VM (1963) Assessment of progress to develop suitable manurial schedule for sugarcane. Ind J Sugarcane Res Dev 8(1):62–64

Blume H (1985) Geography of sugar cane. Verlag Dr. Albert Barlens, Berlin, p 371

Brandes EW (1939) Three generations of cold resistant sugarcane. Sugar Bull 18(4):3–5

Brandes EW, Artschwager E (1958) Sugarcane (*Saccharum officinarum* L.): origin, classification, characteristics, and descriptions of representative clones. USDA Agric. Handbook 122, US Govt. Print Office, Washington DC

Chagas RM, Silveira JAG, Ribeiro RV, Vitorello VA, Carrer H (2008) Photochemical damage and comparative performance of superoxide dismutase and ascorbate peroxidase in sugarcane leaves exposed to paraquat-induced oxidative stress. Pestic Biochem Physiol 90:181–188

Clarkson DT, Hanson JB (1980) The mineral nutrition of higher plants. Annu Rev Plant Physiol 31:239–298

Comai L (2005) The advantages and disadvantages of being polyploid. Nat Rev Genet 6:836–846

Durai R (1997) Study on the stress management in sugarcane. Cooperative Sugar 29:102–104

Earle FS (1928) Sugar cane and its culture. Wiley, New York, p 206

Fageria NK, Virupax C, Baligar JA (2010) Growth and mineral nutrition of field crops, 3rd edn. CRC Press, Boco Raton, p 586

Fahmy-Eman M, Nermin M, El-Gawad A, El-Geddawy IH, Saleh OM, El-Azab NM (2008) Development of RAPD and ISSR markers for drought tolerance in sugarcane (*Saccharum officinarum* L.) Egypt J Genet Cytol 37(1):1–15

Ferreira TH, Gentile A, Vilela RD, Costa GGL, Dias LI, Endres L, Marcelo M (2012) MicroRNAs associated with drought response in the bioenergy crop sugarcane (*Saccharum* spp.) PLoS One 7:e46703. doi:10.1371/journal.pone.0046703

Gentile A, Ferreira TH, Mattos RS, Dias LI, Hoshino AA, Carneiro MH, Souza DM, Calsa T, Nogueira RM, Endres L, Menossi M (2013) Effects of drought on the microtranscriptome of field-grown sugarcane plants. Planta 237(3):783–798

Gentile A, Dias LI, Mattos RS, Ferreira TH, Menossi M (2015) MicroRNAs and drought responses in sugarcane. Front Plant Sci 6:58. (1-13)

Guerzoni JTH, Geraldo BN, Moreira RMP, Hoshino AA, DSilva DD, Bespalhok FJC, Esteves VLG (2014) Stress-induced Δ1-pyrroline-5-carboxylate synthetase (P5CS) gene confers tolerance to salt stress in transgenic sugarcane. Acta Physiol Plant 36(9):2309–2319

Guimarães ER, Mutton MA, Mutton MJR, Ferro MIT, Ravaneli GC, Silva JA (2008) Free proline accumulation in sugarcane under water restriction and spittlebug infestation. Sci Agric 65:628–633

Gunshiam U, Cheong RN, Gillabel J, Merckx R (2014) Effect of conventional versus mechanized sugarcane cropping systems on soil organic carbon stocks and labile carbon pools in Mauritius as revealed by ^{13}C natural abundance. Plant Soil 379:177–192

Gupta V, Raghuvanshi S, Gupta A, Saini N, Gaur A, Khan MS, Gupta RS, Singh J, Duttamanjumdar SK, Srivastava S, Suman A, Khanna JP, Kapur R, Tyagi AK (2010) The water-deficit stress- and red-rot-related genes in sugarcane. Funct Integr Genomics 10:207–214

Hartt CE (1940) The synthesis of sucrose by excised blades of sugarcane: time and temperature. Hawaiian Planter's Rec 44:89–116

Hemaprabha G, Simon S (2010) Identification of two new drought specific candidate genes in sugarcane (*Saccharum spp.*) Electron J Plant Breed 14:1164–1170

Hemaprabha G, Simon S (2012) Genetic diversity and selection among drought tolerant genotypes of sugarcane using microsatellite markers. Sugar Tech 14(4):327–333

https://en.wikipedia.org/wiki/Carbon_sequestration, 21.08.2015

http://hyperphysics.phy-astr.gsu.edu/hbase/biology/phoc.html, 13.05.2016

Iskandar HM, Casu RE, Fletcher AT, Schmidt S, Xu J, Maclean DJ, Manners JM, Bonnett GD (2011) Identification of drought-response genes and a study of their expression during sucrose accumulation and water deficit in sugarcane culms. BMC Plant Biol 11:12–26

Iskandar HM, Widyaningrum D, Suhandono S (2014) Cloning and characterization of *P5CS1* and *P5CS2* genes from *Saccharum officinarum* L. under drought stress. J Trop For Sci 1(1):232–230

Jackson PA (2005) Breeding for improved sugar content in sugarcane. Field Crops Res 92:277–290

Jangpromma NSK, Lomthaisong K, Daduang S, Jaisil P, Thammasirirak S (2010) A proteomics analysis of drought stress-responsive proteins as biomarker for drought-tolerant sugarcane cultivars. Am J Biochem Biotech 6(2):89–102

Jayabal V, Chockalingam S (1990) Studies on drought management in sugarcane. Co-op Sugar 21(8):571–573

Julien MHR, Irvine JE, Benda GTA (1989) Sugarcane anatomy, morphology and physiology. In: Recaud C, Egan BT, Gillaspie AG, Hughes CG (eds) Diseases of sugarcane: major diseases. Elsevier, Amsterdam, pp 1–17

Kanzaria MV, Patel MV (1985) Soils of Gujarat and their management, in soils of India and their management. Fertilizer Association of India, New Delhi, pp 103–129

Kathiresan G, Balasubramanian N (1991) Management of drought in sugarcane during early growth phase. Indian Sugar 41(5):319–324

Khan IA, Bibi S, Yasmin S, Khatri A, Seema N (2013) Phenotypic and genotypic diversity investigations in sugarcane for drought tolerance and sucrose content. Pak J Bot 45(2):359–366

Kido EA, Neto JRCF, de-Oliveira-Silva RL, Pandolfi V, Guimarães ACR, Veiga DT, Chabregas SM, Crovella S, Benko-Iseppon AM (2012) New insights in the sugarcane transcriptome responding to drought stress as revealed by supersage. Sci World J 2012:1–14. doi:10.1100/2012/821062

Krishnamurthi M (1989) Sugarcane genetic resources and their utilization through conventional and unconventional approaches. In: Naidu KM, Sreenivasan TV, Premchandran MN (eds) Sugarcane Varietal Improvement. Sugarcane Breeding Institute, Coimbatore, pp 163–176

Kumar T, Khan MR, Abbas Z, Ali GM (2014) Genetic improvement of sugarcane for drought and salinity stress tolerance using arabidopsis vacuolar pyrophosphatase (AVP1) Gene. Mol Biotechnol 56:199–209

Kurama EE, Fenille RC, Rosa VE, Rosa DD, Ulian EC (2002) Mining the enzymes involved in the detoxification of reactive oxygens species (ROS) in sugarcane. Mol Plant Path 3(4):251–259

Lal C, Yadav BP, Jaiswal CS (2000) Role of drainage and challenges in 21st century. In Proc. 8th ICID International Drainage Workshop, New Delhi, vol I, pp277–288

Levin DA (1983) Polyploidy and novelty in flowering plants. Am Nat 122(1):1–24

Lewis WH (ed) (1980) Polyploidy: biological relevance. Plenum Press, New York, p 583

Li Y, Solomon S (2003) Ethephon- a versatile growth regulator for sugarcane industry. Sugar Tech 5(4):213–233

Lin S, Chen T, Qin X, Wu H, Khan MA, Lin W (2014) Identification of microRNA families expressed in sugarcane leaves subjected to drought stress and targets thereof. Pak J Agric Sci 51(4):925–934

Manoharan ML, Duraisamy K, Vijayaraghavan H (1990) Effect of management practices to improve cane yield and quality under moisture stress conditions. Bharatiya Sugar 15(10):19–31

Markad NR, Kle AA, Pawar BD, Jadhav AS, Patil SC (2014) Molecular characterization of sugarcane (*Saccharum officinarum* L.) genotypes in relation to salt tolerance. The Bio Scan 9(4):1785–1788

Marshall A (2014) Drought –tolerant varieties begin global march. Nat Biotechnol 32(4):308

Mathur RN, Haider IM (1940) A summary of five years of physiological research on sugarcane at the sugarcane Research Station, Shahjahanpur. Proc Intern Soc Sugar Cane Tech 9:11–26

McQualter RB, Dookun-Saumtally A (2007) Expression profiling of abiotic-stress-inducible genes in sugarcane. Proc Aust Soc Sugar Cane Tech 29:878–886

Menossi M, Silva-Filho MC, Vincentz M, Van-Sluys MA, Souza GM (2008) Sugarcane functional genomics: gene discovery for agronomic trait development. Intern J Plant Genomics, 2008:1–11, Article ID 458732, 1–11. doi:10.1155/2008/458732

Ming R, Moore PH, Wu KK, D'Hont A, Glaszmann JC, Tew TL (2006) Sugarcane improvement through breeding and biotechnology. Plant Breed Rev 27:15–118

Molinari HBC, Marur CJ, Daros E, Campos MKF, Carvalho JFRP, Bespalhok FJC, Pereira LFP, Vieira LGE (2007) Evaluation of the stress-inducible production of proline in transgenic sugarcane (*Saccharum* spp.): osmotic adjustment, chlorophyll fluorescence and oxidative stress. Physiol Plant 130:218–229

Naidu KM, Srinivasan TR, Raj SM (1983) (c.f. Shrivastava and Srivastava, 2012)

Nair NV (2011) Sugarcane varietal development programmes in India: an overview. Sugar Tech 13(4):275–280

Nayak N, Das PK, Mahapatra SS, Jena BC (1997) Recent agro-techniques for waterlogged sugarcane in Orissa. Indian Sugar 47(4):265–269

Nieman RH, Clark RH (1976) Interactive effects of salinity and phosphorus nutrition on the concentration of phosphate and phosphate esters in mature photosynthesizing corn leaves. Plant Physiol 57:157–161

Nogueira F, Rosa TS, Vicente E, Menossi M, Ulian EC, Arruda P (2003) RNA expression profiles and data mining of sugarcane response to low temperature. Plant Physiol 132(4):1811–1824

Page RE, Glanville TJ, Truong PN (1986) (*c.f.* Shrivastava and Srivastava, 2012)

Pandey U (1964) Growth and biochemical studies of flood survived canes and varietal resistance to flood injury. Proc Bienn Conf SRDWI 5:340–344

Panje RR, Gill PS, Alam M (1966) Effect of soil cover and trash veins on the productivity of alkaline soils for sugarcane culture. Indian Sugar J 10(1):1–8

Parameswaran P, Ramakrishnan MS (1987) Measures to alleviate drought conditions in sugarcane due to canal closure in wet lands. Indian J Sugarcane Technol 4(2):113–118

Park J-W, Benatti TR, Marconi T, Yu Q, Solis-Gracia N, Mora V, Silva JA (2015) Cold responsive gene expression profiling of sugarcane and *Saccharum spontaneum* with functional analysis of a cold inducible *Saccharum* homolog of NOD26- like intrinsic protein to salt and water stress. PLoS One 10(5):e0125810. doi:10.1371/journal.pone.0125810

Parr JF, Sullivan LA (2005) Soil carbon sequestration in phytoliths. Soil Biol Biochem 37(1):117–124

Parr JF, Sullivan LA (2007) Sugarcane-the champion crop at carbon sequestration. Aust Cane Grower 17:14–15

Parr JF, Sullivan LA, Quirk R (2009) Sugarcane phytoliths: encapsulation and sequestration of a long lived carbon fraction. Sugar Tech 11(1):17–21

Parthasarathy SV (1972) Sugarcane in India. KCP Ltd., Madras, p 804

Patade VY, Suprasanna P (2010) Short term salt and PEG stresses regulate expression of microRNA, miR159 in sugarcane leaves. J Crop Sci Biotechnol 13(3):177–182

Patade VY, Suprasanna P, Bapat VA (2008) Effects of salt stress in relation to osmotic adjustment on sugarcane (*Saccharum officinarum* L.) callus cultures. Plant Grow Regul 55:169–173

Patade VY, Bhargava S, Suprasanna P (2012) Transcript expression profiling of stress responsive genes in response to short-term salt or PEG stress in sugarcane leaves. Mol Biol Rep 39:3311–3318

Patel B, Shrivastava PK, Lad AN, Raman S (2000) Solution of waterlogging and salinity problems through subsurface drainage system a case studying. In: Role of drainage and challenges in 21st century, vol I. Proc. 8th ICID International Drainage Workshop, New Delhi, pp 203–209

Patil ND, Ghonsikar CP (1985) Soil of Maharashtra and their management. In: Soils of India and their management. Fertilizer Association of India, New Delhi, pp 250–265

Perez ZF, Romero ER, Scandaliaris J, Sotillo S (1996) Effect of the drought of 1995 on growth and expected yield of sugarcane. Adv Agroind (Argentina) 16(64):8–11

Perur NG, Mithyantha MS (1985) Soils of Karnataka and their management. In: Soils of India and their management. Fertilizer Association of India, New Delhi, pp 177–207

Prabu GR, Kawar PG, Pagariya MC, Prasad DT (2011) Identification of water deficit stress upregulated genes in sugarcane. Plant Mol Biol Report 29:291–304

Prasada-Rao KK (1989) Land marks in sugarcane research at Anakapalle. In: Prasada-Rao KK (ed) Platinum Jubilee souvenir (1913–1988). Regional Agricultural Research Station, Anakapalle, pp 21–31

Premachandran MN, Prathima PT, Lekshmi M (2011) Sugarcane and polyploidy: a review. J Sugarcane Res 1(2):1–15

Priji PJ, Hemaprabha G (2015) Sugarcane specific drought responsive candidate genes belonging to ABA dependent pathway identified from basic species clones of *Saccharum* sp. and *Erianthus* sp. Sugar Tech 17:130–137

Rajkumar AS, Kambar NS (1999) Drought management in sugarcane. Bharatiya Sugar 24(9):6–9

Ramakrishan SR, Raju JSN, Rao KV, Padmanabham M, Rampandu S (1988) Effect of spacing and nitrogen levels on mineral nutrition, yield and juice quality of late planted (June) rainfed sugarcane. DSTA, 38th Convention, A-43-53, Deccan Sugar Technologists Association, Pune

Raman K, Bhat SR, Tripathi BK (1985) Ratooning ability of sugarcane genotypes under late harvest conditions. Indian Sugar 35:445–448

Ramana TCR, Sreenivasan TV, Palanichami K (1985) Catalogue on Sugarcane Genetic Resources – II *Saccharum barberi*, Jeswiet, *Saccharum sinense*, Roxb. Amend. Jeswiet, *Saccharum robustum* Brandes et Jeswiet ex., *Saccharum edule* Hassk. Grassl. Sugarcane Breeding Institute, Coimbatore

Ramiro DA, Melotto-Passarin DM, Barbosa MA, Santos FD, Gomez SGP, Massola NS, Eric EL, Carrer H (2016) Expression of Arabidopsis Bax inhibitor-1 in transgenic sugarcane confers drought tolerance. Plant Biotechnol J 14:1826–1837. doi:10.1111/pbi.12540

Roach BT, Daniels J (1987) A review of origin and improvement of sugarcane. Proc Copersucar International Sugarcane Breeding Workshop, Sao Paulo, pp 1–32

Rocha FR, Papini-Terzi FS, Nishiyama MY, Venico RJN, Vincentini R, Duarte RDC, de-Rosa VE, Vinagre F, Barsalobres C, Medeiros AH, Rodrigues FA, Ulian EC, Zingaretti SM, Galbiatti JA, Almeida RS, AVO F, Hemerly AS, Silva-Filho MC, Menossi M, Souza GM (2007) Signal transduction-related responses to phytohormones and environmental challenges in sugarcane. BMC Genomics 8:71–93

Rodrigues FA, Laia ML, Zingaretti SM (2009) Analysis of gene expression profiles under water stress in tolerant and sensitive sugarcane plants. Plant Sci 176:286–302

Rodrigues FA, Graça JP, Laia ML, Nhani-Jr A, Galbiati JA, Ferro MIT, Ferro JA, Zingaretti SM (2011) Sugarcane genes differentially expressed during water deficit. Biol Plant 55(1):43–53. ISSN 1573-8264

Rozeff N (2002) Important temperatures for sugarcane production. Sugar J 65(6):8

Rutherford RS (1989) The assessment of proline accumulation as a measure of drought resistance in sugarcane. Proc Ann Conf South African Sugar Technol Assocn 63:136–141

Sato AM, Catuchi TA, Ribeiro RV, Souza GM (2010) The use of network analysis to uncover homeostatic responses of a drought tolerant sugarcane cultivar under severe water deficit and phosphorus supply. Acta Physiol Plant 32:1145–1151

Shaik MM, Hossain MA, Khatoon MM, Nasiruddin KM (2007) Efficient transformation of stress tolerance GLY gene in transgenic tissue of sugarcane (*Saccharum officinarum* L.) Mol Biol Biotech J 5(1&2):37–40

Sharma V (2009) Identification of drought-related quantitative trait loci (QTL) in sugarcane (*Saccharum* spp.) using genic markers. PhD Dissertation, Texas A & M Univ, Texas, pp 64

Shrivastava AK, Srivastava MK (2006) Abiotic stresses affecting sugarcane: sustaining productivity. International Book Distributing Company, Lucknow, p 322

Shrivastava AK, Srivastava S (2012) Sugarcane: physiological and molecular approaches for improving abiotic stress tolerance and sustaining crop productivity. In: Tuteja N, Gill SS, Tiburcio AF, Tuteja R (eds) Improving crop resistance to abiotic stress, vol 2. Wiley-Blackwell, Weinheim, pp 885–922

Shrivastava AK, Srivastava AK, Jain R, Rai RK, Singh P (2009) Compensatory ability in sugarcane. In: Singh SB, Rao GP, Solomon S, Gopalasundaram P (eds) Sugarcane: crop production and improvement. Studium Press, Houston, pp 543–556

Shrivastava AK, Srivastava TK, Srivastava AK, Varucha-Misra SS, Singh VK, Shukla SP (2016) Climate change –induced abiotic stresses affecting sugarcane and their mitigation. ICAR-IISR, Lucknow, p 108

Silva RLO, Ferreira JRCN, Pandol V, Chabregas SM, Burnquist WL, Benko-Iseppon AM, Kido EA (2011) Transcriptomic of sugarcane osmoprotectant during drought stress. In: Vasnthaiah H (ed) Plants and environment. InTech Publishers, p 272. isbn:978-953-307-779-6

Singh S (1985) Natural inductive temperature for primordia initiation and development of floral primordia in sugarcane at Coimbatore. Phyton 45(1):85–91

Singh BN, Lal KN (1935) Limitations of Blackman's law of limiting factors and Harder's concept of relative minimum as applied to photosynthesis. Plant Physiol 10:245–268

Singh M, Ahuja RL, Khanna SS (1985) Soils of Haryana and their management. In: Soils of India and their management. Fertilizer Association of India, New Delhi, pp 130–148

Singh G, Singh OP, Singh RS, Yadav RA, Singh BB (1991) Effect of press mud and fertiliser applied on yield and quality of sugarcane in flood affected conditions. Bhartiya Sugar 16(6):27–30

Smith D (1978) Cane sugar world. Palmer Publications, New York, p 240

Sreenivasan TV, Amalraj VA, Jebadha AW (2001) Catalogue on sugarcane genetic resources IV. *Erianthus* species. Sugarcane Breeding Institute, Coimbatore, p 98

Srivastava SC, Johari DP, Gill PS (1988) Manual of sugarcane production in India. Indian Council of Agricultural Research, New Delhi, p 194

Srivastava S, Pathak AD, Gupta PS, Shrivastava AK, Srivastava AK (2012) H_2O_2-scavenging enzymes impart tolerance to high temperature induced oxidative stress in sugarcane. J Environ Biol 34:657–661

Stebbins GL Jr (1985) Polyploidy, hybridization and the invasion of new habitats. Ann Mo Bot Gard 72:824–832

Stiles W, Cocking EC (1969) An Introduction to the principles of plant physiology, 3rd edn. Methuen, London, p 633

Suman A, Singh KP, Singh P, Yadav RL (2009) Carbon input, loss and storage in sub-tropical Indian Inceptisol under multi-ratooning sugarcane. Soil Tillage Res 104(2):221–226

Tew TL (1987) New varieties. In: Heinz DJ (ed) Sugarcane improvement through breeding. Springer Verlag, New York, pp 559–594

Thiebaut F, Grativol C, Carnavale-Bottino M, Rojas CA, Tanurdzic M, Farinell L, Martienssen RA, Hemerly AS, Ferreira PCG (2012) Computational identification and analysis of novel sugarcane microRNAs. BMC Genomics 13:290–304

Trujillo LE, Menéndez C, Ochogavía ME, Hernández I, Borrás B, Rodríguez R, Coll Y, Arrieta JG, Banguela A, Ramírez R, Hernández L (2009) Engineering drought and salt tolerance in plants using SodERF3, a novel sugarcane ethylene responsive factor. Biotech Appl 26(2):168–171

Ulloa MF, Anderson DL (1991) Sugarcane cultivar response to calcium silicate slag on Everglades Histosols. Paper presented at the ASSCT Annual Meetings. New Orleans pp 111

Vicentini R, Menossi M (2007) Pipeline for macro- and microarray analyses. Braz J Med Biol Res 40(5):615–619

Wahid A, Close TJ (2007) Expression of dehydrins under stress and their relationship with water relations of sugarcane leaves. Biol Plant 51:104–109

Waltz E (2014) Beating the heat. Nature Biotech 32(7):610–613

www.isaaa.org/kc/cropbiotechupdate/,12.05.2016

Yadav RL (1987) Soil fertility improvement through legumes in sugarcane. Fert News 32(10):29–31

Zanca AS, Vicentini RV, Ortiz-Morea EA (2010) Identification and expression analysis of microR-NAs and targets in the bio-fuel crop sugarcane. BMC Plant Biol 110:260–273

Zende NA, Hapase DG (1986) Development of saline and alkaline soils in Maharashtra under sugarcane cultivation, their reclamation and management. Bahratiya Sugar 11(11):41–48

Zingaretti SM, Rodrigues FA, da-Graça JP, Pereira LM, Lourenço MV (2012) Sugarcane responses at water deficit conditions, water stress, Prof. Ismail Md. Mofizur Rahman (ed), In-Tech. Publishers, pp 312. (ISBN: 978-953-307-963-9)

Abiotic Stress Management in Fruit Crops

R.H. Laxman and R.M. Bhatt

Abstract

With nutritional benefits, fruits are important part of human diets. These crops not only play an important role in nutritional security but also offer gainful employment and enhanced income to the farmers' community. However, the abiotic stresses, encountered at critical growth stages, adversely affect their productivity. Further, climate change is likely to increase frequency, intensity, and duration of abiotic stresses. The main abiotic stresses affecting tropical fruit crops in India are the drought/water-deficit, high temperature, and salinity stresses. These stresses cause many morphological, anatomical, physiological, and biochemical changes ultimately impacting both their productivity and quality. Thus, thorough understanding of the adverse influence of abiotic stresses on different crop species is imperative for devising innovative horticultural practices for overcoming the adverse impacts. Timely intervention with appropriate adaptation strategies would help in realizing sustainable yields. Practices like providing irrigation at critical stages, adopting micro-irrigation, use of growth regulators, soil mulching, amendments, and nutrient management need to be implemented for alleviating adverse effects. The advanced irrigation methods like partial root-zone drying (PRD) are another option for limited water conditions. The inclusion of tolerant crops or cultivars and adoption of tolerant rootstocks to graft the choice cultivars would further enable the farmers to overcome adverse effects of abiotic stresses. The focus should be on developing integrated crop-specific adaptation strategies. Integration of all available adaptation options would be the most effective approach in sustaining the production and productivity of fruit crops under abiotic stresses.

R.H. Laxman (✉) • R.M. Bhatt
Division of Plant Physiology and Biochemistry, ICAR-Indian Institute
of Horticultural Research, Hessaraghatta Lake Post, Bengaluru 560089, Karnataka, India
e-mail: laxman@iihr.res.in; rmbt@iihr.res.in

© Springer Nature Singapore Pte Ltd. 2017
P.S. Minhas et al. (eds.), *Abiotic Stress Management for Resilient Agriculture*,
DOI 10.1007/978-981-10-5744-1_18

18.1   Introduction

Horticulture sector, with diverse crops, has been a driving force for nutritive diet and enhanced income in Indian agriculture. Presently, its share in the agriculture GDP is more than 30%. India has witnessed voluminous increase in horticulture production over the last few years both due to substantial increases in area and productivity. During the last decade, the annual growth in area under horticulture sector was about 2.7% and annual production 7.0%. During the year 2013–2014, the total horticulture production in the country was 283.5 million tons from 24.2 million ha area. The production of horticultural crops has outpaced the production of food grains since 2012–2013 (Anonymous 2015a). The special thrust given to horticulture sector through the National Horticulture Mission (NHM) and Horticulture Mission for North East and Himalayan States (HMNEH) has paid rich dividends. The Mission for Integrated Development of Horticulture (MIDH) was launched during the XII 5-year Plan for holistic development of the horticulture sector. These concerted efforts resulted not only in increased production for domestic consumption but also for export.

Horticulture provides employment opportunities, higher income, and nutritional security across various states of the country. The technology-led horticulture during the last two decades has made tremendous impact on production, productivity, and profitability of all horticultural crops. The Government of India has recognized horticulture crops as a means of diversification in agriculture in an eco-friendly manner through efficient use of land and natural resources. Fruits due to their nutritional benefits are highly valuable for humanity, and along with vegetables, they are part of everyday meals. Thus, fruits and vegetables contribute nearly 90% to the total horticulture production in India. Globally, India stands second in the production of fruits and vegetables (Anonymous 2015a). Fruit crops provide not only nutritional security but also livelihood security to the farmers.

In order to realize higher yields, the perfect match between climate of a region and the suitability of a particular fruit species to that region is very essential. The potential yield levels are seldom achieved due to the occurrence of various biotic and abiotic stresses. Worldwide occurrence of environmental stresses is the primary cause of crop losses, with average yield reduction by more than 50% for the major crops (Bray et al. 2000). The majority of fruit crops are peculiar mainly due to perennial nature and deep root system. They undergo vegetative and reproductive phases during different seasons. The abiotic stresses coinciding with these phenological phases play a significant role in determining the duration of phenology and productivity. Further, under climate change and climate variable conditions, fruit crops are likely to face abiotic stresses quiet frequently. Hence, under such circumstances, meeting the increasing demand for fruits becomes challenging. Realizing sustained and enhanced yields under abiotic stress situations primarily depends on implementation of appropriate adaptation strategies. Hence, the adverse effects of abiotic stresses on important fruit crops and available adaptation options are discussed in this chapter.

18.2   Impact of Abiotic Stresses

Though specific agroecological regions are sustaining the cultivation of fruit crops as niche areas, the variability in weather conditions during critical stages of crop growth and development causes heavy yield loss and affect fruit quality. Fruit crops face various abiotic stresses like high temperature, excess and limited moisture, and salinity stresses. These stresses occurring, either at intermittent or terminal stages of crop growth, in an agroecological zone play very significant role in determining phenology, growth, development, and consequently the productivity of horticultural crops. Global warming is likely to increase the frequency, intensity, and duration of excess and limited water and high temperature stresses (Bates et al. 2008). Climate change, with its influence on hydrological cycles leading to changed precipitation pattern, may affect the crop production than increases in temperature. The elevated temperatures would hasten plant transpiration and soil evaporation. These stresses either individually or in combination would significantly influence the production, productivity, and quality of fruit crops.

Environmental stresses during different developmental stages can cause morphological, anatomical, physiological, and biochemical changes (Ahmad et al. 2011). In order to develop timely and appropriate adaptation measures, a better understanding of the overall effects of abiotic stresses on fruit crops is required. Moreover, the interactions of abiotic factors with physiological processes, phenology, growth, and development are extremely important for devising innovative horticultural practices for overcoming the adverse impacts of various abiotic stresses.

18.2.1 Water Stress

Horticultural crops, due to high water requirement, are grown under assured irrigation conditions, and the water-limiting situations adversely affect these crops. However, the timing, intensity, and duration determine the scale of water stress effects. In mango appearance of vegetative flushes is greatly reduced during water stress period. The water stress also causes reduction in number of leaves in a flush, the flush length, and leaf water contents. In mango water stress also plays an important role in induction of flowering mainly through its influence on floral stimulus produced by mature leaves. Under tropical conditions, even though the prevailing temperatures are not as low, water stress for a brief period induces flowering (Scholefield et al. 1986). Through its inhibitory influence on vegetative flushing, water stress may provide more time for accumulation of floral stimulus (Schaffer et al. 1994). The advancement of floral bud break by nearly 2 weeks and floral bud growth and postponement in development of vegetative buds were observed under water stress (Whiley 1986; Nunez-Elisea and Davenport 1994; Schaffer et al. 1994).

Another important fruit crop, grapes, encounters frequent moisture stress conditions. It undergoes several morphological and physiological changes under water stress. Grapevines are considered as relatively tolerant to water stress due to large xylem vessels in comparison to other crops (Serra et al. 2013). The roots keep on growing and exploring deeper soil layers for moisture under water-limiting conditions, but under adequate water supply, these remain confined to topsoil layer (Bauerle et al. 2008). The vines adapt to water scarcity conditions not only by enhancing root length but also reducing the shoot growth (Hardie and Martin 2000). Higher proportion of new roots was observed in different soil layers during dry and hot seasons for increasing the water uptake (Serra et al. 2013).

Field studies have shown that the bananas are quite sensitive to soil moisture stress. The carbon assimilation is affected as stomata close to conserve leaf water (Thomas 1995). Flowering stage is reported to be the most sensitive in banana. In cultivar "Elakki," the lowest yield was obtained when water stress was imposed at flower differentiation stage (Murali et al. 2005). In cv. Robusta, maximum reduction in yield was observed when stress was imposed during a 5-week period immediately after flowering (Hegde and Srinivas 1989). In different cvs. Robusta, Karpuravalli, and Rasthali, water stress at flowering stage caused reduction in bunch weight to the extent of 42.07%, 25.0%, and 18.83%, respectively (Ravi et al. 2013). In papaya under field conditions, water-deficit stress treatment caused 50% reduction in leaves and significantly reduced number of flowers by 86% and fruit by 58%. The growth and development of papaya fruit was also retarded (Masri et al. 1990).

18.2.2 High Temperature Stress

High temperature stress is of concern in tropical and subtropical areas. It causes damages like sunburns on leaves, branches, and stems, leaf senescence and abscission, shoot and root growth inhibition, and fruit discoloration and damage (Almeida and Valle 2007; Wahid et al. 2007). The high temperatures encountered at various stages of crop growth and development affect various physiological processes. The plant carbon fixation through photosynthesis would largely determine the dry matter accumulation and distribution into various plant parts. Reproductive processes are also highly affected by heat stress in most plants (Wahid et al. 2007). High temperature stress disrupts the biochemical reactions fundamental to normal cell functioning, and it primarily affects the photosynthetic functions of higher plants (Weis and Berry 1988). Mango being a tropical tree, though adapted to both tropical and subtropical climatic conditions, endures a wide range of temperatures. The prevailing temperatures determine the vegetative and flowering flushes in mango. Due to episodic vegetative flushes in a mango tree, the interaction of plant and environmental factors controls the synchronization of growth phases. Higher temperatures lead to stronger vegetative bias under sufficient nutrient and water availability (Laxman et al. 2016). Floral induction in mango is temperature dependent (Davenport 2007) and is triggered by temperatures below 16 °C (Schaffer et al. 1994). Floral induction occurred at 15 °C day and 10 °C night temperatures whereas vegetative induction at 30 °C day and 25 °C night temperatures (Whiley et al. 1989, 1991). Panicles that developed during the prevailing low temperatures usually had higher proportion of male flowers (Singh et al. 1974), and the panicles emerging late experiencing higher temperatures had higher percentage of hermaphrodite flowers (Ramaswamy and

Vijayakumar 1992), signifying that the proportion of male and hermaphrodite flowers change with the prevailing temperatures. Thus, the sudden changes in temperatures due to climate variability would influence not only the vegetative and reproductive cycles but also proportion of female flowers in the panicle, leading to effects on productivity.

In wine grapes, each cultivar grows in a suitable range of temperatures, and for each cultivar, it is possible to define climates for premium wine production (Jones 2008). The adaptability of cultivars enables the production of fruit crops over a relatively large range of climates. The high temperatures advance harvest times in grapes with higher sugar concentrations, low acidity, and alterations in aroma compounds. The extreme hot temperatures may affect wine aroma and color through the effects on metabolism (Mira de Orduna 2010). The high temperatures also affect banana growth and production. The leaf production and relative leaf area growth are affected beyond 33.5 °C. The relative growth rate and dry weight increment are sustained till 39.2 °C (Turner and Lahav 1983). Banana can relatively persist under prolonged water stress, but the combined effects of deficit soil moisture along with prolonged prevalence of temperatures beyond 35 °C can reduce banana production (Thornton and Cramer 2012). The prevailing high temperature episodes coinciding with critical phenophases would affect fruit crops to various magnitudes.

18.2.3 Salinity Stress

The area under salt-affected soils in India is 6.74 M ha with approximately 2.95 and 3.79 M ha saline and sodic soils, respectively (Anonymous 2015b). In climate change situations, the crops would further be affected by salinity stress due to accumulation of higher amounts of salts owing to high evaporation. The higher levels of chlorides and sulfates of calcium, magnesium, and sodium present in the soils adversely cause considerable damage to many crops. These dissolved salts in the root zone cause either osmotic stress to roots, or/and when taken up, the salt ions cause toxicity to plants. The accumulation of toxic ions in leaves leads to nutrient imbalance and lower uptake of major nutrients. This results in injury to leaves, inhibition of growth, lack of fruit bearing, and consequently reduces yields.

Studies have shown that the saline conditions are not favorable for successful mango cultivation. The increase in irrigation water salinity caused the reductions in N, K, Ca, and Mg contents in leaves without affecting the contents of P and S. In banana, salt stress-induced necrosis is seen first in leaf margins and subsequently spreads to inner parts of the leaf. The salinity stress causes reduction in pseudostem thickness, delayed flowering, reduced finger size, and low-quality bunches (Ravi and Vaganan 2016). In grapes, also many physiological parameters, growth, and nutrient uptake are affected under salinity stress (Bybordi 2012). The papaya seedling growth was not affected at 2 dS m^{-2}, growth was reduced by 50 per cent at 4 dS m^{-2}, and mortality occurred at salinity levels >6 dS m^{-2} (Makhija and Jindal 1983). Therefore, fruit crops respond differently to salinity stress and are affected to various degrees at different levels of salinity.

18.3   Management Options

Successful cultivation of crops and attaining reasonable yields under abiotic stress situations mainly depends on the available adaptation options. The adaptation efforts would enable us to channelize concerted efforts for the holistic development of horticulture sector empowering marginal and small farmers. Majority of the fruit crops are perennial, possess deep root system, and undergo vegetative and reproductive phases during different seasons in a year. The long time horizon of perennial horticulture crops itself is a challenge. The quick adaptation strategies, like switching over to tolerant cultivars and changing planting dates or season, followed in annual crops are not likely in perennial fruit crops. Hence, the choice of fruit crops should be guided by the suitability of a crop species and their varieties in a particular location. The planting and rearrangement of fruit orchards require long-term consideration. In addition, the preference for a choice variety of fruit delays adoption of a new cultivar than an annual crop. Even the perennial habit slows the process of developing new varieties and limits the options for shifting varieties (Koski 1996). Thus, with these limitations, the adaptation of new varieties of fruit crops takes time and requires long-term considerations. Hence, the abiotic stresses in perennial crops need to be managed mainly through alterations in cultivation practices.

18.3.1 Modification in Cultural Practices

The alterations in cultivation practices help in effective management of abiotic stresses in perennial crops. In mango better tree growth, fruit retention, and fruit size and higher yields are realized under irrigation. The fruit size and yield increased in field-grown mango cv. Hindi with more frequent irrigation (Azzouz et al. 1977). The bigger size fruits and highest yield were obtained with 7-day irrigation interval (Larson et al. 1989). Weekly irrigation during the first 6 weeks in cv. Dashehari reduced fruit drop compared with three weekly irrigation treatments (Singh and Arora 1965). Irrigation during the initial 4–6 weeks succeeding the fruit set is crucial for attaining better fruit size and yield, as cell division and cell wall development take place at initial stage (Whiley and Schaffer 1997). Since the occurrence of water stress immediately after fruit set increases fruit drop, protective irrigation is essential during the fruit development period (Anonymous 2014). Thus, providing irrigation at least during post-fruit set period under water-limiting conditions is very important for realizing sustainable yields.

Under water-deficit conditions, production practices that can improve water use efficiency would help in water saving and bringing more area under irrigation. Plastic mulching helps in reducing soil water evaporation and rainwater impact and provides effective weed control and congenial environment for soil microflora. The use of mulching enhances the production and quality of produce under water-limiting conditions. The production system employing drip irrigation, fertigation, and plastic mulching would help in realizing higher yields. During initial establishment of mango plants in the field, under water scarcity situations, application of 1.25 l of subsoil irrigation per day through pitcher placed one foot below ground and mulching with sugarcane thrash at 1.0 kg basin^{-1} is suggested for better establishment. Rainwater harvesting through opening of circular trenches around trees at a distance of 6 ft and 9 in. width and depth and mulching the trenches with dry mango leaves helps in retaining sufficient moisture in the soil during flowering and fruiting, resulting in higher yield (Anonymous 2014).

In bananas, genotypes belonging to *Musa balbisiana* group are reported to be tolerant to abiotic stresses. In regions with limited water availability, the cultivars like "Saba" (ABB), Monthan, (ABB), Karpuravalli (ABB), and Poovan (AAB) which are relatively drought tolerant than other cavendish clones like "Grand Naine" and "Robusta" could be an option (Ravi and Vaganan 2016). Under water-deficit conditions, 0.1 mM salicylic acid foliar spray at 250 ml plant^{-1} is suggested for banana. Foliar application of kaolinite (5%) during vegetative stage reduces the water loss through transpiration. Basin mulching either with black polythene or with plant residues can be followed to reduce water loss. Application of 5 kg rice husk ash or composted coir pith in the pit at the time of planting is recommended to increase the water-holding capacity in the rhizosphere (Anonymous 2014).

In grapes, providing subsurface irrigation and recommended surface drip irrigation helped in saving 46.8 and 25.9% irrigation water, respectively. The irrigation water saving of 19.1–26.6% is achieved through partial root-zone drying (PRD) technique (Anonymous 2017). Mulching with 400 mm-thick black polyethylene film enhances water use efficiency and crop yield. The application of bagasse at the rate of 10 tons per acre (3″ thick layer covering root-zone strip) and spraying of Anti-stress (permitted biodegradable acrylic polymer) 4–6 ml L^{-1} at 30, 60, and 90 days after foundation pruning and 30 and 60 days after fruit pruning resulted in reduction of 25% irrigation water requirement (Adsule et al. 2013). Crop residue recycling is suggested to build organic carbon reserves to improve soil health and water-holding capacity to cope with dry spells. Salt accumulation in the root zone due to excessive irrigation could be overcome through leaching by impounding excess water and use of soil amendments. Practices like incorporation of crop residues and green manure improves soil organic matter status and inturn soil structure and moisture-holding capacity. In situ moisture conservation using organic or inorganic mulches could be practiced. Intercultural operations need to be reduced to minimize the moisture loss. The measures like contour cultivation and zero tillage help in situ soil moisture conservation. The establishment of farm ponds to harvest the runoff water and utilizing the same to provide protective irrigation in the prolonged dry spell period could be explored.

In salt-affected soils, planting of cooking banana cultivar Saba could be taken up (Ravi and Vaganan 2016). The crop management practices that enhance soil physical properties and nutrient and water availability in the active root zone would be advantageous. The banana cultivar Nendran, which is sensitive to high salinity, showed no salt injury symptoms and recorded highest bunch weight of 10.33 kg when applied with 2 kg Gypsum + 15 kg FYM with 120% K (450 g plant^{-1}). The application of amendment along with higher potassium helped in maintaining higher leaf K:Na ratio and alleviation of sodium salt stress in banana (Jeyabaskaran et al. 2000).

18.3.2 Micro-irrigation

Systematic irrigation scheduling enhances water productivity largely because of improved efficiency and timing of water applications. Through the precise and direct application of water in root zone through drip irrigation, better crop growth and yield can be realized along with considerable savings in water. Drip irrigation method enables judicious use of available irrigation water in fruit orchards. Overall it saves up to 30–70% irrigation water and also helps in realizing higher yields by 25–80% (Sikka and Samra 2010). Various studies have shown that the adoption of micro-irrigation systems increased yield. And productivity of fruit increases by 42.3%. This resulted in improved water use efficiency, and an average irrigation cost has been brought down by 31.9%. This has helped the farmers to introduce new crops. However, in India, only around 8 Mha is under micro-irrigation, but the estimated potential for micro-irrigation is around 69 Mha. Hence there is great potential for adaptation of this technology (Anonymous 2016).

Micro sprinkler irrigation not only helps in water saving to the tune of 20 to 30%, but during summer it helps in reducing temperature in the microclimate and increases the humidity, leading to better growth and yield. Micro-irrigation, because of high cost and intensive management constraints, presently is adopted in few crops. However, it offers a great perspective for water savings due to its advantage of precise application of water at the root zone, and also it is an extremely flexible irrigation method. It could be adapted to almost any crop production situation and climatic conditions. In situations where limited water is available, providing irrigation during critical stages of the crop growth like active growth, flowering, and fruit enlargement is very essential. Micro-irrigation helps in achieving this feat and conservation of water under water-limiting conditions. Hence, appropriate management strategies are needed to solve the production problems.

18.3.3 Adopting Novel Irrigation Methods

In addition to drip irrigation and mulching, novel irrigation methods, like partial root-zone drying (PRD), could be adapted for production of fruit crops under water scarcity conditions. Partial root-zone drying is an irrigation water application technique alternating from one side of the plant to the other. This system purposefully imposes water stress to the plants at specific growth stages by providing limited amounts of plant's daily water use. The production of ABA hormone and other chemical signals in the drying roots presumably reduces stomatal conductance and leaf growth (Gowing et al. 1990) thereby increasing water use efficiency. In mango orchards where water is a limiting factor, PRD may be the key for a sustainable production (Spreer et al. 2007). A frequent response of fruit trees to deficit irrigation (DI) is earliness in flowering. This stress-induced flowering is often explained in terms of a lesser resource competition with vegetative growth effectively restrained by water deficit in evergreen and deciduous fruit trees (Behboudian and Mills 1997). This tree response to DI has been successfully exploited to induce out of season blooming and to increase the levels of flowering in many tropical and subtropical fruit crops. The average yields of 4 years were 83.3, 80.1, 80.8, and 66.1 kg tree^{-1} when irrigated equal to ETc, RDI (regulated deficit irrigation 0.5 ETc), PRD (partial root-zone drying, 0.5ETc), and nonirrigated control, respectively. Further RDI and PRD were at par during normal rainfall years while PRD outyielded RDI during deficient rainfall years. The trees receiving PRD also bore bigger size fruits (Spreer et al. 2009).

Papaya can tolerate certain water deficit without substantial yield reduction. A 30% water deficit induced through PRD and RDI water supply technique did not significantly affect vegetative growth and yield components as compared to full irrigation. Lima et al. (2015) confirmed that PRD technique improved papaya WUE through lower stomatal conductance without affecting the photosynthesis and growth characteristics. Subsurface irrigation in papaya also enhanced water use efficiency (Srinivas 1996). In grape cultivar Thompson Seedless, subsurface irrigation (10 cm depth) produced not only higher yield (12.4 vs. 8.1 t ha^{-1} with surface drip) but also water productivity (28.9 vs. 18.8 kg grapes mm^{-1}) (Sharma et al. 2005). Thus, a shift to these irrigation methods should help in substantial savings in irrigation water.

18.3.4 Choice of Tolerant Rootstocks

In situations where there is a strong consumer preference for a select cultivar that is susceptible and if alternative tolerant cultivars are not available, the option of using rootstocks for better performance needs to be explored. Rootstocks with better root system, having capacity for enhanced water and nutrient uptake, could be used for grafting commercial cultivars to mine water from deeper soil layers. In grapes, cv. Pinot Noir on "101–14 Mgt" rootstock had higher CO_2 assimilation, transpiration rate, and higher water use efficiency than on "3309C" (Candolfi-Vasconcelos et al. 1994). Rootstocks, 110R, 99R, and 1103P, belonging to *Vitis berlandieri* x *Vitis rupestris* crosses, show increased water use efficiency (Satisha et al. 2006, 2007). The rootstocks such as Dogridge, Salt Creek, and *Vitis champini* clone showed tolerance with reduced cytokinin level and increased ABA accumulation at 50% moisture stress compared to irrigated vines (Satisha et al. 2007). Rootstocks 110R, Dogridge, Salt Creek, and B2-56 though enhanced sugar and other compatible solute accumulation showed both moisture and salinity stress tolerance (Jogaiah et al. 2014). Mango rootstocks exhibited differential response during water stress. The cultivars Starch, Peach, and Kensington required 7–9 days, whereas Mylepelian took 16 days to reach negligible photosynthesis rates during water stress. But upon rewatering, Starch, Mylepelian, Peach, and Kensington recovered in 1, 2, 3, and 4 days, respectively (Laxman 2015). Thus, availability of different suitable rootstocks enhances the farmers' ability to manage abiotic stresses.

Although saline conditions have adverse effects on plant height, number of leaves, leaf area, and stem thickness (Ahmed and Ahmed 1997), the mono- and polyembryonic mango cultivars display differential tolerance to salinity stress.

Studies have demonstrated that the polyembryonic genotypes appear to have greater tolerance to salinity compared to monoembryonic types (Jindal et al. 1975). The polyembryonic cultivars Kurakkan and 13-1 have shown tolerance to salinity stress. Among the different polyembryonic mango rootstocks evaluated, Olour and Kurukkan have been identified as tolerant to salt stress (Dubey et al. 2006; Srivastav et al. 2009). The polyembryonic rootstock cvs. Bappakai and Olour (Anonymous 1989) have also been identified as moderately tolerant to salt stress. The Gomera-1 having ability to restrict uptake and transport of Cl and Na ions from root system to the aboveground parts is identified as tolerant to salinity stress (Duran-Zuazo et al. 2003). The tolerant seedlings exhibit physiological tolerance to chloride ion concentrations in leaf tissues.

The rootstocks have the ability to either restrict sodium uptake or, once taken up, sequester sodium in vacuole or older leaves (Paranychianakis and Angelakis 2007). The adverse effects of salinity stress could be successfully alleviated by using tolerant rootstocks. The rootstock Dogridge followed by Salt Creek showed least mortality under saline conditions (8 dS m^{-1}, Yohannes 2006). The cv. Thompson Seedless, grown extensively in India for both domestic consumption and export, when grafted on 110R (*Vitis berlandieri* x *Vitis rupestris*) rootstock showed lower accumulation of sodium ions and sustained the yield over a period of time (Satisha et al. 2010; Sharma and Upadhya 2008). It is also reported that the Thompson Seedless vines, on rootstock 110R, exhibited not only early and uniform sprouting but also increased fruitfulness under saline water (1.8 dS m^{-1}) irrigation (Jogaiah et al. 2013). Vines grafted on B2-56 rootstock which is a clone of 110R accumulated considerably lower Na concentration (Sharma and Upadhya 2008). The rootstocks 1103P and 110R also have shown restriction in uptake of both sodium and chloride in grapevines (Sharma et al. 2011). Hence, grape orchards could be successfully established and grown in salt-affected soils by employing tolerant rootstocks as one of the strategies to overcome the adverse effects.

18.3.5 Choice of Tolerant Crops

In areas where the crops perennially face water and high temperature stresses, the knowledge should be shared with farmers on fruit crops which would be most suitable. In such circumstances, the selection of appropriate fruit species becomes very important. Many fruit crops are endowed with physiological and morphological adaptations and have capacity to withstand adverse effects of water stress. Leaf hairiness, hypostomatous distribution, and sunken stomata are all characteristic features of species that exist in drought-prone regions (Clifford et al. 2002). Ber sheds leaves to avoid extreme water stress during summer. Pomegranate is fairly winter hardy and tolerant to water-deficit and high temperature stresses. It tolerates concentrations up to 40 mM NaCl in irrigation water (Naeini et al. 2006). Fig has adopted to retain high-bound water in the tissue, by having sunken stomata, thick cuticle, and leaf wax coating. Aonla, being a hardy and drought-tolerant subtropical tree, can be grown well under tropical conditions. In salt-affected lands where cultivation of annual field crops is limited, adopting relatively tolerant crops like ber, aonla, guava, grape, karonda, jamun, and phalsa would help in utilization of such lands for horticulture. These crops could be considered as candidate crops to face the challenges of abiotic stresses under climate change conditions.

18.4 Conclusion

Though various adaptation options like cultural practices, advanced irrigation methods, tolerant crops, or varieties and rootstocks are available, the productivity of fruit crops remains low in areas experiencing abiotic stresses. The main reasons are slow pace of adoption by the small and marginal farmers, limited awareness about the potential adverse effects of abiotic stresses, dearth of agroecological zone-based perspective plans, lack of awareness about the risks associated with horticultural crops, and lack of integrated location-specific modules to overcome abiotic stresses. Therefore, focus is required for developing integrated location-specific and crop-specific adaptation strategies for various abiotic stresses. Dissemination of already available adaptation strategies can be taken up through location-specific monitoring networks and creating awareness among farmers on likely climatic risks. The timely availability of planting material of tolerant cultivars needs to be assured through proper institutional mechanism. The institutional support to provide forecast and early warnings needs to be further strengthened. Robust insurance policies linked to climatic risks of a region, recent weather extremes, and weather forecasts are very much essential.

The coping measures need to be further developed with focused research. Multidisciplinary efforts are needed to develop integrated adaptation strategies. There is an immediate necessity to enhance the genetic base through collection of wild and cultivated genotypes having tolerance to abiotic stresses. The identification of traits imparting tolerance to abiotic stresses is an important step in the process of crop improvement and development of tolerant cultivars. Development of transgenic cultivars could be an option for the genetic enhancement approaches. In the case of perennial fruit crops, development of transgenic rootstocks is the best option. Hence, the impacts of abiotic stresses on fruit crops could be overcome by adopting different strategies. An integrated approach with all available options is the most effective for sustaining production, productivity, and quality of fruit crops.

References

Adsule PG, Yadav DS, Satisha J, Sharma AK, Upadhaya A (2013) Good agricultural practices for production of quality table grapes. NRC for Grapes (ICAR), Pune
Ahmad A, Xiao-yu X, Long-chang W, Muhammad FS, Chen M, Wang L (2011) Morphological, physiological and biochemical responses of plants to drought stress. Afr J Agric Res 6(9):2026–2032
Ahmed AM, Ahmed FF (1997) Effect of saline water irrigation and cycocel on growth and uptake of some elements of Taimour and alphonso mango seedlings. Ann Agric Sci Moshtohor 35(2):901–908

Almeida AAF, Valle RR (2007) Ecophysiology of cacao tree. Braz J Plant Physiol 19:425–448

Anonymous (1989) News. Indian Inst Hortic Res 10(4):1–3

Anonymous (2014) Horticultural Advisories for rain deficient or delayed monsoons-2014, ICAR. www.icar.org.in/files/Advisories_Horticultural%20crops.pdf

Anonymous (2015a) Horticultural statistics at a glance 2015, horticulture statistics division, Department of Agriculture, Cooperation & Farmers Welfare, Ministry of Agriculture & farmers Welfare government of India. Oxford University Press, New Delhi

Anonymous (2015b) Vision 2050. Central soil salinity research Institute. Karnal, Haryana

Anonymous (2016) Accelerating growth of Indian agriculture: Micro irrigation an efficient solution. Strategy paper – Future prospects of micro irrigation in India. 2016 Grant Thornton India LLP

Anonymous (2017) Annual Report 2016–17, Department of Agricultural Research and Education-ICAR, Ministry of Agriculture and Farmers Welfare, Govt. of India

Azzouz S, El-Nokrashy MA, Dahshan IM (1977) Effect of frequency of irrigation on tree production and fruit quality of mango. Agric Res Rev 55:59

Bates BC, Kundzewicz ZW, Wu, S, Palutikof JP (2008) Climate change and water. Technical paper of the intergovernmental panel on climate change. Intergovernmental panel on climate change, IPCC Secretariat, Geneva

Bauerle TL, Smart DR, Bauerle W, Stockert CM, Eissenstat DM (2008) Root foraging in response to heterogeneous soil moisture in two grapevines that differ in potential growth rate. New Phytol 179:857–866

Behboudian MH, Mills TM (1997) Deficit irrigation in deciduous orchards. Hortic Rev 21:105–131

Bray EA, Bailey-Serres J, Weretilnyk E (2000) Responses to abiotic stresses. In: Gruissem W, Buchannan B, Jones R (eds) Biochemistry and molecular biology of plants. ASPP, Rockville, pp 1158–1249

Bybordi A (2012) Study effect of salinity on some physiologic and morphologic properties of two grape cultivars. Life Sci J 9:1092–1101

Candolfi-Vasconcelos MC, Koblet W, Howell GS, Zweifel W (1994) Influence of defoliation, rootstock, training system, and leaf position on gas exchange of pinot noir grapevines. Am J Enol Vitic 45:173–180

Clifford SC, Arndt SK, Popp M, Jones HG (2002) Mucilages and polysaccharides in Ziziphus species (Rhamnaceae): localization, composition and physiological roles during drought- stress. J Exp Bot 53:131–138

Davenport TL (2007) Reproductive physiology of mango. Braz J Plant Physiol 19(4):363–376

Dubey AK, Srivastav M, Singh R, Pandey RN, Deshmukh PS (2006) Response of mango (*Mangifera indica*) genotypes to graded levels of salt stress. Indian J Agric Sci 76:670–672

Duran-Zuazo VH, Martinez-Raya A, Aguilar Ruiz J (2003) Salt tolerance of mango rootstock. Span J Agric Res 1:67–78

Gowing DJG, Davies WJ, Jobes HG (1990) A positive root sourced signal as an indicator of soil drying in apple, malus x domestica-Borkh. J Exp Bot 41:1535–1540

Hardie WJ, Martin SR (2000) Shoot growth on de-fruited grapevines: a physiological indicator for irrigation scheduling. Aust J Grape Wine Res 6:52–58

Hegde DM, Srinivas K (1989) Yield and quality of banana in relation to post-flowering moisture stress. South Indian Hort 37(3):131–134

Jeyabaskaran KJ, Pandey SD, Laxman RH (2000) Studies on reclamation of saline sodic soil for banana (cv. Nendran). Proceedings of the international conference on managing natural resources for sustainable agricultural production in the 21st century, New Delhi, vol. 2: 357–360

Jindal PC, Singh JP, Gupta OP (1975) Screening of mango seedlings for salt tolerance. Haryana J Hortic Sci 4:112–115

Jogaiah S, Ramteke SD, Sharma J, Upadhyay AK (2014) Moisture and salinity stress induced changes in biochemical constituents and water relations of different grape rootstock cultivars. Int J Agron. doi:http://dx.doi.org/10.1155/2014/789087

Jogaiah S, Oulkar DP, Banerjee K, Sharma J, Patil AS, Maske SR, Somkuwar RG (2013) Biochemically induced variations during some phonological stages in Thompson Seedless grapevines grafted on different rootstocks. S Afr J Enol Vitic 34:36–45

Jones GV (2008) Climate change and the global wine industry. In: Rae Blair Pat Williams (eds) Proceedings of the 13th AWITC, Sakkie Pretorius, pp 91–98

Koski V (1996) Breeding plans in case of global warming. Euphytica 92(1–2):235–239

Larson KD, Schaffer B, Davies FS (1989) Effect of irrigation on leaf water potential, growth and yield of mango trees. Proc Fla State Hortic Soc 102:226–228

Laxman RH (2015) Unpublished data. ICAR-Indian Institute of Horticultural Research, Bengaluru

Laxman RH, Annapoornamma CJ, Geeta Biradar (2016) Mango. In: Srinivasa Rao NK, Shivashankara KS, Laxman RH (eds). Abiotic stress physiology of horticultural crops. Springer, pp 169–181

Lima RSM, Figueiredoa FAMMA, Martinsa AO, Deusa BCS, Ferraza TM, Gomesa MMA, Sousab F, Glennc DM, Campostrini E (2015) Partial rootzone drying (PRD) and regulated deficit irrigation (RDI) effects on stomatal conductance, growth, photosynthetic capacity, and water-use efficiency of papaya. Sci Hortic 183:13–22

Makhija M, Jindal PC (1983) Effect of different soil salinity levels on seed germination and seedling growth in papaya (*Carica papaya*). Seed Res 11:125–128

Masri M, Razak AS, Ghazalli MZ (1990) Response of papaya (*Carica papaya* L.) to limited soil moisture at reproductive stage. Mardi Res J 18:191–196

Mira de Orduna R (2010) Climate change associated effects on grape and wine quality and production. Food Res Int 43:1844–1855

Murali K, Srinivas K, Shivakumar HR, Kalyanamurthy KN (2005) Effect of soil moisture stress at different stages on yield and yield parameters of "Elakki" banana. Adv Plant Sci 18(2):817–822

Naeini MR, Khoshgoftarmanesh AH, Fallahi E (2006) Partitioning of chlorine, sodium, and potassium and shoot growth of three pomegranate cultivars under different levels of salinity. J Plant Nutr 29:1835–1843

Nunez-Elisea R, Davenport TL (1994) Flowering of mango trees in containers as influenced by seasonal temperature and water stress. Sci Hortic 58:57–66

Paranychianakis NV, Angelakis AN (2007) The effect of water stress and rootstock on the development of leaf injuries in grapevines irrigated with saline effluent. Agric Water Manag 2531:1–8

Ramaswamy N, Vijayakumar M (1992) Studies of the effects of flowering and fruiting behaviour of south Indian mango cultivars in abstract IV. International Mango Symposium, Miami Beach, p 47

Ravi I, Vaganan MM (2016) Abiotic Stress Tolerance in Banana. In: Rao NKS, Shivashankara KS, Laxman RH (eds.). Abiotic Stress Physiology of Horticultural Crops. Springer India, pp 207–222

Ravi I, Uma S, Vaganan MM, Mustaffa MM (2013) Phenotyping bananas for drought resistance. Front Phys 4:Article 9. doi:10.3389/fphys.2013.00009

Satisha J, Prakash GS, Murti GSR, Upreti KK (2006) Response of grape rootstocks to soil moisture stress. J Hortic Sci 1:19–23

Satisha J, Prakash GS, Murti GSR, Upreti KK (2007) Water stress and rootstocks influences hormonal status of grafted grapevines. Eur J Hortic Sci 72:202–205

Satisha J, Somkuwar RG, Sharma J, Upadhyay AK, Adsule PG (2010) Influence of rootstock on growth, yield and fruit composition of Thompson Seedless grown in the Pune region of India. S Afr J Enol Vitic 31:1–8

Schaffer B, Whiley AW, Crane JH (1994) Mango. In: Schaffer B, Andersen PC (eds) Handbook of environmental physiology of fruit crops, vol 2, Sub tropical and tropical crops. CRC Press, Boca Raton, pp165–197

Scholefield PB, Oag DR, Sedgley M (1986) The relationship between vegetative and reproductive development in mango in northern Australia. Aust J Agric Res 37:425–433

Serra I, Strever A, Myburgh P, Deloire A (2013) Review: the interaction between rootstocks and cultivars (*Vitis vinifera* L.) to enhance drought tolerance in grapevine. Aust J Grape Wine Res. doi:10.1111/ajgw.12054

Sharma J, Upadhya AK (2008) Rootstock effect on Tas A-Ganesh (*Vitis vinifera* L.) for sodium and chloride uptake. Acta Hortic 785:113–116

Sharma J, Upadhyay AK, Adsule PG (2005) Effect of drip water application at sub-surface on grapevine performance- a case study. J Appl Hortic 7(2):137–138

Sharma J, Upadhyay AK, Bande D, Patil SD (2011) Susceptibility of Thompson Seedless grapevines raised on different rootstocks to leaf blackening and necrosis under saline irrigation. J Plant Nutr 34:1711–1722

Sikka AK, Samra JS (2010) Rainfed horticulture perspective and priorities. In: Chadha KL, Singh AK, Patel VB (eds) Recent initiatives in horticulture. Horticultural Society of India, New Delhi

Singh RN, Arora KS (1965) Some factors affecting fruit drop in mango (*Mangifera indica* L.) Indian J Agric Sci 35:196

Singh RN, Majumder PK, Sharma DK, Sinha GC, Bose PC (1974) Effect of de-blossoming on the productivity of mango. Sci Hortic 2:399–403

Spreer W, Nagle MC, Neidhart S, Carle R, Ongprasert S, Mueller J (2007) Effect of regulated deficit irrigation and partial root zone drying on the quality of mango fruits (*Mangifera indica* L., cv. 'Chok Anan'). Agric Water Manag 88(1–3):173–180

Spreer W, Ongprasert S, Hegele M (2009) Yield and fruit development in mango (*Mangifera indica* L. cv. Chok Anan) under different irrigation regimes. Agric Water Manag 96:574–584

Srinivas K (1996) Plant water relations, yield, and water use of papaya (*Carica papaya* L.) at different evaporation-replenishment rates under drip irrigation. Trop Agric 73:264–269

Srivastav M, Dubey AK, Singh AK, Singh R, Pandey RN, Deshmukh PS (2009) Effect of salt stress on mortality, reduction in root growth and distribution of mineral nutrients in Kurukkan mango at nursery stage. Indian J Hortic 66:28–34

Thomas DS (1995) The influence of the atmospheric environment and soil drought on the leaf gas exchange of banana (*Musa* spp). PhD thesis, The University of Western Australia

Thornton P, Cramer L (2012) Impacts of climate change on the agricultural and aquatic systems and natural resources within the CGIAR's mandate. CCAFS Working Paper 23. CGIAR Research Program on Climate Change, Agriculture and Food Security (CCAFS), Copenhagen, Denmark

Turner DW, Lahav E (1983) The growth of banana plants in relation to temperature. Aust J Plant Physiol 10:43–53

Wahid A, Gelani S, Ashraf M, Foolad MR (2007) Heat tolerance in plants: an overview. Environ Exp Bot 61:199–223

Weis E, Berry JA (1988) Plants and high temperature stress. Soc Exp Biol 42:329–346

Whiley AW (1986) Crop management review. Proc of the first Australian mango research workshop, Melbourne, pp 186–195

Whiley AW, Schaffer B (1997) Stress physiology. In: Litz RE (ed) The mango: botany, production and uses. CAB International, New York, pp 147–174. 198 Madison Avenue

Whiley AW, Rasmussen TS, Saranah JB, Wolstenholme BN (1989) Effect of temperature on growth, dry matter production and starch accumulation in ten mango (*Mangifera indica* L.) cultivars. J Hortic Sci Biotechnol 64:753–765

Whiley AW, Rasmussen TS, Wolstenholme BN, Saranah JB, Cull BW (1991) Interpretation of growth responses of some mango cultivars grown under controlled temperature. Acta Hortic 291:22–31

Yohannes DB (2006) Studies on salt tolerance of Vitis spp. Ph.D thesis submitted to University of Agricultural Sciences, Dharwad, India, p 132

Impact of Climate Change on Vegetable Production and Adaptation Measures

19

Prakash S. Naik, Major Singh, and J.K. Ranjan

Abstract

Climate change influences vegetable production worldwide. However, its nature and impact vary, depending on the degree of climate change, geographical region, and crop production system. Possible impact of climate change may be visualized by change in productivity with reference to quality of crops; changes in agricultural practices like use of water, fertilizers, and pesticides; and environmental influences particularly in relation to the frequency and intensity of soil drainage which may lead to loss of nitrogen through leaching, soil erosion, and reduction of crop diversity. Vegetables are in general more sensitive to environmental extremes such as high temperatures and soil moisture stress. CO_2, a major greenhouse gas, influences growth and development as well as incidence of insect pests and diseases of vegetable crops. Under changing climatic situations, crop failures, shortage of yields, reduction in quality, and increasing pest and disease problems are common, and they render the vegetable cultivation unprofitable. Agriculture production needs to be adapted to the changing climate by mitigating its impact. Unless measures are undertaken to adapt to the effects of climate change on vegetable production, nutritional security in developing countries will be under threat.

19.1 Introduction

Global vegetable production has made unprecedented growth especially on per capita basis, which has increased more than 60% over the last 20 years. However, the projected growth in the world's population to 9.1 billion by 2050 adds an extra

P.S. Naik • M. Singh (✉) • J.K. Ranjan
ICAR-Indian Institute of Vegetable Research, Varanasi 221 305, Uttar Pradesh, India
e-mail: naikps1952@gmail.com; singhvns@gmail.com; jkranjan2001@yahoo.co.in

© Springer Nature Singapore Pte Ltd. 2017
P.S. Minhas et al. (eds.), *Abiotic Stress Management for Resilient Agriculture*,
DOI 10.1007/978-981-10-5744-1_19

challenge for food and nutritional security worldwide, and it has been reported that more than one billion people are undernourished (FAO 2009), and over one third of the burden of disease in children below 5 years of age is mainly due to undernutrition (Black et al. 2008). Vegetables being the major source of nutrition in terms of vitamins and minerals have to play a crucial role in meeting this challenge. As per FAO statistics 2015, the world vegetable production during 2011 was 1087.59 million tonnes from an area of 56.69 million hectares with average productivity of 19.18 tonnes ha^{-1}. The report suggests that vegetables cover 1.1% of the world's total agricultural area in which Europe and Central Asia contribute 12% of the total global area and 14% of global production. The leading vegetable-producing countries in the world are developing countries which are highly vulnerable to climate change. This issue becomes more pertinent when it relates to the fight against poverty, hidden hunger, and livelihoods of billions of people in developing countries. For providing food and nutritional security to the burgeoning population, global vegetable production needs a sustainable growth. However, climate change is becoming one of the most serious environmental threats to impact agriculture in a long way. Agriculture is highly climate dependent and global climate change will certainly impact agriculture and consequently affect the world's food supply.

Climate change is a complex phenomenon occurred due to changing climatic scenario over time most probably by natural variability or by man-made consequences. Climate change per se is not necessarily harmful but the nuisance arises from extreme weather situation that are difficult to predict are the main concern (FAO 2001). By 2100, increase in atmospheric CO_2 will be recorded from 360 ppm to 400–750 ppm and elevation in sea level from 15 to 95 cm as per Intergovernmental Panel on Climate Change (IPCC 2007a, b). This rise in CO_2 level will subsequently increase the global average temperature from 0.8 °C in 2001 to 1.4–5.8 °C by 2100. Variability in atmospheric air temperature alters the amount and distribution of precipitation; wind patterns resulting into weather extremes, viz., flooding, drought, hailstorms, high temperature, and freezing stress; change in ocean currents; acidification; and forest fires in addition to rate of ozone depletion (Minaxi et al. 2011; Kumar 2012). The current fertilizer use efficiency that ranges between 2 and 50% in India is likely to be reduced further with increasing temperatures. Huge fertilizer uses for boosting agricultural production will in turn lead to higher emission of greenhouse gases. Small fluctuation in atmospheric temperature and rainfall pattern affects both the production and quality of fruits and vegetables ultimately with resultant implication in domestic and international trade. Similarly, a high temperature stimulates the outbreak of new and aggressive pests and diseases along with the weeds which altogether invade our standing crop impeding its performance and yield. Studies focusing on climate change and its consequence in different crops are increasingly becoming the major areas of scientific concern. Scientist and researchers all over the world are engaged in developing different adaptation strategies and improved variety for answering the question of an hour. However, tackling this immense challenge must involve both adaptation to manage the unavoidable and mitigation to avoid the unmanageable while maintaining a focus on its social dimensions.

19.2   Impact of Climate Change on Vegetable Production

Change in productivity due to the impact of climate change with reference to quantity and quality of crops may be visualized by changes in agricultural practices like use of water, fertilizers, and pesticides and environmental influences particularly in relation to the frequency and intensity of soil drainage which may lead to loss of nitrogen through leaching, soil erosion, and reduction of crop diversity. Climate change is basically associated with extremities of four factors, viz., temperature, drought, salinity, and CO_2 concentration. The first three factors are causing environmental stress leading to crop losses worldwide, reducing average yields for most of the crops by more than 50% (Boyer 1982; Bray et al. 2000). Vegetable crops are more sensitive to climate extremes such as elevated temperatures and limited soil moisture which are the major causes for its lower yields and poor quality. These effects will be further magnified by climate change.

19.2.1 Impact of Increasing Temperature

Temperature is the major abiotic stress factor limiting the production and quality of horticulture crops, mainly the vegetables. Generally, most of the vegetable crops require low temperature for their cultivation, and hence average productivity of temperate countries is much higher than tropical regions where high temperature is habitually established. In tropical region where thermal tolerance of crops is already near its limit, even a slight increase in temperature will result in drastic fall in crop productivity. Increasing temperature will reduce the irrigation water availability, decrease soil fertility, and increase soil erosion and flooding and salinity incursion, which are major restrictive factors in sustainable vegetable production (Wheeler and vonBraun 2013; Anwar et al. 2015). A warmer climate is likely to increase the frequency of high day or night temperature stress events that can negatively affect flowering, fruit set, and quality. Analysis of climate trends in tomato-growing locations reported that severity and frequency of above-optimal temperature episodes will increase in the coming decades at an alarming rate (Bell et al. 2000). This is may be due to the fact that vegetative and reproductive processes in tomatoes are strongly modified by temperature alone or in conjunction with other environmental factors (Abdalla and Verderk 1968). Elevated temperature affects the plant developmental processes by disturbing its normal biosynthetic pathways like photosynthesis and translocation processes essential for normal cellular functioning in plants (Weis and Berry 1988). In tomato, if temperatures exceed 32.2 °C during critical flowering and pollination periods (Sato et al. 2001), it causes significant losses in tomato productivity due to reduced fruit set and smaller and lower-quality fruits (Stevens and Rudich 1978). The symptoms of tomato fruit set failure at high temperatures include bud drop, abnormal flower development, poor pollen production, dehiscence and viability, ovule abortion and poor viability, reduced carbohydrate availability, and other reproductive abnormalities (Hazra et al. 2007). Higher temperature during pre-anthesis can influence developmental changes in the anthers,

Table 19.1 High-temperature injury symptoms in different vegetable crops

Crop	Injury symptoms
Snap bean	Brown and reddish spots on the pod; spots can coalesce to form a water-soaked area
Cabbage	Outer leaves showing a bleached, papery appearance; damaged leaves are more susceptible to decay
Lettuce	Damaged leaves assume papery aspect; affected areas are more susceptible to decay; tipburn is a disorder normally associated with high temperatures in the field; it can cause soft rot development during post-harvest
Muskmelon	Characteristic sunburn symptoms: dry and sunken areas; green color and brown spots are also observed on rind
Bell pepper	Sunburn: yellowing and, in some cases, a slight wilting
Potato	Black heart: occurs during excessively hot weather in saturated soil; symptoms usually occur in the center of the tuber as dark-gray to black discoloration
Tomato	Sunburn: disruption of lycopene synthesis; appearance of yellow areas in the affected tissues, no fruit set after 35°C day temperature

Source: Moretti et al. (2010)

particularly irregularities in the epidermis and endothecium, lack of opening of the strontium, and poor pollen formation (Sato et al. 2002). However, in pepper pre-anthesis period is not sensitive to high-temperature exposure as it did not affect pistil or stamen viability, but high post-pollination temperatures inhibited fruit set, suggesting that fertilization is sensitive to high-temperature stress at later stage (Erickson and Markhart 2002). Numerous primary metabolic processes are severely affected by high-temperature stress, viz., photosynthesis, respiration, plant water relations, and cellular membrane stability and enzymatic activities. Secondary metabolic pathways like phytohormone biosynthetic pathways and synthesis of secondary metabolites essential for defense mechanism in plants are altered by fluctuating temperature. The symptoms of high-temperature stress injury vary from crop to crop (Table 19.1). In the past few years, the impact of climate change is seen and experienced worldwide as in Sambalpur, India, and increased temperatures have delayed the onset of winter resulting in the decline of cauliflower yields (Pani 2008). Earlier growers used to harvest 1 kg of inflorescences head but now weight has reduced to 0.25–0.30 kg each. Such reductions in yield hiked production costs of other vegetables like tomatoes, radish, and other native Indian vegetable crops.

Nevertheless, there may be some positive impacts of climate change on crop productivity which is expected to rise slightly in mid- to high latitudes for mean temperature increase up to 3 °C coupled with enhanced CO_2 concentration. Slightly higher temperature increases rate of development in plants, and thus a short life cycle, though less productive, can be beneficial for escaping drought and frost. Late-maturing cultivars could benefit from faster development rate. In colder regions, global warming could lead to lengthening of growth period and optimal assimilation at elevated temperatures.

19.2.2 Impact of Water Stress

Changing climatic scenario and distribution of amount of rainfall leads to a serious consequence of water stress. Water stress comprises of two types of events depending upon the amount of rainfall, viz., the water-deficit stress/drought stress and flooding stress. Vegetables, being succulent products by definition, generally consist of more than 90% water). Thus, water act as a very important factor for determining yield and quality of vegetables. Drought is a meteorological term drastically reducing vegetable productivity worldwide due to insufficient water supply. Drought stress disturbs the plant water status by disturbing the soil water potential by increasing the solute concentration in plant rhizospheric and surrounding soil leading to an osmotic flow of water out of plant cells. The processes ultimately decrease the plant water potential and increase the solute accumulation in plat cell thereby disturbing the cellular membrane and various membrane-bound processes and enzyme activities, namely, photosynthesis and respiration processes. The impact of drought stress in plants depends mainly on its duration, intensity, plant species, and soil properties. In India only about 41% area is irrigated and the remaining 59% is rainfed. By utilizing maximum irrigation potential of the country, still about 50% of the area will depend on rainfed farming system. Under such circumstances, increase in temperatures and changes in rainfall patterns are likely to reduce agricultural productivity in rainfed areas.

19.2.3 Impact of Salinity

Vegetable production is a source of 40% of the world's food) but increasing soil salinity at an alarming rate is becoming the major issue for irrigated crops particularly vegetables. In hot and dry environments, high evapotranspiration results into substantial water loss, thus leaving salt around the plant roots which interferes with the plant's ability to uptake water and thus reduces crop productivity. According to the US Department of Agriculture (USDA), onions are more sensitive to saline soils due to its shallow root system, while cucumbers, eggplant, peppers, and tomatoes are moderately sensitive. Plant sensitivity to salt stress is reflected in loss of turgor, growth reduction, wilting, leaf curling and epinasty, leaf abscission, decreased photosynthesis, senescence, respiratory changes, loss of cellular integrity, tissue necrosis, and ultimately plant death (Jones 1986; Cheeseman 1988). Crops cultivated along coastal regions are highly prone to salinity due to poor water quality with huge salts in irrigated water. Seasonal variations are observed for soil salinity, mostly elevated during dry spells and minimized with rainfall where freshwater flushing is prevalent.

19.2.4 Impact of Elevated CO$_2$ Concentration

Carbon dioxide level in the atmosphere has hiked approximately 35% from prein-dustrial era to 2005 (IPCC 2007a, b). Its concentration in the atmosphere is a key factor for plant growth and physiological processes. Increased atmospheric CO$_2$ is reported as a modifier for net photosynthesis, biomass production, sugar and organic acid contents, stomatal conductance, fruit firmness, seed yield, light, water, nutrient use efficiency, and plant water potential in vegetables (Bazzaz 1990; Cure and Acock 1986; Idso and Idso 1994). The magnitude of the "CO$_2$ fertilization effect" varies tremendously among plant species and even within varieties of same species (Wolfe 1994). Major vegetables come under C$_3$ plants, which respond positively under elevated CO$_2$ concentrations and ambient temperature compared to C$_4$ plants (Patterson et al. 1999), but this effect is found to be reversed during elevated temperatures. However, attaining maximum CO$_2$ benefits often requires more fertilizer (to support bigger plants), optimum temperatures, unrestricted root growth, and excellent control of weeds, insects, and diseases. High CO$_2$ concentration can have additional direct effect on plants that is reducing transpiration (water loss) per unit leaf area by causing partial closure of leaf stomata. However, often this water-conserving response is minimized or not observed at the whole-plant or field (ET) level because high CO$_2$-grown plants have more total leaf (transpirational surface) area. Thus, CO$_2$ fertilization effect cannot be paid off for other detrimental effects occurring due to other abiotic stresses (Luo and Mooney 1999).

Crops show reduced nutritional quality and higher carbon to nitrogen ratio in plant foliage when grown under elevated CO$_2$ concentration. This also stimulates enhanced feeding activities in few herbivores and by pests which lead to greater crop damage (Hill and Dymock 1989; Porter et al. 1991); however, this theme remains controversial. Clark (2004), while working on tropical forests, argued that increasing atmospheric CO$_2$ has negligible effect on biomass production rates. He stressed that the growth of tropical forests is not carbon limited, since higher temperature increases respiration rate and other metabolic processes resulting in increased atmospheric CO$_2$, which reduces forest productivity. There will also be associated changes in the movement of crop pests and diseases, thereby changing the dynamics of their interactions (Harrington and Woiwod 1995; Bale et al. 2002). It has been reported that many of the supreme invasive and noxious weeds respond more positively to increasing CO$_2$ than do most of our cash crops (Ziska 2003). Weeds are much more difficult to control with herbicides at higher CO$_2$ levels as we anticipate will occur in the coming decades (Ziska et al. 1999). It is mostly speculation at this time as to which crops and regions will benefit and which will be worse off by CO$_2$ elevation in the future with regard to weed and pest control. Some leaf-feeding insects cause more crop damage to plants under high CO$_2$ conditions (Hamilton et al. 2005). It remains difficult to predict future rainfall patterns, but wetter summers would tend to favor many foliar pathogens (Coakley et al. 1999). Climate change also influences the ecology and biology of insect pests.

19.3   Adaptation Measures

Farmers from the tropical part of world need new and innovative technologies for adapting and mitigating the undesirable effects of changing climatic scenario on agricultural productivity, particularly vegetable production and quality. It is now becoming a challenging task before farmers from developing countries of the tropics. Novel technologies developed by scientist all over the world through plant stress physiology research can potentially contribute to mitigate the threats from climate change on vegetable production. However, farmers in developing countries are usually smallholders with fewer options and must rely heavily on resources available in their farms or within their communities. They need adaptive tools to manage the adverse effects of climate change on vegetable productivity and quality.

19.3.1 Crop Management Practices

Climate change will alter the environmental conditions for crop growth and require adjustments in management practices at the field scale. These practices need to be explored and utilized for efficient crop management. A number of technologies have been developed at AVRDC – The World Vegetable Center – to alleviate the production challenges of vegetable crops grown under stressful environment.

19.3.2 Efficient Water Management

The efficient water management and quality of irrigation water determine the vegetable yield and quality. Further, the optimum frequency and amount of applied water are functions of climate and weather conditions, crop species, variety, stage of growth, rooting characteristics, soil structure and water holding capacity, irrigation system, and cultural practices (Phene 1989). Lavishly or inadequate water causes abnormal plant growth, predisposes plants to infection by pathogens, and causes nutritional disorders. Under limited water supply, soil water conservation and proper irrigation schedule are essential agronomical practices for maintaining crop yield potential. There are several irrigation methods and the choice depends on the crop, water availability, soil characteristics, and topography. Irrigation practices like overhead, surface, and drip systems could be implemented during water-scare situation. Surface irrigation methods are utilized in more than 80% of the world's irrigated lands, yet its field-level application efficiency is often 40–50% (Von Westarp and Chieng 2004). Drip irrigation delivers water directly to plants through small plastic tubes which prevent water losses due to minimal runoff and deep percolation, and water savings of 50–80% are achieved when compared to surface irrigation. Thus, by implementing the drip irrigation practice, more plants can be irrigated per unit of water and with less labor. In Nepal, drip irrigation practices did not record significant difference in cauliflower yield; however, its long-term use is found to be beneficial as per the cost and labor are concerned (Von Westarp and Chieng 2004). The water-use efficiency in chili and pepper was significantly higher by drip irrigation compared to furrow irrigation, with higher efficiencies observed with high delivery rate drip irrigation regimes (AVRDC 2005). As per the report of AVRDC, in drought-tolerant crop like watermelon, yield differences between furrow and drip were not significant; however, the incidence of fusarium wilt was reduced when a lower drip irrigation rate was used. Thus, drip irrigation is the most preferred due to its low cost, labor-saving, and efficient water management as more plants are to be grown per unit of water and simultaneously increasing the farmers' incomes.

19.3.3 Cultural Practices

Various cultural practices such as mulching, use of shelters, and raised beds are helpful for protection against high temperatures, heavy rains, and flooding. They also conserve soil health in terms of soil moisture and nutrient conservation required for crop production. The organic and inorganic mulches are commonly used for the production of high-valued cucurbitaceous vegetable crops, viz., gynoecious parthenocarpic cucumber hybrid, gynoecious bitter gourd, and summer squash specially zucchini type either in open or in protected condition. This adaptive measure ultimately minimizes the evaporation losses from soil optimizing its temperature, reducing the soil erosion, hampering the weed growth, and protecting the fruits from direct soil contact. Additionally, the organic materials as mulch help in enhancing soil fertility and improving soil structure and other properties. In India, mulching improved the growth of bottle gourd, round melon, ridge gourd, and sponge gourd compared to the non-mulched controls (Pandita and Singh 1992). Yields were the highest when polythene and sarkanda (*Saccharum* spp. and *Canna* spp.) were used as mulching materials. Planting cucurbitaceous vegetables in raised beds can ameliorate the effects of flooding during the rainy season.

19.3.4 Organic Farming

Organic farming emits less greenhouse gases than conventional agriculture (Mader et al. 2002; Pimentel et al. 2005; Reganold et al. 2001) and enhances carbon sequestration in the soil. Organic farming practices preserve soil fertility and maintain or increase organic matter that can reduce the negative effects of climate extremes while increasing crop productivity (ITC and FiBL 2007; Niggli et al. 2008). The organic farming practices that can mitigate the effect of climate change include replacement of chemical fertilizers through organic amendments, crop residue recycling, incorporation of legume in crop rotations, crop diversification, avoidance of burning of crop waste and residues, as well as more use of organic mulches, bio-inoculants, and organic growth-promoting substances as vegetable production strategies. These practices will help to build up soil organic matter and will offer sustainable carbon credit generation (Bellarby et al. 2008; Niggli et al. 2008).

Crop diversification and an increase of soil organic matter will enhance the nutrient buffer capacity and the microbial activity; both will strengthen soil fertility and will enhance resilience against extreme weather events.

19.3.5 Grafting Technique

Changing climate increases in incidence of biotic and abiotic stresses which enforces formers to use more pesticides which become major concern for health and environment. Vegetables are prone to soilborne diseases for this development of resistant variety is the best solution in the disease-prone areas. Grafting method can overcome some of these barriers to make potential genetic stocks available to the farmers. This technique quickly and directly and basically correlates the traits of two plants, one providing a root system (rootstock) and the other providing a shoot (scion). Though the major concern of grafting is to avoid soilborne diseases such as fusarium wilt in *Cucurbitaceae* (cucumber, melon, etc.) and bacterial wilt in *Solanaceae* (tomato, pepper, etc.), it is also practiced for a number of other purposes and has several advantages (Lee et al. 2010) such as yield increase, shoot growth promotion, disease tolerance, nematode tolerance/resistance, low-temperature tolerance, high-temperature tolerance, enhanced nutrient uptake, enhanced water uptake, high salt tolerance, wet soil tolerance, heavy metal and organic pollutant tolerance, quality changes, extended harvest period, etc. Grafting as an intensive labor work is majorly practiced on cucurbits and members of the solanaceous vegetables such as brinjal and tomato. Among grafting-practiced vegetables, certain rootstocks have been identified which possess tolerance/resistance against stresses (Table 19.2). Grafting on these resistant/tolerant rootstocks protects the crop from ill effects of soilborne stresses.

19.4   Developing Climate-Resilient Genotypes/Varieties

Availability of potential and diverse genetic resource material is a key element for farmers as well as breeders to meet the ever-changing population demands and to mitigate the changing climate. Presently, breeders have plenty of genotypes with good yield potential with other yield-contributing traits. But there is a need to screen and diagnose all available genetic materials against different biotic and abiotic stresses to identify potential genetic stock which can tolerate a wide range of climatic conditions.

19.4.1 Breeding for Stress Tolerance

Identification of tolerance/resistance is a major activity for initiating breeding for stress in any crop. After screening the available genetic material, wild species is a best option against breeder to identify the improvement process. Wild and related

Table 19.2 Rootstocks for cucurbitaceous crops and some related characteristics

Rootstock	Cultivar	Major characteristics	Possible disadvantage
Watermelon			
Bottle gourd (*Lagenaria siceraria* L.)	Dongjanggoon, Bulrojangsaeng, Sinhwachangjo (Korea), FR Dantos, Renshi, Friend, Super FR Power (Japan)	VRS, FT, LTT	New *Fusarium* race, susceptible to anthracnose
Squash (*Cucurbita moschata* Duch.)	Chinkyo, No. 8, Keumkang(Korea)	VRS, FT, LTT	Inferior fruit shape and quality
Interspecific hybrid squash (*Cucurbita maxima* Duch. × C. moschata* Duch.)	Shintozwa, Shintozwa #1, Shintozwa #2, Chulgap, (Japan, China, Taiwan, Korea)	VRS, FT, LTT, HTT, SV	Reduced fertilizers required; quality reduction may result
Pumpkins (*Cucurbita pepo* L.)	Keumsakwa, Unyong, Super Unyong	VRS, FT, LTT	Mostly for cucumbers
Winter melon (*Benincasa hispida* Thunb.)	Lion, Best, Donga	GDR	Incompatibility
Watermelon [*Citrullus lanatus* (Thunb.) Matsum. et Nakai]	Kanggang, Res. #1, Tuffnes (Japan), Ojakkyo(Syngenta)	FT	Not enough vigor and disease resistance
African horned (AH) cucumber (*Cucumism etuliferus* E. Mey. Ex)	NHRI-1	FT, NMT	Medium to poor graft compatibility
Cucumber			
Fig leaf gourd (*Cucurbita ficifolia* Bouché)	Heukjong (black seeded figleaf gourd)	LTT, GDT	Narrow graft compatibility
Squash (*Cucurbita moschata* Duch.)	Butternut, Unyong #1, Super Unyong	FT, FQ	Affected by *Phytophthora*
Interspecific hybrid squash (*Cucurbita maxima* Duch. × C. moschata* Duch.)	Shintozwa, Keumtozwa, Ferro RZ,64-05 RZ, GangryukShinwha	FT, LTT	Slight quality reduction expected
Bur cucumber (*Sicyos angulatus* L.)	Andong	FT, LTT, SMT, NMT	Reduced yield
AH cucumber (*Cucumis metuliferus* E. Mey. ex Naud)	NHRI-1	FT, NMT	Weak temperature tolerance

(continued)

Table 19.2   (continued)

Rootstock	Cultivar	Major characteristics	Possible disadvantage
Melon			
Squash (*Cucurbita moschata* Duch.)	Baekkukzwa, No. 8, Keumkang, Hongtozwa	FT, LTT	*Phytophthora* infection
Interspecific hybrid squash (*Cucurbita maxima* Duch. × *C. moschata* Duch.	Shintozwa, Shintozwa #1, Shintozwa #2	FT, LTT, HTT, SMT	*Phytophthora* infection, poor fruit quality
Pumpkin (*Cucurbita pepo* L.)	Keumsakwa, Unyong, Super Unyong	FT, LTT, HTT, SMT	*Phytophthora* infection
Melon (*Cucumis melo* L.)	Rootstock #1, Kangyoung, Keonkak, Keumgang	FT,	FQ *Phytophthora* problem

Cultivars vary greatly depending upon countries, growing types, years, and grafting methods; *VRS* vigorous root systems, *FT Fusarium* tolerance, *LTT* low-temperature tolerance, *ST* strong vigor, *HTT* high-temperature tolerance, *GDT* good disease tolerance, *GDR* good disease resistance, *NMT* nematode tolerance, *SMT* high soil moisture tolerance, *FQ* fruit quality modification (Source: Lee et al. 2010)

species which grows under variable environmental conditions are the untapped reservoir of genepool. This selection process needs to accumulate the superior alleles at multiple loci in a specific genotype, so further improvement can be done by applying different breeding techniques including tissue culture and transgenic.

Here selection of species which is already accumulated for marked climatic change becomes more easy to subject to variable environmental conditions. This also helps breeding techniques to identify genes or gene combinations which confer such resilience. As onions are more prone to salinity, there is an urgent need develop a variety which sustains against saline condition and gives potential yield and ultimately enriches the onion growers. In European gene banks, alliums with cold and frost resistance are also reported. In case of cucurbits, these crops are more adaptable to warm season, and any increase in temperature will largely affect its productivity. Hence, there is an urgent need to develop varieties of cucurbitaceous vegetable crops with general adaptation to hot and humid tropical environments. AVRDC has developed tomatoes and Chinese cabbage with general adaptation to hot and humid tropical environments and low-input cropping systems since the early 1970s by developing tolerant breeding lines for heat and disease resistant.

Tissue tolerance to severe dehydration is not common in crop plants but is found in species native to extremely dry environments (Ingram and Bartels 1996). Some *Cucurbita* sp. possess some xerophytic characters essential to adopt under water scarcity condition. Drought stress is a major environmental factor affecting plant physiological growth and overall development. For development of drought-resistant species, root system of particular crops plays a vital role; *Citrullus colocynthis* a drought-tolerant cucurbit species have a deep root system. Differences in gene expression during drought were studied using cDNA-AFLP. Two genes,

CcrbohD and *CcrbohF*, encoding respiratory burst oxidase proteins were cloned using RACE. Real-time RT-PCR analysis showed that expression of *CcrbohD* was rapidly and strongly induced by abiotic stress imposed by PEG, ABA, SA, and JA treatment. *CcrbohD* has great promise for improving drought tolerance of other cucurbit species.

Besides tackling drought stress, severity of diseases and pests among stressed genotypes as associated action as well as in normal climate change is a major concern. Identification of resistant/tolerant gene/genotype in genepool of crop, its wild species, and related species will be the best option to curb the situation. Large gene pool is available in *Cucurbitaceae* family which can be utilized for the development of pest-tolerant/pest-resistant varieties or elite breeding lines in cucumber, muskmelon, watermelon, and other important cucurbitaceous vegetable crops.

19.4.2 Climate-Proofing Through Genomics and Biotechnology

Stress tolerance in plants is a complex phenomenon involving many genes (Wang et al. 2003). Genome sequencing of some of the vegetable crops including cucumber, muskmelon, watermelon, potato, tomato, and pepper has opened the way to genetic manipulation of genes associated with tolerance to environmental stresses. Even though this technology is rapid and promising, it is not found to be cost-effective on large scale. The use of molecular markers as a selection tool provides the potential for increasing the efficiency of breeding programs by reducing environmental variability, facilitating earlier selection, and reducing subsequent population sizes for field testing. However, its use in vegetable crop is limited, but progess is being initiated in this line for identifying the QTLs and candidate genes for stress tolerance QTLs for drought tolerance have been identified in tomato (Martin et al. 1989). In plants RNA and protein expression profiling alters during stress; reports show approximately 130 drought-responsive genes have been identified using microarrays (Seki et al. 2001; Reymond et al. 2000). These genes are involved with transcription modulation, ion transport, transpiration control, and carbohydrate metabolism. *DREB1A* and *CBF* and *HSF* genes are transcription factors implicated in drought and heat response, respectively (Sung et al. 2003; Sakuma et al. 2002). The *ERECTA* gene regulates plant transpiration efficiency in *Arabidopsis thaliana* (Masle et al. 2005), and the *NHX* and *AVP1* genes are associated with ion transport (Zhang and Blumwald 2001). There are many more genes implicated with stress response, and the current challenge is to identify the ones that confer a tolerant phenotype in the crop of interest. Complete analysis of the genes is already being reported in *A. thaliana*, but overexpression of only few candidate genes contributes for stress-tolerant phenotype in vegetable crops (Zhang et al. 2004). In tomato, expression of *AVP1*, a vacuolar H+ pyrophosphatase from *A. thaliana*, resulted into improved tolerance mechanism for drought stress (Park et al. 2005). The engineered tomato has a stronger, larger root system which allows efficient use of limited water.

The control plants suffered irreversible damage after 5 days without water as opposed to transgenictomatoes which began to show water stress damage only after 13 days but recovered completely as soon as water was supplied. The *CBF/DREB1* genes have been used successfully to engineer drought tolerance in tomato and other crops (Hsieh et al. 2002).

19.4.3 Exploring Climate-Resilient Vegetable Crops for the Future

One of the important steps toward combating the ill effects of climate is to look for those species which may adapt to climatic changes in a better way and make the dynamics of carbon sequestration faster. There is a need to diversify our food basket with new species which have the potential to combat the challenges of food and nutrition even under adverse climatic conditions. The crops for the future are to be identified and promoted in an emerging production scenario. Crops with either C4 cycle or CAM pathway, like amaranth, Indian spinach, opuntia, portulaca, etc., may be promoted as they make carbon sequestration faster under high CO_2 level. Other potential vegetable crops which may be grown on marginal lands with low inputs should be promoted.

19.5 Conclusions

The chapter reveals that global climate change puts forth a series of new challenges and opportunities for researcher globally for sustainable agriculture production under the erratic climatic events. However, it is well understood that factors contributing for food security are likely to be multiplied under climate change. The limited reasons for climate change are known today; nonetheless, as per the available information, anthropogenic activities like industrialization and mechanization may contribute up to some extent. The elevated temperature resulting into two detrimental abiotic stresses, namely, drought and soil salinity, is an important factor limiting vegetable production. Among the major gases contributing the global warming, CO_2 concentration in the atmosphere can enhance plant growth and development as well as the pest and disease outbreak. Under changing climatic situations, crop failures, shortage of yield, reduction in quality, and increasing pest and disease problems are common, and they can render the vegetable cultivation unprofitable. Agriculture practices need to be adapted to the changing climatic scenario along with options for mitigating its effects. Unless measures are undertaken to adapt to the effects of climate change on vegetable production, nutritional security in developing countries will be under threat.

References

Abdalla AA, Verderk K (1968) Growth, flowering and fruit set of tomato at high temperature. The Neth J Agric Sci 16:71–76

Anwar MR, Li LD, Farquharson R, Macadam I, Abadi A, Finlayson J, Ramilan T (2015) Climate change impacts on phenology and yields of five broadacre crops at four climatologically distinct locations in Australia. Agric Syst 132:133–144

AVRDC (2005) Annual report. AVRDC – The World Vegetable Center, Shanhua

Bale JS, Masters ID, Hodkinson C, Awmack TM, Bezemer VK, Brown J et al (2002) Herbivory in global climate change research: direct effects of rising temperature on insect herbivores. Glob Chan Biol 8:1–16

Bazzaz FA (1990) The response of natural ecosystems to the rising global CO_2 levels. Ann Rev Ecol Syst 21:167–196

Bell GD, Halpert MS, Schnell RC, Higgins RW, Lowrimore J, Kousky VE, Tinker R, Thiaw W, Chelliah M, Artusa A (2000) Climate assessment for 1999. Supplement June 2000. Bull Amer Meteorol Soc 81(6)

Bellarby J, Foereid B, Hastings A, Smith P (2008) Cool farming: climate impacts of agriculture and mitigation. Greenpeace International, Amsterdam, 43p

Black RE, Allen LH, Bhutta ZA, Caulfield LE, Onis M de, Essati M, Mathers C, Rivera J (2008) Maternal and child undernutrition: global and regional exposures and health consequences. Lancet 371:243–260

Boyer JS (1982) Plant productivity and environment. Science 218:443–448

Bray EA, Bailey-Serres J, Weretilnyk E (2000) Responses to abiotic stresses. In: Gruissem W, Buchannan B, Jones R (eds) Biochem & molecular biol of plants. ASPP, Rockville, pp 1158–1249

Cheeseman JM (1988) Mechanisms of salinity tolerance in plants. Plant Physiol 87:57–550

Clark DA (2004) Sources or sinks? The responses of tropical forests to current and future climate and atmospheric composition. Phil Trans R Soc 59:477–491

Coakley SM, Scherm H, Chakraborty S (1999) Climate change and plant disease management. Annu Rev Phytopathol 37:399–426

Cure JD, Acock B (1986) Crop responses to carbon dioxide doubling: a literature survey. Agric For Meteorol 38:127–145

Erickson AN, Markhart AH (2002) Flower developmental stage and organ sensitivity of bell pepper (Capsicum annuum L.) to elevated temperature. Plant Cell Environ 25(1):123–130

FAO (2001) http://www.fao.org

FAO (Food and Agricultural Organization of the United Nations) (2009) State of food insecurity in the world: economic crises—impacts and lessons learned. FAO, Rome

Hamilton JG, Dermody O, Aldea M, Zangerl AR, Rogers A, Berenbaum MR, Delucia E (2005) Anthropogenic changes in tropospheric composition increase susceptibility of soybean to insect herbivory. Environ Entomol 34(2):479–485

Harrington R, Woiwod IP (1995) Insect crop pests and the changing climate. Weather 50:200–208

Hazra P, Samsul HA, Sikder D, Peter KV (2007) Breeding tomato (Lycopersicon esculentum mill) resistant to high temperature stress. Int J Plant Breed 1(1):31–40

Hill MG, Dymock JJ (1989) Impact of climate change: agriculture/horticulture systems. DSIR entomology division submission to the New Zealand climate change Programme. Department of Scientific and Industrial Research, Auckland

Hsieh TH, Lee JT, Charng YY, Chan MT (2002) Tomato plants ectopically expressing Arabidopsis CBF1 show enhanced resistance to water deficit stress. Plant Physiol 130:618–626

Idso KE, Idso SB (1994) Plant responses to atmospheric CO_2 enrichment in the face of environmental constraints: a review of the past 10 years' research. Agric For Meteorol 69:153–203

Ingram J, Bartels D (1996) The molecular basis of dehydration tolerance in plants. Annual Revi Plant Physiol Plant Mol Biol 47:377–403

IPCC (2007a) Climate change, Fourth assessment report. Cambridge University Press, Cambridge

IPCC (2007b) Appendix I: glossary. In: Parry ML, Canziani OF, Palutik of JP et al (eds) Climate change: impacts, adaptation and vulnerability. Contribution of working group II to the fourth assessment report of the intergovernmental panel on climate change. Cambridge University Press, Cambridge, pp 79–131

ITC (International Trade Centre UNCTAD/WTO) and FiBL (Research Institute of Organic Agriculture) (2007) Organic farming and climate change. ITC, Geneva

Kumar SV (2012) Climate change and its impact on agriculture: a review. Int J Agric Environ Biotech 4(2):297–302

Lee J-M, Kubota C, Tsao SJ, Bie Z, Hoyos-Echevarria P, Morra L, Oda M (2010) Current status of vegetable grafting: diffusion, grafting techniques, automation. Sci Hortic 127:93–105

Luo Y, Mooney HA (eds). (1999) Carbon dioxide and environmental stress. Academic Press, New York

Mader P, Fliessbach A, Dubois D, Gunst L, Fried P, Niggli U (2002) Soil fertility and biodiversity in organic farming. Science 296:1694–1697

Martin B, Nienhuis J, King G, Schaefer A (1989) Restriction fragment length polymorphisms associated with water use efficiency in tomato. Science 243:1725–1728

Masle J, Gilmore SR, Farquhar GD (2005) The ERECTA gene regulates plant transpiration efficiency in *Arabidopsis*. Nature 436:866–870

Minaxi RP, Acharya KO, Nawale S (2011) Impact of climate change on food security. Int J Agric Environ Biotech 4(2):125–127

Moretti CL, Mattos LM, Calbo AG, Sargent SA (2010) Climate changes and potential impacts on postharvest quality of fruit and vegetable crops: a review. Food Res Int 43:1824–1832

Niggli U, Schmid H, Fliessbach A (2008) Organic farming and climate change. International Trade Centre (ITC), Geneva, p 30

Pandita ML, Singh N (1992) Vegetable production under water stress conditions in rainfed areas. In: Kuo CG (ed) Adaptation of food crops to temperature and water stress. AVRDC, Shanhua, pp 467–472

Pani RK (2008) Climate change hits vegetable crops. Indian Express. Available from: http://www.expressbuzz.com

Park S, Li J, Pittman JK, Berkowitz GA, Yang H, Undurraga S, Morris J, Hirschi KD, Gaxiola RA (2005) Up-regulation of a H-pyrophosphatase (H-PPase) as a strategy to engineer drought –resistant crop plants. PNAS 102:18830–18835

Patterson DT, Westbrook JK, Joyce RJV, Lingren PD, Rogasik J (1999) Weeds, insects and diseases. Clim. Change 43, 711_727.luo

Phene CJ (1989) Water management of tomatoes in the tropics. In: Green SK (ed) Tomato and pepper production in the tropics. AVRDC, Shanhua, pp 308–322

Pimentel D, Hepperly P, Hanson J, Douds D, Seidel R (2005) Environmental, energetic and economic comparisons of organic and conventional farming systems. Bioscience 55(7):557–582

Porter JH, Parry ML, Carter TR (1991) The potential effects of climatic change on agricultural insect pests. Agric For Meteorol 57:221–240

Reganold JP, Glover JD, Andrews PK, Hinman HR (2001) Sustainability of three apple production systems. Nature 410:926–930

Reymond P, Weber H, Damond M, Farmer EE (2000) Differential gene expression in response to mechanical wounding and insect feeding in *Arabidopsis*. Plant Cell 12:707–720

Sakuma Y, Liu Q, Dubouzet JG, Abe H, Shinozaki K, Yamaguchi-Shinozaki K (2002) DNA-binding specificity of the ERF/AP2 domain of *Arabidopsis* DREBs, transcription factors involved dehydration- and cold-inducible gene expression. Biochem Biophys Res Comm 290:998–1009

Sato S, Peet MM, Gardener RG (2001) Formation of parthenocarpic fruit and aborted flowers in tomato under moderately elevated temperatures. Sci Hortic 90:243–254

Sato S, Peet MM, Thomas JF (2002) Determining critical pre- and post-anthesis periods and physiological process in *Lycopersicon esculentum* Mill. exposed to moderately elevated temperatures. J Exp Bot 53:1187–1195

Seki M, Narusaka M, Abe H, Kasuga M, Yamaguchi-Shinozaki K, Carninci P, Hayashizaki Y, Shinozaki K (2001) Monitoring the expression pattern of 1300 Arabidopsis genes under drought and cold stresses by using a full-length cDNA microarray. Plant Cell 13:61–72

Stevens MA, Rudich J (1978) Genetic potential for overcoming physiological limitations on adaptability, yield, and quality in tomato. Hort Sci 13:673–678

Sung DY, Kaplan F, Lee KJ, Guy CL (2003) Acquired tolerance to temperature extremes. Trends Plant Sci 8:179–187

VonWestarp S, Chieng SS (2004) A comparison between low-cost drip irrigation, conventional drip irrigation, and hand watering in Nepal. Agric Water Manag 64:143–160

Wang WX, Vinocur B, Altman A (2003) Plant responses to drought, salinity and extreme temperatures: towards genetic engineering for stress tolerance. Planta 218:1–14

Weis E, Berry JA (1988) Plants and high temperature stress. Soc Expt Biol 42:329–346

Wheeler T, vonBraun J (2013) Climate change impacts on global food security. Science 341:508–513

Wolfe DW (1994) Physiological and growth responses to atmospheric CO_2 concentration. In: Pessarakli M (ed) Handbook of plant and crop physiology. Marcel Dekker, New York

Zhang HX, Blumwald E (2001) Transgenic salt-tolerant tomato plants accumulate salt in foliage but not in fruit. Nat Biotechnol 19:765–768

Zhang JZ, Creelman RA, Zhu JK (2004) From laboratory to field: using information from Arabidopsis to engineer salt, cold, and drought tolerance in crops. Plant Physiol 135:615–621

Ziska LH (2003) Evaluation of the growth response of six invasive species to past, present, and future atmospheric CO_2. J Exp Bot 54:395–406

Ziska LH, Teasdale JR, Bunce JA (1999) Future atmospheric carbon dioxide may increase tolerance to glyphosate. Weed Sci 47:608–615

Part IV

Mitigation Options in Animal Husbandry

Nutritional Management: Key to Sustain Livestock in Drought-Prone Areas

20

N.P. Kurade, B. Sajjanar, A.V. Nirmale, S.S. Pawar, and K.T. Sampath

Abstract

Frequent drought situations with below monsoon rainfall have severe impacts on the availability of nutritional resources for livestock. Droughts are usually accompanied with events of higher temperatures leading to multiple abiotic stresses. Nutritional strategies to overcome feed shortages include improved drought-tolerant fodder varieties including *rabi* crops for producing more roughage, preserving fodder by hay and silage making during lush season, use of grass from reserved forests, enrichment of straw for complete feed blocks and total mixed rations (TMR), lopping of tree leaves for fodder, growing fodder through hydroponics etc. Providing shelters through organization of cattle camps is proving as one of the lifelines for survival of livestock and farmers in the drought situations. Judicious use of novel/unconventional feed resources needs to be exploited to manage nutritional requirements of animals during scarcity by alleviating the effect of anti-nutritional factors (ANF). The use of antistress minerals, vitamins and herbal supplements needs to be further explored to mitigate the harmful effects of drought on livestock health and production.

N.P. Kurade (✉) • B. Sajjanar • A.V. Nirmale • S.S. Pawar
ICAR-National Institute of Abiotic Stress Management,
Baramati, Pune 413115, Maharashtra, India
e-mail: nitin.kurade@icar.gov.in; bksvet@gmail.com; avinash.nirmale@icar.gov.in; sachin.pawar@icar.gov.in

K.T. Sampath
ICAR-National Institute of Animal Nutrition and Physiology,
Adugodi, Bangalore 560 030, Karnataka, India
e-mail: ktsampath@sify.com

© Springer Nature Singapore Pte Ltd. 2017
P.S. Minhas et al. (eds.), *Abiotic Stress Management for Resilient Agriculture*,
DOI 10.1007/978-981-10-5744-1_20

20.1   Introduction

Abiotic stress is defined as the negative impacts of nonliving factors on the living organisms under specific environment. The present classification of abiotic stress as edaphic, drought and atmospheric is based on the basic requirements of plants for their growth and production. Although edaphic factors have influenced the whole biosphere, they do not directly influence animal growth and production. The nutritional requirements of animals are mostly fulfilled through plants which are the source of proteins, carbohydrates, fats, vitamins and minerals. Edaphic factors which directly influence plants have only indirect impact on animals. The effects of climate change on animal performance result mainly from alterations in the nutritional resources besides the direct impact of heat stress and water scarcity during drought cycles (Valtorta 2010). Drought is an extended period of water scarcity due to reduced or absence of rainfall including inefficient water resource management. Water scarcity is the frequent climatic condition adversely affecting agriculture/ livestock production in drought-prone areas of India. Among all other natural hazards, drought sets slowly over months to years and affects comparatively large areas. Therefore, the edaphic, atmospheric and drought factors influence livestock by altering nutrient availability. The inappropriate environmental factors influence directly as well as indirectly by altering supply of nonliving requirements to livestock in the specific production system. Thus, nutritional resource management is a key factor for abiotic stress management in livestock in addition to genetic improvement.

20.2   Drought-Prone Areas in India

The severity of drought has increased in India over the last two centuries (Anonymous 2010). The Rainfed Area Development Programme (RADP) prioritized the districts with less than 60% cultivated area under irrigation and further identified them as arid (31), semi-arid (133) and sub-humid (175) agroecosystems. Recently, extensive studies conducted under National Innovations in Climate Resilient Agriculture (NICRA) have led to identification of 100 districts in peninsular India highly prone to drought (Prasad et al. 2012). Due to adverse impact of drought, vast agricultural land remains uncultivated and leads to severe forage crisis for animals.

Drought-affected areas of India can be categorized into two tracts. The rectangular area that extends from Ahmedabad to Kanpur then to Jalandhar comprising of 0.6 million km^2 desert, and semiarid area comes under the first tract. The rainfall of this region is 400–750 mm. The eastern slope of Western Ghats up to a distance of 300 km from the coast represents the second tract, and that is a rain shadow area of Western *Ghats*. The rainfall in this region is highly erratic and is also less than 750 mm. In addition to these two tracts, there are isolated pockets of drought-prone areas such as Kattabomman, Nellai and Coimbatore districts in Tamil Nadu,

Kachchh and Saurashtra regions, Lalitpur-Janshi region, Mirzapur plateau, Kalahandi region and Odisha and Purulia district of West Bengal (Sarkar 2011).

20.3 Nutritional Resources in Drought-Prone Areas

During periods of drought cycles, there is an overall shortage of feeds and fodder for livestock in the area. Unavailability of forages and low-quality forages result in shortage of dry matter and overall required nutrients for animals. Animals need to be supplemented under such circumstances to maintain production levels. Protein- and energy-deficient status reduces dry matter intake and not only affects the performance but also adds stress to animals.

In a survey conducted in villages of Chopan and Ghorawal in Sonbhadra district, Uttar Pradesh, the nutrient intake through different feed ingredients was not sufficient to fulfil the standard animal requirements (Sagar et al. 2013). Pantgne et al. (2002) reported that feeding practices followed by farmers, in general, are not as per balanced feeding requirements of their animals. Moreover, the farmers from Middle Gangetic Plain did not fed dairy animals as per standard requirements as the feeding practices followed were traditional with deficit amount of DM, CP and TDN (Singh et al. 2008).

There is wide variation in data/estimates of fodder production and availability in the country. Fodder production and its utilization depend on the cropping pattern, type of livestock and climatic and socioeconomic factors. At present, there is an overall net deficit of 61.1% greens, 21.9% dry fodder/crop residues and 64% feeds in the country (Anonymous 2010). Although there are variable estimates of feed and fodder availability in the country (Tables 20.1 and 20.2), all of the estimates point towards overall deficient status of feed and fodders for livestock even in the absence of drought. The projected deficit of green and dry fodder appears to be aggravated during the near future.

20.4 Impact of Drought on Livestock

Besides direct impact of scarcity of water, drought also results in scarcity of nutritional resources and exposure of animals to adverse climatic conditions mainly heat stress (Fig. 20.1). Scarcity of water and forages during drought can increase the risks of poisonings in animal and nutritional deficiencies. The impaired water

Table 20.1 Status of feed and fodder (DM basis) in India (NIANP 2005)

Feeds	Available (million tonnes)	Requirements (million tonnes)	Deficit (%)
Dry fodder	365	412	11
Concentrate	34	47	28
Green fodder	126	193	35
Total	526	652	19

Table 20.2 Supply and demand scenario of forage and roughages (1995–2025) (in million tonnes)

Year	Supply		Demand		Deficit as % of demand (as actual)	
	Green	Dry	Green	Dry	Green	Dry
1995	379.3	421	947	526	568 (59.95)	105 (19.95)
2000	384.5	428	988	549	604 (61.10)	121 (21.93)
2005	389.9	443	1025	569	635 (61.96)	126 (22.08)
2010	395.2	451	1061	589	666 (62.76)	138 (23.46)
2015	400.6	466	1097	609	696 (63.50)	143 (23.56)
2020	405.9	473	1134	630	728 (64.21)	157 (24.81)
2025	411.3	488	1170	650	759 (64.87)	162 (24.92)

Source: Based on X Five Year Plan Document, Government of India. Figures in parentheses indicate the deficit in percentage

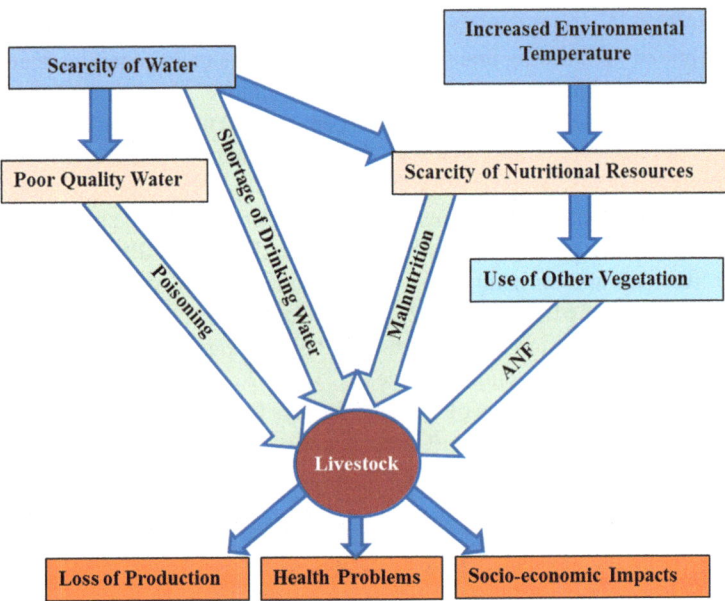

Fig. 20.1 Impact of drought on livestock

quality, feed quality, nutritional deficiency and increased incidence of plant poisonings are some drought-related risks to cattle productivity and health (Poppenga and Puschner 2014). Due to reduced availability of fodder, animals are forced to consume other vegetations or nonconventional feedstuffs with increased risk of exposure to anti-nutritional factors. In addition, the rains after long drought spell result in sudden changes in the grazing conditions for cattle, leading to additional health risks. The conditions most commonly associated with this change are bloat, certain deficiency problems, plant poisonings and clostridial diseases. Other impacts of drought are changes in production systems which include migration of livestock farmers to surplus areas, sale of animals for slaughter and shifting from large ruminant-based systems to small ruminant systems.

20.5   Nutritional Strategies for Livestock

There are many challenges for sustaining livestock wealth due to recurrent drought or delayed monsoon-like situations in drought-prone areas of the country. Efforts regarding suitability of fodder species for increasing production, alternate fodder sources and optimization for their use, storage and transport are critical for mitigation of drought conditions in livestock. Further research on optimizing nutrient availability and utilization by the different livestock species in target areas are warranted for sustainable livestock production in drought-prone areas. Special attention needs to be provided for recommendation of plant- and animal-based mixed production system where forage needs of the animals are fulfilled. Several innovations that have happened during the last few decades, which can help in enhancing productivity of fodders/feed, need to be adopted.

20.5.1 Drought-Tolerant Fodder Varieties

Efforts have been made for increasing the fodder yield of cultivated fodder crops on agricultural lands as well as on wastelands and community pastures (Hegde 2010). The strategy should include selection and breeding of high-yielding and stress-tolerant as well as short duration fodder crops and varieties. Importance may be given to improve the yields through sustainable production practices, efficient conservation practices and strengthening the value chain of dairy and meat producers by providing various critical services required to improve productivity and sustain livelihood. For this, joint efforts of various government and non-governmental agencies are important. A comprehensive review of the improved fodder crop varieties released/notified during the last three decades, fodder production systems and packages of practices for important fodder crop and intensive forage sequences recommended for different regions has been provided in the *Handbook of Agriculture* (2010).

20.5.2 Hay and Silage Making

Haymaking and ensiling are the only methods to conserve forage on a large scale. Haymaking is important in drier climates. However, the trend to conserve forage as silage has been proportionately increased, as compared to the proportion dedicated to hay (Wilkinson et al. 1996). Ensiling has many advantages than haymaking. Forage can be conserved in large quantities in less time, forage conservation is not dependent on weather and thirdly, mechanization is well suited to silage making. However, the feeding value of the resultant forage is reduced relative to that of the original crop which is a disadvantage associated with silage making (Charmley 2000). Silage can be made of crop residues or agricultural and industrial by-products and excess forages. During silage making, forages are preserved by natural or artificial acidification, for use as animal feed in periods when feed supply is inadequate (Mannetje 1999). According to Charmley (2000), there is a genuine possibility that in the future, silages will have superior feeding value than that of the original crop. Physical treatments may help to break down obstacles to improve intake and digestibility of the silage. Expected silage fermentation can be used to augment rumen function. More research efforts, besides popularization of technique, to improve silage intake and utilization using locally available forages are required to overcome the scarcity of fodder during drought cycles in different regions of India.

Making silage from drought-damaged crops needs to be assessed in drought-prone areas in India for nutritional management. Availability of sugar cane tops/whole sugarcane in the drought-prone areas needs to be exploited for an effective drought stress management option. During drought period, plant growth is diminished and nitrates can accumulate in the plant. Soil nitrates are normally absorbed by plants and utilized for the synthesis of plant proteins. Elevated nitrate in annual forages can also occur in summer if subjected to drought stress. Weeds commonly found in corn fields, viz. ragweed, pigweed, lamb's-quarter, nightshade and johnsongrass can also accumulate nitrates in toxic levels under drought conditions. Elevated nitrate levels in corn plants and also in corn silage can possibly be toxic to cattle (Ajaib et al. 2002; Wright et al. 2015).

20.5.3 Complete Feed Blocks

The crop residues have low and variable nutritional value and are bulky and fibrous. Besides this, availability of crop residues varies with location and season. In some areas, there may be deficiency of crop residues, while in some other areas, they are available in large quantity but are mostly wasted. Under emergency situations, complete feed technology has been used to save the animals from hunger and death. Based on the productivity levels of animals, the densified total mixed ration blocks (DTMRBs) or the densified total mixed ration pellets (DTMRPs) of different formulations can be made using different ingredients, comprising of vitamins, minerals and feed additives. Thus, the straw-based densified complete feed as block or pellet technology could play a vital role in providing need-based rations to livestock in drought-prone areas. The technology may provide a means to increase livestock production in the drought-prone areas apart from having other benefits such as decrease in environmental contaminants (including methane emission), increase in earnings of farmers, decrease in labour cost and time for feeding and saving of transportation cost of straw. The block making technology also offers the potential to remove regional disparity in feed availability, if the block or pellet making setups are used as 'feed banks' in regions having excess availability of crop residues. It could be used to provide balanced feed to livestock under crisis situations created by natural disasters such as drought and human conflicts (FAO 2012). However, further research is required on reducing energy cost of straw transportation and feed densification. Monitoring the quality of the processed feed for the availability of essential nutrients to avoid its dilution by the addition of more of non-nutritional feed additives is also required. The use of complete ration in mash form for feeding is more advantageous with regard to feed intake, nutrient utilization, body weight gain and feed conversion efficiency in cattle compared to conventional feeding system (Sharma et al. 2010).

20.5.4 Urea-Molasses Treatment

It is generally recommended to avoid urea treatment during periods of drought. However, urea treatment of poor-quality fodder if done judiciously under controlled condition is beneficial to sustain the periods of drought.

20.5.5 Novel Feed Resources or Alternate Feeds

Search for alternate feed resources and research for its judicious use need to be carried out for the different agroecological systems. Several newer feed resources have been evaluated and found useful for feeding. Incriminating factors have been identified in unconventional feeds and methods for their detoxification have been evolved. Protein cakes after oil extraction from seeds of neem, castor, karanj, palm and mahua have been evaluated and found suitable after detoxification to use for feeding. However, largely this technology is not yet adopted by end users. Over the years, in some regions of Karnataka, Kerala and Assam, the areca (*Areca catechu*) cultivation as a commercial crop has replaced the traditional paddy and other cereal crops to a large extent due to its higher economic returns. There is a gradual shift in the cropping pattern from cereals to more remunerative fruit and horticultural crops. This results in generation of huge quantity of fruit and vegetable residues. Presently, such residues are not effectively used and either composted or dumped in landfills causing environmental pollution. There is a need to develop suitable methods to convert waste to wealth and contribute to value-added feed resources. Some of the potential fruit residues that can be used in feeding are summarized below.

In India, annually about 1.74 million tonne of apple is produced, and waste consists of peels, seed and pulp, which represents 25–35% of fresh apple. The apple pomace on dry matter basis consists of 4.72% crude protein and 48–60% total sugar. It is a good source of energy (75%) and can replace 30% maize grain in ruminant ration. The residue contains high moisture and can be dried in sunlight or, at 65% moisture level, can be made as silage for preservation to use as feed.

Grape pomace is a by-product of wine industry comprising grape pulp, skin, stem and seeds accounting for about 20% of the grape fruit. The annual production is about 1.6 million tons. Grape pomace contains 11% crude protein, 27% crude fibre and 5% lignin on dry matter basis. Due to lower nutritive value, grape pomace can't be used in higher quantity in ruminant feeding.

Majority of the mango is consumed freshly, and only about 2% is processed and generates 40–50% waste. The waste includes peels, juice extraction waste, seeds and kernel. The waste has more sugar and moisture and hence needs to be dried and made silage for preservation. The silage of peels and juice extraction waste is a good energy (70%). The by-products of orange, mandarin orange and lemon are peel and rag. About 40–50% waste is generated from citrus juice industry and contains soluble sugars. It can be made as silage along with dry fodder and used as cattle feed. About 30–40% of whole banana is available for livestock feeding. The banana fruit waste like peels contains more moisture, low protein and more soluble sugars. This can be preserved as silage along with dry fodder and fed to cattle with other feeds. The pineapple wastes include leafy crown, peels and the pomace of juice extraction. The waste can be chaffed and dried to use as bran. Otherwise, after draining the excess moisture, the waste can be made as silage and can be used as a fodder source for cattle/sheep.

20.5.6 Hydroponics

Available literature points towards use of hydroponic fodder production which dates back to the 1930s.This method is gaining renewed attention for producing fodder for sheep, goats and other livestock. The quantity and quality of shoots produced are affected by many factors such as grain quality, soaking time, fodder variety and treatments, temperature, humidity, nutrients used, depth and density of seeds in troughs and the occurrence of mould (Sneath and McIntosh 2003). The technology of hydroponic systems is changing rapidly with systems producing yields never before realized. Water saving techniques and methods that may help to improve productivity are desired for drought-prone areas; hence, hydroponic technique merits closer consideration (Al-Karaki and Al-Hashimi 2012). Hydroponically produced fodder was found to enhance the efficiency of water use as compared to growing fodder in soil. Bradley and Marulanda (2000) reported that green fodder produced in hydroponic system requires only about 10–20% of the water required to produce the same amount of crop in soil culture. More research efforts regarding water saving options including the use of treated waste water for hydroponic green fodder production need to be carried out for applying in drought-prone areas of India.

20.6   Nutrient Management During Drought Period

During periods of scarcity, nutrient management of individual animals is important to sustain health and production or even save the life of animals. Mineral/vitamin supplementation based on the requirement may be provided under the guidance of experts. For example, selenium and vitamin E have beneficial effect on alleviating stressed conditions. The harmful effects of heat stress which is more common during water scarcity can be minimized by supplements of vitamin E and selenium which have protective role on lipid peroxidation (Sahin et al. 2002).

Table 20.3 Strategies suggested by the National Disaster Management Authority (NDMA) and required interventions

Strategies suggested by NDMA	Interventions needed
Assessment of need for fodder will be done well in advance. If a deficit is identified, ways and means to fill the gap will be explored including supplies from the nearest area, within the Mandal, within the district or in the nearby state	Timely and realistic assessments and ensuring availability of fodder. Exploring alternative sources
Raising of fodder in government as well as farmers' lands with buy-back arrangements for fodder cultivated will be promoted	Initiatives of government agencies with area-specific programmes required
Use of tank bunds for fodder cultivation	Suitable guidelines with area-specific varieties needed
Utilizing the period between crops for fodder cultivation	Suitable guidelines with area-specific varieties needed
Distribution of fodder produced within a state in nearby areas	Initiatives of government and non-governmental agencies
Establishment of fodder banks	
Conserving fish and aquaculture during droughts	Need to be coupled with water conservation practices
Utilizing the assistance of Ministry of Railways in transport of fodder and drinking water from unaffected areas to those affected	Initiatives of government agencies required
Organizing online availability of information relating to demand and supply of fodder	
Undertaking market intervention to keep the prices reasonable	
Intensification of water conservation measures in the villages	

During drought cycles, deficit forage conditions are further aggravated in drought-prone areas with severe negative impact on livestock population. To address these problems, several strategies and relief measures being adapted by the National Disaster Management Authority (NDMA) are listed in Table 20.3.

20.7 Conclusions

Most of the abiotic stressors affect livestock through their diet, and their impacts may be managed by providing appropriate nutrition. Therefore, nutritional management is an important aspect of abiotic stress mitigation in livestock. Defining the availability of various nutritional resources, strategic management through various nutritional technologies and judicious use of alternative feed resources may help to sustain the livestock population in drought-prone areas. The role of livestock farmers and initiatives of government and non-governmental agencies are crucial for abiotic stress management in livestock in drought-prone areas of India.

References

Ajaib S, Tiwana US, Tiwans MS, Puri KP (2002) Effect of application method of level of nitrogen fertilizer on nitrate content in oat fodder. Indian J Anim Nutr 17:315–319

Al-Karaki GGN, Al-Hashimi M (2012) Green fodder production and water use efficiency of some forage crops under hydroponic conditions. ISRN Agronomy 2012, Article ID 924672, 5 pages http://dx.doi.org/10.5402/2012/924672

Anonymous (2010) National Disaster Management Guidelines: Management of Drought. A publication of the National Disaster Management Authority, Government of India. ISBN 978-93-80440-08-8, New Delhi

Bradley P, Marulanda C (2000) Simplified hydroponics to reduce global hunger. Acta Hortic 554:289–295

Charmley E (2000) Towards improved silage quality– a review. Water resources systems division http://www.nih.ernet.in/rbis/india_information/draught.htm

FAO (2012) Crop residue based densified total mixed ration – A user-friendly approach to utilise food crop by-products for ruminant production, Walli TK, Garg MR, Harinder, Makkar PS, FAO Animal Production and Health Paper No. 172. Rome

Hegde NG (2010) Forage resource development in India. In: Souvenir of IGFRI Foundation Day, November 2010

Mannetje L (1999) Introduction to the conference on silage making in the tropics. In: Mannetje L (ed) Proc. FAO e-Conf. on Trop. Silage. FAO Plant Prod. and Protect. Paper 161. 1 Sept. – 15 Dec. 1999. Paper 1.0: 1–3

NIANP (2005) Feed base. National Institute of Animal Nutrition and Animal Physiology, Adugodi, Bangalore – 560 030.

Pantgne DD, Kulkarni AN, Gujar BV, Lalyankar SD (2002) Nutrient availability of milch Marathwari buffaloes in their home tract. Indian J Anim Nutr 19:41–46

Poppenga RH, Puschner DVM (2014) Drought related poisoning and nutritional risks to cattle, School of Veterinary Medicine. University of California-Davis, Davis, p 10

Prasad YG, Venkateswarlu B, Ravindra Chary G, Ch S, Rao KV, DBV R, VUM R, Subba Reddy G, Singh AK (2012) Contingency crop planning for 100 districts in peninsular India. Central Research Institute for Dryland Agriculture, Hyderabad, p 302

Sagar V, Anand RK, Dwivedi SV (2013) Nutritional status and reproductive performance of dairy cattle and buffaloes in sonbhadra district of Uttar Pradesh. Int J Sci Nat 4(3):494–498

Sahin K, Sahin N, Yaralioglu S, Onderci M (2002) Protective role of supplemental vitamin E and selenium on lipid peroxidation, vitamin E, vitamin a, and some mineral concentrations of Japanese quails reared under heat stress. Biol Trace Elem Res 85(1):59–70

Sarkar J (2011) Chapter 4 Drought, its impacts and management: scenario in India. In: Shaw R, Nguyen H (eds) Droughts in Asian monsoon region (community, environment and disaster risk management, 8). Emerald Group Publishing Limited, Bradford, pp 67–85

Sharma D, Tiwari DP, Mondal BC (2010) Performance of crossbred female calves fed complete ration as mash or block vis-à-vis conventional ration. Indian J Anim Sci 80(6):556–560

Singh VK, Singh P, Verma A, Mehra UR (2008) On farm assessment of nutritional status of lactating cattle and buffaloes in urban, periurban and rural areas of middle Gangetic Plains. Livest Res Rural Dev 20(8):Article #130. http://www.lrrd.org/lrrd20/8/singh20130.htm

Sneath R, McIntosh F (2003) Review of Hydroponic Fodder Production for Beef Cattle. Department of Primary Industries: Queensland 84. McKeehen, p 54

Valtorta SE (2010) Animal production in a changing climate: impacts and mitigation, National Institute of agricultural technology Rafaela Experimental Station 2300 – Rafaela, Santa Fe Argentina, pp 12

Wilkinson JM, Wadephul F, Hill J (1996) Silage in europe – a survey of 33 countries. Chalcombe Publications, Welton

Wright DL, Mayo DE, Jowers HE (2015) Using drought-stressed corn for silage, hay, or grazing. SS-AGR-274. UF/IFAS Extension, University of Florida, Gainesville, Agronomy Department, p 3

Mitigation Options for GHG Emissions from Ruminants

21

Raghavendra Bhatta and P.K. Malik

Abstract

Livestock are one of the major contributors as well as sufferers of climate change. The adverse impact of climate change on livestock sector is now ubiquitary; however, its intensity is stratified. Livestock production among various agricultural sectors is considered one of the major fronts that is accountable for large greenhouse gas emission. The demand for livestock products is expected to accelerate that would essentially come from more livestock. The increasing livestock numbers would cost larger GHG emission. Livestock production and excrement storage contribute three major greenhouse gases, namely, carbon dioxide, methane and nitrous oxide into the atmosphere. The contribution of CO_2 in GHG emission from livestock is almost negligible due to its continuous cycling into the biological system. India alone contributes 10 Tg methane to the global pool every year that arises from the enteric fermentation. Major livestock species (cattle and buffalo) in the country is held accountable for 85–90% of the annual enteric methane emission. Countries such as India and China are expected to have maximum increase in enteric methane emission in the world during the next 20 years. Temperature, humidity and storage conditions are major deciding factors for the extent of emission from excrement. The GHG emission from manure management depends on the storage conditions. The anaerobic storage of the dung leads to its decomposition and subsequent CH_4 production, whilst aerobic storage results into N_2O emission. This chapter dealt with the GHG generated from livestock production including enteric fermentation and excrement management. Ameliorative and preventive measures are discussed in this chapter for reducing the emission of greenhouse gases that originates from livestock production.

R. Bhatta (✉) • P.K. Malik
ICAR-National Institute of Animal Nutrition and Physiology,
Bangalore 560 030, Karnataka, India
e-mail: ragha0209@yahoo.com; malikndri@gmail.com

© Springer Nature Singapore Pte Ltd. 2017
P.S. Minhas et al. (eds.), *Abiotic Stress Management for Resilient Agriculture*,
DOI 10.1007/978-981-10-5744-1_21

443

21.1   Introduction

Worldwide annual emission of greenhouse gases (GHGs) was estimated about 49 gigatonnes for the year 2005 (WRI 2011). Among the nations, China, the USA and 27 member states of the European Union (WRI 2011) contribute the maximum quantum, whilst India contributes only 4.25% of the total GHG emission (Fig. 21.1). Livestock are an integral component in agriculture sector that globally support the livelihood for more than one billion. About 1/8 of energy and 1/3 of protein requirement of the populace are fulfilled by this sector. Due to the fast changing food habits, the requirement for livestock products such as milk, meat and eggs would increase substantially. To meet this additional requirement, the livestock numbers have to proportionally be multiplied or the production has to be intensified. FAO (2006) in an estimate revealed that the bovine and ovine population is to be 2.6 and 2.7 billion more, respectively, by the end of 2050. Livestock production and climate change are connected through such a complex mechanism where adversity of one has the adverse impacts on another. The threat of climate change to livestock sector is ubiquitary, but the intensity of adversity is stratified depending on the prevailing agroclimatic situations. Livestock is being influenced by climatic variations both directly and indirectly through alterations in comfort (stress), compositional and fodder biomass availability, health, etc. In other words, livestock are both culprit and sufferer due to climate change.

21.1.1 Livestock and GHG Emissions

Livestock majorly emit carbon dioxide (CO_2), methane (CH_4) and nitrous oxide (N_2O) greenhouse gases (GHGs) into the surrounding atmosphere. However, due to continuous cycling of CO_2 in biological systems, it is not being considered whilst

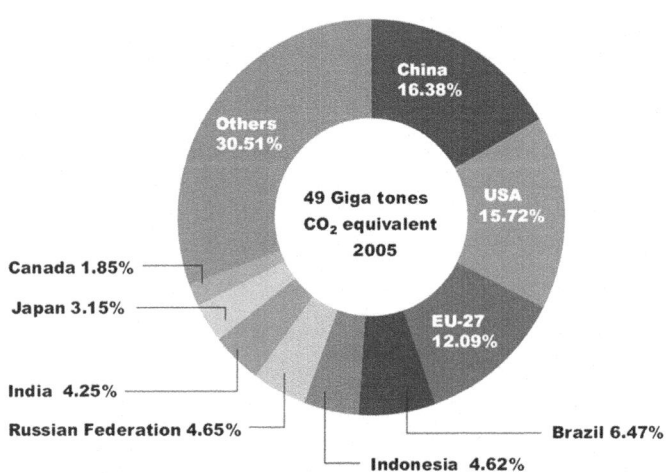

Fig. 21.1 Nation-wise greenhouse gas emissions (WRI 2011)

Table 21.1 Global GHG emission from various agricultural subsectors

Agricultural subsectors	Primary GHG emission	GHG emission (%)
Soil	N_2O	40
Enteric fermentation	CH_4	27
Rice cultivation	CH_4	10
Manure management	CH_4	7
Energy related	CO_2	9
Others	CH_4, N_2O	7
Total		100

quantifying the GHG emissions from livestock (FAO 2006; EPA et al. 2006). Agriculture including livestock is one of the sectors accountable for large emission of greenhouse gases in the atmosphere. On the whole, agriculture contributes about 14–15% of the global GHG (WRI 2005). Soil alone contributes about 40% of the total GHG emission in the form of N_2O from agricultural sector, whilst livestock sector constitutes 34% of GHG emission primarily as methane (27% enteric and 7% manure). Rice cultivation contributes 10% to the total GHG emission from agriculture, and the contribution from other subsectors including biomass burning is about 16% (Table 21.1). Both N_2O (46%) and CH_4 (45%) equally contribute to the total GHG emission from agricultural sector, whilst CO_2 only makes 9% of the total emission from this sector.

Livestock production is held accountable for the larger CH_4 emission that arises from enteric fermentation and excrement disposal, whilst N_2O emission is primarily from the manure management. Ruminants are one of the major sources of CH_4 emission in many countries; however, the emission from manure management due to different disposal systems is comparatively far less than the enteric fermentation. The confined animal operation where manure is primarily disposed in liquid form leads to the CH_4 emission. On the other hand, the disposal of excrement in aerobic environment likely represents the nitrogen loss in many forms including N_2O. Globally, livestock only account for 9% of CO_2 emission, but this sector contributes 35 and 65% of CH_4 and N_2O from anthropogenic sources, respectively. Livestock in Latin America emit the highest enteric CH_4 emission followed by Africa and China (Table 21.2). India with an emission figure of 8–10 Tg is also a significant contributor to the global enteric methane emission. These emissions are projected to increase substantially in the next 20 years (Table 21.3), whilst the increase in CH_4 from manure management may not be much. The enteric methane emission during the next 20 years is expected to increase by 0.7 and 0.8% compounded annual growth rate for India and China, respectively (Table 21.3). China will also have substantial increase in N_2O emission from excrement by the end of 2030 (Fig. 21.2).

21.1.2 Enteric CH_4: A Necessity for Ruminants

Degradation of complex structural carbohydrates is being accomplished in the rumen that accommodates diverse kind of microbes for the vested purpose. These

Table 21.2 Region-wise enteric methane emission

Region	Enteric methane emission (M tonnes CO_2-eq year^{-1})	Emission (%)
Africa	280	14.5
Latin America	460	23.8
North America	136	7.06
Non-EU former Soviet Union	97	5.03
Europe	188	9.83
India	218	11.3
China	259	13.4
Rest of Asia	175	9.09
Australia/New Zealand	88	4.57
Middle East	27.3	1.42
World	1929	100

Modified from Frank O'Mara (2011)

Table 21.3 CH4 emission (Mt CO_2 eq) and projections for the livestock sector

Major emitter		Projection			CAGR
	2010	2020	2030		
Enteric fermentation					
India	265	283	301		0.7
Brazil	225	237	234		0.2
China	162	179	191		0.8
USA	132	138	143		0.4
Pakistan	73	90	111		0.1
ROW	1088	1223	1365		1.04
Manure management		Projection			
	2010	2020	2030		
India	35	39	42		0.9
Brazil	10	11	12		0.8
China	79	83	87		0.5
USA	43	43	43		0.0
France	8	8	8		−0.4
ROW	166	178	192		0.8

Reproduced and modified from EPA (2013)

microbes work in a syntrophic fashion where the end metabolites of one group are used by other group of microbes as substrate. H_2, a major end by-product of the fermentation, is essentially removed from the anaerobic vat (rumen) to maintain the prerequisite favourable conditions for the rumen microbes. Fermentative H_2 is primarily used for the reduction of CO_2 into CH_4 which later on belch out in the surrounding atmosphere via oral route. Archaea, a distinct group of microbe in the rumen, are responsible for the reduction of CO_2 into CH_4. These microbes previously considered under bacterial domain, but recently they are classified (Woese et al. 1990) as under separate domain Archaea. Most of the Archaea species in the rumen primarily use H_2 as basal substrate; however, there are other species too that

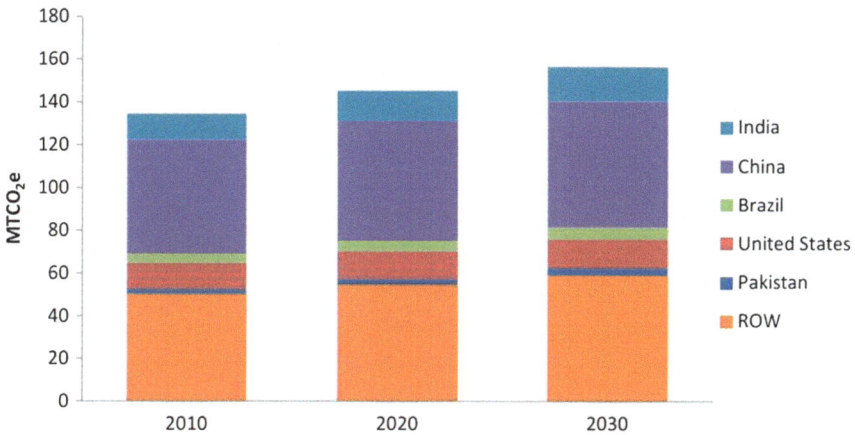

Fig. 21.2 Projections for N_2O emission from livestock sector (2010–2030; EPA 2013) Adapted from MAC Report 2013

use other substrate for methanogenesis. Thus, H_2 may be referred as one of the crucial metabolites, and its partial pressure in the rumen is a deciding factor for determining the extent of methanogenesis (Hegarty and Gerdes 1998). Due to the significant energy loss, rumen methanogenesis is said to be one of the necessary but wasteful processes.

21.1.3 Enteric CH_4 Emission: Indian Scenario

As per latest livestock census (19th census, Govt. of India), India has 512 million livestock, wherein cattle and buffaloes dominate and constitute 60% of the total population. On an average, Indian livestock annually emit 8–10 Tg enteric CH_4 wherein cattle and buffalo aggregately emit >90% of the total emission (Table 21.4). However, small ruminants such as sheep and goats contribute only 7.7% of the enteric methane emission in the country. The rest of the emission comes from the species which are very less in numbers and only confined to specific geographical locations, i.e. yak, mithun, etc. There is a disparity among the cattle species also: crossbred cattle emit more methane than their indigenous counterparts (46 vs 25 kg $h^- yr.^{-1}$).

21.1.4 Rumen Methane Amelioration: Challenges and Opportunities

The challenges related to the enteric methane emission and amelioration are summarised below (Table 21.5) for the easy understanding and to achieve the long-term methane reduction from livestock:

Table 21.4 Relative enteric methane emission by Indian livestock

Species	Methane emission (%)
Cattle	
Indigenous	31.3
Crossbred	14.9
Total from cattle	46.2
Buffalo	45.1
Sheep	2.6
Goat	5.06
Yak	0.03
Mithun	0.11
Horse/pony	0.08
Mule	0.04
Donkey	0.04
Camel	0.17
Pig	0.57
Total	100

Source: Bhatta et al. (2015)

- Lack of reliable and validated country-/region-specific database
- Improper methodologies for methane prediction
- Feed and fodder shortage
- Feed and fodder quality
- Low productivity
- Twinkling methane reduction attributes
- Complexity of rumen microbial system
- Unexplored community of rumen archaea

The enteric CH_4 emission from rumen is H_2 centric, and therefore, the successful reduction through any strategy under investigation should qualify the criteria of minimising H_2 production (Malik et al. 2015). Reducing the numbers of unproductive/low-productive animal, dipping of H_2 production in the rumen to obstruct the supply, alternating disposal of H_2 away from methanogenesis, targeting H_2-utilising microbes for indirect inhibition of methanogenesis and targeting rumen archaea for direct inhibition are most likely the opportunities for minimising the enteric methane emission (Table 21.5).

21.1.5 Manure Management and GHG Emission

Manure is a valuable resource that is obtained from livestock and serves as primary reservoir for various nutrients required by a crop at different physiological stages. The emission of CH_4 from manure management due to aerobic storage in our conditions is very less (0.1 Tg year^{-1}). Nevertheless, it is one of the listed sources for methane emission. The excrement storage in anaerobic conditions is being practised in the developed world where dung is mixed with water and stored as lagoons; it

Table 21.5 Ameliorative measures for enteric CH_4 mitigation

Measures	Opportunities/limitation	Remarks
Reducing livestock numbers	Due to a high number of low-producing or non-producing ruminants, CH_4 emission per kg of livestock product is high. Killing of such livestock is not possible due to the ban on cow slaughter	Low-productive animals should be graded up with rigorous selection for improving their productivity and less enteric methane emission
Feeding of quality feeds	Feed interventions are the best option for methane amelioration. The uninterrupted availability is a question mark. Area under pasture and permanent fodder production is declining or stagnant since the last three decades. Livestock are getting their fodders from 7 to 8% of the arable area	Improving the quality of fodders' availability seems unrealistic under the ever-increasing human population and food-feed-fuel competition scenario
Ionophores	Selective inhibition of microbes and failures to achieve the reduction in the long term are big issues. Animals turn back to normal level of emission after short time. Their use is banned in many European countries	May be tried in rotation as well as in combination for sustaining the reduction in long term
Ration balancing	Ration balancing with feed resources available at farmer's doorstep will improve the productivity with concurrent methane reduction at low input level	Farmers should be made aware about the importance of ration balancing and monetary advantages
Removal of protozoa	Removal of ciliate protozoa from the rumen results in lower CH_4 production. May witness less fibre digestibility. It is practically impossible to maintain protozoa-free ruminants	In spite of complete removal, partial defaunation may be achieved for enteric methane reduction without affecting the fibre digestion
Reductive acetogenesis	Thermodynamics favour methanogenesis in the rumen. The affinity of acetogens for H_2 substrate is considerably lower than methanogenesis. It cannot work until unless methanogens are targeted in the rumen	Reductive acetogenesis may be promoted by simultaneously targeting the rumen archaea. This will ensure less methane with additional acetate availability for the host animal
Plant secondary metabolites	Under the quality fodders' deficit scenario, the use of PSM as methane-mitigating agents is a good option. Dose optimisation and validation of methane migration potential in vivo on large scale are mandatory before recommendation	Inclusion at safe level without affecting the feed fermentability may be a viable option for enteric methane amelioration. Combined action of PSM on in vivo methane emission needs to be assessed

(continued)

Table 21.5 (continued)

Measures	Opportunities/limitation	Remarks
Nitrate/sulphate	Nitrate and sulphate hold the potential to reduce methane emission to a greater extent. These reductive processes are thermodynamically more favourable than methanogenesis. The end product from this productive process will not have any energetic gain for the animal. Intermediary products are toxic to the host animal	Probably slow-releasing sources for these compounds will reduce the toxicity chances caused by intermediary metabolites. Safe level of inclusion must be decided and tested on large number of animals by considering all the species accountable for methane emission
Active immunisation	This approach holds the potential for substantial methane reduction provided methanogenic archaea of the rumen is explored to a maximum extent for identifying the candidate target for the inclusion in vaccine	Information on species and biogeographic variation in methanogenic archaeal community should be explored for considering this approach for enteric methane amelioration
Disabling surface proteins	It is well established that methanogens adhere to the surface of other microbes for H_2 transfer through surface proteins. Identifying and disabling of these surface proteins will certainly reduce enteric methane emission by cutting supply of H_2	This is unexplored area and needs some basic and advance research for exploring the possibility
Bio-hydrogenation	Restricting H_2 supply to methanogens with alternate use of bio-hydrogenation decreases enteric methane amelioration. The use of fat/lipids at high level depresses fibre digestion. Of the total, about 5–7% of H_2 is utilised in this process	This approach is not practical due to high cost of fat/lipids and fibre depression at high level of use

serves as one of the potential sources for CH_4 emission. On the contrary, excrement storage in an open environment favours the large emission of N_2O from the decomposition. Apart from N_2O, nitrogen from excrement is also lost in other forms (NH_3 and NO_X). According to one estimate, India alone is accountable for 9.6% of the CH_4 emission from manure. The emission of CH_4 from excrement in the country is projected to remain constant (9.5–10.2%) during the next 40 years (Table 21.6). Indian livestock on an average contribute 3.9% of the world's total N_2O emission.

Excrement management is a deciding factor for determining the extent of GHG emissions that arise from microbial decomposition. Anaerobic storage and decomposition of the dung lead to larger emission of CH_4. These conditions mostly prevail when manure is stored as lagoons by mixing of excrement with water and urine. Temperature and the duration are two major factors that greatly affect the quantum

Table 21.6 Estimate and projected emissions of methane and nitrous oxide from manure management

	Manure methane (kg × 10⁵)		
	Estimated	Projected	
Methane	2010	2025	2050
World	11,414	12,849	15,046
India	1096	1221	1543
% of total	9.6	9.5	10.2
Nitrous oxide	Manure nitrous oxide (kg × 10⁵)		
	Estimated	Projected	
	2010	2025	2050
World	383	445	516
India	15.3	17.5	21.4
% of total	3.9	3.9	4.1

Source: Patra (2014)

of CH_4 produced. However, dung processing as stacks or piles favours the aerobic degradation and, hence, generates very less CH_4. Volatile solid (VS) is the major degradable fraction of excrement that decomposes and converts into CH_4 during anaerobic decomposition. The VS content/excretion either may be determined or retrieved from the literature. It can also be determined from the feed intake if the average daily VS excretion rates are not known. Feed intake for cattle and buffalo can be estimated using the 'enhanced' characterisation method. The VS content of dung represents the fraction of diet consumed but not digested and thus excreted as faecal material. The VS excretion rate is calculated using the following equation of Dong et al. (2006):

VOLATILE SOLID EXCRETION RATES

$$VS = \left[GE \bullet \left(1 - \frac{DE\%}{100} \right) + \left(UE \bullet GE \right) \right] \bullet \left[\left(\frac{1 - ASH}{18.45} \right) \right]$$

CH_4 emission factor from VS excretion rate may be worked out using the following equation:

$$EF_{(T)} = \left(VS_{(T)} \bullet 365 \right) \bullet \left[B_{o(T)} \bullet 0.67 kg\,/\,m^3 \bullet \sum_{S,k} \frac{MCF_{S,k}}{100} \bullet MS_{(T,,S,,k)} \right]$$

where:

EF (T) = annual CH_4 emission factor for livestock category T, kg CH_4 animal⁻¹ yr.⁻¹

VS (T) = daily volatile solid for livestock category T, kg dry matter animal⁻¹ day⁻¹

Bo (T) = maximum methane producing capacity for manure produced by livestock category T, m³ CH_4 kg⁻¹ of VS excreted

0.67 = conversion factor

MCF (S,k) = methane conversion factors for each manure management system S by climate region k, %

MS (T,S,k) = fraction of livestock category T's manure handled using manure management system S in climate region k, dimensionless

N_2O emissions from manure are a combined result of both nitrification and denitrification process. Nitrogen, carbon content and storage duration are three major factors that decide the quantum of N_2O emission from manure. Nitrification, a necessary step, is likely to occur if there is adequate supply of oxygen during excrement storage. In the process of denitrification, nitrite and nitrates are transformed into N_2O and dinitrogen (N_2). N_2O generation from the manures requires either nitrite or nitrate for the conversion to oxidised forms of nitrogen. A common equation used for quantifying the N_2O emission from manure is given below.

Direct N_2O emission from manure management:

$$N_2O_{D(mm)} = \left[\sum_{S} \left[\sum_{T} \left(N_{(T)} \bullet Nex_{(T)} \bullet MS_{(T,S)} \right) \right] \bullet EF_{3(S)} \right] \bullet \frac{44}{28}$$

where:

N_2OD (mm) = direct N_2O emissions from manure management, kg N_2O yr.$^{-1}$

N (T) = number of head of livestock species/category T in the country

Nex (T) = annual average N excretion kg animal^{-1} yr.$^{-1}$

MS (T, S) = fraction of total annual nitrogen excretion for each livestock species/category T that is managed in manure management system S in the country, dimension less

EF3 (S) = emission factor for direct N_2O emissions from manure management system S in the country, kg N_2O-N/kg N in manure management system S

S = manure management system

T = species/category of livestock

44/28 = conversion of (N_2O-N) (mm) emissions to N_2O (mm) emissions

21.1.6 Preventive and Ameliorative Measures for Minimising GHG Emissions from Excrement

For mitigating the CH_4 and N_2O emissions from the manure management, precautionary/ameliorative measures are compiled and presented in Table 21.7.

Table 21.7 Precautionary/ameliorative measures for reducing GHG emissions from manure management

GHG	Measures
CH_4	Handle manure as a solid or deposit it on pasture rather than storing it in a liquid-based system such as a lagoon (this may increase N_2O emission)
	Capture CH_4 from manure decomposition to produce renewable energy
	Avoid adding straw to manure which serve as a substrate for anaerobic bacteria
	Apply manure to the soil as soon as possible because prolonged storage of manure encourages anaerobic decomposition that results in increased CH_4
	Avoid manure application when the soil is extremely wet, as this leads to anaerobic conditions and increased methane emissions
	Improve animal's feed conversion efficiency by either feeding quality feeds or processing to decrease GHG emissions
	Cover lagoons which involve placing an impermeable floating cover (e.g. plastic cover) over the surface of the tank or lagoon to capture GHGs
N_2O	Apply manure shortly before crop growth to allow for the maximum amount of available nitrogen to be used by the crop
	Avoid applying manure in the late fall and winter because these conditions lead to high emission of N_2O and high nitrogen loss in the spring (Anonymous 2004)
	Avoid applying manure when the weather is hot and windy, or before a storm, because these conditions can increase N_2O emissions
	Implement soil and water management practices such as improving drainage, avoiding soil compaction, increasing soil aeration and using nitrification inhibitors for nitrogen gas production instead of nitrogen oxide (Anonymous 2004)
	Spread manure evenly around the pasture
	Maintain healthy pastures by implementing beneficial management grazing practices to help increase the quality of forages
	Include low-protein levels and the proper balance of amino acids in the diet to minimise the amount of nitrogen excreted, particularly in urine. Use phase feeding to match diet to growth and development
	Move fresh manure to a covered storage facility to reduce adding moisture, which reduces the amount of N_2O emission
	Storage in below-ground facilities with lower temperatures reduces microbial activities

21.2 Conclusion

Enteric fermentation and manure management are two integral components associated to livestock production that contribute to the methane and nitrous oxide emission. These two gases have very high global warming potential than that of CO_2. Apart from an accelerating factor in global warming, enteric methanogenesis also deprives the host animal from a substantial fraction of the biological energy. Developing countries like India cannot afford such a high loss of biological energy. Ameliorative measures for the enteric methane mitigation must be devised by considering local feed resources availability, seasonal variation, etc. Researchers should focus on the approaches that may persist for long and achieve 18–25% reduction in enteric methane emission. The problem of GHG emission (methane and nitrous oxide) is not uniform across the countries and varies according to the storage conditions of the excrement. Due to storage of dung as heap in the country, methane emission from manure is not alarming, and hence, our full focus should be on tackling nitrous oxide from excrement.

References

Anonymous (2004) Manure management and greenhouse gases. Alberta Bulletin No. 11, p. 1–4. available at: http://www1.agric.gov.ab.ca/$department/deptdocs.nsf/all/cl10038/$file/GHGBulletinNo11Manuremanagement pdf ? Open Element

Bhatta R., Malik P. K. and Prasad C. S. (2015) Chapter 15: Enteric methane emission: status, mitigation and future challenges- an Indian perspective. In: Malik PK, Bhatta R, Takahashi J, Kohn RA, Prasad CS (eds.) Livestock production and climate change. Publisher CABI, Oxfordshire. ISBN-13: 978 1 78064 432 5, pp. 229–244

Dong H, Mangino J, McAllister TA, Hatfield JL, Johnson DE, Lassey KR, de Lima MA, Romanovskya A (2006) Emission from livestock and manure management (Chapter 10). In: IPCC guidelines for national greenhouse gas inventories. Available online at http://www.ipcc-nggip.iges.or.jp/public/2006gl/pdf/4_Volume4/V4_10_Ch10_Livestock.pdf

EPA (2013) Global Mitigation of Non-CO_2 Greenhouse Gases: 2010-2030. Washington, DC

EPA, Holtkamp J, Hayano D, Irvine A, John G, Munds Dry O, Newland T, Snodgrass S, Williams M (2006) Inventory of U.S. greenhouse gases and sinks: 1996–2006. Environmental protection agency, Washington, DC. http://www.epa.gov/climatechange/emissions/downloads/08_Annex_1-7.pdf

FAO (2006) Livestock's long shadow. Food and Agriculture Organisation, Rome

Hegarty RS, Gerdes R (1998) Hydrogen production and transfer in the rumen. Recent Adv Anim Nutr 12:37–44

MAC Report (2013) V Agricultural sector. Available at: https://www.epa.gov/sites/production/files/2016-06/documents/mac_report_2013-v_agriculture.pdf

Malik PK, Bhatta R, Takahashi J, Kohn RA, Prasad CS (2015) Livestock production and climate change. CABI book published by CAB International UK & USA, 396 p

O'Mara FP (2011) The significance of livestock as a contributor to global greenhouse gas emissions today and in the near future. Anim Feed Sci Technol 166-167:7–15

Patra AK (2014) Trends and projected estimates of GHG emissions from Indian livestock in comparison with GHG emissions from world and developing countries. Asian Australas J Anim Sci 27:592–599

Woese CR, Kandlert O, Wheelis ML (1990) Towards a natural system of organisms: Proposal for the domains Archaea, Bacteria, and Eucarya. Evolution (NY) 87:4576–4579

WRI (2005) Navigating the numbers: greenhouse gas data and international climate policy. World Resources Institute. Available at: http://pdf.wri.org/navigating_numbers.pdf

WRI (2011) Climate Analysis Indicators Tool (CAIT), version 9.0, Washington DC

Mitigation of Climatic Change Effect on Sheep Farming Under Arid Environment

22

S.M.K. Naqvi, Kalyan De, Davendra Kumar, and A. Sahoo

Abstract

Livestock is the integral part of agricultural systems all over the world. However, in India, climate change has become a serious concern for ensuring nutritional security for the growing population. Small ruminants, especially pastoral farming, serve a major livelihood option and are embedded deep in the culture of resource-poor small and marginal farmers of arid and semiarid western India. The breeds of these arid and semiarid region are well adapted to the local climatic condition and have amalgamated themselves to very harsh climatic factors in the region. The local native animals of this region have their own adaptive mechanism of altering physiological, neuroendocrine, biochemical, cellular, and molecular process to encounter the stress; still, they need to endure stressful conditions due to high temperature, low feed, and water scarcity. All these constraints expose the sheep production into heat stress, nutritional stress, water stress, walking stress, and their combinations. All the stress factors affecting sheep production directly and indirectly and ultimately lead to compromised performance, lower efficiency, and increased mortality and affect the immune system. Giving the poor farmer's economic security, under changing climatic scenario, sheep production has to be sustainable by combating the detrimental effect of different mitigation strategies. In the present chapter, the mitigation strategies have been discussed which include genetic improvement, breeding management, grazing management, nutritional management, utilization of unconventional feed resources, antioxidant supplementation, water management, shelter management, and disease management. Basically, all these strategies are

S.M.K. Naqvi (✉) • K. De • D. Kumar • A. Sahoo
ICAR-Central Sheep and Wool Research Institute,
Avikanagar, Malpura, Rajasthan 304 501, India
e-mail: naqvismk@gmail.com; kalyande2007@gmail.com;
davendraror@gmail.com; sahooarta1@gmail.com

© Springer Nature Singapore Pte Ltd. 2017
P.S. Minhas et al. (eds.), *Abiotic Stress Management for Resilient Agriculture*,
DOI 10.1007/978-981-10-5744-1_22

based on a physical modification of the environment, genetic modification, and improved nutritional management. To get optimum production under changing climatic scenario, holistic approach is needed as per the environmental conditions and available resources.

22.1 Introduction

In the twenty-first century, worldwide climate change is a critical challenge to the mankind as well as the animal kingdom. The developing countries like India are more vulnerable to this threat due to their dependence on natural resources and high poverty levels. High temperature hinders productive and reproductive performance, impairs metabolic health and metabolic status, and weakens immune response. A pertinent increase of drought may affect forage and crop production (Nordone et al. 2010). These situations will likely worsen further under the ensuing changing climatic scenario. Under such situation, sheep production may play a critical role for the development of sustainable and environmentally sound production system (Ben-Salem 2010).

Sheep husbandry is one of the sustainable livelihood options for the people living in arid and semiarid regions. But nature has been a little harsh toward these regions by providing less vegetative resources, climatic variability, and water scarcity. In India, the arid and semiarid region comprise almost 53.4% of the total land area. These regions cover mostly rangelands. Therefore, the poor landless small and marginal farmers of the region rear their livestock through a low-input pastoral system of management (Naqvi et al. 2013b). Nearly 21% of the sheep belong to arid and semiarid region, and they play an important role in the rural economy with its multifaceted utility of wool, meat, milk, skin, and manure. Out of 41 well-recognized breeds of India, 11 breeds survive/exist/sustain/belong to arid and semiarid region. The harsh climatic condition compels the breeds of these areas to make some behavioral, physiological, and morphological adjustment to adopt to the climatic variability. Therefore, the native breeds of these regions have become well adapted to scarce feed supply and shortage of drinking water and able to survive in extreme heat, can migrate long distances, and show resistance against tropical diseases. Nonetheless, the adaptation mechanism supports their survivability, but with a compromised production efficiency.

22.2 Impact of Climate Change on Sheep Production

The pronounced impacts of climate change on sheep production are apparent in breed composition, population and distribution, feed and fodder scarcity, shrinkage of grazing land, reproductive and productive disorders, spread of diseases, poor performances, consumer demand and market trend for wool and meat, etc. Some of the important attributable climate-driven changes are discussed here.

Table 22.1   Current changes in the sheep population in India and top five sheep-producing states

Year	Population (million)	State	Population (million)
1992	50.78	Andhra Pradesh	26.40 (40.5)
1997	57.49	Karnataka	9.6 (14.7)
2003	61.47	Rajasthan	9.1 (13.9)
2007	71.56	Tamil Nadu	4.8 (7.3)
2012	65.07	Jammu and Kashmir	3.4 (5.2)

Figures in parenthesis refer to percent increase in two decades (Source: Livestock Census 2012)

22.2.1 Sheep Population and Distribution

A reduction in the sheep population from 74 million in 2010 to 65 million in 2012 (Livestock Census 2012) raised an alarm on its future status. The reasons may be multifarious, an increasing demand for meat and higher slaughter rate to preference shift toward the rearing of other livestock species. The effect of climate change on the distribution of sheep population is visualized by the recent change in the sheep population from Rajasthan to Andhra Pradesh (Table 22.1). The sheep population in Rajasthan was 10.62 million in 1997, which declined to 10.05 million in 2003 livestock census with an overall reduction of 5.72%. On the other hand, the sheep population in Andhra Pradesh increased by 13.9%.

The frequent drought and famine situations and continuous declining of grazing resources, both in terms of quality and quantity, could be the reasons for the decline of sheep in Rajasthan during the period. In the present climate-changing scenario, migration becomes hard and harsh for the farmers. Flock size is declining sharply over the period; the average flock size was 100–120 in 1990, which has reduced to 70–80 in 2000 and 35–40 in 2008 in the semiarid region of Rajasthan (Shinde and Sejian 2013).

22.2.2 Breed Composition and Distribution

Farmers of arid and semiarid regions always prefer sheep breeds which can withstand thermal and nutritional stresses and able to walk long distances during migration. The evolution of indigenous sheep breeds in different agroecological niches through natural selection over the time period also supports these characteristic changes. Farmers are taking initiative for adopting breeding strategies to cope with the changing climate. Kheri sheep, developed by crossing Malpura with Marwari, are hardy, produce carpet-type wool, have better walking efficiency, and thrive well under the migratory system (Shinde and Sejian 2013). Magra breed found in and around Bikaner, known as Bikaneri Chokla, produce excellent carpet wool with a fiber diameter of 32–35 μ and lustrous character. However, their population got reduced considerably with the intermixing of breeds during migration.

22.2.3 Grazing Land

In arid and semiarid regions of the country, 45–50% of the land is utilized for grazing purposes, and in extreme arid regions of Rajasthan, 90% of the land is utilized for grazing purposes. The area of common property resources (CPR) has declined by 26–63% during the last three decades. This CPR is also now in a state of dwindling and produce only 0.2–0.3 Mg ha^{-1} of dry fodder under normal rainfall, which reduced to 40–50% with changing climate due to decline and erratic distribution of rain (Sankhyan et al. 1999).

22.3 Multiple Stress Factors and Their Effect

22.3.1 Heat Stress

Heat stress is the foremost effect of climate change in the arid and semiarid region in sheep production. Heat stress can disrupt the physiology and productive performance of an animal. The increase in body temperature caused by heat stress has direct, adverse consequences on cellular function. Production losses in domestic animals are largely attributed to increase in maintenance requirement associated with sustaining a constant body temperature and altered feed intake (Indu et al. 2014). Heat stress also hampers the male and female reproduction. Heat stress affects estrus percent and duration, conception rate, lambing rate, and birth weight of lambs (Maurya et al. 2004). In addition, it influences plasma estradiol and progesterone concentration in Malpura ewes (Sejian et al. 2011). Further, a reduction in superovulatory response and embryo production occurred in Bharat Merino ewes (Naqvi et al. 2004) while it reduced the feed intake and body weight (Sejian et al. 2010a).

22.3.2 Nutritional Stress

Another potential stressor in this region is nutritional scarcity which is increasing due to the scarcity of good quality and quantity of fodder. Poor and marginal farmers of semiarid tropics depend a lot on livestock and agriculture for their livelihood. In extensive production system, the well-being of livestock entirely depends on the herbage. Livestock of this region mainly remained undernutrition during late spring and summer due to increased energy output for thermoregulation and concurrent reduction in energy intake (Sahoo et al. 2013). This can result in impairing reproduction and production efficiency of grazing animals (Ali 2008). Sheep and goat flocks migrate in search of feeding and grazing resources from one place to another and from the lower hills (800–900 m above MSL) up to the high alpine pastures (3600–4800 m above MSL) and back again as the climate determines the availability of fodder and forage (Sahoo 2013). But the magnitude of stress was severe when only 60% or less of their nutritional requirement are available (Sejian et al. 2014).

Thermal stress and feed scarcity are the major predisposing factors for the low productivity of ewes under hot semiarid environment. Moberg (2000) described a hypothetical scheme of how two stressors can summit together, and their total impact might be severe on biological functions. The combined thermal and nutritional stress had a severe impact on biological functions, though the native breeds possessed considerable adaptive mechanisms to overcome these stresses (Sejian et al. 2010b).

22.3.3 Water Stress

Water scarcity during summer is a serious problem in this region. Therefore, during this period, availability of good quality drinking water is reduced for all the species. Drinking water is often a limiting factor for livestock in grazing areas of the semiarid region (Sahoo et al. 2015a). Some breeds of small ruminants could survive up to 1 week with little or even no water in hot arid and semiarid regions (Nejad et al. 2014; Chedid et al. 2014), but water shortage affected animal's physiological homeostasis leading to loss of body weight, low reproductive performance, and a decreased resistance to diseases (Barbour et al. 2005). The feed intake, average daily gain, physiological responses, hematological parameters, and reproduction of ewes impaired during water stress (Kumar et al. 2016). Well-adapted Malpura ewes have the capacity to adjust their physiochemical response and reproduction comfortably up to 20% of water restriction during hot summer, and the impact increased with the magnitude of water restriction (Kumar et al. 2016).

22.3.4 Walking Stress

During the summer, another key constraint of arid and semiarid tropical environments is their low biomass productivity (Sahoo et al. 2015b). The availability of feed in the rangeland reduced substantially and most of the time, it is not available in the dry season. As biomass density (feed) per unit area remained very low, the sheep tries to increase the grazing time each day as well as disperse more widely. Therefore, other than thermal stress and feed scarcity, the animals need to walk long distances for grazing (Naqvi et al. 1991) and are exposed to exercise stress during walking in such an environment. These stresses lead to alterations in the process of homeostasis and metabolism. The changes in respiration rate, heart rate, rectal temperature, plasma cortisol, thyroxine, and triiodothyronine showed that native sheep have the capability to adapt to long-distance walking. The negative effect on growth performance shows that productive performances are compromised while trying to adapt to long-distance walking (Sejian et al. 2012b).

22.3.5 Multiple Stresses

Under field condition, most of the time, the above stresses, i.e., heat stress, nutritional stress, water stress, and walking stress, occur in combinations and simultaneously and causing multiple stress. Multiple stress affects body weight, respiration rate, pulse rate, rectal temperature, sweating rate, triiodothyronine, thyroxine, cortisol, hemoglobin, packed cell volume, glucose, and total protein (Sejian et al. 2013). Along with that, both conception and lambing rate also reduced significantly under multiple stress (Sejian et al. 2012a, b). Hence, selection of adapted animal breeds is very important for sustaining animal production under this challenging environment. The breeds of this region have developed adaptive capacity to survive in these adverse conditions.

22.4 Free Ranging and Metabolizable Energy Requirement

In free ranging, if the food availability is scarce, the animals are forced to travel to meet their energy requirements, and this extra energy drain can be an important contributor to the metabolizable energy (ME) need for maintenance (Sahoo 2013). Sheep have adapted well to environmental conditions prevailing in arid lands, being able to obtain an adequate diet even when forage is scarce. Anyhow, the energy requirements in open range may increase severalfold over values assessed for restrained animals due to walking stress in search of scarce feed resources in rugged and otherwise inaccessible terrain. According to NRC (1981), the energy requirement of goat increased 25% under grazing conditions due to increased muscular activity. The energy requirement increased 50% in semiarid rangeland pasture and in slightly hilly land due to higher muscular activity for less availability of biomass in these regions. The requirement of energy increased up to 75% in case of long-distance travel on sparsely vegetated grassland and/or mountainous transhumance pasture. Both the increased energy expenditure of eating and the energy expended in walking would account to this increased maintenance requirement. The heat production (HP) attained $401 \text{ kJ kg}^{-0.75} \text{ d}^{-1}$ for restrained goats. The increase in HP for unrestrained over restrained goats was 43.1%, and daily distance traveled (say 2 km) accounted for 9% of the extra HP.

22.4.1 Sheep Diseases

Other than direct effect of climate change, sheep diseases and resistance are also affected by reduction in natural grazing land and increasing metabolic disorders. Poor-quality green grass due to vitamin A deficiency increases the incidence of corneal opacity and night blindness. The deficiency of vitamins A, D, and B_1 and minerals, viz., calcium, phosphorus, zinc, and copper, is not only prevailing unabated but also is registering an upward trend in the recent past (Shinde and Sankhyan 2007). Copper, cobalt, selenium, zinc, and iodine are some of the trace mineral deficiencies resulting in anemia, retarded growth, and reproductive disorders. Apart from these, deficiencies of vitamins A, D_3, E, B_1, and C were also identified in the flock. The nutritional stress increases the case of pregnancy toxemia and neonatal death due to poor milk yield and immunity, prone to many infectious diseases. The increased morbidity and mortality and declined production under climate change would lead to economic losses to farmers (Singh et al. 2010). Under the traditional system, breeding of sheep is associated with acacia and khejri pods in arid and semiarid regions. The climate change and its effect on tree pods would adversely affect the reproduction in sheep and goats. A poor vegetation cover of CPR would provoke soil ingestion because of their close grazing habit, e.g., soil ingestion increased up to 39.1, 15.6, and 46.4% of dry matter intake in sheep during drought and famine conditions. Soil that accumulates in the rumen and reticulum impairs the digestion and production resulting in mortality under prolonged exposure (Shinde et al. 2005).

22.4.2 Marketing and Economics

The climate change is leading to shrinkage of grazing lands and scarcity of feed and fodder from grazing lands which are directly hampering production performance of sheep. As a consequence, farmers are being forced to sell their lambs early (3–4 months age weighing only 12–14 kg) instead of regular practice of selling lambs of 20–22 kg weight (9–12 months old) (Shinde and Sankhyan 2010). Earlier, flocks were managed completely on grazing resources. However, the grain supplementation to young stock has now become a common practice as a protection against the vagaries of nutritional stress. This along with climate change induced poor health and increased susceptibility to diseases which are adding to the cost of production (Shinde and Sejian 2013).

22.5 Mitigation Strategies

In changing climatic situation of arid and semiarid region, keeping in view the poor farmer's economic security, sheep production has to be sustainable by combating the deleterious effect of different stressors. Thermal variability challenges the animal's ability to maintain energy, thermal, water, and hormonal and mineral balance. Reducing stress on sheep requires multidisciplinary approaches that emphasize animal nutrition, housing, and animal health management. The effect of hot climate can alleviate through suitable management strategies like provision of shade, diminishing the ground reflection, suitable shelters, restriction of feeding during hotter parts of the day, postponement of shearing to cooler season, and control of mating so that late pregnancy occurs in comparatively cooler season (Naqvi et al. 2013b).

22.5.1 Genetic Improvement

The increasing demand for meat urges a serious need of sturdy, heat-tolerant, disease-resistant, and relatively adaptable breeds to the adverse conditions (Moran et al. 2006). In such situation, some of the indigenous breeds are able to cope better than the crossbred. The native sheep breed like Malpura, Chokla, Marwari, and Magra are well adopted in arid and semiarid region and can tolerate heat and nutritional stress. Disease- and parasite-resistant sheep breed need to propagate in this region as the resistant line does not require drenching and reduces the problem of drug resistance, drenching cost, and drug residues in meat and milk (Swarnkar et al. 2009). Pattanwadi and Malpura sheep yield 1.2–1.4 and 0.6–0.7 liter of milk daily which can provide extra security and economic support in the dry regions. The crossbreeding of Awassi sheep that is well adapted to dry and hot conditions, with Pattanwadi or Malpura, to improve milk yield of native sheep appears to be a better option. Fat tail sheep breeds can withstand the harsh climate and can serve as a source of income for poor farmers (Shinde and Sejian 2013).

22.5.2 Breeding Management

In Rajasthan and Gujarat, the farmer used to tie the prepuce of rams with cotton threads to avoid matings during undesirable seasons. Most sheep breeding takes place in July–August, i.e., immediately after the onset of the monsoon and some of it in March–April, when stubble grazing and *Acacia* and *Prosopis* pods are available to the animals. Though 80–100% of animals exhibit estrus throughout the year, considering lambing percent and lamb survival and growth, breeding in March–April and August–September is preferable (Acharya 1982).

22.5.3 Grazing Management

The sheep mainly graze either on pastures, wasteland, fallow lands, or stall fed to meet their nutritional requirements as per their physiological state and levels of production. The climate variability is changing the pattern of land utilization, deforestation, degradation of pasture, and range lands, which ultimately are increasing the gap between availability and requirement of nutrients. In such condition, trees and shrubs provide green biomass of moderate to high digestibility and protein content when other feed reserves are scarce. There are several options for making effective use of shrubs and tree foliage. In many parts of the country, small ruminants are maintained on top feeds than conventional fodder resources. Sheep and goat browse on tree leaves. The tree leaves are also harvested, sun dried, and stored at proper stage; thereafter, they are supplemented during the lean summer months, at the time of feed scarcity to maintain the small ruminant production. Although lopping of standing trees in the forest is prohibited, removal of fallen tree leaves is allowed.

It is estimated that 300–350 million Mg dry fallen leaves and grass is available from the forests which has better CP value than the crop residues. Almost 43 million Mg of this resource, if processed, can be effectively used in livestock feeding. However, fallen tree leaves cannot maintain the sole feeding of sheep. Therefore, plantations of palatable trees like subabul (*Leucaena leucocephala*) can become a good alternative. Mainly locally available species should be preferred which produce leaf residues during lean period (Sahoo et al. 2013).

22.5.4 Nutritional Management During Migration

In Rajasthan, almost 0.5 million sheep are on permanent migration following established migratory routes and seasons (Acharya 1985). From western districts of Rajasthan, almost 1.0 million migrates for 6–9 months to the neighboring states. Most flocks begin migration between October and February and follow set migratory routes and return by May to June or before the onset of the monsoon. The sheep flocks are grazed on uncultivated land during the monsoon, and when the kharif crops are harvested, the animals are permitted to graze on crop stubbles. During later part of the year, beginning in September/October, most nonmigratory flocks graze on the harvested fields and reserve forests in their migratory tracts. During the extreme summer months, the flocks graze during the cooler hours of the day. About 60% of the flocks are penned in open fields away from the house; the rest are penned in temporary courtyards made out of thorny bushes or earth near the house (Naqvi et al. 2013b).

The sheep and goats fulfill their maintenance requirements from grazing during migration and thus considered to be not under stress (Sahoo 2013). However, nutrient requirements are higher during critical physiological stages, viz., last quarter of lactation and early part of lactation. In view of poor pasture quality and high stocking density of grazing lands, such animals remain underfed. Similarly, adequate nutrition of lambs/kids through their dam during pre-weaning phase is important. Healthy lambs of comparatively higher birth weight suckling from optimum fed ewes grow faster and attain finishing weights at an early age. Therefore, the supplementation of ewes/does on pasture is important for economic fat lamb production.

22.5.5 Utilizing Unconventional Feed Resources

The cactus species like prickly pear cactus (*Opuntia ficus-indica*) has been propagated successfully as an alternative to provide biomass and water to sheep during summer scarcity (CSWRI 2013–14). The approach is applicable elsewhere in dry arid region, where feed scarcity along with water scarcity is severe. Lopping of fodder trees like khejri (*Prosopis cineraria*), ardu (*Ailanthus* spp.), and neem (*Azadirachta indica*) serves as the best option during the harsh periods of the year.

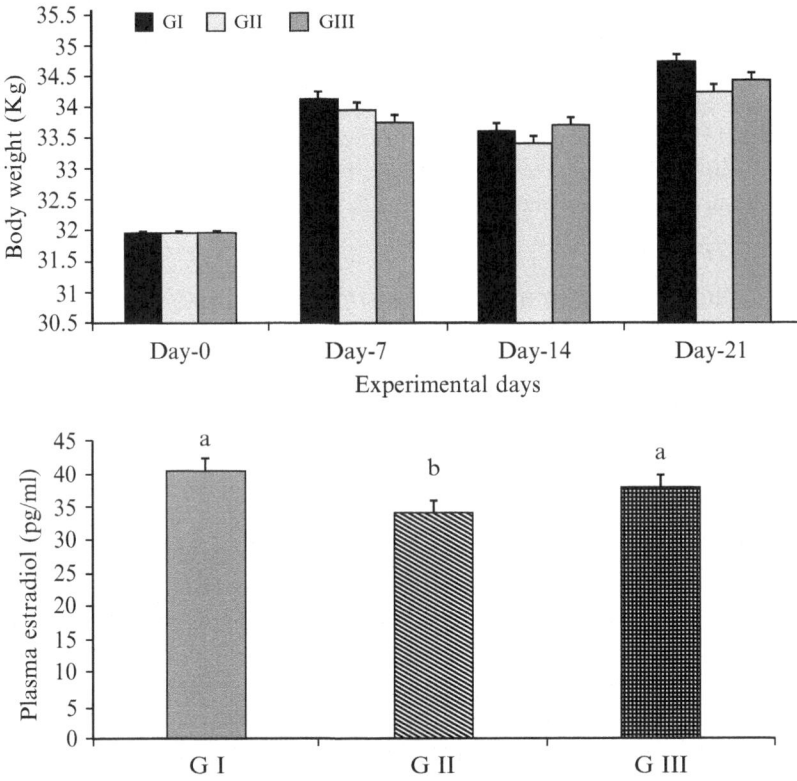

Fig. 22.1 Effect of antioxidant supplementation on body weight and plasma cortisol levels during summer (Source: Sejian et al. 2014) *GI*, control; *GII*, heat stress; *GIII*, heat stress + antioxidant supplementation

22.5.6 Antioxidant Supplementation

Heat stress stimulates excessive production of free radicals (Bernabucchi et al. 2002; Sivakumar et al. 2010). In such situation, the deficiency of dietary trace element affects physiological function and particularly on reproduction. The dietary and tissue balance of antioxidant nutrients is important in protecting tissues against free radical damage. Antioxidants such as vitamins C and E are free radical scavengers, which protect the body defense system against excessively produced free radicals during heat stress and stabilize the health status of the animal. Free radicals and reactive oxygen species play a number of significant and diverse roles in reproductive biology (Agarwal et al. 2006). Mineral mixture and antioxidant like zinc, cobalt, chromium, and selenium and vitamin E supplementation in the feed protected the ewes from adverse effects of heat stress (Fig. 22.1). The adverse effect of heat stress on the productive and reproductive efficiency of Malpura ewes (Tables 22.2 and 22.3) was alleviated by mineral mixture and antioxidant supplementation (Sejian et al. 2014).

Table 22.2 Effect of abiotic stresses on physiological, blood biochemical, hormonal, and reproductive responses of sheep

Parameters	Thermal, nutritional, and combined stress				Walking stress		Water stress		Multiple stress	
	Control	Thermal	Nutritional	Combined	Control	Stressed	Control	Stressed	Control	Stressed
Growth performance										
Initial body weight (kg)	33.8 ± 2.6	33.5 ± 1.9	34.7 ± 1.7	34.9 ± 1.5	27.4 ± 2.4	27.1 ± 2.43	38.8	38.9	30.9 ± 1.6	31.5 ± 1.1
Final body weight (kg)	39.7 ± 2.7ᵃ	35.2 ± 1.5ᵃᵇ	30.4 ± 1.5ᵇ	30.0 ± 1.4ᵇ	32.3 ± 0.6ᵃ	30.1 ± 0.64ᵃᵇ	38.5	37.6	32.6 ± 0.2ᵃ	29.5 ± 0.5ᵇ
ADG (g)	169.1 ± 0.01ᵃ	47.7 ± 0.1ᵇ	−122.6 ± 0.1ᶜ	−138.0 ± 0.1ᶜ	140.0 ± 1.5ᵃ	86.0 ± 1.50ᵇ	−7.4	−37.1	50.3 ± 1.01ᵃ	−56.6 ± 1.4ᵇ
Physiological response										
Respiration (breath min⁻¹)	40.1 ± 2.4ᵃ	130.8 ± 5.6ᵇ	29.8 ± 1.2ᶜ	107.5 ± 5.0ᵇ	53.2 ± 1.3ᵃ	69.3 ± 1.3ᵇ	67.3 ± 2.6	62.9 ± 2.6	58.2	68.7
Pulse rate (beats min⁻¹)	74.6 ± 2.8aᵇ	80.1 ± 2.5ᵃ	67.5 ± 2.9ᵇ	69.4 ± 3.6aᵇ	76.0 ± 2.8	79.8 ± 3.5	71.4 ± 1.3	68.6 ± 1.3	72.4	59.6
Rectal temperature (°F)	102.0 ± 0.1ᵃ	103.3 ± 0.1ᵇ	101.7 ± 0.2ᶜ	102.9 ± 0.1ᵈ	101.8 ± 0.1ᵃ	102.6 ± 0.1ᵇ	101.9 ± 0.1	101.8 ± 0.1	102.1	102.6
Blood biochemical										
Hb (g dl⁻¹)	11.9 ± 0.4ᵃ	9.5 ± 0.3ᶜ	10.8 ± 0.3ᵇ	8.6 ± 0.2ᵈ	9.7 ± 0.1ᵃ	11.0 ± 0.1ᵇ	11.3 ± 0.5ᵃ	12.3 ± 0.5ᵇ	9.9 ± 0.1	10.9 ± 0.1
PCV (%)	41.5 ± 1.5ᵃ	31.3 ± 1.6ᵇ	35.8 ± 1.7ᵇ	26.5 ± 2.1ᶜ	34.2 ± 0.8ᵃ	39.5 ± 0.8ᵇ	41.6 ± 1.8ᵃ	44.0 ± 1.8ᵇ	30.4 ± 0.8	35.8 ± 0.8
Glucose (mg dl⁻¹)	52.1 ± 2.4ᵃ	47.8 ± 1.6ᵃᵇ	44.2 ± 1.9ᵇ	43.0 ± 2.5ᵇ	56.9 ± 1.1ᵃ	47.3 ± 1.1ᵇ	51.2 ± 1.4ᵃ	46.3 ± 1.4ᵇ	45.4 ± 0.5	39.4 ± 0.5
Total protein (g dl⁻¹)	8.9 ± 0.3ᵃ	8.0 ± 0.4ᵃᵇ	7.9 ± 0.3ᵃᵇ	7.1 ± 0.3ᵇ	7.7 ± 0.3	7.8 ± 0.3	–	–	7.3 ± 0.1	6.4 ± 0.1

(continued)

Table 22.2 (continued)

	Thermal, nutritional, and combined stress				Walking stress		Water stress		Multiple stress	
Total cholesterol (mg dl⁻¹)	52.3 ± 1.9[a]	43.0 ± 2.3[b]	42.7 ± 2.2[b]	35.6 ± 2.3[c]	70.9 ± 5.1	67.8 ± 5.3	65.4 ± 1.6[a]	55.5 ± 1.6[b]	9.9 ± 0.1	10.9 ± 0.1
ACP (KA units)	2.1 ± 0.1[a]	1.5 ± 0.1[bc]	1.8 ± 0.1[ab]	1.2 ± 0.2[c]	6.6 ± 0.1	1.3 ± 0.1	–	–	–	–
ALP (KA units)	5.9 ± 0.3[a]	5.1 ± 0.4[b]	5.6 ± 0.3[ab]	4.1 ± 0.3[c]	7.8 ± 1.3	7.9 ± 1.3	–	–	–	–
Hormonal profile										
T_3 (nmol L⁻¹)	1.7 ± 0.01[a]	1.3 ± 0.0[b]	1.4 ± 0.0[b]	1.1 ± 0.01[c]	2.3 ± 0.1[a]	1.8 ± 0.1[b]	–	–	2.4 ± 0.1	1.4 ± 0.1
T_4 (nmol L⁻¹)	76.2 ± 4.2[a]	58.9 ± 3.2[b]	63.0 ± 3.6[b]	46.0 ± 5.1[c]	150.4 ± 6.3[a]	135.7 ± 6.3[b]	–	–	58.4 ± 0.1	24.4 ± 0.1
Cortisol (nmol L⁻¹)	18.6 ± 1.3[d]	77.0 ± 5.2[a]	8.5 ± 1.1[e]	46.4 ± 3.6[b]	16.1 ± 1.1[a]	28.1 ± 1.1[b]	87.5 ± 5.9[a]	69.3 ± 5.9[b]	14.8 ± 1.1	31.0 ± 1.1
Insulin (microIU mL⁻¹)	47.4 ± 2.9[a]	39.8 ± 3.2[ab]	35.2 ± 1.4[bc]	26.4 ± 2.8[c]	–	–			–	–
Estradiol (pg mL⁻¹)	14.6 ± 1.0[a]	12.1 ± 0.7[b]	12.8 ± 0.9[b]	10.0 ± 0.7[c]	31.5 ± 10.0	29.4 ± 9.0	27.7 ± 2.3[a]	19.2 ± 2.5[b]	–	–
Progesterone (ng mL⁻¹)	3.3 ± 0.6[c]	4.5 ± 0.3[ab]	4.0 ± 0.3[bc]	5.2 ± 0.3[a]	5.9 ± 2.8	5.7 ± 2.9	3.3 ± 1.7	3.6 ± 1.4	–	–

Reproductive performance

Ewes in heat (%)	85.7[a]	57.1[b]	85.7[a]	71.4[ab]	–	–	–	100	85.7	66.67[a]	41.7[b]
Estrus duration (hr)	38.0 ± 2.4[a]	23.4 ± 3.3[b]	28.5 ± 5.7[bc]	18.8 ± 3.8[bd]	–	–	–	34.3 ± 6.0	28.0 ± 6.4	32[a]	14.4[b]
Estrus cycle length (days)	18.2 ± 0.3[b]	20.3 ± 0.7[ab]	18.0 ± 0.3[b]	22.3 ± 1.7[a]	–	–	–	17.0 ± 0.4	17.0 ± 0.5	18.2 ± 0.3[b]	23.6 ± 1.5[a]
Conception rate (%)	71.43[a]	42.86[ab]	57.14[ab]	28.57[b]	–	–	–	–	–	83.3[a]	50[b]
Lambing rate (%)	71.43[a]	42.86[ab]	57.14[ab]	28.57[b]	–	–	–	–	–	83.3[a]	50[b]

Source: Sejian et al. (2010a, b; 2011; 2012a, b; 2013); Kumar et al. (2016)

Thermal stress, by keeping ewes in climatic chamber at 40 °C and 55% RH for 6 h/day between 1000 and 1600 h; *nutritional stress*, by providing restricted diet, i.e., 30% intake of control; *combined stress*, by inducing both thermal and nutritional stress; *walking stress*, by compelling ewes to walk 14 km in two spans between 0900 and 1500 h; *water stress*, by providing restricted water, i.e., 40% intake of control; *multiple stress*, by inducing thermal, nutritional, and walking stress at a time

Table 22.3 Effect of mineral antioxidant supplementation on reproductive performance of Malpura ewes

Parameters	Control	Heat stress	Heat stress + antioxidant supplementation
Estrus (%)	85.71	85.71	100
Duration (h)	34.29 ± 6.10	24.86 ± 5.77	36.86 ± 2.42
Length (days)	14.00 ± 0.01	14.33 ± 0.31	14.29 ± 0.18

Source: Sejian et al. (2014)

22.5.7 Water Management

Water scarcity during summer is a serious problem in this region. Water management of the sheep in the desert area is costly affair. During peak summer, literally no groundwater is available in the areas falling under Thar Desert and even the adjoining the same. Watering sheep in nearby ponds is essential for their good health, and for this, watering twice a week is sufficient to sustain health and productivity of sheep. Magra and Marwari breeds are genetically adapted to the desert conditions (Mittal and Ghosh 1983) and are able to produce and reproduce normally even with prolonged periods of drinking naturally occurring saline waters (TSS up to 3500 mg l^{-1}). The native sheep breeds have also the capability to overcome water shortage of about 40% for a month with little effect on their physiology and rejuvenation (Table 22.4; De et al. 2015b). Watershed management is a very effective strategy for long-term supply of water in the areas prone to drought. Integrated watershed management (IWM) provides a framework to integrate natural resource management with community livelihoods in a sustainable way. Several approaches can be initiated starting from desert rainwater harvesting for saving more water and its utilization. Development of watershed will not only help to fertilize the region but also for better management and upkeep of the livestock and sheep in particular (Naqvi et al. 2017).

22.5.8 Shelter Management

The sheep keepers of arid and semi-arid region generally construct different types of structure within their limited resources which are mostly primitive to semi-migratory type. In the southern and the northeastern part of Rajasthan, the flocks are kept in mud hut only in the night and the rest of them in an open field under grazing condition. These huts are mostly attached to the outer side of the owner's house. Roofs are commonly thatched with long rough grass. In the northern and western part of Rajasthan, the flocks are mostly migratory and are kept in open fields during the night. They usually flock back to their native village after 2 days of grazing for drinking water (De et al. 2013).

Table 22.4 Effect of rehydration on water-restricted Malpura ewes

Stage of watering	Body weight (kg)		FI (DMI g $W^{-0.75}$ d)		Hemoglobin (g%)		Packed cell volume (%)		Glucose (mg dL^{-1})		Cortisol (nmol L^{-1})	
	G1	G2	G1	G2	G1	G2	G1	G2	G1	G2	G1	G2
Ad libitum	38.8	38.8	57.5[xy]	54.7[xy]	12.5	12.95	38.7	34.9[y]	50.0[a]	45.3[b]	21.3[y]	22.3[y]
Restriction	38.4	37.5	54.1[y]	51.5[y]	11.8[b]	14.8[a]	41.1	47.3[x]	49.4	45.9	55.4[ax]	34.7[bx]
Rehydration	38.8	38.9	62.6[x]	60.4[x]	12.1	14.1	32.7	35.9[y]	49.2	47.4	29.3[y]	20.2[y]

G1, 20% water restricted; G2, 40% water restriction; Source: De et al. (2015b)

Table 22.5 Meteorological data of different housing during winter

Weather parameters	Control	Bamboo dome	Thermocol insulated
Maximum temperature	8.5 ± 0.7^c	14.6 ± 0.5^a	11.8 ± 0.6^b
Minimum temperature	25.0 ± 0.3^a	23.9 ± 0.4^{ab}	22.6 ± 0.4^b
Relative humidity (%)	59.2 ± 4.5	66.4 ± 2.0	56.4 ± 3.7
Temperature–humidity index^2	12.5 ± 0.9^b	16.1 ± 0.6^a	14.0 ± 1.0^{ab}
Wind velocity (m s^{-1})	5.58 ± 0.4	5.58 ± 0.4	5.58 ± 0.4

Source De et al. (2015a)
a,b, c values within a row with different superscripts differ significantly at P<0.05

The traditional sheep housing depends on the local customs and availability of materials. The sheep houses are mostly constructed in one corner of the main family house, an overhang attached to the roof of the house of open yard with no roof or at the basement under the family house or a separate house of thatched roofs. But mostly the traditional sheds are insufficient in drainage and ventilation. It is generally recommended that smallholder producers who keep a few animals should construct a shed attached to their main house using locally available inexpensive materials to reduce the cost of construction. Generally, thatched roof houses are often sufficient. New knowledge about the animal response to the environment continues to be developed to reduce the effect of climate change (De et al. 2014). Keeping this in view, different models of sheds have been tried. A bamboo dome is one of such sheds, where lambs are kept to protect from severe cold during the night (Table 22.5). Generally, lambs are kept inside the dome in the shed that maintains higher minimum temperature at night and provides comfortable microclimate (De et al. 2015a). Cold- and heat-protected sheds have also been tried to protect the weaner or grower lambs from winter as well as summer. In this special type of shed, the floor is usually at lower level than the outside, and its roof is prepared from thermocol-insulated PV sheet. These types of shed maintain lower maximum temperature and higher minimum temperature than asbestos-roofed shed (De et al. 2015a). Yagya-type shed is also constructed to protect animals during extreme summer. It has pagoda-style roof which provides better ventilation. The wall is double walled with a hollow space in the middle, which is filled up with sand and the sand remains moist by continuous water drip. This innovative strategy protects the animal from direct hot wind as well as provide extra evaporative cooling that keep the microclimate inside the shed comfortable (De et al. 2017). Another option is preparing a mat of locally available dry grasses at the open side wall and sprinkling water three to four times in a day on that mat to provide evaporative cooling. Similarly, canopy curtain may be used in the open side of the wall to provide protection in the chilling cold nights. Increased roof height building with ridge ventilation is constructed to protect from direct solar radiation as well as better ventilation during the summer months.

22.5.9 Disease Management

Sheep in the arid region are relatively less susceptible to diseases due to less humidity that is an impediment for the pathogen development. Most mortalities occur due to liver-fluke infestation, while other causes include enterotoxaemia, anthrax, and foot-and-mouth disease. Sheep pox also results in serious mortality, although it usually becomes epidemic once in 3–4 years. So far health care has been given little attention, and the sick animals are generally treated using indigenous medicines. Animals infested with gastrointestinal parasites, though not a major menace during dry periods, must be treated after 1 week of onset of monsoon and thereafter on case-to-case basis. Nutritional stress also leads to chronic worm problems and thus nutritional supplementation is necessary (Naqvi et al. 2013a). However, considering the overall poor facilities for prevention and control of sheep diseases, a comprehensive plan (e.g., health, disaster reduction) needs to be developed to prevent proliferation of diseases with climate change. In endemic regions, vaccination can be done as a protective measure. By combining improved empirical data and refined models with a broad view of the production system, robust projection of disease risk can be developed (Swarnkar and Singh 2013).

22.6 Conclusion

In India, a major portion of the rural poor community depends greatly on small ruminants for their survival. Like other agricultural systems, small ruminant production system is also likely to be affected directly or indirectly by climate change. This would impact the performance and profitability in sheep production system by lowering feed intake and nutrient utilization and production. Although sheep husbandry, itself, is a climate-smart agriculture, still there is a scope to make it better and sustainable under the pressure of climate change and increasing demand. For this purpose, the present science and technology has to concentrate more on thematic issues related to climatic adaptation, dissemination of new understanding in rangeland ecology, and holistic understanding of pastoral resource management. The issues need to be addressed are early warning system, multiple stress research, simulation model, water availability, exploitation of genetic potential of native breeds, suitable breeding program, and nutritional intervention. However, the present existing condition concentrates on reduction of the magnitude of climate change in the long term, i.e., mitigation and adaptation. In this process, the local livestock farmers should have a key role in determining what adaptation and mitigation strategies existing husbandry system will support to make the production sustainable in the changing climate scenario. Furthermore, integration of new technologies into the research and proper transfer of those into the field offers many opportunities in the development of climate-smart sheep production system.

References

Acharya RM (1982) Sheep and Goat Breeds of India. FAO Animal Production and Health Paper 30, FAO of United Nations, Rome

Acharya RM (1985) Small ruminant production in arid and semi-arid Asia. In: Timon VM, Hanrahan JP (eds) Small ruminant production in the developing countries. Proceedings of an Expert Consultation held in Sofia, Bulgaria

Agarwal A, Gupta S, Sikka S (2006) The role of free radicals and antioxidants in reproduction. Current Opinion in Obst and Gyn 18:325–332

Ali A, Hayder (2008) Seasonal variation of reproductive performance, foetal development and progesterone concentrations of sheep in the subtropics. Reprod Dom Anim 43: 730–734

Barbour E, Rawda N, Banat G, Jaber L, Sleiman FT, Hamadeh S (2005) Comparison of immuno-suppression in dry and lactating Awassi ewes due to water deprivation stress. Vet Res Commun 29:47–60

Ben-Salem H (2010) Nutritional management to improve sheep and goat performances in semiarid regions. Rev Bras Zootec 39:337–347

Bernabucchi V, Ronchi B, Lacetera N, Nardone A (2002) Markers of oxidative status in plasma and erythrocytes of transition dairy cows during hot season. J Dairy Sci 85:2173

Chedid M, Jaber LS, Giger-Reverdin S, Duvaux-Ponter C, Hamadeh SK (2014) Review: water stress in sheep raised under arid conditions. Canadian J Anim Sci 94:243–257

De K, Ramana RC, Kumar D, Sahoo A (2013) Shelter Management- a means to resist extreme climatic variables: climate Resilient small ruminant production, pp 75–83. (eds) Sahoo A, Kumar D, Naqvi, SMK. National Initiative on Climate Resilient Agriculture (NICRA), Central Sheep And Wool Research Institute

De K, Kumar D, Singh AK, Sahoo A, Naqvi SMK (2014) Seasonal variation of physiological response in ewes of farmers' flocks under semi-arid tropical environment. Biol Rhythm Res 45:397–405

De K, Kumar D, Kumar K, Sahoo A, Naqvi SMK (2015a) Effect of different types of housing on behavior of Malpura lambs during winter in semi-arid tropical environment. J Vet Behav 10:237–242

De K, Kumar D, Singh AK, Kumar K, Sahoo A, Naqvi SMK (2015b) Resilience of Malpura ewes on water restriction and rehydration during summer under semi-arid tropical climatic conditions. Small Rumin Res 133:123–127

De K, Kumar D, Singh AK, Kumar K, Sahoo A, Naqvi SMK (2017) Effect of protection against hot climate on growth performance, physiological response and endocrine profile of growing lambs under semi-arid tropical environment. Trop Anim Health Prod 49(6):1317–1323

Indu S, Sejian V, Naqvi SMK (2014) Effect of short term exposure to different environmental temperature on physiological adaptability of Malpura ewes under semi-arid tropical environment. J Vet Sci Med Diagn 3:1–3

Kumar D, De K, Singh AK, Kumar K, Sahoo A, Naqvi SMK (2016) Effect of water restriction on physiological responses and certain reproductive traits of Malpura ewes in semi-arid tropical environment. J. Vet Behav 12:54–59

Livestock Census (2012) Nineteenth livestock census of India, department of animal husbandry, dairying and fisheries, ministry of agriculture, government of India, Krishi Bhawan, New Delhi

Maurya VP, Naqvi SMK, Mittal JP (2004) Effect of dietary energy level on physiological responses and reproductive performance of malpura sheep in the hot semi-arid regions of India. Small Rumi Res 55:117–122

Mittal JP, Ghosh PK (1983) Long-term saline drinking and female reproductive performance in Magra and Marwari sheep of the Indian desert. J Agric Sci 101:751–754

Moberg GP (2000) Biological responses to stress. Implications for animal welfare. In: Moberg GP, Mench JP (eds) Biology of animal stress. CAB International, Wallingford, pp 1–21

Moran DS, Eli-Berchoer L, Heled Y, Mendel L, Schocina M, Horowitz M (2006) Heat intolerance: does gene transcription contribute? J Appl Physiol 100:1370–1376

Naqvi SMK, Hooda OK, Saxena P (1991) Some plasma enzymes of sheep under thermal, nutritional and exercise stresses. Indian Vet J 68:1045–1047

Naqvi SMK, Maurya VP, Gulyani R, Joshi A, Mittal JP (2004) Effect of thermal stress on superovulatory response and embryo production in Bharat merino ewes. Small Rumi Res 55:57–63

Naqvi SMK, Sejian V, Karim SA (2013a) Effect of feed flushing during summer season on growth, reproductive performance and blood metabolites in Malpura ewes under semiarid tropical environment. Trop Anim Health Prod 45:143–148

Naqvi SMK, De K, Gowane G (2013b) Sheep production system in arid and semi-arid regions of India. Ann Arid Zone 52(3&4):1–9

Naqvi SMK, De K, Kumar D, Sejian V (2017) Climate changes, water use and survival during severe water deprivation. In: Sejian V, Bhatta R, Gaughan J, Malik PK, Naqvi SMK, Lal R (eds) Sheep production adapting to climate change. Springer, Singapore, pp 173–187

Nejad JG, Lohakare JD, West JW, Sung KI (2014) Effects of water restriction after feeding during heat stress on nutrient digestibility, nitrogen balance, blood profile and characteristics in Corriedale ewes. Anim Feed Sci Technol 193:1–8

Nordone A, Ronchi B, Lactera N, Ranieri MS, Bernabucci U (2010) Effect of climate changes on animal production and sustainability of livestock systems. Livestock Sci 130:57–69

NRC (1981) Nutrient requirements of goats. National Research Council, National Academic Press, Washington, DC

Sahoo A (2013) Nutritional issues in grazing and migratory sheep and goats. In: Pattanaik AK, Jadhav SE, Dutta N, Das A, Kamra DN, Dass RS (eds) Clinical nutrition approaches for health and productivity of farm animals. CAFT in Animal Nutrition, Indian Veterinary Research Institute, Izatnagar, pp 174–181

Sahoo A, Kumar DK, Naqvi SMK (2013) Strategies for sustaining small ruminant production in arid and semi-arid regions. In: Climate Resilient Small Ruminant Production, (Eds) Sahoo A, Kumar D, Naqvi, SMK. National Initiative on Climate Resilient Agriculture (NICRA), Central Sheep and Wool Research Institute, p. 20–34

Sahoo A, Naqvi SMK, Kalyan D (2015a) Annual progress report, ICAR-Central sheep and wool research Institute. Avikanagar, Rajasthan

Sahoo A, Bhatt RS, Tripathi MK (2015b) Stall feeding in small ruminants: emerging trends and future perspectives. Indian J Animal Nutrit 32:353–372

Sahoo A, Sankhyan SK, Sharma SC (2015c) Pasture in small ruminant production: issues and solutions. In: Eco-responsive feeding and nutrition: linking livestock and livelihood. Pattanaik AK, Verma AK, Jadhav SE, Dutta N, Saikia BN (eds). Thematic Papers, Proceedings of 9th Biennial Animal Nutrition Association Conference, January 22–24, 2015a. Guwahati, p 17–27

Sankhyan SK, Shinde AK, Karim SA (1999) Seasonal changes in biomass yield, nutrient intake and its utilization by sheep maintained on community grazing land. Indian J Animal Sci 69:617–620

Sejian V (2013) Climate change: impact on production and reproduction, adaptation mechanisms and mitigation strategies in small ruminants: a review. The Indian J Small Rumt 19(1):1–21

Sejian V, Maurya VP, Naqvi SMK (2010a) Adaptability and growth of Malpura ewes subjected to thermal and nutritional stress. Trop Anim Health Prod 42:1763–1770

Sejian V, Maurya VP, Naqvi SMK (2010b) Adaptive capability as indicated by endocrine and biochemical responses of Malpura ewes subjected to combined stresses (thermal and nutritional) in a semi-arid tropical environment. Int J Biometeorol 54:653–661

Sejian V, Maurya VP, Naqvi SMK (2011) Effect of thermal, nutritional and combined (thermal and nutritional) stresses on growth and reproductive performance of Malpura ewes under semi-arid tropical environment. J Animal Physiol & Animal Nutrit 95:252–258

Sejian V, Maurya VP, Kumar K, Naqvi SMK (2012a) Effect of multiple stresses (thermal, nutritional, and walking stress) on the reproductive performance of Malpura ewes. Vet Med Int. doi:10.1155/2012/471760

Sejian V, Maurya VP, Naqvi SMK (2012b) Effect of walking stress on growth, physiological adaptability and endocrine responses in Malpura ewes in a semi-arid tropical environment. Int J Biometeorol 56:243–252

Sejian V, Maurya VP, Kumar K, Naqvi SMK (2013) Effect of multiple stresses on growth and adaptive capability of Malpura ewes under semi-arid tropical environment. Trop Anim Health Prod 45:107–116

Sejian V, Bahadur S, Naqvi SMK (2014) Effect of nutritional restriction on growth, adaptation physiology and estrous responses in Malpura ewes. Anim Biol 64:189–205

Shinde AK, Sankhyan SK (2007) Mineral profile of cattle, buffaloes, sheep and goats reared in humid southern-eastern plains of semi-arid Rajasthan. Indian J Small Rumin 13(1):39–44

Shinde AK, Sankhyan SK (2010) Nutritional stress and early disposal of lambs for mutton production in semi-arid region of Rajasthan. Proceeding of National Seminar on Stress Management in Small Ruminant Production and Product processing 29:9–31

Shinde AK, Sejian V (2013) Sheep husbandry under changing climate scenario in India: an overview. Indian J Animal Sci 83(10):998–1008

Shinde AK, Sankhyan SK, Singh NP, Verma DL (2005) Soil ingestion and its effect on dry matter intake and digestibility in sheep grazing on native ranges of semiarid region of Rajasthan. Indian J Small Rumin 11(1):37–42

Singh RK, Sanjay K, Sanjay B, Rajender K (2010) Changing diseases pattern in small ruminants vis -a-vis climate change. In: Karim SA, Joshi A, Sankhyan SK, Shinde AK, Shakyawar DB, SMK N, Tripathi BN (eds) Climate change and stress management: sheep and goat production. Satish Serial Publisher, New Delhi, pp 566–586

Sivakumar AVN, Singh G, Varshney VP (2010) Antioxidants supplementation on acid base balance during heat stress in goats. Asian-Aust J Animal Sci 23:1462–1468

Swarnkar CP, Singh D (2013) Effect of climate change on livestock Diseases. In: Sahoo A, Kumar D, SMK N (eds) Climate Resilient Small Ruminant Production. National Initiative on Climate Resilient Agriculture (NICRA), Central Sheep And Wool Research Institute, Jaipur, pp 53–68

Swarnkar CP, Singh D, Sushil K, Mishra AK, Arora AL (2009) Study on Malpura sheep selected for resistance to Haemonchus contortus. Indian J Animal Sci 79:577–581

Policy Support: Challenges and Opportunities in Abiotically Stressed Agroecosystem

23

K. Palanisami, T. Mohanasundari,
and Krishna Reddy Kakumanu

Abstract

Abiotic stress causes more than 50% losses in crop productivity and hence got major concerns for food and nutritional security. Two case studies addressing the impacts of abiotic stress on agricultural sector, adaptation measure taken up and needed policy options are included. The first case addresses the impact of climatic variables in Godavari River basin of Telangana where the impact of climate change on yield of paddy, groundnut and maize crops had been assessed using the Just-Pope production function. Climate change has serious effect on groundnut (with high yield variation of 69–90%), rice (with moderate yield variation of 23–38%) and maize (with negligible yield variation). Case two discusses about different adaptation strategies followed by agro-silvipastoral farmers to manage the abiotic (drought) stress in Tamil Nadu where among the 17 strategies identified, 12 were indexed as important strategies undertaken. At the time of severe drought stress, farmers used cotton waste as livestock feed, gave vaccination and added shade to protect the livestock from cold and heat stress. Majority of the farmers are only medium adopters, and hence there is an increasing need for creating awareness among the farmers on latest stress management practices by strengthening the extension services and capacity building programmes.

K. Palanisami (✉) • K.R. Kakumanu
International Water Management Institute, Hyderabad, India
e-mail: palanisami.iwmi@gmail.com; kakumanuk@gmail.com

T. Mohanasundari
Tamil Nadu Agricultural University, Coimbatore 641003, Tamil Nadu, India
e-mail: yogana2007@gmail.com

© Springer Nature Singapore Pte Ltd. 2017
P.S. Minhas et al. (eds.), *Abiotic Stress Management for Resilient Agriculture*,
DOI 10.1007/978-981-10-5744-1_23

23.1 Introduction

Agriculture sector in general and crop production in particular are more sensitive to different kinds of stresses that occur in different periods of time and space. These stresses are normally classified into biotic and abiotic stresses. Biotic stress in agriculture encompasses pests and diseases of crops, inimical parasites and microbial infections in animals as well as zoonotic disease-based health problems in animals and humans. Abiotic stresses are "suboptimal environmental conditions caused by non-living factors that are harmful to a living organism". Some types of primary abiotic stresses include drought, salinity, cold, heat, etc. Abiotic stress-related factors affect severely the agricultural production and the livelihood of farmers especially in tropical and subtropical countries where larger proportion of work force is directly depending on climate-sensitive agriculture sector. Among the countries, India is more challenged with multitude of several abiotic and biotic stresses as a result of unfavourable climate and soil conditions resulting in salt stress, low and high temperature stress, flooding stress, chemical stress, oxidative stress and other related stress types. Droughts encompass the global ecosystem as a whole, but the impacts may vary from region to region. According to Miyan (2015), the increasing biophysical vulnerability contexts and intensity in the Asian LDCs cause adverse effects on food security, human health, biodiversity, water resources, hydroelectric power generation, streams, perennial springs and livelihood. Drought is also responsible for increasing pollution, pests and diseases and forced migration and famine. In recent years, there has been a general increase in extreme events including floods, droughts, forest fires and tropical cyclones in the Asian continent (Grover et al. 2003). Climate change-resultant abiotic environment especially changes in hydrological cycles (Rowntree 1990) and temperature regimes may alter the composition of agroecosystems; (Sutherst et al. 1991; NACCAP 2012).

Abiotic stresses, which cause more than 50% losses in crop productivity, are the major concerns for food and nutritional security of additional 0.4 billion Indians by 2050 (Wang et al. 2007). This loss is caused mainly by high temperature (40%), salinity (20%), drought (17%), low temperature (15%) and other forms of stresses (Ashraf et al. 2008). Further it is estimated that only 9% of the world area is conducive for crop production, while 91% is afflicted by various stressors (NIASM 2015).

Agroecosystem environment is largely governed by interactions between abiotic (temperature, humidity, rainfall, soil factors, pollutants, etc.) and biotic (crop plants, weeds, insect pests, pathogens, nematodes, etc.) components. The abiotic stress factors modulate the effects of biotic stresses and are most harmful when they occur in combination (Mittler 2006), greatly influencing crop growth and productivity to the extent of 80% (Oerke et al. 1994; Theilert 2006). Thus, climate-induced changes may affect our ability to expand the food production area as required to feed the burgeoning population of more than ten billion people projected for the middle of the next century. Therefore, understanding abiotic stress responses in plants, animals and fishes and enhancing stress resilience are the most demanding areas in agricultural research. In this context, Bennet et al. (2003) had indicated the possible

Table 23.1   Management of abiotic stress – possible interventions and likelihood progress

Abiotic stress	Possible interventions	Likelihood of progress (in 5 years)
Saline soils and water (presence of salts and of sodium salts)	Increase salt tolerance of rice	High
	Improve leaching; apply gypsum, improve various agronomic practices	High
Waterlogging	Improve water and salt management	High
	Increase tolerance for waterlogging	Medium
	Improve water management	Medium
Water pollution (agrochemicals, industrial waste products)	Improve dosage of agro-chemicals and waste water treatment	Low
Acid sulphate soils (low pH, toxic anaerobic conditions in root zone)	Short duration varieties and seedling vigour	High
	Improve water management	Medium
High and low temperatures	Increase heat and cold tolerance at flowering	Medium
	Improve cooling mechanism for leaves	High
Dry periods and droughts	Improve irrigation infrastructure	Medium
	Increase drought tolerance	High
	Short duration varieties and seedling vigour	High
Floods	Increase tolerance for submergence	High
	Improve water management at river basin level	Low
Air pollution	Improve environmental quality (industrial and urban waste gases)	Low

Source: Bennett (2003)

interventions and the likelihood of major progress that can be done for the management of abiotic stresses (Table 23.1).

Impact of climate change on crop production has been well documented in the works of Palanisami et al. 2015. It is observed that the Indian climate has also undergone significant changes showing increasing trends in annual temperature with an average of 0.56 °C rise over last 100 years (IPCC 2007; Rao et al. 2009; IMD 2010). Warming was more pronounced during post monsoon and winter season with increase in number of hotter days in a year (IMD 2010). Even though there was slight increase in total rainfall received, the number of rainy days has decreased. The rainfed zone of the country has shown significant negative trends in annual rainfall (De and Mukhopadhyay 1999; Lal 2003; Rao et al. 2009). The semiarid regions of the country had maximum probability of prevalence of droughts of varying magnitudes (20–30%), leading to sharp decline in water tables and crop failures (Lal 2003; Rao et al. 2009; Samra 2003). By the end of next century (2100), the temperature in India is likely to increase by 1–5 °C (De and Mukhopadhyay 1999; Lal 2003; IPCC 2007; IMD 2010). According to the estimates of NATCOM (2004), there will be 15–40% increase in rainfall with high degree of variation in its distribution. Apart from this, the country is likely to experience frequently occurring extreme events like heat and cold waves, heavy tropical cyclones, frosts, droughts and floods (NATCOM 2004; IPCC 2007).

Already, the productivity of Indian agriculture is limited by its high dependency on monsoon rainfall which is most often erratic and inadequate in its distribution (Chand and Raju 2009). The country is experiencing declining trend of agricultural productivity due to fluctuating temperatures (Samra and Singh 2004; Aggarwal 2008; Joshi and Viraktamath 2004), frequently occurring droughts and floods (Samra 2003), problem soils and increased outbreaks of insect pests (Joshi and Viraktamath 2004; Srikanth 2007; Dhawan et al. 2007; IARI News 2008; IRRI News 2009) and diseases. These problems are likely to be aggravated further by changing climate which put forth major challenge to attain food security.

Intensive agriculture practices to meet the demands of ever increasing population have caused land degradation problems and also consequently increased the magnitude of abiotic stressors. Further, agricultural intensification through modifications to the environment (like increasing use of irrigation, agrichemicals) and the expansion of farming into undisturbed lands affect natural ecosystems. Hence, in the context of global climate change, it is important to address the abiotic stresses threatening sustainability of agricultural production systems. Hence, understanding abiotic stress responses in crop plants, insect pests and their natural enemies is an important and challenging area in future agricultural research and education. In this context, there is a need to develop simple and low-cost biological methods for the management of abiotic stress, which can be used easily on short-term basis. Also, there is much scope for abiotic stress management and improving the adaptation mechanisms. This chapter includes salient issues that underpin the economics of addressing the impacts of abiotic stress on agricultural sector, adaptation measure that can be taken up and some policy options. Two case studies are discussed below with respect to the impacts and adaptation aspects of abiotic stresses.

23.2 Case Study 1: Impact of Temperature and Precipitation on Crop Yields

Given the importance of abiotic stresses in agriculture, this case study mainly addresses the impact of temperature (both max and min) and precipitation in Godavari River basin of India under varying climate change scenarios.

23.2.1 Approach and Methodology

More accurate region-specific predictions for changes in temperature and rainfall are needed to capture the impact of climate change. Gosain (2011) has applied data from Providing Regional Climates for Impact Studies (PRECIS), a regional circulation model (RCM) for projecting climate changes in Godavari basin. PRECIS is the Hadley Centre portable regional climate model, developed for a grid resolution of

Table 23.2 Projected changes in climatic variables

Change from baseline to	Mean temperature (°C)		Mean precipitation (Per cent)		
	Kharif	Rabi	Kharif	Rabi	Overall
Mid-century (2021–2050)	1.93	2.22	12.5	17.6	13.6
End-century (2071–2098)	4.03	4.28	13.0	53.4	17.8

Source: Calculations based on those reported by Gosain (2011)

0.44° × 0.44°. It captures important regional information on summer monsoon rainfall missing in its parent GCM simulations. The changes in temperature and precipitation (from base line period, 1960–1990) predicted for mid-century (2021–2050) and end-century (2071–2098) (Gosain 2011) were used in getting the projected change in temperature and precipitation for the *Kharif* (June to November) and *Rabi* (December to April) season in study region (Table 23.2).

Based on these changes, two scenarios are formulated, one for the mid-century and the other one for end-century. The mid-century scenario for *Kharif* season is an increase of 1.93 °C and an overall increase of 13.6% in precipitation. This scenario is denoted by 1.93 °C/13.6% for *Kharif*, and for the *Rabi* season, the scenario is 2.22 °C/13.6%. Similarly for the end-century, the scenarios for *Kharif* and *Rabi* are 4.03 °C/17.8% and 4.28 °C/17.8%. It should be noted that in all these scenarios only the annual change in precipitation (and not seasonal changes) is considered. The reason is annual precipitation reflects inter-seasonal water accumulation. These predicted changes are used in the mean and variance functions to predict the climate change-induced average yield and variability in yield.

23.2.2 Study Area, Data and Variables

The Sri Ram Sagar Project (SRSP) in Godavari River basin was selected for the analysis. The SRSP covers four neighbouring districts, viz. Adilabad, Karimnagar, Nizamabad and Warangal. The project is located at Nizamabad, and it augments the irrigation needs of these districts besides providing drinking water to Warangal town. The crop data for the present study consisted of 39 years (1970–2008) panel data on yield of three important crops, viz. rice, maize and groundnut. The data were collected from various sources, publications and websites. The yield data for the crops were collected from season and crop reports of erstwhile Andhra Pradesh and also from the website www.andhrapradeshstat.com. The climate variables were annual precipitation and average seasonal temperature. Meteorological data were collected from various publications and also from the website www.indiawaterportal.org. The annual precipitation data (time series) was collected, which reflects both precipitation falling directly on a crop and inter-seasonal water accumulation within a year (Isik and Devadoss 2006). The temperature data collected is the average temperature observed over the *Kharif* (June to November) and *Rabi* (December to April) seasons.

23.2.3 The Model

In the present study, we focus on the yield and its variability in the context of climate change. Following Isik and Devadoss 2006, we assume that the relation between yield (also known as yield or production per hectare) y_{it} of a crop at district i during year t and the climatic variables x_{it}, viz. precipitation and temperature, is given by the Just-Pope stochastic production function (Just and Pope 1978).

$$y_{it} = f\left(x_{it};\beta\right) + \omega_{it} h\left(x_{it};\delta\right)^{0.5} \tag{23.1}$$

where ω_{it} is the stochastic term with mean zero and variance σ^2, β and δ are the production function parameters to be estimated using historical data. The independent variables (x_{it}) used for the estimations include a constant, annual precipitation (P), temperature (T), trend (t) and three district dummy variables. The expected crop yield is given by $E(y_{it}) = f(x_{it}; \beta)$, and crop variability is given by $V\left(y_{it}\right) = \sigma_\omega^2 h\left(x_{it};\delta\right)$. Hence the functions $f(x_{it}; \beta)$ and $h(x_{it}; \delta)$ are called *mean* and *variance* functions, respectively. Estimation of the above production function can be considered as estimation with heteroscedastic errors as in the following equation (Saha et al. 1997; Kumbhakar 1997):

$$y_{it} = f\left(x_{it};\beta\right) + u_{it} \tag{23.2}$$

where $u_{it} = \omega_{it} h(x_{it}; \beta)^{0.5}$ with $E(u_{it}) = 0$ and $Var(u_{it}) = \sigma^2 h(x_{it}; \delta)$. There are two approaches suggested in many studies to estimate the mean and variance functions of the Just-Pope production function. They can be estimated using feasible generalized least squares or the maximum likelihood method. For example, Barnwal and Kotani (2010) applied the first method. However, Saha et al. (1997) have shown that the estimators under the maximum likelihood method are consistent and more efficient than the feasible generalized least squares method. Hence in our study, maximum likelihood method has been used. Following Ranganathan (2009) and Isik and Devadoss 2006, the following quadratic form is assumed for the mean function:

$$f\left(x_{it};\beta;,d\right) = \beta_0 + \beta_1 P + \beta_2 T + \beta_3 t + \beta_4 P^2 + \beta_5 T^2 + \beta_6 PT + \sum_{i=1}^{i=3} d_i D_i \tag{23.3}$$

where, D_i, i = 1,2,3 are the district dummy variables taking values 1 and 0. The variance function $\sigma_\omega^2 h\left(x_{it};,\delta;,\eta\right)$ with $\sigma_\omega^2 = 1$ was assumed to have exponential form:

$$h\left(x_{it};,\delta;,\eta\right) = \exp\left(\delta x_{it} + \eta D\right) = \exp\left(\delta_0 + \delta_1 P + \delta_2 T + \delta_3 t + \sum_{i=1}^{i=3} \eta_i D_i\right) \tag{23.4}$$

23.2.4 Results and Discussions

A summary statistics of rice yield, precipitation and temperature of the four districts (average for 39 years, (1970–2008) is given in Table 23.3. The major share of annual precipitation is from *Kharif* season in all the districts. Temperature during the *Kharif*

Table 23.3 Summary statistics of yield of rice and climate variables

	Kharif season (June to November)			Rabi season (December to April)		
District	Yield (kg ha⁻¹)	Precipitation (mm)	Temperature (°C)	Yield (kg ha⁻¹)	Precipitation (mm)	Temperature (°C)
Adilabad	1543	941.4	27.3	2179	43.1	26.5
Karimnagar	2437	905.8	27.0	2523	56.1	26.2
Nizamabad	2241	888.2	26.7	2242	47.0	26.5
Warangal	2167	817.8	27.8	2039	50.5	26.7

Source: Palanisami et al. (2015)

Table 23.4 Just-Pope production function for rice: parameter estimates

	Kharif		Rabi	
Mean yield	Coefficient	Std. error	Coefficient	Std. error
Precipitation (R)(in mm)	7.2104**	3.5083	2.4397**	0.995
Temperature (T)(in °C)	2245.9150**	1015.6600	2284.363**	1098.4
Trend(year)	42.4363***	2.3439	42.7376***	2.88
R^2	−0.0013***	0.0003	−0.0016***	0.00038
T^2	−40.1724	50.3485	−43.6677	57.13
R*T	−0.1519	0.1944	0.0528	0.19
Adilabad	−710.6665	59.2652	159.4246**	79.32
Karimnagar	119.0369**	62.4787	427.0203***	67.17
Nizamabad	5.2939	95.3290	215.0166	74.92
Constant	−31671.7	39543.3500	−30787.6	42010.98
Variability in yield				
Precipitation (R)	−0.0014**	0.0006	−0.0004	0.0007
Temperature (T)	0.6296**	0.2775	0.2830	0.246
Trend	0.0267**	0.0126	0.0315**	0.0138
Adilabad	1.0743**	0.3636	−0.3060	0.402
Karimnagar	0.8547**	0.4534	−0.6833*	0.41
Nizamabad	1.9225***	0.4638	−0.2294	0.37
Constant	−6.1009	7.6631	3.8732	6.69
Likelihood fun.	−1096.8		−1106.4	

*Significant at 10% level; ** Significant at 5% level; ***Significant at 1%t level.
Source: Palanisami et al. (2015)

season is slightly higher than that of Rabi season. The most of the coefficients of the climate variables and their square terms are significant for both mean and variance functions (Table 23.4). The coefficient of trend is positive and highly significant in the two seasons showing the technological advancement in rice production in the four districts of Telangana. The percentage losses are computed based on normal yield (Table 23.5). The normal yield is the average yield during the last 5 years ending 2008–2009. For the first climate change scenario, i.e. an increase of 1.93 °C in temperature and 13.6% increase in precipitation, the expected loss in yield during *Kharif* season varies from 1.9 to 9.4%. The highest loss corresponds to Warangal

Table 23.5 Impact of climate change on rice yield in the two seasons (kg ha^{-1})

Climate change (temperature/ rainfall variation)	Parameter	Adilabad district	Karimnagar district	Nizamabad district	Warangal district	All districts
Kharif season						
Mid-century 1.93 °C/13.6%	Normal yield	2262	3115	3226	3009	2972
	Max yield	2616	3445	3332	3326	3180
	MC-predicted yield	2140	3056	3028	2726	2747
	Percent loss	5.4	1.9	6.1	9.4	7.6
	Standard deviation	616	511	763	455	575
End-century 4.03 °C/17.8%	EC-predicted yield	1395	2401	2438	1989	2065
	Percent loss	38.3	22.9	24.4	33.9	30.5
	Standard deviation	1160	964	1439	860	1086
Rabi season						
Mid-century 2.22 °C/13.6%	Normal yield	2460	3338	3214	2929	2985
	Max yield	2544	3374	3260	3255	3108
	MC-predicted yield	2248	3129	2975	2882	2814
	Percent loss	8.6	6.2	7.4	1.6	5.7
	Standard deviation	536	458	828	371	523
End-century 4.28 °C/17.8%	EC-predicted yield	1596	2550	2355	2274	2200
	Percent loss	35.1	23.6	26.7	22.4	26.3
	Standard deviation	998	853	1542	692	975

Source: Palanisami et al. (2015)

district and the lowest to Karimnagar. The standard deviation ranges from 511 to 763 kg. The second climate change scenario produces greater percentage of losses and variability in yield. The percentage loss varies from 22.9 to 38.3% and yield in Adilabad and Warangal district area is expected to suffer maximum losses. A similar conclusion can be drawn for the yield losses and variability for Rabi season also. The variability in yield for end-century is more than that for mid-century. Thus it can be concluded that climate change induces not only loss in yield but also greater variability in yield of rice. This conclusion coincides with the results of (Ranganathan 2009; Barnwal and Kotani 2010).

Parameter estimates of the fitted Just-Pope production functions for maize and groundnut are given in Table 23.6. As in the case of rice, coefficients of most of the climate variables are significant for the two crops in mean function as well as in variance function. Coefficients of precipitation, temperature and temperature square are significant for maize with temperature having negative effect on the mean yield. For groundnut, temperature has positive significant effect. Trend has positive

Table 23.6   Just-Pope function parameters for maize and groundnut

	Maize		Groundnut	
Mean yield	Coefficient	Std. error	Coefficient	Std. error
Precipitation(R)(in mm)	1.458***	0.624	−0.140	5.488
Temperature (T)(in °C)	−1684.180**	901.492	4621.546**	2272.849
Trend(year)	73.988***	4.854	20.096***	3.243
R²	−0.001	0.001	0.000	0.000
T²	29.238***	7.480	−86.010***	31.296
R*T	−0.005	0.364	−0.002	0.193
Adilabad	−443.900***	117.295	−237.896***	69.177
Karimnagar	317.455***	123.596	−40.902	71.334
Nizamabad	168.429	149.899	78.230	130.970
Constant	24377.590	68948.460	−61385.950	40394.420
Variability in yield				
Precipitation (R)	−0.001	0.001	0.000	0.001
Temperature (T)	0.133*	0.075	0.281**	0.130
Trend	0.035**	0.014	0.029	0.022
Adilabad	0.443	0.350	0.656	0.455
Karimnagar	−0.136	0.420	0.347	0.473
Nizamabad	0.368	0.404	1.576	0.501
Constant	8.376	7.709	2.566	7.871
Likelihood fun.	−1182.2		−1081.3	

Source: Palanisami et al. (2015)

significant effect on the two crops. Interaction between precipitation and temperature is not significant for the two crops. Since the coefficients of temperature in the variance function for the two crops are positive and significant, temperature is a risk-increasing factor for the two crops where increase in temperature results in higher variability in yield.

Table 23.7 presents the climate change impact on the two crops. The percentage of loss in yield for maize is small for the first scenario in all the districts. The maximum loss will be about 8% for Warangal district. Surprisingly, for the scenario 4.1 °C/17.8%, the percentage loss seems to decrease, and the maximum loss will be 5.5% for Nizamabad district. However, variability in yield increases by about 10–12%. Thus we can conclude that climate change may not have considerable impact on maize yield in the four districts. However, impact of climate change on groundnut production will be considerable. For the first scenario, the percentage loss varies from 13.8 to 25.2. Nizamabad district will have maximum loss. The standard deviation in yield ranges from 292 to 383 kg. The second scenario will have more damaging effect with the percentage loss varying from 69 to 90% while the standard deviation ranges from 387 to 802 kg. Thus we can conclude that groundnut production will be very much affected by climate change.

Table 23.7 Impact of climate change on maize and groundnut yield (kg ha^{-1})

CC-scenario	Parameter	Adilabad	Karimnagar	Nizamabad	Warangal	Average
Maize						
	Normal yield (kg ha^{-1})	3340	4185	4162	3999	3922
	Max yield (kg ha^{-1})	3248	4010	3861	3692	3703
Mid-century 2.05 °C/13.6%	MC-predicted yield (kg ha^{-1})	3249	4022	3866	3687	3708
	% loss in yield	2.7	3.9	7.1	7.8	5.5
	Standard deviation	785	587	763	667	696
End-century 4.1 °C/17.8%	EC-predicted yield (kg ha^{-1})	3321	4064	3932	3788	3778
	% loss in yield	0.6	2.9	5.5	5.3	3.7
	Standard deviation	889	665	865	757	789
Groundnut						
	Normal yield (Kg ha^{-1})	1344	1602	1865	1412	1556
	Max yield (Kg ha^{-1})	1203	1400	1519	1441	1391
Mid-century 2.05 °C/13.6%	MC-predicted yield (kg ha^{-1})	1072	1325	1394	1217	1254
	% loss (base/ normal yield)	20.2	17.3	25.2	13.8	19.4
	Standard deviation	382	319	606	292	383
End-century 4.1 °C/17.8%	EC-predicted yield (kg ha^{-1}/)	140	490	485	224	338
	% loss (base/ normal yield)	89.6	69.4	74.0	84.1	78.3
	Standard deviation	506	422	802	387	507

Source: Palanisami et al. (2015)

In summary, climate change will have very serious effect on groundnut, moderate effect on rice and negligible effect on maize. Further, stronger climate change will induce higher variability in yield in all the crops. In this context, the following management options have been examined to address the impact of climate change on yield of selected crops:

- System of Rice Intensification (SRI) (with 20% reduction in water use)
- Machine Transplanting (MT) (with 15% reduction in labour use)
- Alternate Wetting and Drying (Maize Water Management (MWM) (with 10% reduction in water use)

Adoption of the water- and labour-saving technologies helps the rice production in the project AWD (with 10% reduction in water use) area. It is observed that in all the cases, the SRI resulted in higher production, gross income and water saving compared to alternate drying and wetting and machine transplanting. Nonetheless, adoption of SRI is less due to its challenges in sowing, cono weeding, etc. As an alternate strategy, machine transplanting can help the rice production releasing the labour to cover additional area under rice. It is understood that in the future, the labour scarcity is expected to reduce the area under rice as it will constraint the transplanting operations. Hence machine transplanting helps to ease the labour scarcity to the extent of 20–25%.

23.3 Case Study 2: Drought Stress Management in Dryland Agrosilvipastoral System

Abiotic stress is to be either managed through mitigation or abatement strategies, while biotic forces are tackled mechanically/chemically or biologically. A wide range of adaptation and mitigation strategies are required to cope with the severe impacts of abiotic stress. Efficient resource management and crop/livestock improvement for evolving better breeds can help to overcome abiotic stresses to some extent. Hence, this case study focuses on the adaptation strategies followed by the dryland farmers to manage the drought stress that accounts for 17% of the crop loss.

23.3.1 Study Area, Data and Methodology

This case study investigates the strategies followed by the livestock-based integrated farmers to manage the drought stress in the dry land area of Tamil Nadu. The study covered Tiruppur district in Tamil Nadu which is a rain shadow region with a rainfall of 600 mm per annum. The district often suffers from severe drought stress due to the increased frequency of drought and erratic pattern of rainfall. As a resilience mechanism, the district leads in mixed farming with animal husbandry as one of the key enterprises where in several locations unique agrosilvipastoral farming is followed. Under this silvipastoral system, Velvel (*Acacia leucophloea*) is allowed to grow in rainfed lands with naturally emerging perennial grass called Kolukattai grass (*Cenchrus ciliaris*) which encourages sheep rearing and became a popular subsidiary occupation in the area. However, the adoption level of this practice is also not increasing over the years in spite of erratic and uncertain rains. The study analysed the adoption aspects of these adaptation strategies by covering 180 farmers from six blocks of the district who were selected using multistage purposive sampling.

23.3.2 Adoption Index

The farmers' were categorized into two groups based on their adoptability: adopt (score 1) and non-adopt (score 0). Data were tabulated using frequency distribution and were analysed descriptively. The adoption level of the respondents was measured by making use of adoption index (Karthikeyan 1994 in Rahman 2007).

$$\text{Adoption Index} = \frac{\text{Respondent Total Score}^*}{\text{Total Possible Score}} 100$$

23.3.3 Extent of Adoption

$$\text{Extent of Adoption} = \frac{\text{Number of respondants who had adopted the practice}^*}{\text{Total number of practices}} 100$$

where:

Respondents total score = total number of practices adopted by a farmer multiplied by the respective practice weight age and summated.

Total possible score = total number of practices recommended, multiplied by the respective practice weight age and summated.

23.3.4 Results and Discussions

Among the 17 strategies identified, viz. change in sowing dates, change in cropping pattern, summer ploughing, deepening of exiting bore well and/or drilling of new bore wells, barbed wire fencing, usage of drought-tolerant varieties, crop insurance and usage of drip irrigation, shifted to nonfarm activity and construction of water harvesting, waste management, purchase of feed and fodder, sell and reduce the herd size, provide shade, lease in more lands, livestock insurance and reviewed vaccination; 12 were indexed as important strategies followed by the sample farmers (Table 23.8). The change in planting dates and the change in cropping pattern were given with high score since the majority of the farmers cope with the drought stress. When the precipitation is lesser than the normal level, they changed their cropping system from agricultural crops to forage crops, as these crops require less water and can be sustained even in the drought conditions. Deepening of exiting bore well or drilling of new bore wells has been followed by the large farmers. Usage of drip irrigation to the crops like coconut was also observed. Marginal farmers and many agricultural labourers were adversely affected by the impact of climate change and therefore shifted to other business like finance, sweet stalls, etc. Some farmers constructed few water-harvesting structures such as farm ponds and drainage channels in their field with the help of the government schemes and on their own to collect the rainwater which recharge their wells. Although many drought-tolerant varieties

Table 23.8 Adoption indices of strategies followed in abiotic stress management

Adoption strategies	Adoption index
Change in cropping pattern	82
Usage of drought-tolerant variety	17
Usage of drip irrigation	35
Deepening of existing wells or drilling new bore wells	75
Shift to nonfarming	29
Change in planting dates	93
Crop/livestock insurance	3
Investing in water-harvesting structures	26
Providing shade	37
Waste management to supplement fodder	83
Reduction in number of livestock	77
Providing livestock vaccination	59

Note: All parameters are assumed to carry equal weight.
Source: Mohanasundari (2015)

Table 23.9 Different Adoption level of farmers (n = 180)

Extent of adoption in crops	Frequency	Percentage
Low (<33 score)	38	21.11
Medium (33–74)	103	57.22
High (>74)	39	21.67
Total	180	100.00

Source: Mohanasundari (2015)

are available, farmers are not aware of those, and hence usages are also very limited. In spite of being aware and avail the crop/livestock insurance schemes, farmers are not interested to avail them because of the low compensation paid and the complex procedures involved.

Farmers use cotton waste as livestock feed from Tiruppur textile industry to supplement the fodder. To sustain their livestock at the time of severe drought stress, farmers purchase the waste at very low cost. When the farmers are unable to meet the fodder demand for livestock at the extreme drought situations, they are reducing the number of livestock by selling. On the other hand, to avoid some seasonal and climatic disease infections, regular vaccination is given for the livestock. Farmers built some sort of infrastructures to provide shade which protects the livestock from severe cold and heat stress. The majority of farmers (57%) were medium level adopters. And about 21% of farmers are low adopters and 22% of farmers' high level adopters (Table 23.9).

23.4   Conclusions and Way Forward

The key message is that water is the key constraint in rice production in the long run, and land put under current fallow due to water scarcity will be a key issue to be dealt with. Implementation of various rice water- and labour-saving technologies will minimize the reduction in rice production between 20–25% under the medium-term and long-term basis. Hence, simply implementing the water management technologies will address the rice production constraints without making any structural interventions such as construction of new storage structures. Already field-level studies in the project area had shown that water-saving technologies will have a higher rate of return in rice production systems (Palanisami et al. 2011). The key question is how and what scale these technologies should be introduced and what kind of institutional and capacity-building mechanisms are needed to achieve this.

The results of the agrosilvipastoral study convey that most of the farmers do not have awareness about the drought-resistant varieties, and even in farmers with some awareness, they are not insuring their crop or livestock for losses. Drilling bore wells in drought prone regions is not a good option due to increasing well failure. However, agrosilvipastoral farmers adopted few strategies such as change in crop pattern, change in planting dates and micro irrigation for fodder to cope with the increasing drought-related stresses. But still, there is an increasing need for creating awareness among the farmers on latest practices and strengthening the extension services. Given the importance of climate change and its impact on agriculture, it is important that abiotic stress-related interventions need to be prioritized. The following are considered important.

Irrigation Ecosystems
- Water management technologies should be piloted in selected locations of the project, and based on the success of these technologies, the upscaling mechanisms should be initiated.
- A cluster approach (covering a group of villages in a location) will be more useful in upscaling the water management technologies, and farmers will be free to interact and follow up with the relevant technologies.
- Labour-saving technologies such as machine transplanting have proved to increase the rice area and production in all the climate scenarios studied. Hence, given the future labour scarcity in rice production, machine transplanting package should be organized at village level through the involvement of local community. A custom hiring unit can be established in the cluster of villages, and farmers can easily forecast their requirement for paddy seedlings and planting in a given time schedule.
- The existing government programmes with the agriculture departments should include the water management technologies in their programme.
- Adequate capacity-building programmes in technology upscaling and mainstreaming should be established. The expertise with the agricultural university research stations and KVKs should be explored for strengthening the capacity-building programmes.

- As the transaction cost of adoption of the adaptation strategies is comparatively high, it is important to address how these costs could be reduced for quick adoption by the farmers (Palanisami et al. 2015).
- There is need for a greater number of dedicated laboratories which deal solely with the production of abiotic stress-tolerant transgenic crops and sharing the results with SAUs.

Rainfed Ecosystem

- In the case of rainfed situation, availability of adequate credit, yield-increasing technology packages to suit drought situations, creating opportunities for off-farm employment, conducting further research on the crop and livestock combination package, introduction of crop and livestock insurance product and investing in water-harvesting structures in dry lands are important components for upscaling agro-silvipastoral systems.

23.5   Areas for Policy Support

Technical and policy instruments complement each other for reduction of harmful impacts and thereby build climate change resilience among crops. Therefore, it is imperative that various dimensions of adoption of different management strategies that have been discussed in the paper are taken into account in implementing techno-economic interventions. As a way forward, the following issues need to be considered in the long-term planning process (Kareemulla and Rama Rao (2013):

- Land degradation and implications – socioeconomic medium and long term. Over the years, land degradation is becoming an issue, and policies that help to manage the land and water ecosystems should be developed along with implementation procedures.
- Community actions for mitigation and coping mechanisms. Already in several locations, community-based mitigation strategies to address abiotic stresses have been identified and tested. It is important to find pathways to upscale them.
- Public policies for communities and regions affected by abiotic stress. Guidelines to develop policy frameworks that are relevant to address the abiotic stresses to suit different agro-ecological environment need to be developed and practiced.
- Relationship of abiotic stress on poverty and resource-poor farmers. As discussed in the paper, in the long run, agriculture production may be affected due to climate change impacts, and it is highly warranted that poverty alleviation programmes with adequate social safety nets particularly in rural sector need to be introduced. Resource-poor farmers should be supported with needed inputs and technology backups to sustain their farming and shared values. The concept of smart villages with package of affordable and appropriate practices is more relevant now.

References

Aggarwal PK (2008) Climate change and Indian agriculture: impacts, adaptation and mitigation. Indian J Agric Sci 78:911–919

Ashraf M, Athar HR, Harris PJC, Kwon TR (2008) Some prospective strategies for improving crop salt tolerance. Adv Agron 97:45–110

Barnwal P, Kotani K (2010) Impact of variation in climatic factors on crop yield: a case of rice crop in Andhra Pradesh, India, Economics and management series, EMS-2010-17. International University of Japan, Minamiuonuma

Bennet EM, Carpenter SR, Peterson GD, Cumming GS, Zurek M, Pingali P (2003) Why global scenarios need ecology. Front Ecol Environ 1(6):322–329

Chand R, Raju SS (2009) Instability in Indian agriculture during different phases of technology and policy. Indian J Agric Econ 64:187–207

De US, Mukhopadhyay RK (1999) Severe heat wave over the Indian subcontinent in 1998 in perspective of global climate. Curr Sci 75:1308–1315

Dhawan AK, Singh K, Saini S, Mohindru B, Kaur A, Singh G, Singh S (2007) Incidence and damage potential of mealy bug, *Phenacoccus solenopsis* Tinsley, on cotton in Punjab. Indian J Ecol 34:110–116

Gosain AK, Rao S (2011) Analysis of climate change scenarios in the Godavari river basin. Draft, Indian Institute of Technology, Delhi

Grover A, Aggarwal PK, Kapoor A, Agarwal SK, Agarwal M, Chandramouli A (2003) Addressing abiotic stresses in agriculture through transgenic technology. Curr Sci 84(3):355–367

IARI News (2008) Brown plant hopper outbreak in rice. 24:1–2

IMD, Annual Climate Summary (2010) India Meteorological Department, Pune. Government of India, ministry of earth sciences, pp 27

IPCC (2007) Climate change- impacts, adaptation and vulnerability. In: Parry ML, Canziani OF, Palutikof JP, vander Linden PJ, Hanson CE (eds) Cambridge University Press, Cambridge, UK, pp 976

IRRI News (2009) Pest outbreaks in India. Rice today, 6. www.ricenews.irri.org

Isik M, Devadoss S (2006) An analysis of the impact of climate changes on crop yields and yield variability. Appl Econ 38:835–844

Joshi S, Viraktamath CA (2004) The sugarcane woolly aphid, *Ceratova cunalanigera* Zehntner (Hemiptera: Aphididae): its biology, pest status and control. Curr Sci 87:307–316

Just RE, Pope RD (1978) Stochastic specification of production functions and economic implications. J Econ 7:67–86

Kareemulla K, Rama Rao CA (2013) Socio-economic and policy issues in abiotic stress management. Crop stress and its management: perspectives and strategies. pp 565, DOI 10.1007/978-94-007-2220-0_18

Karthikeyan C (1994) Sugar factory registered growers – An analysis of their involvement and impact. Unpub. MSc (Ag) Thesis, TNAU, Coimbatore

Kumbhakar SC (1997) Efficiency estimation with heteroscedasticity in a panel data model. Appl Econ 29:379–386

Lal M (2003) Global climate change: India's monsoon and its variability. J Env Studies Policy 6:1–34

Mittler R (2006) Abiotic stress, the field environment and stress combination. Trends Plant Sci 11(1):15–19

Miyan MA (2015) Droughts in Asian least developed countries: vulnerability and sustainability. Weather Climate Extremes 7:8–23

Mohanasundari T (2015) Adaptation of agrosilvipastoral system to the changing climate scenario – An economic analysis in Western Tamil Nadu. Unpublished Ph. D Thesis Department of Agricultural Economics, Tamil Nadu Agricultural University, Coimbatore

NACCAP (2012) Climate change impacts on pest animals and weeds. Communicating climate change. National Agriculture and Climate Change Action Plan (NACCAP), Bureau of meteorology, department of international journal of scientific and research publications, Volume 2, Issue 11, 14 ISSN 2250–3153

NATCOM (2004) India's initial national communication to the United Nations framework- convention on climate change. Ministry of Environment and Forests, pp 268

NIASM (2015) Vision 2015. National institute of abiotic stress management, Malegaon, Baramati 413 115. Pune

Oerke EC, Debne HW, Schonbeck F, Weber A (1994) Crop production and crop protection. Elsevier, Amsterdam

Palanisami K, Kakumanu KR, Suresh-Kumar D, Chellamuthu S, Chandrasekaran B, Ranganathan CR, Giordani-Mark (2011) Do investments in water management research pay? Evidences from Tamil Nadu. India, Water Policy

Palanisami K, Ranganathan CR, Kakumanu KR, Nagothu US (2015) Climate change and agriculture in India: studies from selected river basins. Routledge Publishers, New Delhi

Rahman S (2007) Adoption of improved technologies by the pig farmers of Aizawi district of Mizoram, India. Livestock Research for Rural Development, 19, Article #5. Retrieved from http://www.Irrd.org/Irrd19/1/rahm19005.htm

Ranganathan CR (2009) Quantifying the impact of climatic change on yields and yield variability of major crops and optimal land allocation for maximizing food production in different agroclimatic zones of Tamil Nadu: An econometric approach, Working paper, Research Institute for Humanity and Nature, Kyoto

Rao GGSN, Rao AVMS, Rao VUM (2009) Trends in rainfall and temperature in rainfed India in previous century. In: Aggarwal PK (ed) Global climate change and Indian agriculture case studies from ICAR network project. ICAR Publication, New Delhi, pp 71–73

Rowntree PR (1990) Estimate of future climatic change over Britain. Weather 45:79–88

Saha A, Havenner A, Talpaz H (1997) Stochastic production function estimation: small sample properties of ML versus FGLS. Appl Econ 29:459–469

Samra JS (2003) Impact of climate and weather on Indian agriculture. J Indian Soc Soil Sci 51:418–430

Samra JS, Singh G (2004) Heat wave of March 2004: impact on agriculture, Indian council of agricultural research, New Delhi, 2004, p 32

Srikanth J (2007) World and Indian scenario of sugarcane woolly aphid. In: Mukunthan N, Srikanth J, Singaravelu B, Rajula Shanthy T, Thiagarajan R, Puthira Prathap D (eds) Woolly aphid management in sugarcane, vol 154. Extension Publication, Sugarcane Breeding Institute, Coimbatore, pp 1–12

Sutherst RW, Maywald GF, Bottomly W (1991) From CLIMEX to PESKY, a generic expert system for risk assessment. EPPO Bull 21:595–608

Theilert W (2006) A unique product: the story of the imidacloprid stress shield. Pflanzenschutz-Nachrichten Sci Forum Bayer 59:73–86

Wang W, Vinocur B, Altman A (2007) Plant responses to drought, salinity and extreme temperatures towards genetic engineering for stress tolerance. Planta 218:1–14

Inculcating Resilience to Agriculture Under Abiotically Stressed Environments: Way Forward

24

Paramjit Singh Minhas, Jagadish Rane, and Ratna Kumar Pasala

Abstract

Several transformative changes like growing population, changing lifestyles, expanding urbanisation, accelerated land degradation and climate change-induced abiotic stresses are challenging the future food and nutritional security world over especially the low-income countries. All these changes necessitate development of a strategic framework for agricultural innovations which can ensure inclusive and sustainable agricultural growth especially in harsh agroeco-systems afflicted by abiotic stresses. A multidisciplinary and holistic approach to manage the stressed environments should aim at characterisation of abiotically stressed environments; reoriented, novel and scaled-up natural resource management (NRM) technologies for stress mitigation; improved adaptation to stressed environments; and task-oriented capacity building. Augmentation, integration and promotion of the best available tools, approaches and technologies should involve investments and incentives for breeding protocols, regional networks for exploring synergies and dynamic policy support. Several leads for policy support towards successful mitigation and adaptation have been listed for meeting the future challenges of abiotic stresses.

P.S. Minhas (✉) • J. Rane • R.K. Pasala
National Institute of Abiotic Stress Management, Indian Council for Agricultural Research, Baramati, Maharashtra, India
e-mail: minhas_54@yahoo.co.in; jagrane@hotmail.com; pratnakumar@gmail.com

© Springer Nature Singapore Pte Ltd. 2017
P.S. Minhas et al. (eds.), *Abiotic Stress Management for Resilient Agriculture*,
DOI 10.1007/978-981-10-5744-1_24

24.1   Introduction

The agriculture led to the domestication of many plant and animal species and consequently to the exploitation of natural resources to support them. The art and then the science of agriculture gradually evolved to meet the food demand of the society. The recent advances in agriculture have evolved so strongly that the food grain production could be doubled in just four decades (1960–2000), whereas it took almost 10,000 years for production to touch one billion tons by 1960. This inspiring feat of unprecedented increase in food grain production, referred as 'green revolution', resulted from the fertiliser- and irrigation-responsive genetically improved crop varieties combined with other improved agronomic practices in addition to appropriate policy support (Khush 2001). Similar analogies can be extended to other components of food and nutritional security such as milk, egg, meat, fruits, vegetables and fibre. However, several transformative changes are occurring world over which include growing population, changing lifestyles, expanding urbanisation, accelerated land degradation and climate change. Therefore it is emerging that the past gains in the food productions have started showing the signs of fatigue and can even come to a halt if we fail to adopt new approaches to enhance food production in the context of stagnation in genetic gain in crops and adverse effect of emerging abiotic stresses as consequence of land degradation and climate change. Despite all these constraints, there is a scope to keep pace with desired quantity and quality demands of food for ever-increasing population. One of the obvious alternatives is systematic efforts to envision challenges and opportunities to formulate strategies for minimising the impacts of abiotic stresses featuring drought, high/low temperature, acidity/salinity, nutrient deficiencies/toxicities, extreme weather events, etc. This needs special attention by scientists, academicians, administrators and policymakers for developing an inclusive growth and sustainable agroecosystem oriented strategic framework to promote innovations in all sectors of agriculture. Thus the way forward for sustainable and abiotic stress resilient agriculture is based on new approaches as earlier described in the different comprehensive chapters of this book by experts from diverse disciplines.

24.2   Strategy

The concept of abiotic stress is not new to agriculture. Perhaps its management co-evolved with modern agriculture. However, benefits of modern agriculture rarely reached ecologically challenged regions and that have huge potential for sustaining agricultural productivity. Since the current mitigation and adaptation options are insufficient to face the challenges for food security, there is a need to change the strategy for addressing abiotic stress in agriculture through research, management, capacity building and policy change so as to promote novel, innovative and rewarding technologies. However, the key to such an approach has to be multidisciplinary and should essentially consider the following aspects:

- Characterisation of abiotically stressed environments
- Reoriented, novel and scaled-up stress mitigation NRM technologies
- Enhanced efforts to improve adaptation to changing environment
- Task-oriented capacity building that can create next-generation highly capable managers, researchers and farmers
- Networks that can connect, share and explore synergies at regional, national and international level
- Dynamic policy support tools to realise benefit of advances in science

24.3 Characterisation of Abiotically Stressed Environments

Unlike industrial sectors, the agriculture is largely dependent on natural resources which are always exposed to aberration in weather parameters and anthropogenic activities. Abiotically stressed environments are the outcome of their influences. The monsoon has been and continues to be the driving factor for agriculture and many other related sectors. Hence in tropical countries like India, the responsibilities to watch, warn, inform and to act during vagaries of monsoon have been a priority. Even the most of the promising technologies to manage agricultural risks are dependent on weather information and remain underexploited. Therefore, a number of government and private agencies are involved in providing weather forecasting services. The emphasis is to develop end-to-end capacity in weather monitoring and to evolve mechanisms for drought monitoring and early warning for extreme weather events. There is a general consensus among scientific and farming community that the existing technologies can go a long way to help farmers if the monsoon forecast becomes more robust. Moreover, with their enhanced ability of detecting, monitoring and forecasting, it is further possible to lessen the impact of drought, floods, hailstorms and other weather extremes. To take advantage of advances in collection, compilation and analysis of data set, in data management capacity and in modelling science, the Department of Tropical Meteorology is now gearing up for better forecast of monsoon. The National Monsoon Mission programme launched by the Ministry of Earth Sciences (MoES) in 2012 is likely to improve monsoon prediction at all temporal and spatial scales through joint efforts of national and international scientific communities. By deciphering the drivers of the variability in the Indian monsoon, the prediction will become more robust, and the success from such efforts will be the game changer for abiotic stress management and extreme weather events including drought.

In addition to weather forecasting science, advances in remote sensing and GIS are providing great opportunities for explaining abiotic stresses at regional level. These advances are making it possible to map abiotic stresses in agroecosystems even at block and village levels because of enhanced resolution of images provided by satellite like IKONOS, QuickBird, etc. These provide the basis for assessing the vulnerability of crop species to different abiotic stresses in isolation or in various combinations. Since the impact of environmental factors widely varies across space and time, the abiotic stress maps are highly critical for making decision on abiotic stress management strategies. These tools can gradually evolve as robust decision support system for managing abiotic stresses if further supplemented adequately with information on edaphic factors derived from soil maps, cropping patterns and their responses to stressors, existing regional to local models and GIS-based statistical tools.

Since the extreme weather events are becoming more frequent, to make future agriculture remunerative, risk-free and sustainable, the dynamic characteristics of atmospheric stressors have to be understood in detail. These should include extremes of precipitation, viz. drought, flood, submergence and hailstorm; extremes of temperatures, viz. heat, cold and frost; extremes of radiation, viz. high, low, UV and cloudy days; extremes of wind, viz. cyclone, sand, dust storm, etc., in addition to adverse effect of ozone. Since the knowledge of different dimensions of these factors is elementary for managing abiotic stresses, it is necessary to generate information on various processes that lead to energy and gas exchanges from earth surface, source or sink potential of various ecosystems under varied climatic conditions for developing climate-smart mitigation strategies.

In the changing climatic scenario, a large number of strategies to reduce yield gaps are being evaluated by employing various crop simulation models. Since these models are relatively inexpensive, it is possible to explore the specific management options including those for mitigation and adaptation to counter the adverse effects of abiotic stresses often induced or amplified by climate change. Optimisation of these models together with impact assessment tools can help in understanding and defining the abiotic stress environments to be targeted and simulated for assessment of stress tolerance in germplasm.

Though the recent advances have enhanced the understanding of physical and physiological impacts of various abiotic stressors, much remains to be done regarding quantifying its impacts and long-term implications on agriculture. The types and level of stresses must be properly quantified for future references. In addition, stresses arising out of increased atmospheric aerosol and decreasing in available light need extra attention as change in land use pattern and crop residue burning has changed the way we have been dealing these aspects in the past.

24.4 Reoriented, Novel and Scaled-Up Stress-Mitigating NRM Technologies

Though human beings do not have full control of natural phenomenon, there is a large scope for minimising the adverse effects by addressing the secondary causes of atmospheric, edaphic and water-related stresses. There are many proposed practices for management of land and water which can help in mitigation of abiotic stresses. The benefits of these practices ensure basically with increase in organic matter, replenishment of nutrients, improvement in soil structure, reduction in erosion of soil, increase in infiltration of water, increase in water-use efficiency and also uptake of nutrient (Winterbottom et al. 2013). Among these, the most promising to combat abiotic stresses include conservation agriculture (CA), Integrated

Farming Systems (IFS), integrated soil fertility management, rainwater harvesting, agroforestry, etc.

Conservation agriculture (CA) combines three environmentally friendly agronomic practices to reduce tillage, retain crop residues on surface or maintain cover crops and to rotate or diversify crops in a remunerative cropping sequence. Hence CA can save resources and optimise agricultural system for sustainable intensification, enhanced economic benefits, improved natural resources and more efficient external inputs (Jat et al. 2015). CA systems have worked in all kind of environments/ecologies and helped millions of farmers through arresting land degradation, improving input use efficiency, adapting and mitigating climate extremes/abiotic stresses and improving farm profitability (Kasam et al. 2014). CA can benefit agroecosystem through its capacity to sequester atmospheric CO_2 in the soil organic matter and also reduce emissions of GHG through efficient management of input use. CA when adapted along with other component technologies for efficient soil and water management can serve as foundation for mitigation of abiotic stresses. This warrants for synergies to capitalise on different elements of CA for the stressed environments.

The other approach for stress mitigation and building resilience is resorting to 'Integrated Farming Systems' which spread the stress-induced risks amongst the multi-enterprises and thus provide for a better buffering capacity. Since the abiotic stresses show differential impacts on the crop/tree/animal/fishery enterprises, such a diversification of farming enterprises can at least provide for life-saving returns to the farmers. Thus Integrated Farming Systems can be an ultimate solution for stabilising farmers' income and can act as safeguard for extremes of weather events/abiotic stresses. Stress-specific IFS will be guided by long-term weather events, demography, and socio-economic and agricultural landscape conditions which are essential to find out the best combination of crops, crop varieties, water-saving protocols, livestock, land use plans for grain, fodder, dual-purpose crops, agroforestry/horticulture, etc. For instance, the past experiences show that in regions with rainfall of 500–700 m, the livestock-based IFS should rely on water-use efficient grasses, trees and bushes that can provide fodder, fuel and shelter. In rainfall zones of 700–1000 mm, crop-horticulture-livestock systems that perform better integration of paddy with fisheries are ideal where rainfall is more than 1100 mm (Joshi et al. 2005). Therefore, future efforts are required to identify more economical and sustainable IFS modules for the stressed environments.

Nutrient deficiencies and low factor productivity are the matters of great concern in the abiotically stressed environments. The new decision tools like Nutrient Expert, GreenSeeker, remote sensing coupled with ICTs, etc., are becoming very handy for quick recommendations on SSNP (site-specific nutrient management) and thus improving nutrient use efficiencies. Such decision-making support tools using modern communication platforms are the need of the hour for stressed environments where the present nutrient use efficiencies are dismally low and the overall fertiliser use is very low and highly imbalanced, i.e. skewed towards nitrogen only. It is possible to offset the abiotic stresses and to minimise GHG emissions from agriculture through integrated nutrient management practices which involve judicious use of mineral fertilisers and amendments for soil productivity improvement through application of organic manure/compost/leaf/residues of crops and reduction of salinity/alkalinity/acidity through application of gypsum/lime/rock phosphate, etc.

Effective management of rainwater is also very critical for stress mitigation and successful agriculture. Without due attention to conservation of water and soil, a major part of the rainfall water is lost as runoff from denuded farm fields' simultaneously stimulating soil erosion. Thus slowing water runoff with in situ water harvesting practices can be of great help to farmers for adjusting to fluctuations in rainfall. Though the suitability of the practices depends upon the topography of the field, temporal and spatial distribution of rainfall and type of soil, crop, etc., a variety of simple low-cost practices like compartmental bunds, conservation furrows, broad bed and furrow, ridge and furrow, ridge planting, set furrow, inter-plot rainwater harvesting, etc., have been developed to effectively retain and collect rainwater in the fields well before it gets lost through runoff. These if customised for specific abiotic stresses can be of great help to offset their impacts. Moreover, the Integrated Watershed Management Systems (IWMS) have been advocated on a large scale where surplus water is diverted to storage structure and used as such or in conjunction with groundwater for meeting the water requirements at critical crop stages. Drip irrigation is emerging as a viable solution for improving the use efficiency of this water and can further be utilised to offshoot the other abiotic stresses.

Similarly the understanding of biology and management of horticultural plantations have greatly improved under abiotically stressed environments where the shallowness of soils/hard subsoil layers restricts access to stored soil moisture (e.g. in regions with drought) and nutrients and allows episodic events such as prolonged flooding/waterlogging, drought, etc. To overcome these constraints, the on-site soil and water conservation technologies like trenching, contour/strip planting and graded farrows/ridges are being advocated to enhance crop/tree growth, while the off-site techniques include storage of runoff, transport of canal/drain water through multistage pumping, water tankers and switching to drip irrigation. It is generally argued that successful cultivation of fruit trees requires a major shift in planting techniques, site preparation for planting and post-planting management since the initial establishment and growth of fruit tree saplings are highly critical for fruit orchards. Thus the aim should be to create favourable niches in the ambient where their roots are located. The typical examples are that of auger hole planting in alkali land and sub-surface planting and furrow irrigation in saline lands (Minhas and Dagar 2016). With such innovative planting techniques and the site management, the orchard establishment can be boosted in other stressed environments also.

Microclimate modification research needs focus as it is possible to protect crops from abiotic stresses such as low temperature stress in subtropical parts of country where night-time irrigation (flooding) can save crops like oranges and grapes from freezing. Mulch can be used to protect soil from freezing and plant from low temperature root damage. For high temperature stress, frequent sprinkling of water on foliage can prevent plants from huge evapotranspiration loss. Proper row arrangement can be done so that cold air/hot air cannot directly pass on to crops. Soil and land management practices such as ridge and furrow sowing can prove better germination of seeds and also can prevent crop lodging. Several hydrogels are also being advocated for moisture retention in soil where water deficits are frequent but need field validation. Use of antitranspirant can save plants from excess transpirational water loss. Some reflectants can prevent detrimental temperature which affects crop growth and yield. Irrigation management schedules need to be tailored for different crop-agroecology combinations for better crop growth taking into consideration water requirements at critical growth stages and capacity of soil to meet the evaporation and transpirational water demand. Irrigation requirements of most of the cultivated crops have been standardised, but these need further optimisation under stressed environments, e.g. there is a need for additional irrigation to meet leaching requirements. Water-saving technologies like deficit irrigation/partial root drying, etc., often fail to enhance yield but have the benefits of improving water productivity, reducing pest-disease incidences, irrigation costs and labour requirement, etc. Therefore, all these technologies need to be fine-tuned for stressed conditions.

Protective structures used in vegetable cultivation can lessen the impact of extreme temperatures, hailstorm, waterlogging, etc., through sufficient protection for both the crop and soil. Even high-value crops such as grapes are highly vulnerable to disastrous hailstorms which necessitate protection technology. Protective structure can also provide suitable microclimate for plant growth, but research is needed to make this structure economic, durable and more productive.

While plant has inherent mechanisms to respond to the stresses, these mechanisms can be facilitated by exogenous application of bioregulators to alleviate stress-like drought, salinity and high temperature (Srivastva et al. 2016). The future prospects for bioregulators lie in identification of the best combination of crop or varieties and agroecosystem as well as optimised dose, frequency of application and appropriate stage for efficiently mitigating the stress. Moreover, the opportunities should be explored to use the wide range of microorganisms prevailing in problematic soils on surfaces of plants and in marine ecosystems by employing appropriate techniques. The useful microbes or their products in isolation or in combination should be the target for facilitating nutrient cycling, tolerance to abiotic stresses, enhanced fixation of atmospheric nitrogen and promotion of plant growth under harsh agroecosystems. Candidate endophytic isolates identified so far particularly from abiotically stressed ecosystem need to be validated as promising technologies for abiotic stress tolerance.

Nutrition being one of the major factors in mitigating heat stress in livestock, nutrient supplementation and feeding practices for cattle, small ruminants, poultry and fish should be given priority to improve their performances under heat stress. Anti-nutritional factors associated with abiotic stresses need special attention to enhance productivity of livestock and fisheries. Nano(bio-)technological innovations have potential to cater to the nanofeed demand in addition to deliver other effective nanoparticle-based products for climate-resilient agriculture. Stress mitigation strategies for livestock can be developed through research for robust prediction of animal growth, body composition, feed requirement and waste output in future climate.

The contingency plans particularly in the event of natural disaster such as drought have been formulated and being propagated. The lack of appreciation from end users often points at the need for validation of these technologies as it has been demonstrated in some real-time contingency plans. There should be robust mechanisms to get feedback from farmers on utility of these technologies so that they can be scaled up or modified to make it acceptable or to be replaced by other options.

24.5 Enhanced Efforts to Improve Adaptation

Approaches for improved adaptation to abiotic stresses in agriculture include growing crops and their varieties which are relatively tolerant to specific stress and a shift to intercropping, crop diversification, etc., which help to cope with the adverse influence of stress/extreme weather event. So far there has been a notion that selections of crop cultivars for high yield potential in crop breeding programme would indirectly incorporate traits associated with tolerance to abiotic and biotic factors. However, the recent slowed down genetic gains vis-a-vis productivity of crops indicate that this empirical approach may not be sufficient to meet the requirement of stress resilience if future food demands are to be realised. Hence focus should be on complementary effects of trait-based approaches to breed stress-resilient crops, livestock and fish. In this regard, there is a need for better insight into the biological mechanisms underlying stress responses that can support classical and emerging modern technologies in production, breeding and biotechnology. Metagenomics, which has redefined the concept of genome analysis, should be utilised to accelerate the process of gene discovery through gene mining in stress-resilient microbes for conferring tolerance to stresses. Recent developments in genomics widen scope for mechanistic insights into organisms through classical genetics and even the difficult phenotypes can be tackled. Use of genomics for abiotic stress tolerance should continue for realising their benefit for end user as it has enhanced our understanding about evolution of the genes and genomes. *Omics* approaches are overlapping and dependent on each other, and integration of all these approaches is necessary to enhance stress tolerance in cultivars.

Dissecting genetic basis of different traits through QTL analysis and gene identification accelerates the breeding process for stress-ready crops. Marker-assisted selection (MAS) serves as an effective tool to introgress positive alleles in desirable genetic background. Thus combining crop breeding with various genomic tools would spur the efforts of multi-stress-tolerant cultivars. This approach has been successful in inducing both drought and submergence tolerance in rice. Such leads should be utilised in breeding programme to improve stress tolerance in other crops including vegetables and fruit crops. The exact location of the QTLs and closely linkages of markers determine the effectiveness of MAS. Many of the vegetables are highly sensitive to post-flowering rise in temperature which severely affects their productivity. Leads from rice reveal that it is possible to focus traits associated with inherent diurnal adjustment to escape peak temperatures during fertilisation for the purpose of genetic improvement in high temperature tolerance. Identification of abiotic stress-resilient root stocks can help improve inherent tolerance of horticultural and vegetable crops, and the strategy may make the orchards more resilient to climate change. Focus on deep-rooting species of Cucurbitaceae such as *Citrullus colocynthis* can pay more dividends in drought-prone areas.

The phenotyping is the most important among all the factors that affect the accuracy in the detection of QTL locations, and hence it needs very reliable selection indices and screening platform. This has now led to various options for high-throughput phenotyping. Many of them rely on image-based tools and mechanisation. Image analysis to interpret the responses of plants to stresses is a great challenge. In addition, data management and statistical tools are essential to tackle huge set of data generated through phenomics. There is a lot of scope for research in these areas to facilitate genomic-based gene discovery.

As in any other abiotic stresses, breeding for salt tolerance relies on effective protocols for screening, existence of genetic variability for a target trait and techniques to transfer the genes to the desired cultivars of crops. Hence future research should emphasise all the three dimensions of breeding for plant's tolerance to drought, high temperature, salinity, waterlogging, etc. Molecular approaches for complementing the conventional plant breeding for stress tolerance need to be accelerated. For further improvement in plant's tolerance to salt and waterlogging tolerance in crops, future research should focus on opportunities to regulate multiple Na^+ transporters by deep insight into expression of associated genes and signalling cascades. At present, it is not clear as to which of the transport processes that we know so far could be integrated to develop salt- and waterlogging-tolerant crops. Similarly, basic research on stomatal regulation should be translated into tools and management practices for optimising water use.

For the productivity improvement in crops, a foremost focus has been on photosynthesis and chloroplast. However, the process of respiration is highly responsive to abiotic stresses, and hence mitochondria need attention. For effectively controlling the process of respiration, it is necessary to enhance research efforts for elucidation of the different ways mitochondria are functioning inside the cell. Interventions of technological advances can facilitate for system biology approach for accomplishing this task. This is going to be a challenge for the next decade.

Emerging phenomic techniques are providing for enormous opportunities to search stress tolerance genes in a wide range of cultivated and wild species of plants in gene banks. Across the world there are about estimated 1,750 plant gene banks which have maintained about 7.4 million accessions as ex situ collections by employing different approaches. National gene banks have 89% of these collections and 130 of these gene banks possess more than 10,000 accessions (FAO 2010). Countries like India have as many as 346,000 germplasm of various crops, and these germplasm can be a treasure of desirable traits for drought tolerance. In addition, a large number of mutant lines with induced genetic variation exist for different crop species, and new techniques have emerged to create new genetic variants for gene function studies by employing molecular tools. In addition, recently emerged CRISPR/Cas9 system has a great potential to allow convenient and highly efficient methods for editing many genes in plants in a desired way for conferring stress tolerance in plants (Ma et al. 2016).

Identification of heat and hypoxia tolerance traits in fish, identifying the molecular pathways affected by atmospheric stress and marker assistant selection of fast growing and tolerant species through transcriptomic and proteomic studies. With the emergence of heat stress as one of the major problems in Indian poultry industry, our primary area of focus should be to explore innovative approaches, including genetic marker-assisted selection of poultry breeds for increased heat tolerance and disease resistance for better productivity. Application of modern molecular techniques in poultry breeding has great potential to improve poultry productivity in a sustainable manner. The best choice of animals and breeds is highly essential for livestock sectors, while research in fisheries sensitive to low temperatures needs appropriate combination of species.

There is also possibility to improve income and livelihood of farmers in harsh abiotically stressed agroecologies through introduction of appropriate crops and other components such as livestock in farming system context. Desirable changes in components of cropping and farming system need research to enhance the profit for farmers from marginal lands. To replace water-guzzling crops and to ensure equitable distribution of natural resources such as water, it is essential to explore underutilised and nonconventional crops that can provide export opportunities for farmer and hence profitable farming in water scarcity zone. This needs research on all aspects of crop production, breeding and also marketing.

To make agriculture in abiotically stressed agroecologies profitable, it is essential to develop modules in the agribusiness framework including as many components as a small farmer can afford and manage. These modules can contain ruminants, poultry, medicinal plants, horticulture and other profitable commodities in addition to traditional crops. Perhaps this should have its base in demonstrated research output from research on integrated farming research; however, it should go beyond the traditional thoughts to make it an agribusiness to attract rural youth. Research on this aspect can provide immediate dividends as profits will drive the agriculture. Such an approach for cluster of farmers can be successful when milkroute model can be adapted to collect the produce from farmers at reasonable price, and supply of input and market intelligence are integrated in this module by agencies involved in the village level trade.

24.6   Ways to Accomplish the Task

It is universally accepted that abiotic stresses impact agriculture through complex process often by acting in combinations. This is also evident from the meagre progress that has been achieved so far to address this issue. However, advances in scientific tools provide new opportunities for accelerating the research for understanding the stress environments as well as mitigation and adaptation options. There is a need to integrate all available information in repositories accessible for technology developers, disseminators and end users. This information should be effectively used for capacity building to create new generation of agricultural managers, policymakers and researchers. There should be accelerated efforts to achieve synergies across the disciplines and institution working on this aspect.

24.6.1 Repository of Information on Abiotic Stresses

With our capacity to handle huge data that is increasing with advances in computer science and data management tools, the volume of cured set of all the data relevant to a various abiotic stresses across diverse agroecologies will be the key for decision-making on mitigation and adaptation options for stress management. Expected database for abiotic stress management should have information about the abiotic stress environment including edaphic and atmospheric factors at smallest possible units scaling down even from block level to village level. It should have information about the land use plans and contingency plans to deal with abiotic stresses such as drought. The repository also should reveal the current status of tolerance to abiotic stresses in prevailing cultivars, species, etc. Such information can be obtained from multilocation trials. All the information related to technologies evolved and which are evolving should be accessible to stakeholders including young educated farmers who can contribute to enrich the repository with their feedbacks. Mobile-based applications can facilitate this process and can provide new avenue for research in effective employment of such tools for extending the technologies and decision-making tools for farmers.

Genomics as a game changer has evolved significantly in the recent past to enhance our ability to observe and dissect the minute changes in plant phenotype governed by many genes located at different places in a chromosome of an organism. There are ample accesses to methods that can help in analysing quantitative traits and hence detecting even dozens of specific gene combinations at many loci associated with critical plant traits such as days to flowering or tolerance to stresses. A huge data set on plant genomics is available as an outcome of whole genome sequencing routinely carried out in several plants which include cultivated species. Genomic tools allow direct analysis of allelic diversity and microheterogeneity in chromosome organisation through insight into hundreds or thousands of distinct genomes of inbred lines or ecotypes. This huge set of genomic information can open new avenues to test hypothesis about epigenetics, transposons which are often referred as jumping genes, rearrangement of genome, the amplification/duplications/losses of genes in addition to conservation and change in both promoter region and transcription unit. This needs to be connected to explain development or physiology under abiotic stress environment.

There is an opportunity to rediscover and revisit old hypothesis with new tools and knowledge in place. This can help in bridging the major gaps in our understanding about in vivo regulation of cellular metabolism as metabolic activities are organised to meet the specific needs and desired functions of a particular cell or tissue. This can be achieved if the remarkable advances in capacity for in vivo measurement of metabolic processes are integrated with 'omics' data.

24.6.2 Enhanced Focus on Traits of Interest

Several definitions exist for stress tolerance in the crop plants (Levitt 1972; Blum 2005; Gilbert and Medina 2016); however, the ultimate aim is the performance in terms of grain yield or biomass irrespective of schools of thought. For instance, any mechanism acting at cell, tissue, plant or crop canopy level to impart survival and performance of plants under limited soil moisture can have great impact on improving drought tolerance in crop plants. Nonetheless it is well acknowledged that the stress tolerance mechanisms are complex and governed by many traits/genes. Since the current results of yield and yield component based selections in breeding are not as remunerative as that of past decades, trait-based approaches are being pursued for genetic improvement. The value of stress-related trait can be realised only if it is possible to manipulate it either genetically or by agronomic means to reduce the losses in crop yield caused by the stress such as drought (Purcell and Specht 2012). Such traits can also be identified with the assumption that there will always be more to know about physiology of tolerance to drought, but we know more than enough to act (Passioura 2012; Purcell and Specht 2012). Some of the aspects that need to be emphasised for further improvement in crops for drought tolerance are explained in the following sections.

The stomatal role in water relations of plants (Lawson and Blatt 2014) is well established, but how this knowledge can be used in improving efficiency of water uptake and dehydration avoidance in different crop plants is to be elucidated. While about genetic variations in traits such as stomatal density, we do not have fare idea about dynamics of stomatal aperture in the diurnal cycle, which can serve as trait for screening. Some insights into the processes such as the production of signal molecules by roots that makes the leaves respond by closing the stomata, but we do not know the underlying genetic variability of those perceived signals to close the stomata that can be exploited to improve drought tolerance of many crops with some exception. We also do not know the genetic variability and possibilities of genetic manipulations for capacity of cells to mop up free radicals produced during dehydration or produce molecules that preserve their ability to hold on to water (Pennisi 2008). The role of aquaporins, protein channels located in the plasma membrane, is being increasingly highlighted (Meng et al. 2016); however, it is not clear if aquaporins or expression of genes coding them is the limiting factor in present cultivars of crops for their performance in drought environment. These examples clearly reveal that substantial resources and time have to be devoted to translate the basic knowledge generated into screening tools for drought tolerance assessment in germplasm/breeding lines.

As opined by Gilbert and Medina (2016), there are some basic questions which remain unanswered for many of the crop plants in the context of different drought scenarios. It is necessary to understand drought response determining features of interactions among stress factors such as extreme temperature, vapour pressure deficit and solar radiation which all occur together under natural conditions. Further, it is necessary to evaluate the present cultivars of crops for the following:

- Temperature threshold for drought tolerance which will emerge from experiments to assess impact of extreme temperatures on drought responses. These experiments should also consider effect of plasticity of plant/root/leaf on response to and recovery after drought.
- Effect of the frequency, duration and intensity of drought imposition on recovery of plants in general and specifically of tropical and subtropical orchard crops
- Intrinsic features affected by drought that limit recovery – for example, loss of hydraulic conductance, signalling factors associated with stomatal regulation, damage to photosynthetic system, depletion of carbohydrates stored in vegetative part, reduction in functional leaf area and or reduction in sink size, i.e. potential fillable seeds, etc.

24.6.3 Protocols and Methods of Screening

To answer these questions mentioned in the previous section, it is essential to distinguish stress imposed by drought from stress imposed by water deficit usually in experiments conducted in controlled environments such as greenhouse or growth chamber. If relevance of a particular trait is established, even water-deficit treatments in controlled environments can reveal meaningful information about the plants' response to drought that are common in field environments. Further search for answers to these questions should culminate in identification of useful traits for crop improvement and also associated genes.

This can be carried out within the framework of model proposed by Passioura (2012) wherein yield under moisture deficit environment is defined as function of water uptake, transpiration efficiency and biomass partitioning efficiency. While the search for promising traits associated with plant's tolerance to drought is in progress, the shift from water-use efficiency (WUE) to effective use of water (EUW) has been suggested for genetic gain in crop productivity under drought (Blum 2009). Emphasis on EUW is based on the argument that many traits that contribute to WUE come at the cost of plants performance under favourable condition particularly with respect to genetic improvement and management practices. It is essential to know when such traits and genes are useful for plant breeders (Passioura 2012).

Implication of trait research will be wider if we differentiate the symptomatic and acclimatise responses of plants to drought. This can help in deciding the relevance of a trait in the context of crop growth stages and type of soil or other influencing factors. Further, it is possible to identify trait or associated genes which can impart drought tolerance by creating or simulating the drought scenario of targeted agroecosystem (Tardieu 2012). If we are aiming for productivity enhancement, we may focus on those traits that promote opportunistic strategies, i.e. less reduction under stress but more yields under sufficient conditions. On the other hand, conservative traits associated with stable yield, i.e. may be less in moisture sufficient conditions, may be useful for chronically drought-affected regions (Pennisi 2008).

Trait identification and characterisations are crucial for getting benefit of advances in the field of genomics which has allowed us to know more about genes with high-throughput methods (Huang et al. 2013; Pallotta et al. 2014; Valluru et al. 2014). Further, for accelerating crop improvement, we should know the gene functions. Considering the fact that the stress response of plant is a consequence of a series of gene action, it is necessary to know the phenotype-genotype relationship in high-throughput mode (Tester and Langridge 2010) to decipher gene and gene network associated with drought tolerance. Techniques used in the past for characterising plant responses are not sufficient for understanding functions of genes and developing reliable functional maps. The need for overcoming these inadequacies has led to the emergence of 'phenomics' as a new branch of biological sciences (Furbank and Tester 2011; Yang et al. 2013; Klukas et al. 2014; Fahlgren et al. 2015, Rahaman et al. 2015). Phenomics relies on non-invasive image-based techniques and automation to characterise responses of a large number of plants so that the data generated can improve the power of prediction of association between the gene and genotypes (Rahaman et al. 2015). Background noise in these methods was the major concern to ensure precision of characterisation of phenotype; however, methods for minimising or assessing such noise are also emerging (Biscarini et al. 2016). This area of science is expected to help in tailoring crops for specific drought-stressed agroecosystem and also can accelerate identification of promising genotypes out of germplasm accessions of crops and their wild relatives.

24.6.4 Task-Oriented Capacity Building

Abiotic stress studies have been a small component of course curriculum of state agricultural universities and deemed universities supported by centre. Taking into the complexity in understanding the nature, time of occurrence and magnitude of stress, the present educational system is not sufficient for abiotic stress management. As the agriculture as a profession is now transforming from subsistence to business mode, there will be demand for well-qualified managers for agribusiness to assess, predict and manage abiotic stresses. Hence, capacity building should be oriented to create next-generation highly capable managers, researchers and farmers. While the undergraduate courses sufficiently cover the basic of agricultural science, the postgraduate courses are to be reoriented to include problem identification and problem-solving capacity with respect to different abiotic stresses. They should be trained to use the latest *omics* tools for accomplishing this task in research institutes. These young brains should be learners as well as contributors to the abiotic stress repository to make it dynamic and impactful. The products emerging in future from agricultural universities should have experts to manage atmospheric stresses, drought and edaphic stresses, while each of them may occur alone or in combination with others.

In the academic field, there is now an increasing shift from conventional division concept to school of thought approach, which specifically addresses abiotic stress responses in plants and other commodities. For example, biologists with a broad view of cellular biology are now equipped with the tools to look at the general behaviour of preferred plant-model systems and to ensure efficient selection of the key genes regulating the entire plant processes. This is now providing boost for system biology approach that requires the integration of interdependent disciplines. They are bioinformatics and computational biology which are increasingly employed in model networks, functional genomics essential for developing and implementing tools for analysis of biological systems in the high-throughput mode and interdisciplinary team of biologists to test the relevant hypothesis and to generate proof of concept. These research areas have to be handled in coordinated manner for achieving the targets of a centre of excellence in systems biology. Similar analogy can be extended to build capacity in understanding the stress environment. The capacity building should meet the needs of this system to carry forward benefits of basic research for abiotic stress management.

The capacity building should cater to the needs of farmers who may wish to consult for expertise in analysis and prediction of abiotic stresses, assessment of plant responses through advanced phenotyping methods for large germplasm, fish and veterinary nutrition and stress management, advisors for soil health and productivity in harsh areas, land use planner, use of bioregulators to mitigate stress, etc., in addition to cropping system and market advisors with thorough policy support system knowledge. In addition to postgraduates, there should be regular training on abiotic stress management for those who are in action in agroecologically challenged environment. Thus the capacity building strategy should address requirement of research, abiotic stress management in field, policy- and decision-making as well as farmers. Such education can be made cheaper by engaging IT tools particularly for those in service and business.

24.6.5 Networks for Exploring Synergies

In fact, several research and developmental organisations are working on various aspects of abiotic stresses, but their efforts are too inadequate considering the magnitude of the problem. Looking at the past scenario, it comes out that these organisations have been working in isolation and within their disciplinary boundaries. In order to alleviate the effects of multiple stressors, a holistic multidisciplinary approach to build up systems perspectives is the need of the hour to get the best combination of mitigation and adaptation technologies for a particular agroecosystem that is often featured by multiple stressors and that needs to be defined with greater precision.

Having known the complexity of abiotic stresses in agriculture largely determined by the nature of agroecologies and the plant responses, the global scientific community is now ready to establish collaboration across the borders. This should provide opportunity to make the best use of research investment already made and to translate them into the products useful for management of abiotic stresses. Hence, it is highly beneficial to establish the links with centre of excellence in advancing the science for benefit of agriculture in ecologically challenged environments.

Earlier efforts on abiotic stresses were perceived not as remunerative as that on products for favourable environments by private sector. Now, there is change in trend with increasing involvement of private firms in commodities resilient to drought, high temperature and salinity. This should provide additional dimension for future research network for management of abiotic stresses. There is an opportunity to utilise existing institutional mechanisms to evaluate the germplasm, advanced lines of crops, conservation agriculture, multi-enterprise farming and bioregulator technology specifically to promote stress mitigation at different locations and adaptation to abiotic stresses. Such efforts can get boosted by involvement of progressive farmers and private stakeholders.

24.6.6 Dynamic Policy Support Tools

Lack of requisite investment and sufficient market intelligence especially in the regions afflicted by abiotic stressors are the major causes of ineffective R&D institutions and infrastructure that all together failed to determine pragmatic prices of inputs as well as natural resources. This necessitates policy change for shaping technology development, dissemination and marketing to sustain agricultural outputs with efficient use of resources. It is possible to promote the adoption of available technologies such as integrated soil fertility management, multiple uses for enhancing water productivity, carbon-sequestering practices, etc., by identifying constraints in technology. It remains a great challenge to develop technology management strategy particularly for enhancing adoption of proven techniques for efficient use of natural resources, water and energy. This makes it essential to assess the costs and benefits of other options emerging or likely to emerge from advances in science and innovations in institutional and marketing strategies. There is a need for relook at scientific applications of natural resource technologies such as conservation agriculture, organic agriculture, precision agriculture as well as new sciences such as of omics, nanotechnology, etc. For a pragmatic policy support, it is imperative to analyse potential of all these technologies to pay-offs necessarily in terms of productivity and food security. While accomplishing this it is necessary to take into account the spatial and temporal variability in crops grown, features of soils and other agroecological and socio-economic aspects as well as impact of existing policies at regional and national level. Some leads in this direction are listed below:

- The present abiotic stress/disaster management policies are skewed towards crisis management with possible solution in relief measures, employment opportunities, etc., which all become a component of process that recurs at the next crisis event, while the permanent damage that drought/other events leave through land degradation has yet to receive due attention. Policies should promote location-specific measures to prevent the permanent damage to agroecosystem.
- Since the technologies are to be tailored for the finest level of agroecologies in future, it is necessary to have a real as well as virtual web-based platform for interface among agricultural officers, policymakers, farmers and scientist to build a larger database and to derive appropriate decision for development and implementation of technologies for abiotic stresses.

- Enhancing coping strategies of the country to extreme weather events by establishing 'risk management and mitigation centre' at the block level which should work in close coordination with line departments.
- There is a need to place greater emphasis on future crops to be designed specifically for the smallest components of agroecosystems based on possible acceptance and impact. This needs compilation and analysis of huge set of data on the prevailing farming system, soils, weather, etc.; the similar activity presently being carried out at mega environment needs to be scaled down to macro environment at village levels in the immediate future.
- Priority diversification of farming enterprise through Integrated Farming Systems to provide for life-saving returns and quick recovery in case of extreme weather events.
- The flow of exotic germplasm has been constrained by recently evolved regulations with additional concerns on national biodiversity. Lack of germplasm exchange can also reduce the inflow of technologies that emerge from investment and efforts in other organisations. There should be provisions to promote exchange of germplasm and promising technologies in pipeline for management of abiotic stresses mentioned above through research collaboration with transparency in utility of material and benefits for the country.
- Priorities given to resilience to climate change at national level has now provided significant support for identification of traits and genes relevant to stress tolerance in the events of drought, flood, salinity and nutrient stress. State-of-the art facilities for applying omics approach are created but not sufficiently complemented by providing adequate capacity building. Thereby latter should be implemented on priority
- Molecular marker approaches are promising tools for drought and submergence tolerance, however, advantage of transgenic approaches and several technologies that have been patented so far are yet to be realised for improving food and fodder crops for stress tolerance. Freedom to operate (FTO) that may emerge from negotiations between the institutes and promotions of confined field trials can facilitate validation of these technologies at the earliest.
- Innovative technologies for mitigating the abiotic stressors are rarely adopted despite clear demonstration of benefit, e.g. for dryland systems. It makes provisions for doubts that if some utility claims of novel and innovative technologies are influenced more by interest and wishes than real-time utility. However, some of the technologies have been successfully demonstrated as real-time contingencies. Adoption of such technologies needs policy support.
- Policy initiatives for efficient risk and crisis management of natural disasters should aim at access to microcredit and insurance services in addition to banking, communication and information services before, during and after a disaster event as predictions about these events are gradually becoming more robust.
- Enhanced optimism among farmers is critical to promote technologies for stress-prone areas – this can come from streamlining and strengthening the channels of technology transfers and supports to reinforce the demonstrated technology. Adequate human resources (HRs) induction and training could be fostered.

Investments in viable income-generating activities could be fostered together with the active promotion of communities' self-help and solidarity mechanisms.

- Occurrence of calamities like drought demoralises farmers who may not be willing to venture into any unrealised technologies. The common lands including forest land can be utilised for demonstrating the technologies like fodder production taking into consideration probabilities of rainfall deficit. They can serve as in situ fodder banks particularly in drought-prone areas.

- Market surplus often occurs when the drought-prone areas get benefit of a favourable monsoon. Hence enhanced food production does not lead to enhanced profit because of reduction in prices of commodity. The yield losses due to stresses like drought get often compensated by improved quality of product particularly in dryland crops. This quality deserves premium value if agriculture produce reaching the market gets segregated. Hence, policy for segregated procurement of produce is to be evolved. Further, village level crop plan compatible with value chain can maximise the profit for farmers in each of the drought cycle.

- The slow growth of purchasing power is the major constraint in ensuring the food security to the people in the rain-fed ecosystems. Drought-resistant cultivars of crops, cost-effective IFS, efficient techniques to harvest rainwater, adoption of moisture conservation techniques and intercropping are crucial for stabilising and improving the agricultural production in dry land.

- Economic and durable expansion of irrigation and reasonable prices for key inputs such as water, energy and fertiliser coupled with rationalisation of minimum support prices for all the agricultural produces can enhance resources essential for community-based investment for prevention of soil degradation, augmentation of irrigation capacity and development of rural infrastructure such as roads, processing units and market.

- Funding policies should promote collective efforts by research institutes for addressing different components of abiotic stress management taking into consideration the magnitude of the task and complexities underlying the occurrence, nature and management of these stresses. These policies should be based on clear definition of the gaps existing in pipeline from the point of technology generation to technology outreach for tangible impact.

- The choice between 'risk aversion' and 'loss aversion' is critical for a farmer who prefers high-yielding local genotypes over drought-tolerant one to avoid loss aversion. Policies to support risks should be appropriately combined with technologies.

- While drought-proofing tools are evolved and being propagated after persistent effort, farmers are not appropriately covered for risks due to highly unpredictable events like hailstorms, etc. High-value agricultural enterprises like export-oriented fruit orchard are to be supported for adopting protective structures that cost high at initial stage which small farmers cannot afford. In addition, postharvest technology promotions should consider alternative use of damaged produce at the harvest.

References

Biscarini F, Cozzi P, Casella L, Riccardi P, Vattari A, Orasen G et al (2016) Genome Wide Association Study for traits related to plant and grain morphology, and root architecture in temperate rice accessions. PLoS One 11(5):e0155425

Blum A (2005) Drought resistance, water-use efficiency, and yield potential–are they compatible, dissonant, or mutually exclusive? Crop Pasture Sci 56:1159–1168

Blum A (2009) Effective use of water (EUW) and not water-use efficiency (WUE) is the target of crop yield improvement under drought stress. Field Crop Res 112:119–123

Fahlgren N, Feldman M, Gehan MA, Wilson MS, Shyu C, Bryant DW, Baxter I (2015) A versatile phenotyping system and analytics platform reveals diverse temporal responses to water availability in Setaria. Mol Plant 8(10):1520–1535

FAO (2010) The second report on the state of the world's plant genetic resources. Rome, Italy: FAO; 370 p.

Furbank RT, Tester M (2011) Phenomics–technologies to relieve the phenotyping bottleneck. Trends Plant Sci 16:635–664

Gilbert ME, Medina V (2016) Drought adaptation mechanisms should guide experimental design. Trends Plant Sci 21:639–647

Huang R, Jiang L, Zheng J, Wang T, Wang H, Huang Y et al (2013) Genetic bases of rice grain shape: so many genes so little known. Trends Plant Sci 18:218–226

Jat ML, Jat RK, Sidhu HS, Parihar CM, Sapkota TB, Jat HS, Gathala MK, Saharawat YS, Singh Y (2015) Conservation agriculture and soil healh vis-a-vis nutrient management: what is business as usual? Extended summaries, national dialogue on nutrient management for improving soil health, Sept 28–29, 2015, TAAS, ICAR, CIMMYT, IPNI, CSISA, KAI, New Delhi, pp 19–22

Joshi PK, Jha AK, Wani Suhas P, Sreedevi TK, Shaheen FA (2005) Impact of watershed program and conditions for success: a meta-analysis approach. Global theme on agro-ecosystems. Report no. 46. Patancheru 502 324, Andhra Pradesh, India; International Crops Research Institute for the Semi-Arid Tropics, Hyderabad

Kasam A, Friedirch T, Sims B, Kiezle J (2014) Sustainable intensification and conservation agriculture. In: Proc of tech session 1 on conservation agriculture: building sustainability, 6th world congess on conservation agriculture, June 22–25, Winipeg

Khush GS (2001) Green revolution: the way forward. Nat Rev Genet 2(10):815–822. Retrieved from http://dx.doi.org/10.1038/35093585

Klukas C, Chen D, Pape JM (2014) Integrated analysis platform: an open-source information system for high-throughput plant phenotyping. Plant Physiol 165:506–518

Lawson T, Blatt MR (2014) Focus on water: stomatal size, speed and responsiveness impact on photosynthesis and water use efficiency. Plant Physiol 164:1556–1570

Levitt J (1972) Responses of plants to environmental stresses. Academic Press, New York

Ma X, Zhu Q, Chen Y, Liu Y (2016) CRISPR/Cas9 platforms for genome editing in plants: developments and applications. Mol Plant 9:961–974

Meng LL, Song JF, Wen J, Zhang J, Wei JH (2016) Effects of drought stress on fluorescence characteristics of photosystem II in leaves of *Plectranthus scutellarioides*. Photosynthetica 54(3):414–421. doi:10.1007/s11099-016-0191-0

Minhas PS, Dagar JC (2016) Synthesis and way forward: agroforestry for waterlogged saline soils and poor-quality waters. In: Dagar JC, Minhas PS (eds) Agroforestry for management of waterlogged saline soils and poor-quality waters, Adv Agrofor series 13. Springer-Verlag, New Delhi

Pallotta M, Schnurbusch T, Hayes J, Hay A, Baumann U, Paull J et al (2014) Molecular basis of adaptation to high soil boron in wheat landraces and elite cultivars. Nature 51:88–91

Passioura J (2012) Phenotyping for drought tolerance in grain crops: when is it useful to breeders? Funct Plant Biol 39:851–859

Pennisi E (2008) The blue revolution, drop by drop, gene by gene. Science 320:171–173

Purcell LC, Specht JE (2012) Physiological traits for ameliorating drought stress. In: Roger B, James ES (eds) Agronomy monograph 16. Soybeans: improvement, production, and uses. American Society of Agronomy, Madison, pp 569–620

Rahaman MM, Chen D, Gillani Z, Klukas C, Chen M (2015) Advanced phenotyping and phenotype data analysis for the study of plant growth and development. Front Plant Sci 6:619–630. doi:10.3389/fpls.2015.0061

Srivastva AK, Ratnakumar P, Minhas PS, Suprasanna P (2016) Plant bioregulators for sustainable agriculture: integrating redox signaling as a possible unifying mechanism. Adv Agron 137:237–278

Tardieu F (2012) Any trait or trait-related allele can confer drought tolerance: just design the right drought scenario. J Exp Bot 63:25–31

Tester M, Langridge P (2010) Breeding technologies to increase crop production in a changing world. Science 327:818–822

Valluru R, Reynolds MP, Salse J (2014) Genetic and molecular bases of yield-associated traits: a translational biology approach between rice and wheat. Theor Appl Genet 127:1463–1489

Winterbottom R, Reij C, Garrity D, Glover J, Hellums D, Mcgahuey M, Scherr S (2013) Improving land and water management. Working paper, instalment 4 of creating a sustainable food future. Washington, DC: World Resources Institute, Accessible at: http://www.worldresourcesreport. org

Yang W, Duan L, Chen G, Xiong L, Liu Q (2013) Plant phenomics and high-throughput phenotyping: accelerating rice functional genomics using multidisciplinary technologies. Curr Opin Plant Biol 16:180–187

Index

© Springer Nature Singapore Pte Ltd. 2017
P.S. Minhas et al. (eds.), *Abiotic Stress Management for Resilient Agriculture*,
DOI 10.1007/978-981-10-5744-1

513

Printed by Printforce, the Netherlands